T0217298

Applied and Numerical Harmonic Analysis

More information about this series at http://www.springer.com/series/4968

Jon H. Davis

Methods of Applied Mathematics with a Software Overview

Second Edition

 Birkhäuser

Jon H. Davis
Kingston, ON, Canada

ISSN 2296-5009   ISSN 2296-5017   (electronic)
Applied and Numerical Harmonic Analysis
ISBN 978-3-319-82805-3   ISBN 978-3-319-43370-7   (eBook)
DOI 10.1007/978-3-319-43370-7

Mathematics Subject Classification (2010): 00A69, 30-01, 33-xx, 34-01, 35-01, 35Qxx, 42Axx, 42Cxx, 46Cxx, 65Mxx, 44-01, 97R20

Printed on acid-free paper

This book is published under the trade name Birkhäuser, www.birkhauser-science.com
The registered company is Springer International Publishing AG
The registered company address is: Gewerbestrasse 11, 6330 Cham, Switzerland

To Ethan and Nathaniel

ANHA Series Preface

The *Applied and Numerical Harmonic Analysis* (*ANHA*) book series aims to provide the engineering, mathematical, and scientific communities with significant developments in harmonic analysis, ranging from abstract harmonic analysis to basic applications. The title of the series reflects the importance of applications and numerical implementation, but richness and relevance of applications and implementation depend fundamentally on the structure and depth of theoretical underpinnings. Thus, from our point of view, the interleaving of theory and applications and their creative symbiotic evolution is axiomatic.

Harmonic analysis is a wellspring of ideas and applicability that has flourished, developed, and deepened over time within many disciplines and by means of creative cross-fertilization with diverse areas. The intricate and fundamental relationship between harmonic analysis and fields such as signal processing, partial differential equations (PDEs), and image processing is reflected in our state-of-the-art *ANHA* series.

Our vision of modern harmonic analysis includes mathematical areas such as wavelet theory, Banach algebras, classical Fourier analysis time-frequency analysis, and fractal geometry, as well as the diverse topics that impinge on them.

For example, wavelet theory can be considered an appropriate tool to deal with some basic problems in digital signal processing, speech and image processing, geophysics, pattern recognition, biomedical engineering, and turbulence. These areas implement the latest technology from sampling methods on surfaces to fast algorithms and computer vision methods. The underlying mathematics of wavelet theory depends not only on classical Fourier analysis but also on ideas from abstract harmonic analysis, including von Neumann algebras and the affine group. This leads to a study of the Heisenberg group and its relationship to Gabor systems and of the metaplectic group for a meaningful interaction of signal decomposition methods. The unifying influence of wavelet theory in the aforementioned topics illustrates the justification for providing a means for centralizing and disseminating information from the broader, but still focused, area of harmonic analysis. This will be a key role of *ANHA*. We intend to publish with the scope and interaction that such a host of issues demands.

Along with our commitment to publish mathematically significant works at the frontiers of harmonic analysis, we have a comparably strong commitment to publish major advances in the following applicable topics in which harmonic analysis plays a substantial role:

Antenna theory *Prediction theory*

Biomedical signal processing *Radar applications*

Digital signal processing *Sampling theory*

Fast algorithms *Spectral estimation*

Gabor theory and applications *Speech processing*

Image processing *Time-frequency and*

Numerical partial differential equations *time-scale analysis*

Wavelet theory

The above point of view for the *ANHA* book series is inspired by the history of Fourier analysis itself, whose tentacles reach into so many fields.

In the last two centuries, Fourier analysis has had a major impact on the development of mathematics, on the understanding of many engineering and scientific phenomena, and on the solution of some of the most important problems in mathematics and the sciences. Historically, Fourier series were developed in the analysis of some of the classical PDEs of mathematical physics; these series were used to solve such equations. In order to understand Fourier series and the kinds of solutions they could represent, some of the most basic notions of analysis were defined, e.g., the concept of "function." Since the coefficients of Fourier series are integrals, it is no surprise that Riemann integrals were conceived to deal with uniqueness properties of trigonometric series. Cantor's set theory was also developed because of such uniqueness questions.

A basic problem in Fourier analysis is to show how complicated phenomena, such as sound waves, can be described in terms of elementary harmonics. There are two aspects of this problem: first, to find, or even define properly, the harmonics or spectrum of a given phenomenon, e.g., the spectroscopy problem in optics, and, second, to determine which phenomena can be constructed from given classes of harmonics, as done, for example, by the mechanical synthesizers in tidal analysis.

Fourier analysis is also the natural setting for many other problems in engineering, mathematics, and the sciences. For example, Wiener's Tauberian theorem in Fourier analysis not only characterizes the behavior of the prime numbers but also provides the proper notion of spectrum for phenomena such as white light; this latter process leads to the Fourier analysis associated with correlation functions in filtering and prediction problems, and these problems, in turn, deal naturally with Hardy spaces in the theory of complex variables.

Nowadays, some of the theory of PDEs has given way to the study of Fourier integral operators. Problems in antenna theory are studied in terms of unimodular trigonometric polynomials. Applications of Fourier analysis abound in signal processing, whether with the fast Fourier transform (FFT), or filter design, or the adaptive modeling inherent in time-frequency-scale methods such as wavelet theory. The coherent states of mathematical physics are translated and modulated Fourier transforms, and these are used, in conjunction with the uncertainty principle, for dealing with signal reconstruction in communications theory. We are back to the raison d'être of the *ANHA* series!

College Park, MD, USA   John J. Benedetto

Preface

This book has evolved from courses I have presented on "Methods of Applied Mathematics." Parts of the material have served as a text for a half-year course constructed specifically for engineers, while the book as a whole serves as a text for a yearlong course for students of mathematics and many applied fields of study.

In format and content, the text is broadly organized around the theme of applications of Fourier analysis. The treatment covers both classical applications in partial differential equations and boundary value problems, and a substantial number of topics associated with Laplace, Fourier, and discrete transform theories. The topics covered are useful both in traditional continuum mechanics and mathematical physics areas and in various phases of "modern" applied mathematics such as control and communications. Transform inversion problems really involve the use of complex variable techniques, so I also include substantial material on complex analysis and applications. In addition to the usual transforms, we also cover some more advanced topics in the final chapter. These include short-time Fourier analysis, as well as geometrically based transforms applicable to boundary value problems.

The book provides the student with a firm grasp of the fundamental notions and techniques of the area and, at the same time, conveys a sense of the wide variety of problems in which the methods are useful. We emphasize not only the computational aspects of problem-solving but also the limitations and implicit assumptions inherent in using the formal methods.

The included problems range in difficulty from mechanical exercises to introductions to applications reluctantly left out of the main body of the material. The students must be convinced that the solution of problems is not a spectator sport. I have accumulated convincing statistical evidence that boundary value problems are substantially harder to solve by means of paper and pencil than with the aid of blackboard and chalk.

The semester-length course can cover the topics of Fourier series, boundary value problems, and Laplace, Fourier, and z-transforms. These subjects are treated at an introductory level in Sections 2.1, 2.2, 2.7, 2.8, 3.1–3.4, 3.6–3.7, 6.1–6.8, 7.1–7.3, 7.8–7.9, and 8.1–8.2 in the text. Students taking this course require the equivalent of the standard two-year calculus/differential equations sequence and an introductory linear algebra course as background.

The yearlong course covers in greater or lesser detail the full range of material in Chapters 2–4 and 6–8. The deeper mathematical background of these students is exploited to give (in addition to basic problem-solving skills) a treatment of convergence issues, an introduction to Sturm–Liouville theory, and a complex variable approach to transform inversion. The discrete transforms (discrete Fourier, z-, finite and fast Fourier) are also treated in more depth than in the half-year course.

Students in the year-length course have had a second level linear algebra course, analysis at the level of Fulks, *Advanced Calculus* or Marsden and Tromba's *Vector Calculus*, and concurrently take a course in complex analysis. It is not necessary to take such a concurrent course, however, in order to handle the bulk of the material. Complex analysis is seriously used only in connection with transform inversion integrals in Sections 7.5 and 7.6. A treatment of complex analysis at the required level is given in Chapter 5. This chapter contains more material than is required for the purpose of supporting the rest of the text but includes the basic topics, including the residue theorem and principle of the argument.

Computation

Some applied mathematics topics at the level of this text are complex, at least compared to the experience of many students. Many solutions appear in the form of infinite series rather than as simple numerical answers. In this situation, it is useful to be able to visualize solutions or graph the "special functions" which arise in the course of boundary value problem solutions.

The commercial MATLAB$^{(TM)}$ program is widely used and has strong support for graphic visualization, as well as routines applicable to many of the topics covered in this text. When the first edition of this book was produced, MATLAB$^{(TM)}$ was commonly made available on a site-license basis in university campus student computer labs. Personal copies of the full version of the software were beyond the reach of student budgets.

It is now possible to support the software needs of this text entirely with the use of free software. Appendix B has an introduction to Octave, Maxima, and Python based on the assumption that the reader has been exposed to some programming, realizes that different languages serve different purposes, and is especially interested in generating graphics. That appendix also provides references to sources of further information, as well as some information about the processes for installing and running these free software packages on various hardware and software environments.

We assume that the students are sufficiently "computer literate" that the use of various software tools is not a major stumbling block. We have integrated the use of such programs into the flow of the text and provide blocks of code applicable to the problems under discussion.

As with all software programs, the best advice to the beginner is to "read the
...manual." Search engines are your friend and will usually turn up an instance of
a person with the same naive question you are pondering. A surprising number of
answers to not-so-naive questions also appear.

Acknowledgements

The text was produced using a variety of open-source and freely available software,
including LaTeX, ghostview, and the GNU gpic, m4, gawk, and sed programs, all
running under GNU/Linux. The circuit diagrams were created using the m4 circuit
macros package by D. Aplevich.

Thanks are finally due to the staff at Birkhauser, who guided and suggested a
second edition with a free software emphasis, and to my wife Susan for tolerating
my involvement with another book project.

Kingston, ON, Canada   Jon H. Davis

Contents

1 **Introduction** 1
   1.1   An Overview 1
   1.2   Topics by Chapter 2
   1.3   Applying Mathematics 3
   References 3
   Further Reading 3

2 **Fourier Series** 5
   2.1   Introduction 5
   2.2   Inner Products and Fourier Expansions 16
   2.3   Convergence of Fourier Series 26
   2.4   Pointwise and Uniform Convergence of Fourier Series 38
   2.5   Gibb's Phenomenon and Summation Methods 45
   2.6   Summation Methods 47
   2.7   Fourier Series Properties 56
   2.8   Periodic Solutions of Differential Equations 69
   2.9   Impedance Methods and Periodic Solutions 75
   2.10 Power Spectrum and Parseval's Theorem 77
   References 84
   Further Reading 84

3 **Elementary Boundary Value Problems** 85
   3.1   Introduction 85
   3.2   The One-Dimensional Diffusion Equation 86
   3.3   The Wave Equation 92
   3.4   The Potential Equation 97
   3.5   Discrete Models of Boundary Value Problems 101
   3.6   Separation of Variables 110
   3.7   Half-Range Expansions and Symmetries 127
   3.8   Some Matters of Detail 157
   Further Reading 172

4   Sturm–Liouville Theory and Boundary Value Problems 173
    4.1   Further Boundary Value Problems 173
    4.2   Self-adjoint Eigenvalue Problems 180
    4.3   Sturm–Liouville Problems 190
    4.4   Power Series and Singular Sturm–Liouville Problems 209
    4.5   Cylindrical Problems and Bessel's Equation 222
    4.6   Multidimensional Problems and Forced Systems 238
    4.7   Finite Differences and Numerical Methods 264
    4.8   Variational Models and Finite Element Methods 273
    4.9   Computational Finite Element Methods 279
    References 283
    Further Reading 284

5   Functions of a Complex Variable 285
    5.1   Complex Variables and Analytic Functions 285
    5.2   Domains of Definition of Complex Functions 294
    5.3   Integrals and Cauchy's Theorem 300
    5.4   Cauchy's Integral Formula, Taylor Series, and Residues 307
    5.5   Complex Variables and Fluid Flows 322
    5.6   Conformal Mappings and the Principle of the Argument 334
    Reference 350
    Further Reading 350

6   Laplace Transforms 351
    6.1   Introduction 351
    6.2   Definitions of the Laplace Transform 352
    6.3   Mechanical Properties of Laplace Transforms 356
    6.4   Elementary Transforms and Fourier Series Calculations 366
    6.5   Elementary Applications to Differential Equations 377
    6.6   Convolutions, Impulse Responses, and Weighting Patterns 387
    6.7   Vector Differential Equations 403
    6.8   Impedance Methods 417
    Reference 423
    Further Reading 423

7   Fourier Transforms 425
    7.1   Introduction 425
    7.2   Basic Fourier Transforms 432
    7.3   Formal Properties of Fourier Transforms 438
    7.4   Convolutions and Parseval's Theorem 450
    7.5   Comments on the Inversion Theorem 465
    7.6   Fourier Inversion by Contour Integration 477
    7.7   The Laplace Transform Inversion Integral 499
    7.8   An Introduction to Generalized Functions 514
    7.9   Fourier Transforms, Differential Equations, and Circuits 526
    7.10 Transform Solutions of Boundary Value Problems 535
    7.11 Band-limited Functions and Communications 553
    Further Reading 566

8   Discrete Variable Transforms... 567
  8.1   Some Discrete Variable Models...................................... 567
  8.2   Z-Transforms 575
  8.3   Z-Transform Properties.. 578
  8.4   Z-Transform Inversion Integral 587
  8.5   Discrete Fourier Transforms 602
  8.6   Discrete Fourier Transform Properties 607
  8.7   Some Applications of Discrete Transform Methods 623
  8.8   Finite and Fast Fourier Transforms 638
  8.9   Finite Fourier Properties.. 641
  8.10   Fast Finite Transform Algorithm 649
  8.11   Computing The FFT... 656
  References 664
  Further Reading 664

9   Additional Topics 665
  9.1   Local Waveform Analysis.. 665
  9.2   Uncertainty Principle 666
  9.3   Short-Time Fourier Transforms 675
  9.4   Two-Sided Transforms 685
  9.5   Walsh Functions 690
  9.6   Geometrically Based Transforms 694
  References 707
  Further Reading 707

ERRATUM .... E 1

A   Linear Algebra Overview 709
  A.1   Vector spaces.. 709
  A.2   Linear Mappings... 711
  A.3   Inner Products... 712
  A.4   Linear Functionals and Dual Spaces............................... 713
  A.5   Canonical Forms... 713

B   Software Resources 717
  B.1   Octave 717
  B.2   Python 738
  B.3   Maxima... 756
  B.4   Computing Platforms 760

C   Transform Tables 765
  C.1   Laplace Transforms... 765
  C.2   Fourier Transforms 766
  C.3   Z Transforms... 766
  C.4   Discrete Fourier Transforms 767
  References 768

Index 771

Chapter 1
Introduction

1.1 An Overview

This text is intended to provide an introduction to certain methods of applied mathematics, especially those arising from the area of Fourier analysis. These methods are widely applicable both on an operational basis for the solution of particular problems and on a conceptual basis for the analysis and understanding of models arising in a wide variety of applied contexts.

The topics treated include some that are usually described as in the area of classical mathematical physics, as well as problems arising in a more contemporary engineering mathematics context. These labels may be appropriate for some of the examples and applications. However, the methods discussed have obvious applicability to diverse other models arising in other applications areas. Some of these are described in the problems.

The material below consists of a mixture of analytical results and applications illustrating the results. We include a certain amount of model formulation and analysis, as well as some technical discussion of topics underlying the analytical results.

It is perhaps appropriate to make some remarks about the "level" of the treatment of the topics. The level of rigor in the material is (by design) uneven. Topics are treated on a rigorous level when it is convenient to do so, and when such a treatment is essential or useful for the understanding of the issue under discussion. On some occasions, required results are quoted with attendant (hopefully convincing) argument for their plausibility. Methods are of little use without knowledge of the limits of their applicability, and the creation of novel methods requires knowing where the standard approaches came from.

© Springer International Publishing Switzerland 2016
J.H. Davis, *Methods of Applied Mathematics with a Software Overview*,
Applied and Numerical Harmonic Analysis, DOI 10.1007/978-3-319-43370-7_1

1.2   Topics by Chapter

The second chapter consists of an introduction to the theory and elementary applications of Fourier series. Convergence of such series is discussed on the basis of an inner product space approach. Applications include periodic solutions of ordinary differential equations, impedance methods for electric circuits, and a discussion of the "power spectrum" notion derived from Parseval's Theorem.

The third chapter considers elementary boundary value problems, and might possibly be considered a further application of the Fourier series results of Chapter 2. This chapter also contains some derivations of the standard boundary value problem models, and a discussion of discrete boundary value problem analogue systems.

The fourth chapter treats higher-dimensional, non-rectangular boundary value problems. Included is a treatment of Sturm–Liouville expansions encountered in such models. The chapter also includes some discussion of series solutions and Bessel equations, as well as inhomogeneous boundary value problems.

An introduction to functions of a complex variable occupies Chapter 5. The basic results needed for applications are developed, including Cauchy's integral formula and the Residue Theorem. Complex functions are discussed, including the principle of the argument and conformal mappings. Applications are made to problems of fluid flow in this chapter, and to transform inversion in later chapters.

The sixth chapter introduces Laplace transform methods. In this chapter Laplace inversion is treated on a "tricks and tables" basis. Applications include ordinary differential equations, transient circuit analysis, and an introduction to input-output analysis of linear systems.

Continuous time Fourier transforms appear in Chapter 7. The Fourier inversion is treated both on an introductory algebraic basis, and later by using the inversion integral. The inversion integral for Laplace transforms is introduced as a special case of Fourier inversion. We also include in this chapter an introduction to generalized functions which includes a discussion of the underlying definitions.

Applications of Fourier transforms are made to ordinary differential equations, integral equations, linear systems, communications problems, impedance analysis, and partial differential equations.

Discrete variable transforms are the subject of the eighth chapter. Properties and applications of both the discrete Fourier and z-transforms are included. We also discuss the finite discrete Fourier transform, together with the associated Fast Fourier Transform algorithm of digital signal processing fame.

The final chapter provides an introduction to some transform methods of a more specialized nature than those considered above. The topics include two-sided and Walsh transforms, integral transforms associated with the Sturm–Liouville problems of Chapter 4, and the more recently developed topics in local waveform analysis, including short-time Fourier transforms.

1.3 Applying Mathematics

It goes without saying that applying mathematics requires both knowledge of subject areas of application, and of mathematical ideas and results that can be applied. In this text there are applications (especially in the problems) to a variety of subject areas. The topics are discussed as they appear to the level necessary to understand the discussion and questions.

The text contains some of what might be regarded as "theory," particularly on topics related to analysis. The idea here is that a little knowledge is dangerous, and that making competent applications of the material requires understanding what assumptions are being made in the process. The topics are of sufficient depth that it is possible to make bogus calculations on the basis of inadequate understanding.

Anyone who applies mathematics is eventually struck with wonder that it all seems to fit so well. Why is the universe not random chaos, but instead filled with phenomena that seem to be well described by boundary value problems, or contours of functions of a complex variable?

This seems to be one of the "Deep Questions." This text discusses topics that belong historically to both the areas of mathematical physics and engineering mathematics. The articles [2] and [1] are well-regarded discussions of the issue by authors in both of these areas. The basic unresolved issue is whether the universe "really is" mathematical in nature, and we are merely discovering what is already there, or mathematical structures are constructed out of nothing, and subsequently shape our perceptions of physical reality. Issues of that sort are discussed in some of the books in the references.

References

1. R.W. Hamming, The unreasonable effectiveness of mathematics. Am. Math. Mon. **87**(2) (1980)
2. E. Wigner, The unreasonable effectiveness of mathematics in the natural sciences. Commun. Appl. Math. **13**(1) (1960)

Further Reading

R. Aris, *Vectors, Tensors, and the Basic Equations of Fluid Mechanics* (Prentice-Hall, Englewood Cliffs, 1962)

G. Fauconnier, E. Sweetser (eds.), *Spaces, Worlds, and Grammar* (University of Chicago Press, Chicago/London, 1996)

J.A. Fodor, *The Language of Thought* (Harvard University Press, Cambridge, 1975)

R. Jackendorff, *Languages of the Mind* (MIT Press, Cambridge, 1992)

J.R. Johnson, *Introduction to Digital Signal Processing* (Prentice-Hall, Englewood Cliffs, 1989)

M. Livio, *Is God a Mathematician* (Simon and Schuster, New York, 2009)

J.G. Proakis, *Digital Communications*, 3rd edn. (McGraw-Hill, New York, 1995)

M. Tegmark, *Our Mathematical Universe: My Quest for the Ultimate Reality* (Afred A. Knopf, New York, 2014)

Chapter 2
Fourier Series

2.1 Introduction

A common requirement in various applied mathematical problems is to express a given function as a linear combination of other known functions. The reasons for this vary from analytical necessity to computational convenience.

Expansions in terms of polynomials are familiar from experience with Taylor series of the form

$$f(x) = \sum_{n=0}^{\infty} a_n x^n,$$

in which the unknown constants a_n are determined by values of the derivatives of the given function f.

In many contexts an expansion of the given function in terms of trigonometric functions must be obtained, in preference to the polynomial bias of the Taylor series.

If f is expressible as a finite Fourier series

$$f(x) = a_0 + \sum_{n=1}^{N} a_n \cos(nx) + b_n \sin(nx), \tag{2.1}$$

then it is easy to obtain formulas expressing the constants a_0, a_n, b_n in terms of the given function.

The original version of this chapter was revised. An erratum to this chapter can be found at DOI
http://dx.doi.org/10.1007/978-3-319-43370-7_10

© Springer International Publishing Switzerland 2016
J.H. Davis, *Methods of Applied Mathematics with a Software Overview*,
Applied and Numerical Harmonic Analysis, DOI 10.1007/978-3-319-43370-7_2

The keys to these relations lie in the orthogonality conditions

$$\frac{1}{\pi} \int_0^{2\pi} \sin(mx)\,\sin(nx)\,dx = \begin{cases} 0 & m \neq n, \\ 1 & m = n, \end{cases}$$

$$\frac{1}{\pi} \int_0^{2\pi} \sin(mx)\,\cos(nx)\,dx = 0,$$

$$\frac{1}{\pi} \int_0^{2\pi} \cos(mx)\,\cos(nx)\,dx = \begin{cases} 0 & m \neq n, \\ 1 & m = n\,(\neq 0). \end{cases}$$

Integrating both sides of (2.1) over the interval $[0, 2\pi]$ determines a_0 as

$$a_0 = \frac{1}{2\pi} \int_0^{2\pi} f(x)\,dx.$$

From (2.1) we obtain by multiplying through by $\cos(nx)$

$$f(x)\,\cos(nx) = a_0 \cos(nx)$$

$$+ \sum_{m=1}^{N} a_m \cos(mx)\,\cos(nx) + b_m \sin(mx)\,\cos(nx)$$

and integrating over the interval $[0, 2\pi]$ (using the orthogonality conditions) we obtain

$$a_n = \frac{1}{\pi} \int_0^{2\pi} f(x)\,\cos(nx)\,dx.$$

In a similar fashion we obtain

$$b_n = \frac{1}{\pi} \int_0^{2\pi} f(x)\,\sin(nx)\,dx$$

providing the expressions (known as the Euler equations)

$$a_0 = \frac{1}{2\pi} \int_0^{2\pi} f(x)\,dx. \tag{2.2}$$

$$a_n = \frac{1}{\pi} \int_0^{2\pi} f(x)\,\cos(nx)\,dx$$

$$b_n = \frac{1}{\pi} \int_0^{2\pi} f(x)\,\sin(nx)\,dx$$

for the constants a_0, a_n, b_n.

It is often the case in practice that it is more convenient to use complex exponential function (rather than real trigonometric function) notation.

In the case of a finite Fourier series [1]

$$f(x) = \sum_{-N}^{N} c_n e^{inx},$$

the relevant orthogonality condition is simply that

$$\frac{1}{2\pi} \int_0^{2\pi} e^{inx} e^{-imx} dx = \begin{cases} 0 & m \neq n, \\ 1 & m = n. \end{cases}$$

From this it follows in a similar fashion that the complex analogue of the relations (2.2) is concisely given by

$$c_n = \frac{1}{2\pi} \int_0^{2\pi} f(x) e^{-inx} dx. \tag{2.3}$$

Once the relations (2.2) and (2.3) have been obtained, one is tempted to write expressions of the form

$$f(x) = a_0 + \sum_{n=1}^{\infty} a_n \cos(nx) + b_n \sin(nx)$$

or

$$f(x) = \sum_{n=-\infty}^{\infty} c_n e^{inx}, \tag{2.4}$$

called Fourier series expansions of f for a more or less arbitrary function f defined on the interval $[0, 2\pi]$.

Since the expressions (2.4) are in fact infinite series, there are various questions that arise. Among these are the sense in which the indicated series (2.4) converge (if at all), as well as the class of functions f for which the expansion is appropriate. Some questions along these lines are discussed in the following section.

Aside from these issues, there is the more mundane enterprise of the calculation of the Fourier series expansions (2.4) in the case of concretely given functions f. In principle this is the simple task of evaluating the given integrals. In practice (as would be expected) various methods facilitate the required calculations. Some such techniques arise from various Laplace transform calculations, and are discussed in Chapter 6.

---

[1] We adopt the notation $i^2 = -1$ here.

It is clear that for a given function f the calculations of the c_n and a_n, b_n coefficient sets are closely related. From Euler's formula

$$e^{-inx} = \cos(nx) - i\sin(nx)$$

we obtain

$$c_n = \frac{1}{2\pi} \int_0^{2\pi} f(x)\, e^{-inx}\, dx$$

$$= \frac{1}{2} \frac{1}{\pi} \int_0^{2\pi} f(x)\, [\cos(nx) - i\sin(nx)]\, dx$$

$$= \frac{(a_n - i\, b_n)}{2}$$

for $n \neq 0$, since a_0 is defined with a different coefficient than a_n. Similarly (for $n \neq 0$)

$$a_n + i\, b_n = 2\, c_{-n}.$$

From these relations we obtain

$$a_n = c_n + c_{-n},$$

$$-i\, b_n = c_n - c_{-n},$$

$$b_n = i\, (c_n - c_{-n})$$

and finally

$$c_0 = a_0$$

(since each is evidently the average value of f).

The coefficient formulas and expansions obtained above are valid whether the function in question is real valued or complex valued. Obviously, if f is real valued the coefficients a_n and b_n are all real numbers, while the c_n may assume complex values even in this case. For this reason, the trigonometric expansion (2.1) is often referred to as the *real form* of the Fourier series. The expansion using the complex exponential function is then called the *complex form*.

In the special case of a real valued function, the expression coefficients are related by the simple formulas

$$a_n = 2\,\mathrm{Re}(c_n), \qquad\qquad\qquad (2.5)$$

$$b_n = -2\,\mathrm{Im}(c_n).$$

The above relations are useful in cases where an exponential integral is more convenient than one involving trigonometric terms. Note also that only one integration is involved in the computation of the complex form c_n, while the real form involves evaluation of two integrals. In practice, the complex form is almost always more convenient for calculations.

Example. Define f on the interval $[0, 2\pi]$ by

$$f(x) = \begin{cases} e^x & \text{for } 0 < x < 2\pi \,, \\ f(x + 2\pi) & \text{otherwise.} \end{cases}$$

Compute the real and complex forms of the Fourier series for f. We compute first the complex form of the series.

$$\begin{aligned} c_n &= \frac{1}{2\pi} \int_0^{2\pi} f(x) \, e^{-inx} \, dx \\ &= \frac{1}{2\pi} \frac{e^{(1-in)x}}{(1 - in)} \bigg|_0^{2\pi} \\ &= \frac{1}{2\pi} \frac{(e^{2\pi} - 1)}{(1 - in)}. \end{aligned}$$

Hence, the complex form is

$$f(x) \sim \sum_{-\infty}^{\infty} \frac{1}{2\pi} \frac{(e^{2\pi} - 1)}{(1 - in)} e^{inx}.$$

(We defer the question of equality in the above until the following chapter.) Since

$$c_n = \frac{1}{2\pi} \frac{(e^{2\pi} - 1)(1 + in)}{(1 + n^2)}$$

we obtain, since the given function is real,

$$a_n = \frac{1}{\pi} \frac{(e^{2\pi} - 1)}{(1 + n^2)},$$

$$b_n = \frac{-1}{\pi} \frac{(e^{2\pi} - 1)n}{(1 + n^2)}.$$

Although integration of complex valued functions is formally defined by the separate integration of real and imaginary parts, it may be verified easily that standard integration techniques (and tables) apply.

The real form of the series is therefore

$$f(x) \sim \frac{(e^{2\pi} - 1)}{1} + \sum_{n=1}^{\infty} \left\{ \frac{1}{\pi} \frac{(e^{2\pi} - 1)}{(1 + n^2)} \cos(nx) - \frac{1}{\pi} \frac{(e^{2\pi} - 1)n}{(1 + n^2)} \sin(nx) \right\} .$$

We invite the reader to compute this last result by direct evaluation of the a_n and b_n integrals.

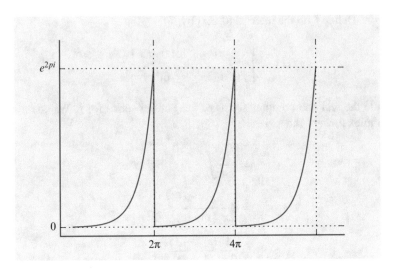

Fig. 2.1 Periodic exponential function

Assuming that equality actually holds in an appropriate sense for the above Fourier series expansions, it is clear that the right-hand sides of the expressions calculated as series above must define functions periodic of period 2π in spite of the fact that the original function was only given on the interval $[0, 2\pi]$. The graph of the function defined by the series therefore must appear as in Figure 2.1. Clearly similar considerations apply in the general case.

One implication of this observation is that the integrals appearing in (2.3) may be evaluated over any convenient interval of length 2π, provided that f is interpreted as being the appropriately defined periodic function. That is, for any x_0,

$$c_n = \frac{1}{2\pi} \int_{x_0}^{x_0 + 2\pi} f(x)\, e^{-inx}\, dx,$$

with similar modifications holding for the $\{a_n, b_n\}$ formulas.

Example. Let f be defined by

$$f(x) = \begin{cases} x & \text{if } 0 \leq x < \pi, \\ x - 2\pi & \text{if } \pi \leq x < 2\pi \end{cases}$$

on the interval $[0, 2\pi]$. Find the real and complex forms of the Fourier series for f.
The 2π-periodic extension of f is given simply by

$$f(x) = \begin{cases} x & \text{if } -\pi \leq x < \pi, \\ f(x + 2\pi) = f(x) \end{cases}$$

(see Figure 2.2). From this it is clear that the interval $[-\pi, \pi]$ is a suitable one for the coefficient evaluation.

$$c_n = \frac{1}{2\pi} \int_{-\pi}^{\pi} x\, e^{-inx}\, dx$$

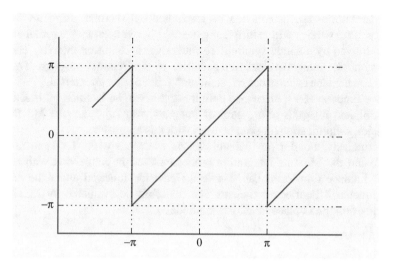

Fig. 2.2 Sawtooth function

$$= \frac{1}{2\pi}\frac{e^{-inx}}{-n^2}(-inx-1)\Big|_{-\pi}^{\pi}$$

$$= \frac{+i(-1)^n}{n}, \ n \neq 0.$$

Obviously, $c_0 = 0$. (This is obtained from Figure 2.2.) The complex form of the series is therefore

$$f(x) \sim \sum_{\substack{n=-\infty \\ n\neq 0}}^{\infty} \frac{+i(-1)^n}{n} e^{inx}$$

The real form of the series follows from the coefficient relation found above. We have

$$a_0 = 0$$

$$a_n = 0$$

$$b_n = \frac{2(-1)^{n+1}}{n}$$

and hence

$$f(x) \sim \sum_{n=1}^{\infty} \frac{2(-1)^{n+1}}{n} \sin(nx).$$

Example. Fourier coefficients can be calculated by Maxima, using the "fourie" package. The package will readily compute the Fourier coefficients for an example function given by a single symbolic formula over a definition interval. The issue is the symbolic integral calculation facilities of the Maxima program. The basic integration handles expressions of "standard" functions on an interval.

If the function has a piecewise definition, this can be handled by breaking up the coefficient integrals into a sum of integrals over sub-intervals. Maxima has expression simplification that can be used with this approach.

Alternatively, Fourier coefficients can be evaluated with "Laplace transform tricks", and the Maxima integration procedure used in connection with calculation of Laplace transforms can handle a Heaviside function under the name of "step_function." Fourier coefficients can therefore be evaluated by Maxima by detouring through Laplace transform methods.

WXMAXIMA

```
-->        load(fourie)
           fourier(x, x, pi)
           foursimp(%)
```

(%o9) $/usr/share/maxima/5.37.3/share/calculus/fourie.mac$

(%t10) $a_0 = 0$

(%t11) $a_n = 0$

(%t12) $b_n = \dfrac{2\left(\dfrac{sin(\pi n)\pi^2}{\pi^2 n^2} - \dfrac{cos(\pi n)\pi^2}{\pi n} \right)}{\pi}$

(%o12) $[\%t10, \%t11, \%t12]$

(%t13) $a_0 = 0$

(%t14) $a_n = 0$

(%t15) $b_n = - \dfrac{2\left(-1 \right)^n \pi}{\pi n}$

(%o15) $[\%t13, \%t14, \%t15]$

```
-->        fourier((x/pi)^2, x, pi)
           foursimp(%)
```

(%t16) $a_0 = \dfrac{1}{3}$

(%t17) $a_n = \dfrac{2\left(\dfrac{sin(\pi n)\pi}{\pi n} - \dfrac{2 sin(\pi n)\pi}{\pi^3 n^3} + \dfrac{2 cos(\pi n)\pi}{\pi^2 n^2} \right)}{\pi}$

(%t18) $b_n = 0$

(%o18) $[\%t16, \%t17, \%t18]$

(%t19) $a_0 = \dfrac{1}{3}$

(%t20) $a_n = \dfrac{4\left(-1 \right)^n}{\pi^2 n^2}$

(%t21) $b_n = 0$

(%o21) $[\%t19, \%t20, \%t21]$

Example. Let the function f be defined by

$$f(x) = \begin{cases} 1 & \text{if } 0 \leq x < \pi, \\ -1 & \text{if } \pi \leq x < 2\pi, \\ f(x + 2\pi) = f(x) & \text{otherwise.} \end{cases}$$

Then (by inspection) $c_0 = 0$, and the complex form expansion coefficients are

$$\begin{aligned} c_n &= \frac{1}{2\pi} \left[\int_0^\pi (1)e^{-inx}\, dx + \int_\pi^{2\pi} (-1)e^{-inx}\, dx \right] \\ &= \frac{1}{2\pi} \left[\frac{e^{-inx}}{-in} \Big|_0^\pi + \frac{(-1)e^{-inx}}{-in} \Big|_\pi^{2\pi} \right] \\ &= \frac{1}{2\pi} \left[\frac{(-1)^n - 1}{-in} + \frac{(-1)^n - 1}{-in} \right] \\ &= \frac{1}{in\pi} [1 - (-1)^n]. \end{aligned}$$

The complex form expansion is

$$f(x) \sim \sum_{\substack{n=-\infty \\ n \neq 0}}^{\infty} \frac{1}{in\pi} [1 - (-1)^n] e^{inx}. \tag{2.6}$$

From the coefficient relations

$$a_n = 0,$$

$$b_n = \frac{2}{n\pi} [1 - (-1)^n],$$

so that the real form is

$$f(x) \sim \sum_{n=1}^{\infty} \frac{2}{n\pi} [1 - (-1)^n] \sin(nx).$$

Example. Define f (periodic of period 2π) by

$$f(x) = \begin{cases} x & \text{if } 0 \leq x < \pi, \\ -x & \text{if } -\pi \leq x < 0, \\ f(x + 2\pi) = f(x) & \text{otherwise.} \end{cases}$$

Then (consider a graph of f)

$$c_0 = \frac{\pi}{2}$$

and

$$c_n = \frac{1}{2\pi} \left[\int_{-\pi}^{0} (-x)e^{-inx} \, dx + \int_{0}^{\pi} (x)e^{-inx} \, dx \right]$$

$$= \frac{1}{2\pi} \left[\frac{-e^{-inx}(-inx - 1)}{-n^2} \bigg|_{-\pi}^{0} + \frac{e^{-inx}(-inx - 1)}{-n^2} \bigg|_{0}^{\pi} \right]$$

$$= \frac{1}{2\pi} \left[\frac{1 + (-1)^n(in\pi - 1)}{-n^2} + \frac{1 + (-1)^n(-in\pi - 1)}{-n^2} \right]$$

$$= \frac{1}{n^2\pi} [(-1)^n - 1].$$

This gives in turn

$$b_n = 0,$$

$$a_n = \frac{2}{n^2\pi} [(-1)^n - 1],$$

$$a_0 = \frac{\pi}{2}.$$

The two forms of the expansion are therefore

$$f(x) \sim \frac{\pi}{2} + \sum_{\substack{n=-\infty \\ n \neq 0}}^{\infty} \frac{1}{n^2\pi} [(-1)^n - 1] \, e^{inx}$$

and

$$f(x) \sim \frac{\pi}{2} + \sum_{n=1}^{\infty} \frac{2}{n^2\pi} [(-1)^n - 1] \cos(nx).$$

Problems 2.1.

1. Let f be defined by

$$f(x) = \begin{cases} 1 & \text{if } 0 \leq x < \pi, \\ 0 & \text{if } \pi \leq x < 2\pi, \\ f(x + 2\pi) = f(x) & \text{otherwise.} \end{cases}$$

Determine both the real and complex forms of the Fourier expansion of f.

2. Repeat Problem 1 for the function defined by

$$f(x) = \begin{cases} x^2 & \text{if } 0 \leq x < \pi, \\ (2\pi - x)^2 & \text{if } \pi \leq x < 2\pi, \\ f(x + 2\pi) = f(x) & \text{otherwise.} \end{cases}$$

3. Repeat Problem 1 for the function h defined by

$$h(x) = \sin\left(\frac{x}{2}\right), \quad 0 < x < 2\pi.$$

4. Consider the infinite series calculated in Problem 1. Evaluate the series at the point $x = \pi$. Does the series sum to $f(\pi)$?

5. Find the real form of the Fourier expansion for the function

$$f(x) = \begin{cases} e^x & \text{if } 0 \le x < 2\pi, \\ f(x + 2\pi) = f(x) & \text{otherwise} \end{cases}$$

by direct evaluation of the real coefficients a_0, $\{a_n\}$, and $\{b_n\}$. The problem can be done this way, but ... .

6. Define f by

$$f(x) = \begin{cases} x & \text{if } 0 \le x < \frac{\pi}{2}, \\ (\pi - x) & \text{if } \frac{\pi}{2} \le x < \pi, \\ 0 & \text{if } \pi \le x < 2\pi, \\ f(x + 2\pi) = f(x) & \text{otherwise.} \end{cases}$$

Find the real and complex forms of the Fourier expansion for f.

7. Repeat Problem 6 for the function f defined by

$$f(x) = \begin{cases} e^{-|x|} & \text{if } -\pi \le x < \pi, \\ f(x + 2\pi) = f(x) & \text{otherwise.} \end{cases}$$

8. Define f by

$$f(x) = \begin{cases} \pi + x & \text{if } -\pi \le x < 0, \\ \pi - x & \text{if } 0 \le x < \pi, \\ f(x + 2\pi) = f(x) & \text{otherwise.} \end{cases}$$

Find the real and complex Fourier expansions for f.

9. Suppose that the function f is exactly expressible as a finite Fourier series, so that

$$f(x) = \sum_{n=-N}^{N} c_n e^{inx}$$

for some fixed finite N. Use the orthogonality conditions for the complex exponential functions to show that

$$\frac{1}{2\pi} \int_0^{2\pi} f(x)\overline{f(x)}\, dx = \frac{1}{2\pi} \int_0^{2\pi} |f(x)|^2\, dx = \sum_{n=-N}^{N} |c_n|^2.$$

2.2   Inner Products and Fourier Expansions

As mentioned in the previous section, the formal derivation of the Fourier expansion formulas sheds little light on the meaning of the expansion formulas. The expansions are formally infinite in length, while (at least in computational practice) one must settle for finite sums. In this case, a notion of closeness of the finite sums (partial sums of the indicated infinite series) to the original function is required.

In the case of Fourier series such a notion of closeness can be readily obtained from the properties of inner product spaces. This approach to the expansion problem has the additional benefit that it applies directly to problems more general than the trigonometric expansion described above. Such problems arise during the solution of partial differential equations describing various physical processes involving non-rectangular physical geometries. These expansions require the use of coefficient calculation formulas that can be readily understood on the basis of an inner product space approach, and are otherwise rather mysterious.

Problems of this type are discussed in some detail in Chapter 4, and are illustrated with examples involving cylindrical and spherical geometries for the underlying physical problem.

Definition (Inner product space). An inner product space is a vector space V (in general with complex numbers as the field of scalars) equipped with an inner product function $(.,.)$ defined from pairs of vectors $V \times V$ and taking values in $\mathbb{C}$, (the complex numbers) satisfying (for all $u, v, w \in V$, $\alpha, \beta \in \mathbb{C}$) the conditions

1. $(\alpha u + \beta v, w) = \alpha (u, w) + \beta (v, w)$,
2. $(u, w) = \overline{(w, u)}$,
3. $(u, u) \geq 0$,
4. $(u, u) = 0$ if and only if $u = 0$.

Associated with an inner product space is a notion of distance, or a norm function, defined by

$$\|u\| = ((u, u))^{1/2}$$

and enjoying the properties of a norm. The proof that the norm $\| \cdot \|$ satisfies the desired triangle inequality

$$\|u + v\| \leq \|u\| + \|v\|$$

can be obtained by use of the Cauchy–Schwarz inequality

$$|(u, v)| \leq \|u\| \, \|v\|.$$

This inequality is applicable in such a wide variety of contexts that it is hard to overstate its usefulness. (See Problem 8, Section 2.10, and Section 7.6 equation (7.20) for typical applications to estimation and convergence estimates.)

Example. Let $V = \mathbb{C}^n$ (n-tuples of complex numbers), and define

$$(x, y) = \sum_{i=1}^{n} x_i \overline{y_i}.$$

Example. Let V denote complex-valued functions f defined on $[0, 2\pi]$ and such that[2]

$$\int_0^{2\pi} |f(x)|^2 < \infty.$$

Then $(f, g) = \frac{1}{2\pi} \int_0^{2\pi} f(x)\overline{g(x)}\, dx$ defines an inner product. The restriction that both f and g be of finite norm guarantees (by the Cauchy–Schwarz inequality) that the inner product function is well defined and produces a finite number for each argument pair f, g.

Example. Let r be a strictly positive continuous real-valued function defined on $[0, 2\pi]$. Then

$$(f, g) = \int_0^{2\pi} f(x)\overline{g(x)}\, r(x)\, dx$$

also defines an inner product on the vector space V of complex-valued functions.

Example. Clearly multidimensional versions of the above are possible. The expression

$$(f, g) = \int_0^1 \int_0^1 f(x, y)\overline{g(x, y)}\, dx dy$$

defines an inner product on functions defined on the unit square and such that

$$\int_0^1 \int_0^1 |f(x, y)|^2\, dx dy < \infty,$$

$$\int_0^1 \int_0^1 |g(x, y)|^2\, dx dy < \infty.$$

---

[2] There is a subtle point (which we largely bypass) involved in this example. Strictly speaking, the indicated integrals should be understood as Lebesgue type integrals rather than conventional Riemann integrals. Further, the vectors should be regarded as equivalence classes of functions whose difference on sets of points is negligible as far as integration is concerned.

The presence of an inner product provides a geometrical notion of orthogonality associated with the vector space in question. The vectors u, v are called *orthogonal* whenever

$$(u, v) = 0$$

holds.

In an inner product space there are distinguished sets of vectors referred to as orthonormal sets.

Definition (Orthonormal set). Given a set of vectors $\{u_1, u_2, \ldots, u_N\}$ in an inner product space V, the set is called *orthonormal* provided that

$$(u_i, u_j) = \delta_{ij} = \begin{cases} 1 \ i = j, \\ 0 \ i \neq j. \end{cases}$$

The geometrical interpretation of such sets is that the elements are all mutually perpendicular and of unit length.

The inner product space appropriate to the discussion of Fourier series is that of complex-valued functions.

Example. Let V be the inner product space of complex-valued square-integrable functions. Then the set of functions

$$\{e^{inx}\}_{-N}^{N} = \{e^{-iNx}, e^{-i(N-1)x}, \ldots, 1, e^{ix}, \ldots, e^{iNx}\}$$

is an orthonormal set with respect to the inner product

$$(f, g) = \frac{1}{2\pi} \int_0^{2\pi} f(x)\overline{g(x)}\, dx.$$

(Note that the above set contains $2N + 1$ functions as elements.)

Example. Let V be defined as in the previous example. Then the set of functions ($2N + 1$ in number)

$$\{\sqrt{2}\, \sin(x), \ldots, \sqrt{2}\, \sin(Nx), 1, \sqrt{2}\cos(x), \ldots, \sqrt{2}\cos(Nx)\}$$

is also an orthonormal set in V with respect to the same inner product.

The partial sums of the Fourier series expansion formally arrived at in Section 2.1 have the form

$$a_0 + \sum_{n=1}^{N} a_n \cos(nx) + b_n \sin(nx)$$

or

$$\sum_{-N}^{N} c_n e^{inx}.$$

An inevitable natural question to consider is the closeness of these partial sums to the function from which the expansion coefficients were calculated.

Since the introduction of an inner product provides a notion of distance between functions, it is natural to use this to measure the degree of error associated with a partial sum. In fact, rather than just considering the formulas for the expansion coefficients derived above in Section 2.1, we may consider the problem (apparently more general) of choosing a set of (complex) constants $\{\alpha_n\}_{n=-N}^{N}$ with the property that the mean-square error

$$\frac{1}{2\pi} \int_0^{2\pi} \left| f(x) - \sum_{n=-N}^{N} \alpha_n e^{inx} \right|^2 dx$$

is as small as possible. Note that this quantity is exactly the square of the norm of the approximation error, given that the norm is defined through the inner product.

It turns out that a solution to this problem produces the Fourier coefficient formulas of the previous section. The minimization could be carried out as a calculus problem, but it is more generally useful to carry out the derivation on the basis of an inner product space argument. This approach provides the sound basis for discussing convergence and error estimates, and is the source of intuition useful in various applications of Fourier series methods.

The central result applicable to this problem (incidentally also to an extremely wide number of problems of interest) is known as the *projection theorem*. The result quoted below is an elementary version (derivable from linear algebra), as only a finite-dimensional projection is involved. This version, however, suffices for the problems under consideration.

Theorem (Projection Theorem). *Let V be an inner product space and W a finite-dimensional subspace of V. Then for an arbitrary vector $v \in V$, there exists a vector $\hat{w}$, the unique best approximation to v as a linear combination of vectors from the approximating subspace W. That is, $\hat{w}$ satisfies the optimality condition*

$$\|v - \hat{w}\| \leq \|v - w\| \ \text{for all} \ w \in W.$$

The best approximant $\hat{w}$ is called the projection of v on W, and is characterized by the condition that the approximation error is orthogonal to the subspace W. That is, if the error vector is

$$e = v - \hat{w},$$

then

$$(e, w) = 0 \ \text{for all} \ w \in W.$$

The best approximant $\hat{w}$ is readily calculated in the case when an orthonormal basis $\{u_1, u_2, \ldots, u_M\}$ is available for the subspace W. Here the formula is

$$\hat{w} = \sum_{i=1}^{M} (v, u_i)\, u_i.$$

This is readily verified by noting that

$$(v - \hat{w}, u_j) = (v, u_j) - \sum_{i-1}^{M} (v, u_i)(u_i, u_j)$$

$$= (v, u_j) - (v, u_j)$$

$$= 0,$$

so that the error is orthogonal to a basis for W, and hence orthogonal to the whole subspace W as required.

Another consequence of the orthogonality condition on $\hat{w}$ is a version of the Pythagorean Theorem. Since

$$v = (v - \hat{w}) + \hat{w} = e + \hat{w},$$

we have

$$\|v\|^2 = ((v - \hat{w}) + \hat{w},\ (v - \hat{w}) + \hat{w})$$

$$= (e, e) + (e, \hat{w}) + (\hat{w}, e) + (\hat{w}, \hat{w})$$

$$= \|e\|^2 + \|\hat{w}\|^2.$$

This leads one to construct the suggestive picture of Figure 2.3 to illustrate the situation.

Example. If one applies the above result to the inner product space of square-integrable functions defined above, and chooses as W the subspace of V obtained as linear combinations of the orthonormal set $\{e^{inx}\}_{n=-N}^{N}$, one obtains the following. The vector v which is to be approximated is identified with the (square-integrable) function f. The projection coefficients are (the inner products)

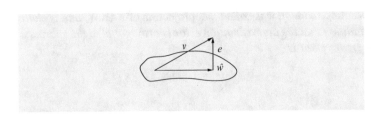

Fig. 2.3 Projection theorem

$$(v, u_n) = \frac{1}{2\pi} \int_0^{2\pi} f(x)\overline{e^{inx}}\, dx$$

$$= \frac{1}{2\pi} \int_0^{2\pi} f(x)e^{-inx}\, dx$$

$$= c_n,$$

and the best approximant is simply the partial sum of the Fourier series

$$\sum_{n=-N}^{N} (v, u_n)\, u_n = \sum_{n=-N}^{N} c_n\, e^{inx}.$$

Entirely similar computations using the trigonometric orthogonal set lead to the real form Fourier series expansion formulas.

In addition to providing a rational derivation of the standard Fourier expansions, the derivation supplies some side benefits.

One of these is the identification of square-integrable functions as the natural functions for which the expansions are appropriate. Such functions correspond physically to processes of finite energy, and so arise naturally in various problems. (See Section 2.10 below).

On a more practical level, one is led directly to a simple formula for the approximation error. From the Pythagorean Theorem we have

$$\|v\|^2 = \|e\|^2 + \|\hat{w}\|^2,$$

while the norm squared of the approximant $\hat{w}$ is readily calculated in terms of an orthonormal basis. Since

$$\hat{w} = \sum_{n=1}^{N} \alpha_n\, u_n,$$

$$\alpha_n = (v, u_n)$$

with $\{u_n\}$ orthonormal, we have

$$\|\hat{w}\|^2 = \left(\sum_{i=1}^{N} \alpha_i u_i, \sum_{j=1}^{N} \alpha_j u_j \right)$$

$$= \sum_{i=1}^{N} \sum_{j=1}^{N} \alpha_i \overline{\alpha_j}(u_i, u_j)$$

$$= \sum_{i=1}^{N} |\alpha_i|^2.$$

This means

$$\|v\|^2 = \|e\|^2 + \sum_{i=1}^{N} |\alpha_i|^2,$$

and hence the error norm is

$$\|e\|^2 = \|v\|^2 - \sum_{i=1}^{N} |\alpha_i|^2.$$

This provides an explicit formula for the size of the error in terms of the given vector v and the currently calculated expansion coefficients

$$\{\alpha_i\}_{i=1}^{N}.$$

Example. For the inner product space V of square-integrable functions, the above relation in explicit terms is

$$\frac{1}{2\pi} \int_0^{2\pi} \left| f(x) - \sum_{n=-N}^{N} c_n e^{inx} \right|^2 dx = \frac{1}{2\pi} \int_0^{2\pi} |f(x)|^2 dx - \sum_{n=-N}^{N} |c_n|^2.$$

Example. The corresponding calculation may be carried out for the real form of the expansion. The resulting formula may be derived directly by taking account of the factors of $\frac{1}{2}$ involved in comparing the $\{c_n\}$ defined above with the corresponding real form coefficients. Alternatively, one may use the relations

$$c_0 = a_0$$

$$c_n = \frac{a_n - ib_n}{2}$$

$$c_{-n} = \frac{a_n + ib_n}{2}$$

to give by direct calculation

$$|c_n|^2 + |c_{-n}|^2 = \frac{|a_n|^2 + |b_n|^2}{4} + \frac{|a_n|^2 + |b_n|^2}{4}.$$

Combining this with the complex case result,

$$\frac{1}{2\pi} \int_0^{2\pi} \left| f(x) - a_0 - \sum_{n=1}^{N} a_n \cos(nx) + b_n \sin(nx) \right|^2 dx$$

$$= \frac{1}{2\pi} \int_0^{2\pi} |f(x)|^2 dx - |a_0|^2 - \sum_{n=1}^{N} \frac{|a_n|^2 + |b_n|^2}{2},$$

we obtain the error expression in terms of a real form Fourier series expansion.

Formally taking the limit as $N \to \infty$ in the above formulas, we obtain (by assuming that the error term approaches 0 as $N \to \infty$) the so-called Parseval relations:

$$\frac{1}{2\pi} \int_0^{2\pi} |f(x)|^2 \, dx = \sum_{n=-\infty}^{\infty} |c_n|^2 \qquad (2.7)$$

$$\frac{1}{2\pi} \int_0^{2\pi} |f(x)|^2 \, dx = |a_0|^2 + \sum_{n=1}^{\infty} \frac{|a_n|^2 + |b_n|^2}{2}.$$

The validity of this limit calculation is discussed in detail below.

These relations provide an interpretation of the partition of signal energy into harmonic components; this interpretation is developed in Section 2.10 below. The Parseval relations (2.7) also provide a means of evaluating the sums of certain infinite series.

Example. Take

$$f(x) = x, \ 0 \le x < 2\pi,$$

then

$$c_n = \frac{1}{2\pi} \int_0^{2\pi} x e^{-inx} \, dx$$

$$= \frac{1}{2\pi} \frac{e^{-inx}(-inx - 1)}{(-in)^2} \Big|_0^{2\pi}$$

$$= \frac{1}{-in}, n \ne 0.$$

$c_0 = \pi$, the average of f. Assuming (as can be shown) that

$$\lim_{N \to \infty} \frac{1}{2\pi} \int_0^{2\pi} \left| x - [\pi + \sum_{\substack{n=-N \\ n \ne 0}}^{N} \frac{-1}{in} e^{inx}] \right|^2 dx = 0,$$

Parseval's relation provides

$$\frac{1}{2\pi} \int_0^{2\pi} x^2 \, dx = \frac{4}{3} \pi^2$$

$$= \sum_{n=-\infty}^{\infty} |c_n|^2$$

$$= \pi^2 + 2 \sum_{n=1}^{\infty} \frac{1}{n^2}.$$

and so lets us evaluate

$$\sum_{n=1}^{\infty} \frac{1}{n^2} = \frac{\pi^2}{6}.$$

Problems 2.2.

1. Use the inner product axioms to show that for all complex scalars β we have

$$(u, \beta\, v) = \overline{\beta}\, (u, v)$$

and

$$\|\beta u\|^2 = |\beta|^2\, \|u\|^2.$$

2. Use the inner product axioms to prove the "parallelogram law"

$$\|x + y\|^2 + \|x - y\|^2 = \|x\|^2 + \|y\|^2.$$

Illustrate this relation with a diagram in the case of a two-dimensional (real) vector space.

3. Show that the functions of two variables u_{mn} defined by

$$u_{mn}(x, y) = \sin(nx)\, \cos(my)$$

are orthonormal with respect to the inner product

$$(u, v) = \frac{1}{\pi^2} \int_0^{2\pi} \int_0^{2\pi} u(x, y)\overline{v(x, y)}\, dxdy.$$

4. Define a sequence of functions on $[0,1]$ by the conditions

$$f_0(x) = 1,\ 0 < x < 1,$$

$$f_1(x) = \begin{cases} 1 & 0 \le x < \frac{1}{2}, \\ -1 & \frac{1}{2} \le x < 1, \end{cases}$$

$$f_2(x) = f_1(2x),\ 0 \le x < \frac{1}{2},$$

$$f_2(x + \frac{1}{2}) = f_2(x),$$

$$f_3(x) = f_2(2x),\ 0 \le x < \frac{1}{2},$$

$$f_3(x + \frac{1}{4}) = f_3(x);$$

alternatively,

$$f_3(x) = f_1(4x), \quad 0 < x < \frac{1}{4},$$

$$f_3\left(x + \frac{1}{4}\right) = f_3(x).$$

In general,

$$f_j(x) = f_1(2^{j+1}x), \quad 0 \le x < \frac{1}{2^{j+1}},$$

$$f_j\left(x + \frac{1}{2^{j+1}}\right) = f_j(x), \quad j = 2, 3, 4, \ldots.$$

a. Graph these functions.
b. Show that these functions are pairwise orthonormal in the sense that

$$\int_0^1 f_j(x) f_k(x) \, dx = \delta_{jk}$$

$$= \begin{cases} 1 & k = j, \\ 0 & k \ne j. \end{cases}$$

(These functions are readily generated by use of digital counter circuits.)
5. Consider the "square wave" functions of Problem 4, above. State a Parseval relation for the expansion of a given function f assumed expressible as a series of square waves:

$$f(x) = \sum_{j=0}^{\infty} d_j f_j(x).$$

Hint: Define d_j through projection, and apply the Pythagorean theorem.
6. Calculate the projections of the functions $\{e^{inx} \, e^{-inx}\}$ on the two-dimensional subspace spanned by $\{\cos(nx), \sin(nx)\}$ (n is fixed). Use the inner product associated with the Fourier series derivation above. Formally deduce the relation between the real and complex Fourier expansions from your projection computations.
7. Use the Parseval relations (2.7) and Fourier coefficient calculations from earlier examples to evaluate

$$\sum_{n=1}^{\infty} \frac{1}{(2n-1)^2},$$

and

$$\sum_{n=1}^{\infty} \frac{1}{n^2 + 1}.$$

2.3   Convergence of Fourier Series

The discussion of Fourier series expansions in the previous section is based entirely on a circle of ideas from linear algebra. This approach provides the required expansion formulas and provides an expression for the mean-square error associated with a partial sum of the series by use of the Projection Theorem.

However, in order to carefully discuss the limiting process leading to the Parseval relation formally derived above, it is necessary to introduce some convergence notions from analysis.

Convergence notions are usually introduced in introductory calculus texts in terms of limits of sequences of real numbers. Recall that convergence of a sequence of real numbers is initially phrased in terms of the existence of a real number L (the limit) with the property that

$$\lim_{n \to \infty} |s_n - L| = 0.$$

The difficulty with this situation is of course that it requires one to guess the value of the limit in order to know that it exists, while one would prefer to be able to deduce existence from the properties of the given sequence itself. (Actual computation of L then proceeds, say, numerically if necessary.)

The fact that this preferred situation is the one which actually holds is the content of the celebrated Cauchy Criterion for convergence (of a real or complex number sequence).

Theorem (Cauchy Criterion). *The sequence $\{s_n\}$ converges if and only if for each number $\epsilon > 0$, there exists an integer $N(\epsilon)$ (depending on ϵ) with the property that*

$$|s_n - s_m| < \epsilon$$

for all $n, m \geq N(\epsilon)$.

The next problem to consider is that of the convergence of a sequence of functions. Given a sequence of functions $\{s_n(\cdot)\}$ defined on some domain (for instance, an interval of the real line), we may obtain a sequence of numbers by selecting a fixed value x_1 in the domain and forming the numerically valued sequence $\{s_n(x_1)\}$. The sequence so constructed may or may not be convergent in the sense discussed above.

Selection of a different value x_2 of the domain produces a different sequence $\{s_n(x_2)\}$ whose convergence may be considered.

Definition (Pointwise Convergence). The sequence $\{s_n\}$ is said to be *pointwise convergent* (on a domain) provided that for each fixed value of x (in the domain) the numerically valued sequence $\{s_n(x)\}$ is convergent.

If the sequence is pointwise convergent (on a domain), then the sequence defines a limit function s for every x in the domain by the rule

$$s(x) = \lim_{n \to \infty} s_n(x).$$

Example. Consider the sequence of functions $\{s_n\}$ defined on the interval $[0,1]$ by

$$s_n(x) = x^n, \ 0 \le x \le 1.$$

This sequence is pointwise convergent to the function s defined by

$$s(x) = \begin{cases} 0 & 0 \le x < 1, \\ 1 & x = 1. \end{cases}$$

Example. Define a sequence of functions $\{s_n\}$ by

$$s_n(x) = \begin{cases} nx & 0 \le x < \frac{1}{n}, \\ n(\frac{2}{n} - x) & \frac{1}{n} \le x < \frac{2}{n}, \\ 0 & \frac{2}{n} \le x \le 1. \end{cases}$$

Then $\{s_n\}$ is pointwise convergent to the zero function. This follows since $s_n(0)$ vanishes for all n, while for each $x > 0$, $2/n$ is eventually smaller than x, and $s_n(x)$ vanishes thereafter, making the limit value 0.

Definition (Pointwise convergence). The formal statement of the condition that the sequence of functions $\{s_n\}$ defined on some interval I should be pointwise convergent to a function s defined on I is obtained by applying the usual definition of limit (of a numerical sequence) for each x in I. This gives the conclusion that for each x in I, and each $\epsilon > 0$, there exists an integer $N(x, \epsilon)$ such that

$$|s_n(x) - s(x)| < \epsilon$$

for all $n \ge N(x, \epsilon)$. Note that the integer $N(x, \epsilon)$ depends on the point x of the interval I. A pointwise convergent sequence of functions may converge much more rapidly for some values of x than for others. This may be verified for the case of the examples above.

While the idea of pointwise convergence may seem one of the most natural notions to apply to a sequence of functions, it is a relatively weak condition with some built-in drawbacks. One of these relates to problems of the numerical evaluation of the limit function. The definition shows that the cut-off index $(N(x, \epsilon))$ required for given accuracy depends on the argument x. From the point of view of computation, it would be more convenient (for purposes of programming) if the cut-off index were independent of x. A second consideration relates to the technical properties of the limit function. It is easy to see (from the example above) that a pointwise convergent sequence of continuous functions $\{s_n\}$ on an interval I may converge to a discontinuous limit. It is also sometimes desired to draw conclusions about integrals (or derivatives) of a sequence of functions. A concept that plays a central role in the treatment of such problems (see [2], for example) is that of uniform convergence. (This concept may be discussed in terms of functions with a largely arbitrary domain of definition although we discuss it below in terms of the interval [0,1].)

Definition (Uniform Convergence). A sequence of functions $\{s_n\}$ defined on the interval $[0, 1]$ is said to be uniformly convergent to the function s if and only if for each $\epsilon > 0$ there exists an integer $N(\epsilon)$ with the property that

$$\sup_{x \in [0,1]} |s_n(x) - s(x)| < \epsilon$$

for all $n > N(\epsilon)$.

The intuitive content of this definition is that (as contrasted with the idea of pointwise convergence) the rate of convergence is independent of the point in the interval chosen for observation.

Example. The sequence $\{s_n\}$ discussed earlier, where

$$s_n(x) = \begin{cases} nx & 0 \leq x < \frac{1}{n}, \\ n(\frac{2}{n} - x) & \frac{1}{n} \leq x < \frac{2}{n}, \\ 0 & \frac{2}{n} \leq x \leq 1, \end{cases}$$

is not uniformly convergent to the zero function, since, in fact,

$$\sup_{x \in [0,1]} |s_n(x) - 0| = s_n(\frac{1}{n}) = 1,$$

which cannot be made arbitrarily small.

Example. Let the interval of definition be $I = [0, 1/2]$, and define

$$s_n(x) = x^n, \ 0 \leq x \leq 1/2.$$

Then $\{s_n\}$ is uniformly convergent to the zero function, since

$$\sup_{x \in [0, \frac{1}{2}]} |s_n(x) - 0| = (\frac{1}{2})^n,$$

and this may be made less than any $\epsilon, 0 < \epsilon < 1$, as long as

$$n > \frac{\ln(\epsilon)}{\ln(\frac{1}{2})}.$$

In analogy with the case of the convergence of a numerically valued sequence, it would be useful to have a criterion for uniform convergence which did not in principle require knowledge of the limit function to establish convergence. Such a result is provided by the so-called Cauchy Criterion for uniform convergence.

Theorem (Cauchy Criterion). *A sequence of functions defined on the interval [0,1] converges uniformly if and only if for any $\epsilon > 0$ there exists an integer $N(\epsilon)$ with the property that*

$$\sup_{x \in [0,1]} |s_n(x) - s_m(x)| < \epsilon$$

for all $m, n \geq N(\epsilon)$.

One of the useful consequences of the notion of uniform convergence is the fact that the limit function of a uniformly convergent sequence of continuous functions defined on the interval $[0, 1]$ is itself continuous. This may be proved in a relatively straightforward fashion, and is left as a problem for this section. The combination of this observation with the Cauchy Criterion immediately above provides a result which provides a motivating example for the convergence framework discussed below.

Theorem (Cauchy Criterion for continuous functions). *A sequence of continuous functions $\{s_n\}$ defined on the interval $[0, 1]$ is uniformly convergent to a continuous limit function s if and only if for each $\epsilon > 0$ there exists an integer $N(\epsilon)$ such that*

$$\sup_{x \in [0,1]} |s_n(x) - s_m(x)| < \epsilon$$

for all $m, n > N(\epsilon)$.

The striking similarity of the two Cauchy criteria quoted above suggests the existence of a unifying underlying formalism. One framework which presents this is that of normed vector spaces.

Definition (Normed vector space). A *normed vector space* V is a vector space (over the real or complex field, R or C) equipped with a norm function $\|\cdot\| : V \mapsto R$ satisfying

1. $\|\alpha v\| = |\alpha| \, \|v\|$ for all scalar α,
2. $\|u + v\| \leq \|u\| + \|v\|$, and
3. $\|v\| = 0$ if and only if $v = 0$.

Example. If $V = R$, then $\|v\| = |v|$ defines a norm.

Example. If $V = R^n$ then $\|v\| = (\sum_{i=1}^{n} x_i^2)^{\frac{1}{2}}$ (the x_i are the coordinates of v) defines a norm.

Example. If V is the space of continuous functions from $[0, 1]$ to R, then

$$\|v\| = \sup_{x \in [0,1]} |v(x)|$$

defines a norm.

The definition of a normed vector space provides a replacement for the notion of "absolute value," and the usual "absolute value manipulations" are readily duplicated in this broader context. Definitions of notions encountered in the framework of real or complex numbers may be readily generalized by replacing "absolute value" by "norm." The most fundamental such notion is that of a limit. This amounts to more than a hollow typographic exercise, since the exercise produces useful generalizations.

Definition (Limit). A sequence $\{s_n\}$ in a normed vector space V is said to have a limit s (in V) if and only if, for any $\epsilon > 0$, there exists an index $N(\epsilon)$ such that

$$\|s_n - s\| < \epsilon$$

for all $n > N(\epsilon)$.

The previous discussion of classical convergence results shows evidence of a common base underlying the Cauchy convergence criteria. This base can be exposed by considering sequences of vectors $\{s_n\}$ in a normed vector space V, and defining a subclass consisting of Cauchy sequences.

Definition (Cauchy sequence). A sequence $\{s_n\}$ in a normed vector space V is a *Cauchy sequence* if and only if, for any $\epsilon > 0$, there exists an integer $N(\epsilon)$ with the property that

$$\|s_n - s_m\| < \epsilon$$

for all $m, n > N(\epsilon)$.

Definition (Complete normed space). A normed linear space V is called *complete* if and only if every Cauchy sequence in V has a limit (in V).

Example. If $V = R$, $\|v\| = |v|$, then V is a complete normed vector space (this is exactly the usual assertion of completeness of the real numbers, often discussed in the appendix of an introductory calculus text).

Example. If V is the set of continuous functions from $[0, 1]$ to R, and

$$\|v\| = \sup_{x \in [0,1]} |v(x)|,$$

then a Cauchy sequence is one for which, given $\epsilon > 0$, there exists $N(\epsilon)$ such that

$$\sup_{x \in [0,1]} |s_n(x) - s_m(x)| < \epsilon$$

for all $n, m > N(\epsilon)$.

Completeness is then the assertion that every Cauchy sequence converges to a limit function continuous on $[0, 1]$. This is exactly the statement of the result given above regarding the Cauchy Criterion for uniform convergence of a sequence of continuous functions defined on the interval $[0, 1]$.

Definition (Banach space). As a cultural remark, we mention that a complete normed space is called a *Banach space*. The examples above are hence all examples of Banach spaces.

It should be clear that it is in general a lot of work to prove that any given normed vector space is complete. In essence, all of the hard analysis (e.g., epsilon-delta argument) is tidily packed in the assertion of completeness for any particular example.

To specialize to the discussion of topics relevant to Fourier series, recall that an inner product space naturally inherits a norm through the inner product,

$$\|v\| = ((v, v))^{1/2}.$$

As a normed vector space, it is subject to the convergence constructs mentioned above. Cauchy sequences are naturally defined using the inner product norm, and one may inquire whether or not the resulting normed space is a complete one, that is, whether or not it is actually a Banach space.

Definition (Hilbert space). An inner product space which is complete with respect to the norm inherited from its inner product is called a *Hilbert space*.

Since Fourier series are associated with expansions in terms of orthonormal vectors, we next consider such problems in the context of Hilbert spaces.

Suppose that V is an infinite-dimensional Hilbert space, and consider an infinite sequence of orthonormal vectors $\{u_i\} = \{u_1, u_2, \ldots\}$ in V. Associated with such a sequence is a sequence of subspaces of V defined by

$$W_N = \text{span } \{u_1, u_2, \ldots, u_N\}.$$

For a fixed $v \in V$, we may construct the orthogonal projection of v on the subspace W_N. This is given by the expression

$$\hat{w}_N = \sum_{n=1}^{N} \alpha_n u_n$$

where the expansion coefficient α_n is computed from

$$\alpha_n = (v, u_n).$$

Since $\hat{w}_N$ is the orthogonal projection of v on the subspace W_N, from the Pythagorean theorem we have (with the error $e_N = v - \hat{w}_N$)

$$\|v\|^2 = \|e_N\|^2 + \|\hat{w}_N\|^2$$

so that

$$\|\hat{w}_N\|^2 \leq \|v\|^2.$$

Since

$$\|\hat{w}_N\|^2 = \sum_{n=1}^{N} |\alpha_n|^2,$$

we obtain

$$\sum_{n=1}^{N} |\alpha_n|^2 \le \|v\|^2.$$

This latter result is known as *Bessel's inequality*. Since v is fixed independent of N, this implies (since the sequence of partial sums is bounded) that the series of nonnegative real terms

$$\sum_{n=1}^{\infty} |\alpha_n|^2 \tag{2.8}$$

is a convergent infinite series, since the sequence of partial sums is evidently monotone and bounded from above.

Associated with the sequence of subspaces W_n is a sequence of vectors $\{\hat{w}_n\}$ in V, consisting of the successive projections of v on W_n. We immediately inquire whether or not such a sequence has a limit.

Because we are assuming that V is a Hilbert space, $\{\hat{w}_n\}$ converges if and only if $\{\hat{w}_n\}$ is a Cauchy sequence with respect to the norm defined by the inner product. The convergence problem reduces to checking whether or not this latter condition holds. To test this, we compute the quantity

$$\|\hat{w}_n - \hat{w}_m\|,$$

assuming $n > m$. Since

$$\hat{w}_n - \hat{w}_m = \sum_{j=m+1}^{n} \alpha_i u_j$$

and the $\{u_j\}$ are orthonormal, we have

$$\|\hat{w}_n - \hat{w}_m\|^2 = \sum_{j=m+1}^{n} |\alpha_j|^2.$$

Since the infinite series (2.8) is convergent, the partial sums form a Cauchy sequence, and hence, given any $\epsilon^2 > 0$, we have

$$\|\hat{w}_n - \hat{w}_m\|^2 = \sum_{j=m+1}^{n} |\alpha_j|^2 < \epsilon^2.$$

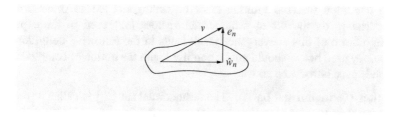

Fig. 2.4 Projection and error vector

for $n, m > N(\epsilon^2)$. Here the index function N is that associated with the fact that the sequence of partial sums of the numerical series (2.8) is a Cauchy sequence. Hence the inequality

$$\|\hat{w}_n - \hat{w}_m\| < \epsilon$$

holds for $n, m > N(\epsilon^2) = N_1(\epsilon)$, so that $\{\hat{w}_n\}$ is indeed a Cauchy sequence in V.

If we call $\{\hat{w}_n\}$ the Fourier expansion sequence of v (since this is what it is in the concrete case motivating these considerations) we have shown by the above that the Fourier expansion sequence $\{\hat{w}_n\}$ is a convergent one in the Hilbert space. This is a consequence only of the definition of completeness of the vector space, and the assumption that the sequence $\{u_n\}$ consists of orthonormal vectors.

In view of the motivation behind the projection procedure defining $\{\hat{w}_n\}$ (see Figure 2.4), we expect that $\{\hat{w}_n\}$ approaches the vector v, although nothing we have said so far guarantees this. From the relation

$$v = e_n + \hat{w}_n,$$

we see that $\hat{w}_n \to v$ if and only if $e_n \to 0$. Since we know that $\{\hat{w}_n\}$ is a sequence with a limit, the relation $v = e_n + \hat{w}_n$ shows that $\{e_n\}$ does also. Taking the limit in the above, denoting $\lim \hat{w}_n = \hat{v}$, and defining

$$\hat{e} = v - \hat{v},$$

it is easy to see (by the construction of $\hat{v}$) that for each fixed basis vector u_j

$$(\hat{e}, u_j) = 0. \tag{2.9}$$

That is, the limiting error is orthogonal to all of the vectors utilized in the expansion.

If the only vector orthogonal to all of the expansion vectors is the zero vector, then we may conclude that the limiting error $\hat{e}$ must vanish, and hence that

$$v = \hat{v} = \lim_{n \to \infty} \sum_{j=1}^{n} \alpha_j u_j = \sum_{j=1}^{\infty} (v, u_j) \, u_j.$$

The condition that the limiting error be orthogonal to the orthonormal set is a restriction on the set of orthonormal vectors $\{u_n\}$ used in the expansion. The importance of this convergence result leads to the following definition. (The terminology used below should not be confused with the notion of completeness of a normed space introduced above.)

Definition (Orthonormal basis). The orthonormal set $\{u_n\}$ is called a *complete orthonormal set* (or an orthonormal basis) for the Hilbert space V if and only if one of the following equivalent conditions holds:

1. There is no nonzero vector $\hat{e}$ in V orthogonal to $\{u_j\}$.
2. For an arbitrary vector v in V the sequence of approximants $\{\hat{w}_N\}$ given by

$$\hat{w}_N = \sum_{j=1}^{N} (v, u_j)\, u_j$$

approaches v.

We emphasize that completeness of a set $\{u_j\}$ is a property inherent in the set itself. There is enough room in an infinite-dimensional Hilbert space to easily construct infinite orthonormal sets which are not complete.

Combining the above definition with the convergence discussion above, we record the result as a theorem.

Theorem (Orthonormal Expansion Theorem). *Let V be a Hilbert space, $v \in V$, and let $\{u_j\}$ be a complete orthonormal set in V. Then*

$$v = \sum_{j=1}^{\infty} (v, u_j)\, u_j.$$

Before specializing the discussion back to the case of Fourier series we derive a Parseval theorem in this context.

Suppose $\{u_j\}$ is a complete orthonormal set, and that

$$v = \sum_{j=1}^{\infty} \alpha_j\, u_j,$$

$$w = \sum_{j=1}^{\infty} \beta_j\, u_j$$

are expansions of vectors $v, w \in V$. The problem is to compute the inner product of v and w in terms of the expansion coefficient sets $\{\alpha_j\}$, $\{\beta_j\}$. This is obtained as follows.

It is easy to show (see the problems below) that if $w_N \to w$ in V, then

$$(v, w) = \lim_{N \to \infty} (v, w_N).$$

(That is, the inner product function is continuous in its vector arguments.) Take

$$w_N = \sum_{j=1}^{N} \beta_j u_j$$

and compute the inner product

$$(v, w_N) = (v, \sum_{j=1}^{N} \beta_j u_j)$$

$$= \sum_{j=1}^{N} \overline{\beta_j}(v, u_j)$$

$$= \sum_{j=1}^{N} \alpha_j \overline{\beta_j}.$$

Since $w_N \to w$, we obtain the conclusion that

$$\lim_{N \to \infty} \sum_{j=1}^{N} \alpha_j \overline{\beta_j} = \sum_{j=1}^{\infty} \alpha_j \overline{\beta_j}$$

exists, and is equal to the desired inner product.

Theorem (Parseval's Theorem). *Let $\{u_j\}$ be a complete orthonormal set in the Hilbert space V, and*

$$v = \sum_{j=1}^{\infty} \alpha_j u_j,$$

$$w = \sum_{j=1}^{\infty} \beta_j u_j.$$

Then

$$(v, w) = \sum_{j=1}^{\infty} \alpha_j \overline{\beta_j} = \sum_{j=1}^{\infty} (v, u_j)\overline{(w, u_j)},$$

and in particular

$$\|v\|^2 = (v, v) = \sum_{j=1}^{\infty} \alpha_j \overline{\alpha_j} = \sum_{j=1}^{\infty} |(v, u_j)|^2.$$

We now illustrate the above results with the example of the standard form of the Fourier series.

The vector space V consists of complex valued functions (strictly, equivalence classes of such) defined on the interval $[0, 2\pi]$, and square-integrable there. The inner product is defined by

$$(f, g) = \frac{1}{2\pi} \int_0^{2\pi} f(x)\overline{g(x)}\, dx.$$

This space is conventionally denoted by the symbol $L^2[0, 2\pi]$, (saving repetition of the paragraph-long description above).

Classical results from analysis (available, for example, in references [1, 3]) assert that the space $L^2[0, 2\pi]$ is complete (i.e., is a Hilbert space), and further that the orthonormal set $\{u_n\}_{n=-\infty}^{\infty}$

$$u_n(x) = e^{inx}, \; n = 0, \pm 1, \pm 2, \ldots,$$

is in fact a complete orthonormal set. These results contain the hard analysis associated with this approach to Fourier series convergence problems.

Taking these results as given, we consider an arbitrary $f \in L^2[0, 2\pi]$, and compute

$$\begin{aligned} (f, u_n) &= \frac{1}{2\pi} \int_0^{2\pi} f(x)\overline{e^{inx}}\, dx \\ &= \frac{1}{2\pi} \int_0^{2\pi} f(x) e^{-inx}\, dx \\ &= c_n. \end{aligned}$$

With some small abuse of the subspace dimension counting convention utilized above, we take the subspace

$$W_N = \text{span}\, \{u_{-N}, \ldots, u_0 = 1, \ldots, u_N\}$$

(of dimension $2N + 1$) and construct the projection of f on W_N. This is just

$$f_N(x) = \sum_{-N}^{N} c_n e^{inx}.$$

From the above general results we conclude that $f_n \to f$ in the sense of convergence in $L^2[0, 2\pi]$ (referred to as *mean-square convergence*), that is,

$$\lim_{N \to \infty} \frac{1}{2\pi} \int_0^{2\pi} \left| f(x) - \sum_{-N}^{N} c_n e^{inx} \right|^2 dx = 0.$$

Further, Parseval's relation (in the square form) gives

$$\frac{1}{2\pi} \int_0^{2\pi} |f(x)|^2 \, dx = \sum_{-\infty}^{\infty} |c_n|^2. \tag{2.10}$$

This discussion gives a rigorous derivation of the Parseval relation discussed formally in the previous section. It is now clear that what is required for the validity of the result is that the expansion vectors should consist of a complete orthonormal set. Since this is the case for the standard Fourier series set $\{e^{imx}\}$, Parseval's result is justified in this case.

Problems 2.3.

1. State and prove a Cauchy Criterion for the pointwise convergence of a sequence of functions $\{s_n\}$ defined on the interval $[0, 1]$.

2. Consider the sequence of functions $\{s_n\}$ defined by

$$s_n(x) = x^n, \ 0 \le x \le 1.$$

   Show that this sequence is pointwise convergent, and explicitly find the required $N(x, \epsilon)$. Show further that this sequence is not uniformly convergent.

3. Show that the limit of a uniformly convergent sequence of continuous functions defined on the interval $[0, 1]$ is itself continuous.
   Hint: The limit function is arbitrarily close to a continuous function s_n in the sequence, and the values of $s_n(x_1)$, $s_n(x_2)$ may be made close by taking x_1 near x_2.

4. The Weierstrass M-test for uniform convergence states that if

   a. $\sup_{x \in [0,1]} |u_n(x)| \le M_n$, and

   b. $\sum_{n=1}^{\infty} M_n < \infty$,

   then the sequence (of partial sums) $\{s_N\}$ given by

$$s_N(x) = \sum_{n=1}^{N} u_n(x)$$

   is uniformly convergent on the interval $[0, 1]$. Prove this result.
   Hint: Apply the Cauchy Criterion for uniform convergence.

5. Use the Cauchy–Schwarz inequality

$$|(u, v)| \le \|u\| \|v\|$$

   to prove that the norm defined through an inner product

$$\|u\| = (u, u)^{1/2}$$

   satisfies the triangle inequality.
   Hint: Show $\|u + v\|^2 \le \|u\|^2 + \|v\|^2$.

6. Prove the Cauchy Criterion for uniform convergence of a sequence of functions uniformly continuous on the interval $[0, 1]$ to a function continuous there.

7. Use the Cauchy–Schwarz inequality to show that the following continuity conditions hold for the inner product function. If $\{u_n\} \to u$, $\{v_n\} \to v$, and w is a fixed vector, then

   a. $(u_n, w) \to (u, w)$,
   b. $(w, v_n) \to (w, v)$,
   c. $(u_n, v_n) \to (u, v)$.

8. Use the continuity of the inner product (as in the proof of the Parseval's Theorem) to prove relation (2.9),

$$(\hat{e}, u_j) = 0 \text{ for all } \{u_j\},$$

   where $\hat{e}$ is the limiting error of the projection sequence $\{\hat{w}_n\}$ defined above.

9. In the Hilbert space $L^2[0, 2\pi]$, find an infinite orthonormal set $\{v_n\}$ which is not a complete set.

10. Assuming the Riesz–Fischer result that $\{e^{inx}\}$ is complete in $L^2[0, 2\pi]$, prove that $\{1, \cos(nx), \sin(nx)\}$ is also complete.

11. Use the product form of Parseval's Theorem to evaluate

$$\sum_{\substack{n=-\infty \\ n \neq 0}}^{\infty} \frac{[(-1)^n - 1]}{n^2(1 + in)}.$$

Hint: Explicitly summing Fourier expansions is usually not possible, so a better approach is to "borrow" coefficient calculations from previously worked examples.

2.4 Pointwise and Uniform Convergence of Fourier Series

We calculated above the Fourier series representation of the function f defined by

$$f(x) = x, \quad -\pi < x \leq \pi, \tag{2.11}$$

and found that the resulting expansion takes the form

$$f(x) = 2 \sum_{n=1}^{\infty} \frac{(-1)^n}{n} \sin(nx). \tag{2.12}$$

When $x = \pi$, the above series consists entirely of zero terms, and hence sums to zero. This discrepancy between the sum of the series and the declared value

$f(\pi) = \pi$ of course has no effect on the mean-square convergence of the series to the function. It does, however, indicate that pointwise convergence is not to be automatically expected in the case of Fourier series expansions.

The fact that for

$$c_n = \frac{1}{2\pi} \int_0^{2\pi} f(x)e^{-inx}\, dx,$$

$$f_N = \sum_{-N}^{N} c_n\, e^{inx},$$

we have only (mean-square convergence)

$$\lim_{N\to\infty} \frac{1}{2\pi} \int_0^{2\pi} |f(x) - f_N(x)|^2\, dx = 0$$

leaves open the possibility that $f_N(x)$ might fail to converge to $f(x)$ for a set of x negligible as far as integration is concerned. The f_N might fail to converge for, say, all x of the form $x = \frac{m}{2^n}$ (numbers manipulated by binary computers) and hence be effectively useless for numerical computation.

In order to obtain results guaranteeing pointwise (or even better, uniform) convergence of the Fourier series, it is necessary to impose some smoothness requirements on the function f. There is a wide variety of such results available; we present below only a result which fits naturally with the inner product approach adopted above.

Definition (Function with finite mean-square derivative). A periodic function $g \in L^2[0, 2\pi]$ with finite mean-square derivative [3] is one for which

1. $g(x + 2\pi) = g(x)$,
2. $g(b) - g(a) = \int_a^b g'(x)\, dx$ (for all real a, b) for some 2π-periodic function g' such that

$$\int_0^{2\pi} |g'(x)|^2\, dx < \infty.$$

Example. The function defined by

$$g(x) = \begin{cases} x & \text{for } 0 \le x \le \pi, \\ -x & \text{for } -\pi \le x \le 0, \\ g(x + 2\pi) = g(x) & \text{otherwise} \end{cases}$$

---

[3] Technically, g coincides in the $L^2[0, 2\pi]$ sense with a 2π-periodic absolutely continuous function with square-integrable derivative. The notion of "absolutely continuous function" is required for the validity of the "fundamental theorem of calculus" in a Lebesgue integral context.

is such a function. This is true although g fails to be differentiable at multiples of π. The definition of g' by the requirements

$$g'(x) = \begin{cases} 1 & \text{for } 0 \le x \le \pi, \\ -1 & \text{for } -\pi \le x \le 0, \\ g'(x + 2\pi) = g'(x) & \text{otherwise} \end{cases}$$

suffices to verify the required conditions of the definition.

These definitions and examples emphasize the idea that the Fourier series are intimately connected with periodic functions (or functions defined on a circle) in spite of the fact that they may arise in problems involving functions defined on an interval. This aspect cannot be side-stepped in convergence discussions, and conversely may be exploited in other situations (see Chapter 3).

Our result on pointwise convergence is that being a function with finite mean-square derivative (as defined above) is a sufficient condition to guarantee pointwise convergence of the Fourier series.

Theorem (Pointwise Fourier series convergence). *Let $g \in L^2[0, 2\pi]$ be a 2π-periodic function with finite mean-square derivative. Then the Fourier series expansion of g converges pointwise to $g(x)$ for all x.*

Proof. What is required is to show that, for each fixed x, the partial sums (of the numerically valued series)

$$s_N(x) = \sum_{n=-N}^{N} c_n e^{inx}$$

converge to the number $g(x)$.

We know that since

$$\int_0^{2\pi} |g'(x)|^2 \, dx < \infty$$

the function g' is entitled to a mean-square convergent Fourier series expansion of the form

$$\sum_{n=-\infty}^{\infty} c_n' e^{inx}.$$

The expansion coefficients are calculated as

$$c_n' = \frac{1}{2\pi} \int_0^{2\pi} g'(x) e^{-inx} \, dx$$

$$= \frac{1}{2\pi} \left[g(x) e^{-inx} \Big|_0^{2\pi} + in \int_0^{2\pi} g(x) e^{-inx} \, dx \right]$$

$$= \frac{1}{2\pi} in \int_0^{2\pi} g(x)\, e^{-inx}\, dx$$

$$= in\, c_n.$$

(Aside: This calculation shows that the Fourier series of a function with finite mean-square derivative may be differentiated on a term by term basis to obtain the Fourier series of the derivative.) Therefore (in the mean-square sense)

$$g'(x) = \sum_{\substack{-\infty \\ n \neq 0}}^{\infty} inc_n\, e^{inx}.$$

To obtain the function g from g', one simply integrates, say over the interval $[0, \zeta]$, where $(0 < \zeta < 2\pi)$.

But the expression

$$g(\zeta) - g(0) = \int_0^{\zeta} g'(x)\, dx$$

can be identified as the inner product of g' with the step function I_ζ defined on $[0, 2\pi]$ by

$$I_\zeta(x) = \begin{cases} 2\pi & \text{for } 0 \leq x < \zeta, \\ 0 & \text{for } \zeta \leq x < 2\pi, \end{cases}$$

since

$$(g', I_\zeta) = \frac{1}{2\pi} \int_0^{\zeta} 2\pi\, g'(x)\, dx = g(\zeta) - g(0).$$

By the product form of Parseval's theorem, the indicated inner product may be calculated in terms of the Fourier expansion coefficients. We have

$$\frac{1}{2\pi} \int_0^{2\pi} g'(x)\, \overline{I_\zeta(x)}\, dx = \sum_{n=-\infty}^{\infty} c'_n \overline{\beta_n}$$

where $\{\beta_n\}$ are the expansion coefficients of I_ζ,

$$\beta_n = \frac{1}{2\pi} \int_0^{2\pi} I_\zeta(x)\, e^{-inx}\, dx$$

$$= \begin{cases} \frac{e^{-in\zeta} - 1}{-in} & \text{for } n \neq 0, \\ \zeta & \text{for } n = 0. \end{cases}$$

Since Parseval's theorem guarantees convergence of the coefficient product infinite series,

$$g(\zeta) - g(0) = \sum_{\substack{n=-\infty \\ n \neq 0}}^{\infty} c'_n \frac{\overline{e^{-in\zeta} - 1}}{-in}$$

$$= \sum_{\substack{n=-\infty \\ n \neq 0}}^{\infty} c'_n \frac{e^{in\zeta} - 1}{in}$$

$$= \sum_{\substack{n=-\infty \\ n \neq 0}}^{\infty} c_n \left(e^{in\zeta} - 1 \right)$$

$$= \sum_{n=-\infty}^{\infty} c_n \left(e^{in\zeta} - 1 \right)^4$$

is a convergent series. This result expresses the $g(\zeta)$ value as an infinite series very close to desired Fourier series expression.

The desired convergence result follows readily as soon as it is verified that

$$\sum_{n=-\infty}^{\infty} c_n$$

is convergent (this allows "splitting" of the series obtained above).

But the convergence of $\sum c_n$ follows from Parseval's Theorem:

$$c_n = in\, c_n \frac{1}{in} \quad n \neq 0$$

$$= c'_n \frac{1}{in} \quad n \neq 0$$

with

$$\sum_{n=-\infty}^{\infty} |c'_n|^2 < \infty$$

(since g' is square-integrable) while

$$\sum_{\substack{n=-\infty \\ n \neq 0}}^{\infty} \frac{1}{n^2} < \infty.$$

---

[4] since $e^{in\zeta} - 1$ vanishes when $n = 0$.

Hence, by the product form of Parseval's Theorem, the series

$$\sum_{\substack{n=-\infty \\ n \neq 0}}^{\infty} c_n = \sum_{\substack{n=-\infty \\ n \neq 0}}^{\infty} c'_n \overline{\frac{1}{-in}}$$

converges. Now knowing that the series can be split into the sum of two series, rearranging the above, we obtain

$$g(\zeta) = \sum_{n=-\infty}^{\infty} c_n e^{in\zeta} + K \tag{2.13}$$

$(K = g(0) - \sum_{n=-\infty}^{\infty} c_n)$ so that the series is within a constant of the required function value.

In fact, $K = 0$, since (by computation of the constant component of each side of the equation) we have

$$\frac{1}{2\pi} \int_0^{2\pi} g(\zeta) \, d\zeta = c_0 + K,$$

i.e.,

$$c_0 = c_0 + K.$$

Therefore $K = 0$ and so

$$g(0) = \sum_{n=-\infty}^{\infty} c_n.$$

This finally shows that for $0 < \zeta < 2\pi$ (where I_ζ is not the 0 function)

$$g(\zeta) = \sum_{n=-\infty}^{\infty} c_n e^{in\zeta}, \tag{2.14}$$

so that the series converges as required, for ζ in the open interval $(0, 2\pi)$. However, for $\zeta = 0, 2\pi$, the series reduces to

$$\sum_{n=-\infty}^{\infty} c_n,$$

shown above convergent to $g(0)$.

This completes the proof that the series converges for each value of ζ in $[0, 2\pi]$ to the corresponding value $g(\zeta)$ of the function.

A closer look at the discussion above shows that the convergence of the Fourier series for a periodic function with a finite mean-square derivative is actually uniform.

This follows from the observation that the argument above actually may be used to show that the series

$$\sum_{n=-\infty}^{\infty} |c_n|$$

is convergent. (This follows since Parseval's Theorem involves the squared magnitude of the coefficients, so that the phase of the c_n may be absorbed into the other factor.)

Since

$$|c_n e^{inx}| = |c_n|,$$

the Weierstrass M-test may be invoked to show that the series (shown pointwise convergent above)

$$\sum_{n=-\infty}^{\infty} c_n e^{in\zeta}$$

is uniformly convergent.

This observation proves the following result.

Theorem (Uniform Fourier series convergence). *The Fourier series of a periodic* $L^2[0, 2\pi]$ *function f with finite mean-square derivative is uniformly convergent to f.*

As might be supposed, the functions defined above are not the only ones for which there is interest in the question of pointwise convergence of the Fourier expansion.

A class sufficiently wide to cover many applications is obtained by allowing the insertion of a finite number of jump discontinuities in functions of the type discussed above.

A result of this general form can be obtained by explicitly analyzing pointwise convergence of a suitable single example where jump discontinuities are present. The sawtooth function (2.11), (2.12) is a convenient case for an ad hoc analysis.

The partial sum of the sawtooth series is simply

$$\sum_{n=1}^{N}(-1)^{n+1}\frac{2\sin(nx)}{n} = -\sum_{n=1}^{N}(-1)^n \int_0^x 2\cos(n\zeta)\, d\zeta$$

$$= -\int_0^x \sum_{n=1}^{N} 2(-1)^n \cos(n\zeta)\, d\zeta.$$

We interchange sum and integral above without trepidation since the sums are finite. The indicated sums may be explicitly evaluated using the usual geometric series formula

$$\sum_{n=1}^{N} \gamma^n = \frac{\gamma[1 - \gamma^N]}{1 - \gamma}$$

and the relation

$$2\cos(n\zeta) = e^{in\zeta} + e^{-in\zeta}.$$

The result is

$$\sum_{n=1}^{N}(-1)^n 2\cos(n\zeta) = \frac{-e^{i\zeta}[1-(-1)^N e^{iN\zeta}]}{1+e^{i\zeta}} + \frac{-e^{-i\zeta}[1-(-1)^N e^{-iN\zeta}]}{1+e^{-i\zeta}}$$

$$= -1 + (-1)^N \frac{e^{i(N+1)\zeta}}{1+e^{i\zeta}} + (-1)^N \frac{e^{-iN\zeta}}{1+e^{i\zeta}}$$

The partial sum of the series is hence given by

$$\sum_{n=1}^{N}(-1)^{n+1}\frac{2\sin(nx)}{n} = \int_0^x \{1-(-1)^N \frac{e^{i(N+1)\zeta}+e^{-iN\zeta}}{1+e^{i\zeta}}\}d\zeta$$

$$= x - \int_0^x (-1)^N \frac{e^{i(N+1)\zeta}+e^{-iN\zeta}}{1+e^{i\zeta}}\,d\zeta$$

The above represents an explicit formula for the partial sum of the series, and it easily follows (see Problem 6 below) that the integral term tends to zero as $N \to \infty$ provided $|x| < \pi$. We conclude then that for $|x| < \pi$, the sawtooth series converges pointwise to the value of the sawtooth function.

2.5 Gibb's Phenomenon and Summation Methods

The previous sections describes a result concerning pointwise convergence of a useful class of Fourier series examples. The result does not provide much in the way of qualitative description of the convergence process.

Some experimentation with plotting partial sums of Fourier series shows that examples from functions with jump discontinuities seem to show considerable "overshoot" in the neighborhood of the discontinuity. This effect is known as "Gibb's phenomenon," after the primary discoverer.

The effect can be clearly seen with the symmetrical square wave example given by

$$f(x) = \begin{cases} -1 & -\pi \le x < 0, \\ 1 & 0 \le x < \pi, \\ f(x+2\pi) & \text{otherwise.} \end{cases}$$

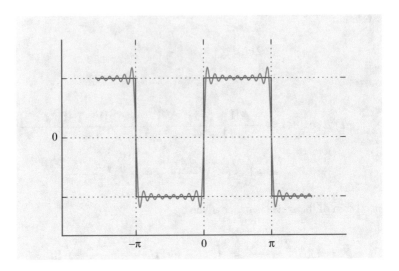

Fig. 2.5 Square wave and Gibb's phenomenon

The Fourier series expansion in this case is

$$f(x) = \sum_{n=1}^{\infty} \frac{2}{\pi} (1 - (-1)^n) \frac{\sin(nx)}{n},$$

and an even partial sum is

$$s_{2n}(x) = \frac{4}{\pi} \left(\sin(x) + \frac{\sin(3x)}{3} + \cdots + \frac{\sin(2n-1x)}{2n-1} \right).$$

A plot of this displays the Gibb's effect in Figure 2.5.

A similar diagram holds for other examples with a jump discontinuity. This particular case is of interest because the simple form of the series makes it possible to explicitly calculate the height of the overshoot. The analytical procedure is a variant on the approach taken above with the sawtooth convergence discussion.

The peak overshoot occurs at a local maximum of the partial sum function $s_{2n}(x)$. To find the critical points, compute

$$s'_{2n}(x) = \sum_{k=1}^{2n} \frac{2}{\pi} (1 - (-1)^k) \cos(kx)$$

$$= \frac{4}{\pi} (\cos(x) + \cos(3x) + \cdots + \cos(2n-1x).)$$

This can be evaluated using Euler's formula

$$\cos(kx) = \frac{e^{ikx} + e^{-ikx}}{2}.$$

We have

$$s'_{2n}(x)$$

$$= \frac{2}{\pi} \left[e^{ix} \left(1 + e^{2ix} + \cdots + e^{(2n-1)ix} \right) + e^{-ix} \left(1 + e^{-2ix} + \cdots + e^{-(2n-1)ix} \right) \right]$$

$$= \frac{2}{\pi} \left[e^{ix} \left(\frac{1 - e^{2nix}}{1 - e^{2ix}} \right) + e^{-ix} \left(\frac{1 - e^{-2nix}}{1 - e^{-2ix}} \right) \right]$$

$$= \frac{2}{\pi \, |1 - e^{2ix}|^2} \left[\left((e^{ix} - e^{-ix})(1 - e^{2nix}) \right) - \left((e^{ix} - e^{-ix})(1 - e^{-2nix}) \right) \right]$$

$$= \frac{8}{\pi |1 - e^{2ix}|^2} \sin(x) \sin(2n\,x).$$

This derivative is positive from $x = 0$ to the first critical point when

$$2nx = \pi,$$

$$x = \frac{\pi}{2n}.$$

The value of the partial sum at the critical point is

$$s_n \left(\frac{\pi}{2n} \right) = \sum_{k=1,3,5\cdots}^{2n-1} \frac{4}{\pi} \frac{\sin(k \frac{\pi}{2n})}{k}$$

$$= \sum_{k=1,3,5\ldots}^{2n-1} \frac{4}{\pi} \frac{\sin(k \frac{\pi}{2n})}{\frac{k\pi}{2n}} \frac{\pi}{2n}.$$

As $n \to \infty$ this tends to[5]

$$\frac{2}{\pi} \int_0^\pi \frac{\sin(x)}{x} \, dx = 1.178979744\ldots,$$

nearly 18% above the value of the function.

2.6 Summation Methods

In spite of the horrendous indication of the Gibb's phenomenon calculation for the square wave example, it is possible to get a more accurate function from the series partial sums by making use of a "summation method." These are often studied as a

---

[5]The sum is "half the terms" of a $2n$ length Riemann sum for the integral.

means for extracting a value from otherwise divergent infinite series, although our interest here is in speeding or smoothing the convergence of a (known convergent) Fourier series.

There are many summation methods, but perhaps the best known (as well as easiest to analyze) one is *Cesaro's method.*

Cesaro's method is simply to average the sequence of partial sums. That is, instead of the sequence of $\{s_n\}$ where

$$s_n = \sum_{j=0}^{n} a_j,$$

we consider the running averages of these $\{S_n\}$ where

$$S_n = \frac{1}{n} \sum_{k=1}^{n} s_k.$$

To say that the $\{a_j\}$ series has a sum means exactly that

$$\lim_{n \to \infty} s_n = S$$

exists. It is easy to see that the Cesaro sums (the running average of the partial sums) have the same limit as the original partial sums, that is, the sum of the original series. This is actually a "limit effect" and really has nothing to do with the fact that our sequence is derived from an infinite series.

To see how this happens, suppose that $\epsilon > 0$ is given, and that $N(\epsilon)$ is the "waiting time" associated with the $\{s_n\}$ sequence, so that

$$|s_n - S| < \epsilon \text{ for } n \geq N(\epsilon).$$

But then,

$$S_n = \frac{1}{n} \sum_{k=1}^{n} s_n$$

$$= \frac{1}{n} \sum_{k=1}^{N(\epsilon)-1} s_n + \frac{1}{n} \sum_{k=N(\epsilon)}^{n} s_n.$$

The first term is of fixed length and tends to zero as $n \to \infty$. The second term is bounded by

$$\frac{n - N(\epsilon)}{n} (S - \epsilon) \leq \frac{1}{n} \sum_{k=N(\epsilon)}^{n} s_n \leq \frac{n - N(\epsilon)}{n} (S + \epsilon).$$

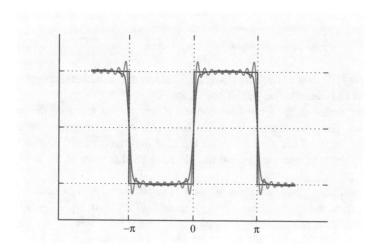

Fig. 2.6 Square wave partial and Cesaro sums

Since ϵ is arbitrary, this shows that

$$\lim_{n \to \infty} S_n = S,$$

so that the Cesaro sum tends to the same limit as the original sequence. The averaging tends to suppress the overshoot of the Gibb's phenomenon. The effect may be seen dramatically in Figure 2.6.

Graphing Partial and Cesaro Sums

Many of the partial differential equations considered in later chapters have closed form solutions in the form of Fourier series. Since we want to be able to generate graphical diagrams of these series it is useful to write some procedures which simplify the process.

Plots with Octave

The plotting functions found in many environments provide a very similar command structure. Examples include Octave, Maxima, and Numerical Python. An introduction to the syntax and facilities of these plotting environments is provided in Appendix B, and that can be consulted if the programs are unfamiliar.

Several issues arise when use of software for plotting series expansions is considered. In the first is simple. The commands

OCTAVE

```
t = -5:.01: 5;
plot(t, f(t), t, g(t));
```

will overlay the plots of two functions f, g. That idiom is sufficient for experimenting with plots of Fourier series partial and Cesaro sums.

Fourier series differ from one another solely in the form of the expansion coefficient. The actual form of the series varies also between complex and real form, and in the limits of the sum. It is also the case that the 0th coefficient often has a different formula from the general one, or may be the value of an indeterminate form.

A partial sum procedure must adopt a standard form, and particular examples must adjust to the convention. Although numerical software typically supports the use of complex numbers, many examples involve real functions, and ending up with a real result involves cancellation of the imaginary component when the complex form expansions are used. In numerical calculations it is more likely (even a near certainty) that a numerically very small imaginary part will be the result. If this is plotted, the result will not be what is expected. One must be aware of this effect when numerically handling Fourier series.

We define the partial sum procedure so that the sum starts with index 1 (like Octave vectors), so the upper limit counts the number of terms. The procedure takes a first argument which is a function that returns the value of the term for a given x and term index n.

The code initializes storage and uses feval to evaluate each term.

OCTAVE

```
% partialsum(term, x, n)
% sum of term(x, 1) +.... + term(x, n)

function y = partialsum(term, x, n)

   tempval = zeros(size (x));

   for kn = 1:n
      tempval = tempval + feval(term, x, kn);
   endfor

   y = tempval;

endfunction
```

A Cesaro sum procedure follows from the interpretation of averaging the partial sums of a series.

OCTAVE

```
% Cesaro(term, x, N)
%
function y = Cesaro(term, x, N)

    tempval = zeros(size (x));

    for kk = 1:N
        tempval = tempval + partialsum(term, x, kk);
    endfor

    y = tempval/N;

endfunction
```

The Gibb's phenomenon illustration is generated from

OCTAVE

```
% squareover
% script to plot squarewave and overshoot

fs = 2048;
t = -10: 1/fs: 10;

plot(t, squarewave(t, 2*pi),...
    t, partialsum('squareterm', t, 17),...
    t, Cesaro('squareterm', t, 17));
```

Plotting Using Python

The python code used to produce plots is quite close to the Octave version. The underlying reason for this is probably that both projects based the structure on the styles of the commercial MATLAB program.

More details about Python use are provided in chapter B. What is evident from comparing this to Octave equivalent above is that the Python has a little more "overhead," because of its explicit control of object namespaces. The other point of difference is that the python matplotlib buffers the images that the code generates, and only places the image on the screen (or in some formatted file) when commanded to do so. That surprises users who are used to octave (and MATLAB(TM)) where images appear as they are generated.

PYTHON

```
import numpy as np
import matplotlib.pyplot as plt

x = np.linspace(-np.pi, np.pi, 1000)
y = np.sin(x)

plt.plot(x, y)
plt.show()
```

The "serious" plotting code is very close to the Octave equivalents, modulo details of python syntax. The utility procedures for setting up the plot fonts, colors, sizes and the like, and generating a usable output file are in the python discussion of chapter B.

The example illustrates the ease of including mathematical text on matplotlib plots. As long as it is enabled in the plot setup parameters, it just takes a raw python string enclosed in double dollar signs, the usual syntax for inline TEXexpressions.

PYTHON

```
#
# Python script to plot squarewave overshoot
#

import matplotlib.pyplot as plt
import numpy as np
from plot_functions import setup_plot, Epsfile

# dict for passing plot parameters

plot_dict = dict()

# TeX symbols for plot

p03 = r'$\pi$'
m03 = r'$- \pi$'
p04 = r'$0$'

#

def partialsum_1x(term, x, n):

    tempval = 0.0
```

```
    for kn in    range(1, 1+n):
        tempval = tempval + term(x, kn)


    return tempval

#

def squareterm(x, n):

    y = (2/(n*np.pi))*(1 - (-1)**n)*np.sin(n*x)

    return y


#

fig_axis = plt.gca()
fig = plt.gcf()

setup_plot(fig, fig_axis)
#
plot_dict['ytick'] = [-1, 0, 1]
plot_dict['yticklabel'] = ["", "0", ""]
plot_dict['xtick'] = [-np.pi, 0, np.pi]
plot_dict['xticklabel'] = [m03, p04, p03]
#
#
Epsfile(plot_dict, 'x', 'squareover.pdf', \
    ['np.sign(np.sin(x))', 'partialsum(squareterm, x, 17)'],\
    ['x = np.linspace(-5, 5, 1000)',\
      'x = np.linspace(-5, 5, 1000)'])

#
#
```

Language Choice

From the examples above it is clear that plots can be generated using either Octave or Python (or Maxima, for that matter). Code blocks for the plots can be presented in either form.

Probably more Octave examples are present in the text, and the primary reason is that the Octave codes are easier to read, because of the "object noise" that is a feature of the Python syntax. Of course, this reflects the fact that package/namespace control is explicit in Python. When you use both languages, you come to appreciate the way python handles modules and packages.

The figure function plots of the text were almost entirely generated with Python, numpy, and matplotlib. The surface plots largely were done with Octave code. The choice is primarily an aesthetic one, with an extra bonus to matplotlib for easy LaTeXmath font handling.

Problems 2.6.

1. Suppose that it is known that the Fourier series for the sawtooth f above converges to f pointwise for $-\pi < x < \pi$, while $f(\pi) = f(-\pi) = 0$. This problem provides the steps required to combine the sawtooth convergence discussion with the previous pointwise results to obtain a result for pointwise convergence in the presence of function discontinuities.

   For the discussion of discontinuities it is useful to use the standard notation for left and right function limits.

   $$f(a + 0) = \lim_{\epsilon \to 0^+} f(a + \epsilon),$$

   $$f(a - 0) = \lim_{\epsilon \to 0^-} f(a + \epsilon).$$

   a. Show that for the sawtooth function $f(x) = x$, $-\pi < x \le \pi$, $f(x + 2\pi) = f(x)$ we have

   $$\frac{1}{2\pi}[f(\pi + 0) - f(\pi - 0)] = -1.$$

   b. Construct a function $j(x)$ (jump to order function) such that

   $$j(a + 0) - j(a - 0) = A,$$

   from a time shifted, scaled version of the sawtooth.
   c. Show that a 2π periodic function which has the properties that

   i. it has two jump discontinuities in each period, and
   ii. otherwise is piecewise linear,

   may be written as the sum of two scaled, time shifted versions of the sawtooth function, and perhaps an additive constant to account for an average value. Construct graphs of several such functions. (This may seem silly, but there is graphical insight in the way the period, slopes, and jumps fit together.)

2. Show that the Fourier series for a piecewise linear periodic function with two jump discontinuities per cycle has the properties that its Fourier series converges

   a. to the function value at a point where the function is continuous, and
   b. to a value at the midpoint of the jump at a point of jump discontinuity.

3. Generalize the previous conclusion to the case of J (for the number of jumps) jump discontinuities.
4. Finally, conclude that a function which can be decomposed as the sum of

    a. a function with a finite mean-square derivative, and
    b. a piecewise linear function with a finite number of jump discontinuities

    has a Fourier series which converges pointwise everywhere to the average of the left- and right-hand limits of the function values. This is a different way of stating the same conclusion as held for the piecewise linear case.
5. Decompose the symmetric square wave into the sum of two sawtooth functions, and one piecewise linear function with a finite mean-square derivative.
6. Carry out the only hard part of the proof of pointwise convergence of the sawtooth expansion. That is, show that for $|x| < \pi$,

$$\lim_{N\to\infty} \int_0^x (-1)^N \frac{e^{i(N+1)\zeta} + e^{-iN\zeta}}{1 + e^{i\zeta}} \, d\zeta = 0.$$

Hint: Integration by parts will move the N to the denominator, where the effect of large N is more apparent.
7. Suppose that f is defined on the interval $[-\pi, \pi]$ as

$$f(x) = \begin{cases} x^3 & 0 \le x < \pi, \\ -x^2 & -\pi \le x \le 0, \end{cases}$$

and that

$$c_n = \frac{1}{2\pi} \int_0^{2\pi} f(x) e^{-inx} \, dx.$$

Does the series

$$\sum_{n=-\infty}^{\infty} c_n e^{inx}$$

converge for $x = \frac{7\pi}{2}$? If so, to what value? What about $x = 19\pi$? (**Hint:** consider Problem 4 above.)
8. What is the Cesaro sum for the non-convergent series

$$\sum_{n=1}^{\infty} (-1)^n?$$

9. Show that the Nth Cesaro sum of

$$\sum_{n=1}^{\infty} a_n$$

is

$$a_1 + (1 - \frac{1}{N})a_2 + (1 - \frac{2}{N})a_3 + \cdots (1 - \frac{N-1}{N})a_N,$$

so that the procedure actually weights the terms with a triangular pattern.

10. The Abel sum of the series

$$\sum_{n=1}^{\infty} a_n$$

is

$$\lim_{r \to 1} \frac{1}{1-r} \sum_{n=1}^{\infty} a_n \, r^n.$$

What is the Abel sum of the evidently divergent series

$$\sum_{n=1}^{\infty} 1?$$

2.7   Fourier Series Properties

The application of Fourier series methods in various problems requires some facility with standard manipulations associated with Fourier series expansions. The purpose of this section is to describe the most useful of these manipulations.

Variable Scaling

The first problem is that of scaling of the independent variable in a problem. For convenience of exposition in the previous sections, the basic length (or time) scale was normalized implicitly to 2π. As this is a rather uncommon measurement in practice, it is necessary to slightly modify the above formulas in the case of a general length scale.

The simplest way to produce the required expansion is, of course, to guess the required result. The argument leading to the correct guess proceeds as follows.

Consider the graphs of the expansion functions associated with the 2π-interval expansions constructed above. It is clear that we can produce a set of functions with the analogous orthogonality properties on the interval $[0, L]$ simply by rescaling the variable. See Figure 2.7.

The required transformations (in the complex expansion case) are given by

$$e^{inx} \mapsto e^{in\,(2\pi/L)\,x},$$

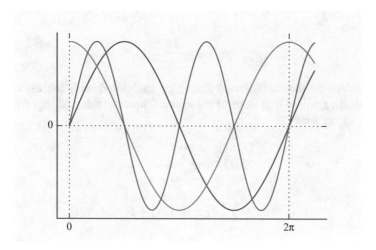

Fig. 2.7 Fourier expansion functions

and in the real form by

$$\sin(nx) \mapsto \sin(n(2\pi/L)x),$$
$$\cos(nx) \mapsto \cos(n(2\pi/L)x).$$

The formulas for calculation of the expansion coefficients may be derived (the hard way) by change of variable in the original formulas. A simpler method is to realize that

$$\left| e^{in(2\pi/L)x} \right|^2 = 1$$

and

$$\frac{1}{T} \int_0^T \left| \sin(n(2\pi/L)x) \right|^2 dx = \frac{1}{2}$$

where T is any multiple of the period L of the integrand. These relations force the conclusion that

$$c_n = \frac{1}{L} \int_0^L f(x) e^{-in(2\pi/L)x} \, dx, \tag{2.15}$$

$$a_0 = \frac{1}{L} \int_0^L f(x) \, dx,$$

$$a_n = \frac{2}{L} \int_0^L f(x) \cos(n(2\pi/L)x), \, dx$$

$$b_n = \frac{2}{L} \int_0^L f(x) \sin(n(2\pi/L)x) \, dx$$

are the appropriate coefficient relations.

The quantity

$$\omega_0 = \frac{2\pi}{L}$$

is referred to as the fundamental frequency associated with the expansion. It represents the (radian) frequency of the lowest frequency sinusoid appearing in the resulting expansions

$$f(x) = \sum_{n=-\infty}^{\infty} c_n e^{in\omega_0 x},$$

$$f(x) = a_0 + \sum_{n=1}^{\infty} a_n \cos(n\omega_0 x) + b_n \sin(n\omega_0 x).$$

(This interpretation assumes that L is the minimal period associated with f. Obviously, if f is really periodic of period say $L/2$, then all odd subscript c_n above vanish, and any rational person would regard $4\pi/L$ as the fundamental frequency.)

Example. Consider first the analysis of the output of a standard half-wave rectifier circuit (Figure 2.8). The output voltage is given by

$$e_0(t) = \begin{cases} \sin(\Omega t) & 0 \le t \le \frac{\pi}{\Omega}, \\ 0 & \frac{\pi}{\Omega} \le t \le \frac{2\pi}{\Omega} = T, \\ e_0(t+T) = e_0(t) & \text{otherwise.} \end{cases}$$

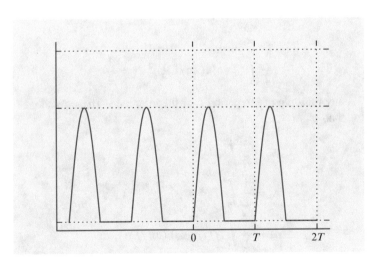

Fig. 2.8 Half-wave rectified signal

The fundamental frequency associated with a Fourier expansion is

$$\omega_0 = \frac{2\pi}{T} = \Omega,$$

the same as the source frequency.

$$
\begin{aligned}
c_n &= \frac{1}{T} \int_0^{\frac{\pi}{\Omega}} \sin(\Omega t)\, e^{-in\omega_0 t}\, dt \\[2mm]
&= \frac{1}{T} \int_0^{\frac{\pi}{\Omega}} \frac{e^{i\Omega t} - e^{-i\Omega t}}{2i}\, e^{-in\omega_0 t}\, dt \\[2mm]
&= \frac{1}{T \cdot 2i} \left[\frac{e^{i(\Omega - \omega_0)t}}{i(\Omega - \omega_0)} - \frac{e^{-i(\Omega + \omega_0)t}}{-i(\Omega + \omega_0)} \right] \Bigg|_0^{\frac{\pi}{\Omega}} \\[2mm]
&= \frac{1}{T\,2i} \left[\frac{e^{i\pi} e^{in\pi \frac{\omega_0}{\Omega}} - 1}{i(\Omega - n\omega_0)} + \frac{e^{-i\pi} e^{-in\pi \frac{\omega_0}{\Omega}} - 1}{i(\Omega + n\omega_0)} \right] \\[2mm]
&= \frac{[e^{(i\pi \frac{\omega_0}{\Omega})^n} + 1]}{T} \frac{\Omega}{\Omega^2 - n^2\omega_0^2} \quad (n \neq \pm 1 \text{ if } \omega_0 = \Omega).
\end{aligned}
$$

If $\omega_0 = \Omega$,

$$
\begin{aligned}
c_1 &= \frac{1}{T} \int_0^{\frac{\pi}{\Omega}} \frac{[1 - e^{-2i\Omega t}]}{2i}\, dt \\[2mm]
&= \frac{1}{T} \frac{1}{2i} \frac{\pi}{\Omega}.
\end{aligned}
$$

Similarly,

$$c_{-1} = \frac{-1}{T} \frac{1}{2i} \frac{\pi}{\Omega}.$$

Using $\omega_0 = \Omega$, the expansion becomes

$$
\begin{aligned}
e_0(t) &= \frac{\pi}{\Omega T} \sin(\Omega t) + \sum_{\substack{-\infty \\ n \neq \pm 1}}^{\infty} \frac{(-1)^n + 1}{T} \frac{\Omega}{\Omega^2 - n^2\Omega^2} e^{in\Omega t} \\[2mm]
&= \frac{1}{2} \sin(\Omega t) + \sum_{\substack{-\infty \\ n \neq \pm 1}}^{\infty} \frac{(-1)^n + 1}{2\pi(1 - n^2)} e^{in\Omega t} \\[2mm]
&= \frac{1}{2} \sin(\Omega t) + \sum_{m=-\infty}^{\infty} \frac{1}{\pi(1 - 4m^2)} e^{2im\Omega t}.
\end{aligned}
$$

Evaluation of c_0 from the above gives $c_0 = \frac{1}{\pi}$.

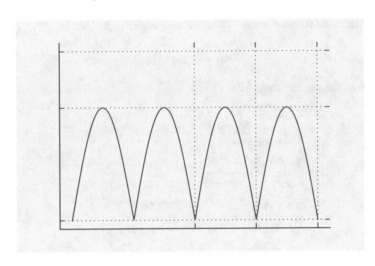

Fig. 2.9 Full-wave rectified signal

A waveform which provides a larger constant (D.C.) component is produced by use of full-wave rectifiers.

The output for this model is presented in Figure 2.9. The output is described by

$$
e_0(t) = \begin{cases} \sin(\Omega t) & 0 \le t \le \frac{\pi}{\Omega}, \\ e_0(t + \frac{\pi}{\Omega}) = e_0(t) & \text{otherwise.} \end{cases}
$$

In this case, the fundamental frequency associated with e_0 is given by

$$
\omega_0 = 2\,\Omega, \tag{2.16}
$$

and hence is twice the source frequency Ω.

The expansion coefficients are given by

$$
c_n = \frac{1}{\frac{\pi}{\Omega}} \int_0^{\frac{\pi}{\Omega}} \sin(\Omega t) e^{-in\omega_0 t}\, dt
$$

which fortunately may be readily evaluated from the calculations made above. Identifying $\frac{\pi}{\Omega}$ with T, and using $\frac{\omega_0}{\Omega} = 2$,

$$
c_n = \frac{2\Omega}{\pi} \cdot \frac{\Omega}{\Omega^2 - 4n^2\Omega^2}
$$

$$
= \frac{2}{\pi} \cdot \frac{1}{1 - 4n^2}
$$

so that the expansion for the full-wave case is

$$e_0(t) = \sum_{n=-\infty}^{\infty} \frac{2}{\pi} \frac{1}{(1 - 4n^2)} e^{i 2n\Omega t}.$$

This result has a constant component twice that of the previous case (as is obvious from Figure 2.9).

Delay Law

Another naturally occurring variation in an independent variable (beyond the scaling considered above) is a shift of origin. This has a simple effect on the associated Fourier expansions.

Given a function f of period L, a function g describable as "f delayed by x_0" (see Figure 2.10) may be defined by

$$g(x) = f(x - x_0).$$

The Fourier expansion coefficients of g are

$$c_n^{shift} = \frac{1}{L} \int_0^L f(x - x_0)\, e^{-in\omega_0 x}\, dx$$

$$= \frac{1}{L} \int_{-x_0}^{L-x_0} f(\zeta) e^{-in\omega_0 (\zeta + x_0)}\, d\zeta$$

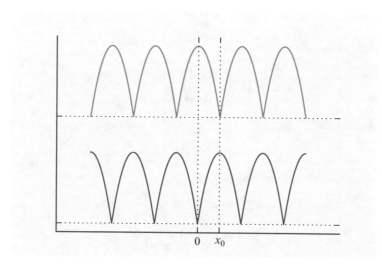

Fig. 2.10 Delayed function

$$= e^{-in\omega_0 x_0} \frac{1}{L} \int_0^L f(x) e^{-in\omega_0 x} \, dx$$

$$= e^{-in\omega_0 x_0} c_n.$$

That is, the expansion coefficients of the "shifted" f are related to those of the original by a multiplication by the delay factor

$$e^{-in\omega_0 x_0}.$$

We will see that similar relations hold in the case of Fourier and Laplace transforms (indeed, all such relations may be viewed as special cases of a more lofty theory of Fourier analysis).

Obviously, related formulas may be derived for the case of real form Fourier expansions. They have little use beyond providing exercise in the use of trigonometric identities.

Example. Suppose that f of period 2 is defined by

$$f(t) = \begin{cases} |t| & -1 \le t < 1, \\ f(t+2) = f(t) & \text{otherwise.} \end{cases}$$

Then we have

$$\omega_0 = \frac{2\pi}{2} = \pi,$$

and

$$f(t) = \frac{1}{2} + \sum_{\substack{n=-\infty \\ n \ne 0}}^{\infty} \frac{(-1)^n - 1}{\pi^2 n^2} e^{in\pi t}.$$

If g is defined by

$$g(t) = \begin{cases} |t-1| & 0 \le t < 2, \\ g(t+2) = g(t), \end{cases}$$

then it is clear that

$$g(t) = f(t-1),$$

so that the expansion coefficients of g are

$$c_n^{\text{shift}} = e^{-in\pi \, 1} c_n$$

$$= (-1)^n \frac{(-1)^n - 1}{\pi^2 n^2},$$

$$c_0^{\text{shift}} = c_0 = \frac{1}{2}.$$

Hence

$$g(t) = \frac{1}{2} - \sum_{\substack{n=-\infty \\ n \neq 0}}^{\infty} \frac{(-1)^n - 1}{\pi^2 n^2} e^{in\pi t}.$$

Differentiation Law

A major use of Fourier series is in the analysis of periodic phenomena in various physical systems. Since this analysis usually involves differential equations, the interaction between differentiation and Fourier expansions is a topic of primary interest.

Since time is the independent variable in many such problems, we adjust our notation accordingly.

We consider a function f of period T, with an expansion

$$f(t) = \sum_{n=-\infty}^{\infty} c_n e^{in\omega_0 t}.$$

Assume that f is a function of finite mean-square derivative [6], and that the derivative function f' is represented as

$$f'(t) = \sum_{n=-\infty}^{\infty} c'_n e^{in\omega_0 t}$$

The problem is to relate these two expansions. But

$$
\begin{aligned}
c'_n &= \frac{1}{T} \int_0^T f'(t)\, e^{-in\omega_0 t}\, dt \\
&= \frac{1}{T} \left[f(t)\, e^{-in\omega_0 t} \Big|_0^T + in\omega_0 \int_0^T f(t)\, e^{-in\omega_0 t}\, dt \right] \\
&= \frac{1}{T}\, in\omega_0 \int_0^T f(t)\, e^{-in\omega_0 t}\, dt \\
&= in\omega_0\, c_n
\end{aligned}
$$

(The "boundary terms" disappear since the functions are periodic.)

---

[6]The technical statement of the required condition is given above for the case of 2π periodic functions.

What this establishes is that if

$$f(t) = \sum_{n=-\infty}^{\infty} c_n e^{in\omega_0 t}$$

and if the derivative of f exists and is square-integrable over a period (has finite energy) then f' has a mean-square convergent expansion [7]

$$f'(t) = \sum_{n=-\infty}^{\infty} in\omega_0 c_n e^{in\omega_0 t}.$$

Note that this is exactly what one obtains by formal differentiation of the series term by term under the summation sign. Differentiation is readily expressed in terms of the Fourier coefficients as multiplication of by the factor *in* ω_0.

Example. The function f is defined by $f(t + 2) = f(t)$,

$$f(t) = \begin{cases} t & 0 \le t \le 1, \\ 2 - t & 1 \le t \le 2. \end{cases}$$

Clearly $f'(t) = \begin{cases} 1 & 0 < t < 1, \\ -1 & 1 < t < 2, \end{cases}$

$$f(b) - f(a) = \int_a^b f'(t)\, dt \text{ for all real } a \text{ and } b,$$

and f' has finite energy. (The nonexistence of $f'(t)$ for integer values of t has no effect on the representability of f as an anti-derivative). f is therefore a function of finite mean-square derivative.

The Fourier expansion of $f(t)$ is given by

$$f(t) = \frac{1}{2} + \sum_{\substack{n=-\infty \\ n \ne 0}}^{\infty} \frac{(-1)^n - 1}{n^2 \pi^2} e^{in\pi t}$$

while

$$f'(t) = \sum_{\substack{n=-\infty \\ n \ne 0}}^{\infty} - \frac{(-1)^n - 1}{in\pi} e^{in\pi t}.$$

This is obtainable by direct differentiation, or more tediously by performing the required integrations.

---

[7] A more careful discussion of this result is in Section 2.3 above.

Example (More properly, a non-example). Let f be a "sawtooth" of period 2π defined by

$$f(t + 2\pi) = f(t),$$

$$f(t) = t, \ -\pi < t \leq \pi.$$

Then the Fourier expansion of f is

$$f(t) = \sum_{\substack{n=-\infty \\ n \neq 0}}^{\infty} \frac{i}{n} (-1)^n e^{int}.$$

Evidently $f'(t)$ exists for all t different from an odd multiple of π, and

$$f'(t) = 1, \ \text{for } t \neq (2n + 1)\pi.$$

However, f is not an anti-derivative of this f' since the value 1 integrates to an ever increasing straight line, not the value of f, so that f fails to be a function of finite mean-square derivative. Formal term by term differentiation leads to the expression

$$\sum_{\substack{n=-\infty \\ n \neq 0}}^{\infty} (-1)(-1)^n e^{int},$$

which has no interpretation as a mean-square convergent Fourier series, since

$$\sum_{n=-\infty}^{\infty} |c_n'|^2 = \sum_{n=-\infty}^{\infty} 1$$

diverges.

This last example probably deserves some further explanation. From Parseval's Theorem we may essentially conclude that if the coefficients of the termwise differentiated series are square-summable (i.e., $\sum |c_n'|^2 < \infty$) then the calculations make sense in the context of mean-square convergence of the series involved. This is close to various other results concerning differentiation of integrals or infinite series, generally paraphrased as "try the formal calculation, and if the results make sense (converge) then the formalities are justified, and the answer is the result in hand."

For the example above, the formal calculation produces an answer with no interpretation as a mean-square convergent series, and the process has no meaning in the context of manipulations with such series. It is possible to discuss such problems in a context of "generalized functions," but this is a topic not pursued here in connection with Fourier series. It amounts to treating the objects under discussion as something other than a function with finite energy, and requires a completely different theoretical framework than the inner product space approach of this chapter.

The alternative framework ("distribution or generalized function theory") is discussed in connection with Fourier transforms in Section 7.8.

To complete this section, we explicitly write out the "differentiation rule" in terms of the real form of the Fourier series. The simplest way to obtain the desired result is to simply invoke the connections between the real and complex forms noted in Section 2.1.

We then conclude that the relevant restrictions are identical to those mentioned above, while the "rule" amounts to formal termwise differentiation of the series:

If

$$f(t) = a_0 + \sum_{n=1}^{\infty} a_n \cos(n\omega_0 t) + b_n \sin(n\omega_0 t),$$

then, assuming that f' is square-integrable (i.e., that f has a finite mean-square derivative),

$$f'(t) = \sum_{n=1}^{\infty} -a_n (n\omega_0) \sin(n\omega_0 t) + b_n (n\omega_0) \cos(n\omega_0 t).$$

The relative simplicity of the complex form formula has the effect that use of the complex form leads to more economical calculations in most problems.

Problems 2.7.

1. Use the shift rule to obtain the complex Fourier coefficients for the function g, where

$$g(t) = \begin{cases} t - \pi & 0 < t < 2\pi, \\ g(t + 2\pi) = g(t) & \text{otherwise,} \end{cases}$$

from those of f

$$f(t) = \begin{cases} t & -\pi < t < \pi, \\ f(t + 2\pi) = f(t) & \text{otherwise.} \end{cases}$$

2. Derive the form of the shift rule appropriate to the real form of the Fourier expansion.
3. Define f by

$$f(t) = \begin{cases} 0 & 0 \le x < 1, \\ 1 & 1 \le x < 2, \\ 0 & 2 \le x < 3, \\ f(x + 3) = f(x) & \text{elsewhere.} \end{cases}$$

Find the real and complex forms of the Fourier expansion for f.

4. Find the real and complex form Fourier expansions for h defined by

$$h(t) = \begin{cases} t & 0 \le t < 3, \\ 0 & 3 \le x < 6, \\ h(t+6) = h(t). \end{cases}$$

5. Suppose that f is a periodic real-valued function which is even, so that

$$f(-x) = f(x),$$
$$f(x+L) = f(x).$$

Show that the $\{c_n\}$ expansion coefficients for f are all real valued, and that all of the $\{b_n\}$ coefficients associated with the real form expansion vanish identically.

6. Show that in the case of a periodic, odd, real-valued function g,

$$g(-x) = -g(x),$$
$$g(x+L) = g(x),$$

the real form Fourier expansion consists entirely of "sine terms" so that

$$g(x) = \sum_{n=1}^{\infty} b_n \sin(n\omega_o x).$$

7. Define g by

$$g(t) = \begin{cases} 0 & 0 \le x < 1, \\ 1 & 1 \le x < 2, \\ -1 & 2 \le x < 3, \\ 0 & 3 \le x < 4, \\ g(x+4) = g(x) & \text{everywhere else.} \end{cases}$$

Find the real and complex Fourier expansions for g.

8. Construct a graph of the function f defined by

$$f(x) = \begin{cases} x/a & 0 < x < a, \\ \frac{(L-x)}{(L-a)} & a < x < L, \\ f(-x) = f(x), \\ f(x+2L) = f(x). \end{cases}$$

Show that the real form of the Fourier expansion for f is given by

$$f(x) = \sum_{n=1}^{\infty} \frac{2L^2}{a(L-a)} \frac{\sin(n\pi \frac{a}{L})}{n^2\pi^2} \sin(\frac{n\pi x}{L}).$$

9. Define a function f of period 2 as

$$f(t) = t^2 (t-2)^2, \, 0 \le t \le 2.$$

Find the real and complex forms of the Fourier expansion of f.
10. Differentiate f above, and compute directly the Fourier series expansions of f'. Verify that these coincide with the result of term by term differentiation of the results of Problem 2.
11. Define f by

$$f(t) = \begin{cases} e^t, & 0 < t < 1, \\ f(t+1) = f(t). \end{cases}$$

Compute the complex Fourier expansion of f. Differentiate termwise the resulting series, and note that, although it appears that

$$\frac{d}{dt}e^t = e^t,$$

the Fourier expansion of f does not reappear after differentiation. Explain this result.
12. A "clipped, full-wave rectified sine wave" is defined by the function f (assume $0 < C < E$)

$$f(t) = \begin{cases} E\sin(\Omega t), & 0 < t < \frac{1}{\Omega} \sin^{-1}(\frac{C}{E}), \\ C, & \frac{1}{\Omega} \sin^{-1}(\frac{C}{E}) \le t \le \frac{\pi}{\Omega} - \frac{1}{\Omega} \sin^{-1}(\frac{C}{E}), \\ E\sin(\Omega t), & \frac{\pi}{\Omega} - \frac{1}{\Omega} \sin^{-1}(\frac{C}{E}) \le t \le \frac{\pi}{\Omega}, \\ f(t + \frac{\pi}{\Omega}) = f(t). \end{cases}$$

a. Make a graph of f, and nod gravely at the appropriateness of the name of this function.
b. Compute the complex form of the Fourier expansion of f.

13. The use of synchronous digital circuit components in various physical systems results in the presence of signals (functions) which are periodic and piecewise constant on time intervals of uniform length.

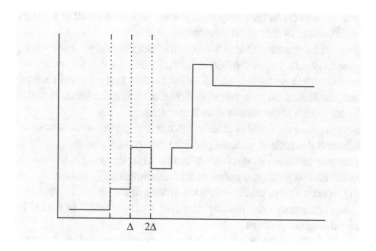

Fig. 2.11 Piecewise constant function

Define a function f by

$$f(t) = \begin{cases} f_0, & 0 < t < \frac{T}{N} = \Delta, \\ f_1, & \Delta < t < 2\Delta, \\ f_2, & 2\Delta < t < 3\Delta, \\ \dots \\ f_{N-1} & (N-1)\Delta < t < N\Delta = T. \end{cases}$$

(See Figure 2.11)

a. Compute the complex form of the Fourier expansion for f.
b. Assume that the numbers $\{f_j\}$ are sample values of some continuous function g, that is,

$$f_j = g(j\Delta).$$

By treating the quantity n/N as a small parameter, use a Taylor series argument to identify the formula of (a) as a "Riemann sum" for the evaluation of the corresponding Fourier coefficient of g.

2.8   Periodic Solutions of Differential Equations

The problems under consideration in this section may be described as the determination of periodic solutions of linear, time-invariant differential equations subject to periodic forcing functions.

Such problems arise in the analysis of linear circuits, as well as in investigations of linear mechanical systems and structures.

As a preface to the discussion, we remark that calculations for such problems often proceed (or are presented) on a formal basis in the following sense. One assumes at the outset (either explicitly or implicitly) that a periodic solution to the problem exists. Based on this premise, "results" are calculated. Needless to say, difficulties may arise if the assumption is violated.

The most obvious difficulties arise when the homogeneous solutions of the governing differential equation include periodic components of frequency coinciding with frequencies included in the forcing terms. This situation is usually referred to as resonance, and, typically, a solution whose amplitude is unbounded may exist. Clearly the search for periodic solutions makes little sense in such a situation. Luckily, such resonances are usually revealed by an attempt to divide by zero in the formal calculation process.

A more subtle problem arises when the homogeneous solution contains either nonresonant oscillatory terms, or terms that are unbounded with increasing time. The formal calculations may then actually produce a candidate for a periodic solution. However, this solution will be unattainable in any physical sense, since the "bad unstable modes" will be excited in virtually any experiment.

This situation does not arise in the analysis of inherently stable physical systems, such as passive circuits; however, circuits containing *active* devices (and mechanisms with analogous energy sources) *are* subject to such pitfalls.

We illustrate the basic method for such problems with the following example.

Example. Consider the circuit of Figure 2.12, and assume that the voltage source $e_0(t)$ is a square wave of period 2 given by

$$f(t) = \begin{cases} \quad 1 & 0 < t \le 1, \\ \quad -1 & 1 < t << 2, \\ e_0(t+2) = e_0(t). \end{cases}$$

Then the fundamental frequency is $\omega_0 = \pi$, and the Fourier series expansion of e_0 is given by

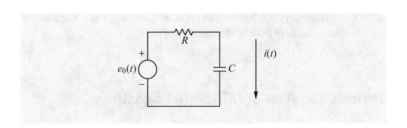

Fig. 2.12 RC circuit

$$e_0(t) = \sum_{\substack{n=-\infty \\ n\neq 0}}^{\infty} -\frac{(-1)^n - 1}{in\pi} e^{in\pi t}.$$

The governing equation of the circuit is

$$RC\frac{dv}{dt} + v = e_0(t). \tag{2.17}$$

We seek a periodic solution v of the same period as the forcing function, expressible in the form (of a mean-square convergent series)

$$v(t) = \sum_{n=-\infty}^{\infty} D_n e^{in\pi t}.$$

This assumption together with the requirement that v satisfy the equation means that $\frac{dv}{dt}$ also has a mean-square convergent expansion. The implication is that $\frac{dv}{dt}$ may be computed by termwise differentiation. If v satisfies (2.17), then

$$\sum_{n=-\infty}^{\infty} RC(in\pi)D_n e^{in\pi t} + \sum_{n=-\infty}^{\infty} D_n e^{in\pi t} = \sum_{\substack{n=-\infty \\ n\neq 0}}^{\infty} -\frac{(-1)^n - 1}{in\pi} e^{in\pi t}.$$

Since two Fourier expansions are equal only when the expansion coefficients are equal, we determine $\{D_n\}$ by equating coefficients of e on both sides of the above.

$$(in\pi RC + 1)D_n = -\frac{(-1)^n - 1}{in\pi},$$

$$1 D_0 = 0.$$

Finally,

$$D_n = -\frac{(-1)^n - 1}{in\pi (in\pi RC + 1)}, \quad n \neq 0,$$

and the solution is

$$v(t) = \sum_{\substack{n=-\infty \\ n\neq 0}}^{\infty} -\frac{(-1)^n - 1}{in\pi (in\pi RC + 1)} e^{in\pi t}.$$

The solution obtained above is often referred to as a *particular solution* in texts on elementary differential equations. The *general solution* to the above differential equation consists of the sum of the above and a multiple of the solution $e^{-t/RC}$

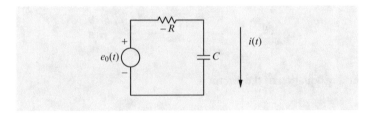

Fig. 2.13 Active RC circuit

associated with the homogeneous equation. The general solution therefore tends to
the periodic solution found above as $t \to \infty$. The solution v above is the whole
solution only when an initial condition problem for (2.17) is chosen so that the
initial value lies on the periodic solution. Obviously the initial condition

$$v(0) = \sum_{\substack{n=-\infty \\ n \neq 0}}^{\infty} -\frac{(-1)^n - 1}{in\pi \, (in\pi \, RC + 1)}$$

is such a choice.

Example. A conventional model of some active circuit devices is a negative resistor.
The active analog of the circuit above is in Figure 2.13, and is governed by the
equations

$$-RC \frac{dv}{dt} + v = e_0(t).$$

An attempt to find a periodic solution to the above produces

$$v(t) = \sum_{\substack{n=-\infty \\ n \neq 0}}^{\infty} -\frac{(-1)^n - 1}{in\pi \, (-in\pi \, RC + 1)} e^{in\pi t}.$$

This *is* a periodic solution of the equation, associated with the initial condition

$$v(0) = \sum_{\substack{n=-\infty \\ n \neq 0}}^{\infty} -\frac{(-1)^n - 1}{in\pi \, (-in\pi \, RC + 1)}.$$

However, solutions of the corresponding homogeneous equation are of the form
$Ae^{+t/RC}$, and the system is therefore unstable. Surely smoke will rise before the
periodic solution calculated above will be observed in practice.

Periodic solutions may also be determined on the basis of the real form Fourier
expansions. Since the algebra involved is several times more involved than the
above, we recoil from this alternative.

Single constant coefficient differential equations of higher order may be handled in a manner similar to the above. All that is required is the ability to differentiate a Fourier series formally more than once.

Single higher-order equation models often arise through a process of elimination of variables in a problem involving several energy storage elements. For such systems, it is often more natural to give a so-called "state variable" formulation of the system equations.

Such formulations are written in "vector-matrix" notation. Homogeneous equations take the form

$$\frac{d}{dt}\mathbf{x} = \mathbf{A}\mathbf{x}$$

where $\mathbf{A}$ is an $n \times n$ matrix, and $\mathbf{x}$ denotes an n-dimensional (column) vector of functions, referred to as the state variables.

In an electrical network, a vector of capacitor voltages and inductor currents is (modulo redundant variables) a natural choice for the state vector. In mechanical systems, position and momentum variables play the same role.

The corresponding inhomogeneous system with a scalar valued forcing function e takes the form

$$\frac{d\mathbf{x}}{dt} = \mathbf{A}\mathbf{x} + \mathbf{b}\,e(t),$$

where $\mathbf{b}$ is an n-vector.

Periodic solutions of such equations may be found by procedures which, modulo minor complications arising from the fact that some variables are vector-valued, are identical to that employed for the scalar example above.

Suppose that

$$e(t) = \sum_{n=-\infty}^{\infty} c_n e^{in\omega_0 t}$$

is a periodic forcing function, and that a solution of the system of equations of the form

$$\mathbf{x}(t) = \sum_{n=-\infty}^{\infty} \mathbf{D}_n e^{in\omega_0 t}$$

is sought. (Since $\mathbf{x}$ is vector valued, so are the $\{\mathbf{D}_n\}$.) Putting this (guess) into the governing equation we obtain

$$\sum_{n=-\infty}^{\infty} (in\omega_0)\,\mathbf{D}_n e^{in\omega_0 t} = \sum_{n=-\infty}^{\infty} \mathbf{A}\mathbf{D}_n e^{in\omega_0 t} + \sum_{n=-\infty}^{\infty} \mathbf{b}\,c_n e^{in\omega_0 t}.$$

Equating coefficients in the usual way,

$$(in\omega_0\,\mathbf{I} - \mathbf{A})\,\mathbf{D}_n = \mathbf{b}\,c_n$$

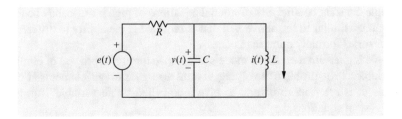

Fig. 2.14 RLC parallel circuit

follows. Assuming that $\mathbf{A}$ has no eigenvalues of the form $in\omega_0$ (i.e., no resonances are present) the above is uniquely solvable:

$$\mathbf{D}_n = (in\omega_0\,\mathbf{I} - \mathbf{A})^{-1}\,\mathbf{b}\,c_n.$$

The periodic solution is

$$\mathbf{x}(t) = \sum_{n=-\infty}^{\infty} (in\omega_0\,\mathbf{I} - \mathbf{A})^{-1}\,\mathbf{b}\,c_n\,e^{in\omega_0 t}.$$

Example. The circuit of Figure 2.14 may be readily analyzed by the above method. The governing equations are

$$L\frac{di}{dt} = v(t),$$

$$C\frac{dv}{dt} = [(e(t) - v(t))/R] - i(t).$$

Taking the inductor current i and the capacitor voltage v as the components of the state vector,

$$\frac{d}{dt}\begin{bmatrix} i \\ v \end{bmatrix} = \begin{bmatrix} 0 & \frac{1}{L} \\ \frac{-1}{C} & \frac{-1}{RC} \end{bmatrix}\begin{bmatrix} i \\ v \end{bmatrix} + \begin{bmatrix} 0 \\ \frac{1}{RC} \end{bmatrix}e(t).$$

We identify

$$\mathbf{A} = \begin{bmatrix} 0 & \frac{1}{L} \\ \frac{-1}{C} & \frac{-1}{RC} \end{bmatrix},$$

$$\mathbf{b} = \begin{bmatrix} 0 \\ \frac{1}{RC} \end{bmatrix},$$

$$\mathbf{x} = \begin{bmatrix} i \\ v \end{bmatrix}.$$

We calculate successively

$$(in\omega_0 \mathbf{I} - \mathbf{A}) = \begin{bmatrix} in\omega_0 & \frac{-1}{L} \\ \frac{1}{C} & in\omega_0 + \frac{1}{RC} \end{bmatrix},$$

$$(in\omega_0 \mathbf{I} - \mathbf{A})^{-1} = \frac{\begin{bmatrix} in\omega_0 + \frac{1}{RC} & \frac{1}{L} \\ \frac{-1}{C} & in\omega_0 \end{bmatrix}}{(in\omega_0)(in\omega_0 + \frac{1}{RC}) + \frac{1}{LC}}.$$

If we assume

$$e(t) = \sum_{n=-\infty}^{\infty} c_n e^{in\omega_0 t}$$

then the periodic solution of the system (circuit) is simply

$$i(t) = \sum_{n=-\infty}^{\infty} \frac{1}{RLC} \frac{c_n}{(in\omega_0)(in\omega_0 + \frac{1}{RC}) + \frac{1}{LC}} e^{in\omega_0 t},$$

$$v(t) = \sum_{n=-\infty}^{\infty} \frac{1}{RC} \frac{(in\omega_0) c_n}{(in\omega_0)(in\omega_0 + \frac{1}{RC}) + \frac{1}{LC}} e^{in\omega_0 t}.$$

Since the example above is a passive circuit, the solution calculated above is physically meaningful. In the general situation, the usual caveats about stability of the solutions of the homogeneous system must be applied to the results of the formal calculations.

2.9 Impedance Methods and Periodic Solutions

The "state variable" methods for periodic solutions discussed are closely related to the so-called impedance analysis methods for electrical circuits (and certain mechanical analogues)

The connection arises on the basis of two simple observations. The first observation is that the individual first-order differential equations in a state variable formulation arise from the application of the Kirchoff laws of circuit analysis. The rate equation for a capacitor voltage arises from the Kirchoff current law, while that for an inductor current arises from the Kirchoff voltage law.

The second observation is that in the treatment of periodic solutions in linear circuits, it suffices (due to the possibility of superposition of solutions) to consider each harmonic component separately.

From the classic circuit relations

$$e(t) = R\,i(t),$$

$$e(t) = L\frac{di}{dt},$$

$$i(t) = C\frac{de}{dt},$$

and the Fourier series differentiation formulas, we deduce that periodic current-voltage pairs are related by (we adopt here the $j = \sqrt{-1}$ convention of circuit analysis)

$$\sum_{n=-\infty}^{\infty} E_n\, e^{jn\omega_0 t} = \sum_{n=-\infty}^{\infty} R\,I_n\, e^{jn\omega_0 t},$$

$$\sum_{n=-\infty}^{\infty} E_n\, e^{jn\omega_0 t} = \sum_{n=-\infty}^{\infty} (jn\omega_0 L)\, I_n e^{jn\omega_0 t},$$

$$\sum_{n=-\infty}^{\infty} I_n\, e^{jn\omega_0 t} = \sum_{n=-\infty}^{\infty} (jn\omega_0 C)\, E_n e^{jn\omega_0 t}.$$

That is, the Fourier expansion coefficients of voltage V_n and current I_n are related by (the impedance relations)

$$E_n = R\,I_n, \tag{2.18}$$

$$E_n = (jn\omega_0 L)\,I_n,$$

$$I_n = (jn\omega_0 C)\,E_n$$

in the cases, respectively, of a resistor, inductor, or capacitor. These relations are illustrated in Figure 2.15.

From the above we deduce that the adoption of the impedance relations (2.18) as a replacement for Ohm's law, and the imposition of the Kirchoff conservation laws must lead directly to the equations of the analogous state variable formulation, and hence provide the same solution.

The treatment of periodic solutions is but one application of impedance methods. Similar (in fact, nearly identical) considerations arise in the application of Fourier and Laplace transform methods to circuit problems. These are mentioned in later sections.

The above treatment is intended mainly to indicate the relation of impedance methods to periodic solution problems, emphasizing "state variable" formulations of the circuit equations. For relations to "loop" or "node" analysis, we refer the interested reader to circuits text.

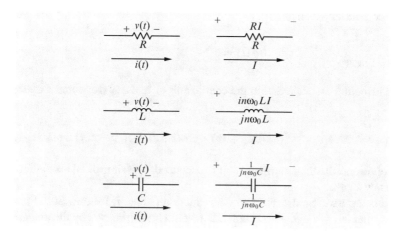

Fig. 2.15 Impedance relations

2.10 Power Spectrum and Parseval's Theorem

The Parseval relation

$$\frac{1}{T} \int_0^T |f(t)|^2 \, dt = \sum_{n=-\infty}^{\infty} |c_n|^2$$

which arises in the discussion of Fourier series convergence is a result which actually has strong physical interpretations (and consequent implications). This fact (or its close relatives) is central in various analyses of signal processing schemes, and related communications problems. Related applications appear in the area of vibration analysis.

At the heart of a physical understanding of Parseval's relation is the intuitive notion of energy or power. If f represents the output of a periodic ideal voltage source, then the average power dissipated in a one-ohm resistor connected across the source is given by

$$\frac{1}{T} \int_0^T |f(t)|^2 \, dt.$$

For this reason, functions f such that

$$\frac{1}{T} \int_0^T |f(t)|^2 \, dt < \infty$$

are often referred to as functions of finite average energy (or power).

In the Fourier expansion of

$$f(t) = \sum_{n=-\infty}^{\infty} c_n e^{in\omega_0 t}$$

the contribution associated with the coefficients c_k and c_{-k} (for some fixed $k > 0$, say) is

$$c_{-k} e^{-ik\omega_0 t} + c_k e^{ik\omega_0 t} = (c_{-k} + c_k) 2 \cos(k\omega_0 t) + i(c_k - c_{-k}) 2 \sin(k\omega_0 t).$$

This represents the total component of f associated with sinusoidal oscillation at the frequency $k \cdot \omega_0$.

If this term were extracted from the function f, then it follows from Parseval's relation that the average power remaining would be diminished by the amount

$$|c_k|^2 + |c_{-k}|^2.$$

An algebraic expression of this is that

$$\frac{1}{T} \int_0^T |f(t) - c_{-k} e^{-ik\omega_0 t} - c_k e^{ik\omega_0 t}|^2 \, dt = \sum_{n=-\infty}^{\infty} |c_n|^2 - |c_k|^2 - |c_{-k}|^2.$$

This observation may be rephrased as the claim that $|c_k|^2 + |c_{-k}|^2$ is the power in f associated with the frequency $k\omega_0$ (called the kth harmonic).

This information may be presented graphically by plotting $|c_k|^2$ against the discrete index k (or $k\omega_0$). Such a graphical display is called the *power spectrum* of the function f, or the power spectral density.

Example. Define f as the "sawtooth" function

$$f(t) = \begin{cases} t & -T/2 < t < T/2, \\ f(t + T) = f(t). \end{cases}$$

Then explicit calculation of the Fourier coefficients gives

$$c_n = (-1)^n \frac{i}{n\omega_0}, \quad n \neq 0,$$

$$c_0 = 0.$$

Construction of the power spectral density reduces to a plot of

$$|c_n|^2 = \frac{1}{\omega_0^2} \frac{1}{n^2} \quad \text{versus } n.$$

The plot is given in Figure 2.16.

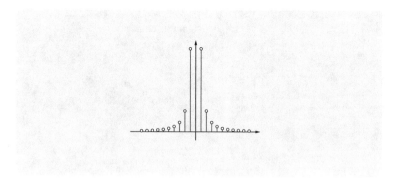

Fig. 2.16 Power spectrum plot

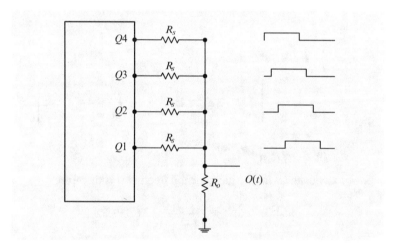

Fig. 2.17 Ring counter output circuit

Example. Wide availability of digital integrated circuit components leads to an interest in (analog) synthesis based on such devices. The power spectrum notion is the natural analysis tool for such arrangements.

A model ring counter based sinusoid generator is illustrated in Figure 2.17, and the resulting waveform is illustrated in Figure 2.18.

Analysis of the indicated resistor network provides the result that

$$O(t) = \frac{1}{4 + \frac{R_s}{R_o}} (Q1(t) + Q2(t) + Q3(t) + Q4(t)),$$

illustrated in Figure 2.18.

The quality of this hardware arrangement as a sinusoid generator is based on the proportion of power associated with the fundamental frequency.

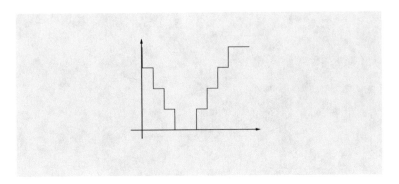

Fig. 2.18 Ring counter approximation

We calculate

$$c_1 = \frac{1}{T} \int_0^T O(t)\, e^{i\omega_0 t}\, dt$$

$$= \frac{1}{T\left(4 + \frac{R_s}{R_o}\right)} \left[\int_0^{T/10} 3\, e^{-i\omega_0 t}\, dt + \cdots + \int_{9T/10}^T 4\, e^{-i\omega_0 t}\, dt \right]$$

$$= \frac{1}{(4 + R_s/R_o)i}\, (.9796) e^{i(.6283)}.$$

The power associated with the fundamental frequency is therefore

$$|c_1|^2 + |c_{-1}|^2 = \frac{1}{(4 + \frac{R_s}{R_o})^2}\, 2\,(.9597).$$

The total average power in $O(t)$ is

$$\frac{1}{(4 + \frac{R_s}{R_o})^2} \frac{1}{10} [3^2 + 2^2 + 1^2 + 1^2 + 2^2 + 3^2 + 4^2 + 4^2] = 6\, \frac{1}{(4 + \frac{R_s}{R_o})^2}.$$

The fundamental frequency makes up

$$\frac{2(.9597)}{6} \times 100 \text{ percent} = 32 \text{ percent}$$

of the total. In fact, a large portion of the remaining power is contained in the constant (D. C.) component. We easily compute

$$c_0 = \frac{1}{10}[3 + 2 + 1 + 0 + 0 + 1 + 2 + 3 + 4 + 4] \frac{1}{(4 + R_s/R_o)}$$

$$= (2) \frac{1}{(4 + R_s/R_o)}$$

so that the fraction of the power of frequencies higher than the fundamental frequency is

$$\frac{6 - 4 - 2(.9597)}{6} = 1.34x10^{-2}.$$

In practice the result may be improved by selecting the resistance array values to minimize the above (rather than to simplify the computations).

Problems 2.10.

1. Find a periodic solution to the differential equation

$$\frac{dx}{dt} + 2x = f(t),$$

where

$$f(t) = \begin{cases} e^t & 0 < t < 1, \\ f(t+1) = f(t). \end{cases}$$

2. a. Produce a state variable model for the circuit of Figure 2.19.
   (f represents a current source.)
   b. From the equations above deduce a single second-order differential equation for the inductor current.
   c. Let f be given by

$$f(t) = \begin{cases} t & 0 < t < 37, \\ f(t+37) = f(t). \end{cases}$$

   Find a periodic solution of the model.

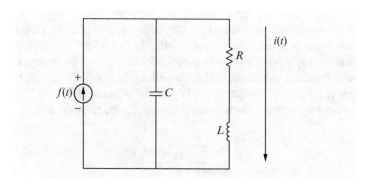

Fig. 2.19 RLC circuit with current source

d. Let f be as above, and compute a periodic solution for the second-order current equation. Verify that the results of that and the state variable model are the same.

3. Reproduce the results of Problem 2a by using impedance methods to analyze the circuit.

4. Analyze the voltage-driven RLC network of Section 2.10 above by means of impedance methods. Explicitly note the correspondence between the resulting linear equations and the two-by-two system determining the Fourier coefficients

$$\left\{ in\omega_0 \begin{bmatrix} 1 & 0 \\ 0 & 1 \end{bmatrix} - \begin{bmatrix} 0 & \frac{1}{L} \\ \frac{-1}{C} & \frac{-1}{RC} \end{bmatrix} \right\} \mathbf{D}_n = \begin{bmatrix} 0 \\ \frac{1}{R} \end{bmatrix} c_n$$

arising in the "state variable" model.

5. Show that the formal periodic solution of the higher-order differential equation

$$a_n \frac{d^n x}{dt^n} + a_{n-1} \frac{d^{n-1} x}{dt^{n-1}} + \cdots + a_0 x = f(t)$$

is given by

$$x(t) = \sum_{n=-\infty}^{\infty} \frac{c_n}{p(in\omega_0)} e^{in\omega_0 t}$$

where

$$f(t) = \sum_{n=-\infty}^{\infty} c_n e^{in\omega_0 t}$$

and

$$p(\lambda) = a_n \lambda^n + a_{n-1} \lambda^{n-1} + \cdots + a_0$$

is the characteristic polynomial of the differential equation. Conclude that complex form periodic solutions are available "instantly" for such problems. (The function $\frac{1}{p(\lambda)}$ is referred to as the transfer function of the system described by the differential equations). What conditions on p are required in order that the formal solution computed above is physically meaningful?

6. How many terms are required in order that the Fourier partial sum

$$\sum_{n=-N}^{N} c_n e^{in\omega_0 t}$$

for the sawtooth function contains 90% of the power of the function?

7. Compute and plot the power spectrum of the function

$$f(t) = \begin{cases} t & 0 < t \le 1, \\ 2 - t & 1 < t \le 2, \\ f(t+2) = f(t). \end{cases}$$

8. Radar receivers (and digital communications devices making use of "matched filters") make receiver decisions based on computed values of expressions of the form

$$\int_0^T w(t) f(t) \, dt$$

which are inner products of the received signal $f(t)$ with the filter waveform $w(t)$.

Suppose that the received signal is approximated by the partial sum of its Fourier series

$$f_N(t) = \sum_{n=-N}^{N} c_n e^{in\omega_0 t}$$

and the required inner product is computed by

$$\int_0^T w(t) f_N(t) \, dt.$$

Given that w is normalized

$$\int_0^T |w(t)|^2 \, dt = 1,$$

we wish to make the calculated error

$$\int_0^T w(t) f(t) \, dt - \int_0^T w(t) f_N(t) \, dt$$

small. Since this quantity scales linearly with the magnitude of the function f it really ought to be normalized. Use the Cauchy–Schwarz inequality to derive a power spectrum condition in order to determine N such that the calculation error

$$\frac{\int_0^T w(t) f(t) \, dt - \int_0^T w(t) f_N(t) \, dt}{\|f\|}$$

is in magnitude less than 1%.

The result is required in order to evaluate the computational requirements of a proposed signal processing scheme.

9. Is the result of Problem 8 dependent on the use of Fourier approximations in any essential way?

10. Suppose that

$$f(t) = sin(\omega_0 t)$$

and that x is the periodic solution of

$$\frac{d^2x}{dt^2} + .01\,\Omega\,\frac{dx}{dt} + \Omega^2 x = f(t).$$

Sketch the variation in the power spectrum of $x(t)$ as the parameter Ω varies (with ω_0 held fixed).

References

1. I.N. Gelfand, C.B. Shilov, *Generalized Functions*, vol. 2 (Academic, New York, 1968)
2. J.E. Marsden, *Elementary Classical Analysis* (W.H. Freeman and Company, San Francisco, 1974)
3. W. Rudin, *Real and Complex Analysis* (McGraw-Hill, New York, 1966)

Further Reading

H. Anton, C. Rorres, *Applications of Linear Algebra* (Wiley, New York, 1979)
C.M. Close, *The Analysis of Linear Circuits* (Harcourt, Brace and World, New York, 1966)
P.M. DeRusso, R.J. Roy, C.M. Close, *State Variables for Engineers* (Wiley, New York, 1965)
K. Hoffman, R. Kunze *Linear Algebra* (Prentice-Hall, Englewood Cliffs, 1960)
A. Kolmogorov, S. Fomin, *Metric and Normed Spaces* (Graylock Press, Rochester, 1957)
Technical Staff, *CMOS Databook* (National Semiconductor Corporation, Santa Clara, 1977)

Chapter 3
Elementary Boundary Value Problems

3.1 Introduction

Partial differential equations arise as models for various physical phenomena. These equations as a group are noticeably less tractable (in virtually every sense) than the analogous ordinary differential equations.

In the theory of ordinary differential equations, it is shown that for equations of the form

$$\frac{du}{dt} = f(u(t), t),$$

weak hypotheses on the functions f suffice to guarantee the existence and uniqueness of solutions. What this means in essence is that most reasonable equations of this form are at least candidates for models of physical systems. In particular, in the case of linear equations a fairly complete analysis is available.

In the case of partial differential equations this happy state of affairs does not persist. The equation

$$-i\frac{\partial u}{\partial x_1} + \frac{\partial u}{\partial x_2} - 2(x_1 + ix_2)\frac{\partial u}{\partial x_3} = f(x_3)$$

(constructed by H. Lewy, in 1957), while linear and innocent in appearance, has the property that solutions often fail to exist, even for forcing functions f which are infinitely differentiable.

Such unpleasant results can often be avoided by concentrating on partial differential equations which have "strong physical origins," and we consider only such problems below.

There is a set of three classical equations, which are prototype models of three different physical situations. Beyond their obvious use in establishing these models, the equations stand as landmarks and examples for the study of partial differential equations themselves.

© Springer International Publishing Switzerland 2016
J.H. Davis, *Methods of Applied Mathematics with a Software Overview*,
Applied and Numerical Harmonic Analysis, DOI 10.1007/978-3-319-43370-7_3

In the sections below, we formulate and solve these standard heat, wave, and potential equations. The mechanics of the solution technique relies heavily on Fourier series. This is not surprising, as such series historically arose in Joseph Fourier's attempt (1822) to find solutions of the heat equation.

The attempt to meet boundary conditions associated with the equations leads naturally to the so-called half-range expansions in the theory of Fourier series.

In addition to the standard equations, we develop below several examples of analogue models. These can be used to motivate and explain the solutions of the partial differential equations. Beyond this, these models arise naturally in various digital or "hybrid" computational schemes for solution of partial differential equations. Analysis of the analogue models for the standard partial differential equations provides explicit error analysis tools for such computational methods.

The chapter concludes with some discussion of the analytic properties of the formal solutions for the standard equations, as well as some results on existence, uniqueness, and continuous dependence of solutions on the given data.

3.2 The One-Dimensional Diffusion Equation

The process of equation derivation seems to be a skill which is regarded largely as witchcraft by the uninitiated, and something akin to common sense by the practitioner. It is a topic about which it is difficult to give useful specific advice ("Isolate the essential effects, ignore the unimportant, etc.,")

Having disclaimed the possibility of giving advice, we make the following comments on the derivations of the standard models below. The essential physical effects behind the models below are local in origin: heat flows in response to the local temperature gradient, particles accelerate in response to local force imbalances, and equilibria are established in response to local force balances. The governing equations are derived by exploiting this to write down the relevant physical law for an (infinitesimal) piece of material isolated from the whole. The process is entirely analogous to the use of free-body diagrams in problems of mechanics. (Conversely, there are variational derivations of the classical partial differential equations analogous to variational formulations in mechanics.)

The basic idealized physical problem in which the diffusion equation arises is that of heat flow through a uniform slab.

We consider an (transversally) infinite slab of thickness L, described by a temperature function u depending on the variables x (denoting distance along the slab), and t (denoting time). See Figure 3.1. The relevant physical law is Newton's law of cooling, expressed as

$$F = -\alpha \frac{\partial u}{\partial x},$$

where F denotes the heat flux per unit cross-sectional area in the material.

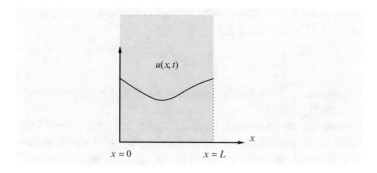

Fig. 3.1 Temperature in a slab

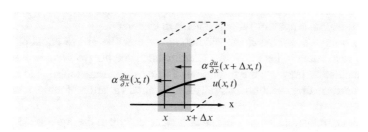

Fig. 3.2 Heat flux through a slice

From the slab of thickness L we extract a slice of thickness Δx centered at x. See Figure 3.2. If we let ρ (assumed constant) denote the density of the slab material, and c the specific heat, then the total heat content at time t, per unit cross sectional area, of the slice above is

$$u(x, t)\Delta x\rho c.$$

This heat content varies only due to the transfer of heat through the faces of the slice. The rate of flow into the slice is

$$-\alpha\frac{\partial u(x - \frac{\Delta x}{2}, t)}{\partial x}$$

while the rate of flow out is

$$-\alpha\frac{\partial u(x + \frac{\Delta x}{2}, t)}{\partial x}.$$

Equating the net flow rate to the rate of change of heat content,

$$\frac{\partial}{\partial t}u(x, t)\Delta x\rho c = -\alpha\frac{\partial u(x - \frac{\Delta x}{2}, t)}{\partial x} - -\alpha\frac{\partial u(x + \frac{\Delta x}{2}, t)}{\partial x}, \tag{3.1}$$

$$\frac{\partial}{\partial t}u(x, t) = \frac{\alpha}{\rho c}\frac{1}{\Delta x}\left(\frac{\partial u(x + \frac{\Delta x}{2}, t)}{\partial x} - \frac{\partial u(x - \frac{\Delta x}{2}, t)}{\partial x}\right).$$

The standard model is now obtained by letting Δx (arbitrary above) approach zero. This gives

$$\frac{\partial u}{\partial t} = \kappa \frac{\partial^2 u}{\partial x^2}. \tag{3.2}$$

The combination $\kappa = \frac{\alpha}{\rho c}$ is often referred to as the diffusion constant of the equation and is intrinsically positive.

The derivation given above is a "fast and dirty" one which sheds little light on the implicit assumptions. More general models employ three spatial dimensions and derive the governing equation by use of the integral theorems of vector calculus. With such a formulation it is easy to incorporate nonuniform material properties, and it is more apparent what is involved in the assumption of uniformity in the transverse dimensions used above. We reserve such a derivation for a treatment of multidimensional boundary value problems in Chapter 4.

One consequence of such considerations is that the above equation may be regarded as a model of heat flow in a laterally insulated bar (of finite lateral extent). Lateral insulation prevents the formation of transverse temperature gradients in an initially uniform cross section, and only longitudinal variation of temperature needs to be considered.

The reader may note that the verbal argument leading to the equation (3.1) above is to some extent apparently dependent on Figure 3.2. It should be verified that, in fact, the resulting equation (3.1) is independent of the assumed temperature profile of Figure 3.2. The algebraic signs of the indicated derivatives in fact "adjust" to provide the intuitively correct result. See the problems below.

The basic equation (3.2) arises as a model for many physical processes besides the case of heat flow. A quantity is said to diffuse (through a medium) if the flux of the quantity through the medium is proportional to the concentration gradient of the quantity in question.

(The physical law governing this mode of transport is referred to as Fick's law of diffusion.) In such a situation with a single spatial dimension (and uniform physical parameters) it is easy to duplicate the argument used above to conclude that the concentration function c is governed by the analogous equation

$$\frac{\partial c}{\partial t} = \kappa \frac{\partial^2 c}{\partial x^2}.$$

This equation arises as a model for the motion of pollutants (particulates, large molecules, etc.) through a gas. The diffusion may be thought of as arising on a microscopic scale from collisions between the molecules of the atmosphere and the particles in question.

The equation (3.2) is only a partial specification of a complete model for the problem under consideration. The equation has arisen on the basis of a local analysis of the phenomenon, and must be supplemented by both "boundary conditions" (corresponding to the arrangements of conditions on the slab faces) and an "initial condition" in the form of an initial temperature distribution.

The simplest imaginable boundary condition is that of specification of a temperature at each side of the slab. In this case, we supplement (3.2) by the conditions

$$u(0, t) = T_0, \ u(L, t) = T_L,$$

where T_0 and T_L are the applied temperatures at $x = 0, x = L$ respectively.

Even with boundary conditions specified, freedom remains to assign an arbitrary initial temperature distribution

$$u(x, 0) = f(x).$$

The complete boundary value formulation is then

$$\frac{\partial u}{\partial t} = \kappa \frac{\partial^2 u}{\partial x^2}, \tag{3.3}$$

$$u(0, t) = T_0,$$

$$u(L, t) = T_L,$$

$$u(x, 0) = f(x).$$

The problem (3.3) may be converted to one with homogeneous boundary conditions:

$$\frac{\partial v}{\partial t} = \kappa \frac{\partial^2 v}{\partial x^2},$$

$$v(0, t) = 0,$$

$$v(L, t) = 0,$$

$$v(x, 0) = g(x),$$

through the transparent device of subtracting an appropriate multiple of a straight line from the solution of (3.3). See the problems below.

It is clear that in addition to the steady boundary conditions of (3.3), one may consider problems for which the boundary temperatures vary with time.

Other boundary conditions of interest arise from the imposition of various "heat transfer" conditions at the slab faces. The simplest of these is the condition that the face be insulated, so that no heat is transferred. Since the heat flux is proportional to the temperature gradient, the relevant condition is that

$$\frac{\partial u}{\partial x} = 0.$$

More complicated conditions arise in problems involving heat transfer to a fluid in contact with the face. Assuming that the heat transfer to the fluid is proportional to the difference between the slab face and fluid temperatures leads (after appropriate steady state adjustments analogous to the above) to a boundary condition of the form

$$\alpha \frac{\partial u}{\partial x} + \beta u = 0$$

at the fluid interface.

Problems 3.2.

1. Construct diagrams analogous to Figure 3.2 illustrating the four possible permutations of temperature slope sign (plus or minus at each side). Verify that in each case equation (3.1) provides an appropriate expression for the rate of change of heat.

2. Show that a reversal of the spatial axis in the heat equation model leaves the basic governing equation the same. Do this in two ways: first by the formal change of variables

$$x_1 = L - x$$

in the governing equation, and second by justifying a new equation (3.2) in this context.

3. Consider the slab model leading to the standard heat equation (3.2), and suppose that in addition to the above hypotheses, the slab generates heat internally (e.g., from radioactive decay) at a rate

$$g(x)$$

per unit length at location x. Derive the governing equation. Do the boundary conditions change? Initial conditions?

4. Suppose that the concentration c of the noxious pollutant glurp is governed in still water by the equation

$$\frac{\partial c}{\partial t} = \kappa \frac{\partial^2 c}{\partial x^2}.$$

a. Consider a river of length L (name negotiable), uniform cross section, and flowing at velocity V_0. Find a partial differential equation for the concentration of glurp in the river (assuming uniformity across the river).

b. Suppose that glurp is (inadvertently, of course) introduced into the river at a rate proportional to $g(x)$ per unit length at location x. Modify your model to take this into account.

   (Here and in the above, you may assume pollution-free headwaters.)

c. Suppose that the river bottom absorbs glurp from the moving stream at a rate proportional to the concentration in the ambient water. Modify your model to take this into account, and also construct an equation giving the evolution of the glurp deposit profile of the river bottom.

5. a. Show that the heat equation

$$\frac{\partial u}{\partial t} = \kappa \frac{\partial^2 u}{\partial x^2}, \quad 0 < x < L$$

has solutions of the form

$$u(x, t) = a + bx.$$

b. Use this observation to find a solution of the heat equation satisfying the boundary conditions

$$u(0, t) = T_0,$$

$$u(L, t) = T_L.$$

c. Use your solution of (b) and the linearity of the heat equation to show that the solution of the problem

$$\frac{\partial u}{\partial t} = \kappa \frac{\partial^2 u}{\partial x^2},$$

$$u(0, t) = T_0,$$

$$u(L, t) = T_L,$$

$$u(x, 0) = f(x), \quad 0 < x < L$$

can be reduced to the solution of the homogeneous (i.e., zero boundary condition) problem of the text.

6. Suppose that a uniform slab of thickness L is in contact with fluids at temperatures T_0 and T_L at the points $x = 0, L$, respectively. If u denotes the slab temperature, suppose that the heat transfer rate between the fluid and slab is given by

$$r_0(T_0 - u(0, t))$$

and

$$r_L(T_L - u(L, t))$$

at the boundary $x = 0, x = L$, respectively. Show that with an appropriate redefinition of variables and constants, the governing model is

$$\frac{\partial v}{\partial t} = \kappa \frac{\partial^2 v}{\partial x^2}$$

with boundary conditions of the form

$$av + b\frac{\partial v}{\partial x} = 0, \quad x = 0,$$

$$cv + d\frac{\partial v}{\partial x} = 0, \quad x = L.$$

7. Diffusion models find application in the design and analysis of semi-conductor electronic devices. Current analysis of such devices is carried out on the basis of models of minority carrier concentration in the material.

   Minority carriers "disappear" from the bulk material according to an exponential decay law associated with recombination.

   A model for this effect is provided by the assertion that the concentration of minority carriers in a semi-conductor is governed (in the case of bulk material and uniform concentration) by the law

   $$c(t) = c_\infty + (c(0) - c_\infty)e^{-\frac{t}{\tau_e}}.$$

   The term c_∞ represents the characteristic equilibrium concentration.

   This model applies to the case of a hypothetical uniform bulk distribution of minority carriers. In the case of a nonuniform concentration, the minority carriers are also subject to diffusion. The constant relating the carrier flux to the concentration gradient is D. Consider a one-dimensional model of the above situation, and derive the model

   $$\frac{\partial c}{\partial t} = \frac{c_\infty - c(x, t)}{\tau_e} + D\frac{\partial^2 c}{\partial x^2}$$

   for the distribution of minority carriers in the material.

3.3 The Wave Equation

A second partial differential equation of a classical type may be obtained from an analysis of the transverse motions of a string under tension.

The physical model under consideration is illustrated in Figure 3.3.

It is assumed that displacements from equilibrium are small, that the string is under uniform tension, and that the string is completely flexible (supports no bending moments).

A model equation may be derived by constructing a free-body diagram for a "small" piece of the string. Such a diagram is illustrated in Figure 3.4.

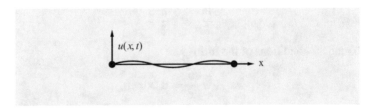

Fig. 3.3 Vibrating string

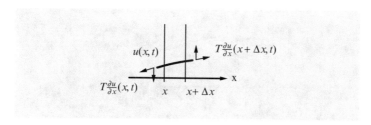

Fig. 3.4 String element force balance

Assuming that the string has a uniform mass density of ρ (per unit length) we obtain

$$\rho \Delta x \frac{\partial^2 u}{\partial t^2} = T \sin(\theta_1) - T \sin(\theta_2).$$

(This assumes small displacements to justify replacing the arc length by Δx. The center of mass acceleration is replaced by $\frac{\partial^2 u}{\partial t^2}$ in anticipation of the inevitable $\Delta x \to 0$ argument.) Recall that for small angles θ,

$$\sin(\theta) \approx \tan(\theta).$$

Making this replacement in the above,

$$\rho \Delta x \frac{\partial^2 u}{\partial t^2} = T \tan(\theta_1) - T \tan(\theta_2),$$

$$\rho \Delta x \frac{\partial^2 u}{\partial t^2} = T \frac{\partial}{\partial x} u(x + \frac{\Delta x}{2}, t) - T \frac{\partial}{\partial x} u(x - \frac{\Delta x}{2}, t)$$

and division by Δx gives (as $\Delta x \to 0$)

$$\frac{\rho}{T} \frac{\partial^2 u}{\partial t^2} = \frac{\partial^2 u}{\partial x^2}, \tag{3.4}$$

which is known as the (one-dimensional) wave equation.

As in the case of the heat equation discussed above, it is necessary to adjoin boundary and initial conditions in order to complete the problem formulation.

The initial conditions for the wave equation are easy to understand. Since the equation was derived on the basis of Newton's law of motion, we expect to prescribe an initial velocity and position for each "mass point" of the model. As the wave equation intuitively describes a continuum of such points, we prescribe two functions for initial position and velocity distributions. These take the form

$$u(x, 0) = f(x), \ 0 < x < L, \tag{3.5}$$

$$\frac{\partial u}{\partial t}(x, 0) = g(x).$$

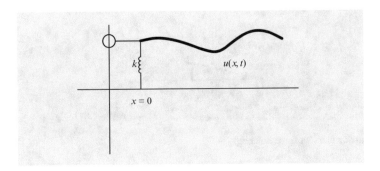

Fig. 3.5 Elastic restraint boundary condition

The boundary conditions for this model are determined by the physical arrangements imposed at the ends of the string.

The simplest of these is the fixed end case. The corresponding boundary conditions are simply that the displacement must vanish at the end in question;

$$u(0, t) = 0,$$

$$u(L, t) = 0.$$

Another variant is that of the "elastically constrained" end condition illustrated in Figure 3.5.

Equating the transverse component of the tension to the spring force, we obtain

$$k u(0, t) = T \frac{\partial u}{\partial x}(0, t)$$

as the appropriate condition.

Intuitively, the fixed end is obtained as $k \to \infty$ in the above; an opposite extreme is obtained if one lets $k = 0$. This corresponds intuitively to an "infinitely soft spring," and is referred to as a free-end condition.

If both ends are free, the conditions are

$$\frac{\partial u}{\partial x}(0, t) = 0, \qquad\qquad (3.6)$$

$$\frac{\partial u}{\partial x}(L, t) = 0.$$

It should be clear that the above boundary conditions may be inter-mixed in the same problem. Similarly, strings of different mass density may be connected back to back. Such a model (or at least the solution thereto) sheds light on scattering problems.

Another source of wave equations may be found in electromagnetic theory. In a uniform, source-free region, Maxwell's equations may be written as

$$\nabla \times \mathbf{E} = -\frac{\partial \mathbf{B}}{\partial t} = \mu_0 \frac{\partial \mathbf{H}}{\partial t}, \qquad (3.7)$$

$$\nabla \times \mathbf{H} = \frac{\partial \mathbf{D}}{\partial t} = \epsilon_0 \frac{\partial \mathbf{E}}{\partial t},$$

$$\nabla \cdot \mathbf{D} = \epsilon_0 \nabla \cdot \mathbf{E} = 0,$$

$$\nabla \cdot \mathbf{B} = \mu_0 \nabla \cdot \mathbf{H} = 0.$$

Here $\mathbf{E}$ is the electric intensity, $\mathbf{D}$ the displacement density, $\mathbf{B}$ the magnetic flux density, and $\mathbf{H}$ the magnetic intensity. Apply the operation $\nabla \times$ to the first equation of equations (3.7) to obtain

$$\nabla \times (\nabla \times \mathbf{E}) = -\frac{\partial}{\partial t} \nabla \times \mathbf{B} = -\mu_0 \frac{\partial}{\partial t} \nabla \times \mathbf{H}$$

$$= -\mu_0 \epsilon_0 \frac{\partial}{\partial t} \frac{\partial \mathbf{E}}{\partial t}$$

$$= \frac{1}{c^2} \frac{\partial^2 \mathbf{E}}{\partial t^2}.$$

Since

$$\nabla \times (\nabla \times \mathbf{E}) = \nabla(\nabla \cdot \mathbf{E}) - \nabla^2 \mathbf{E}$$

and

$$\nabla \cdot \mathbf{E} = 0,$$

we obtain an equation for the electric field in the form

$$\frac{1}{c^2} \frac{\partial^2 \mathbf{E}}{\partial t^2} = \nabla^2 \mathbf{E},$$

$$\frac{1}{c^2} \frac{\partial^2 \mathbf{E}}{\partial t^2} = \frac{\partial^2 \mathbf{E}}{\partial x^2} + \frac{\partial^2 \mathbf{E}}{\partial y^2} + \frac{\partial^2 \mathbf{E}}{\partial z^2}.$$

This is a three-dimensional wave equation for each component of the electric field $\mathbf{E}$. The same equation may be obtained for the magnetic field by starting with the second equation of (3.7).

The boundary conditions associated with the above equations may be considered at length. The simplest condition which applies is that the electric field must vanish inside of a perfectly conducting region. The possibilities of surface charges and currents complicate the formulation of boundary conditions, and we refer the reader to texts on electromagnetic theory for a discussion of these topics at length.

Problems 3.3.

1. Consider the derivation of the equation of a vibrating string above, and suppose that the motion under consideration takes place in a viscous medium. The effect of this viscosity is to add resisting forces proportional to the instantaneous velocity of the string. Derive the governing equation for this case.

2. Suppose that the string under discussion has density that varies with location along the string. Modify the "standard" wave equation for application to this case.

3. Suppose that the string of Problem 1 is also subject to the force of gravity.

   a. Derive the governing equation for this case.
   b. Calculate the steady state displacement for the case of a fixed end model.
   c. Use the solution of (b) to reduce the "gravity present" model to that of Problem 2 above. Conclude that the omission of gravity from the discussion above was justified.

4. Suppose that a string is of density ρ_0 over the interval $(0, L/2)$, and density ρ_1 over $(L/2, L)$. Derive the governing equations and the appropriate "matching" conditions at the junction.

5. Show that if ϕ and ψ are twice continuously differentiable functions of a single variable, then

$$u(x, t) = \phi(x + ct) + \psi(x - ct)$$

   formally satisfies

$$\frac{1}{c^2} \frac{\partial^2 u}{\partial t^2} = \frac{\partial^2 u}{\partial x^2}.$$

   Express the initial position and velocity in terms of ϕ, ψ, and their derivatives.

6. Consider the modeling of the population of a death-free utopia. Let $x > 0$ denote age, and t (as usual) time. If u denotes the population density function, that is,

$$\int_A^B u(\zeta, t)\, d\zeta$$

   is the number of inhabitants between ages A and B, find a partial differential equation governing u.

7. Consider Problem 6, and find a verbal interpretation of the quantity $u(0, t)$. Hence describe plausible boundary conditions for the equation derived there.

8. By sheer force of intellect, find an expression for the solution to the model of Problem 7 assuming that $u(0, t) = b(t)$ is known (or observed).

9. Add a touch of reality by introducing a death rate function r into the models above. Modify the result of Problem 8 above in the case that r is a constant, and then for an age-dependent death rate.

10. The standard wave equation also arises as a model for the longitudinal motions of a simple beam. Consider an elastic beam of length L Let $u(x, t)$ denote the departure from equilibrium of the material points originally at location x.

Hooke's law asserts that the stress in the beam is proportional to the strain $\frac{\partial u}{\partial x}$. This means that the equivalent force acting on a free-body cross section is proportional to $\frac{\partial u}{\partial x}$ at the cross section. Use this information to deduce that

$$\frac{1}{c^2}\frac{\partial^2 u}{\partial t^2} = \frac{\partial^2 u}{\partial x^2}$$

for this model.

3.4   The Potential Equation

A third partial differential equation of classical type is the potential equation. While the heat and wave equations may be described as "equations of motion" (evolution equations) and have associated initial conditions, the equations of this section are associated with conditions of static equilibrium (and hence are accompanied by boundary conditions only).

As in the case of the previous examples, the equation under consideration arises in a variety of problems. One of these is the theory of electrostatics. For simplicity, we restrict attention to two-dimensional models. The electric field has two components (E_x, E_y), and the static condition ensures that the curl of the field vanishes (c.f. Maxwell's equations of the previous section). Since the curl vanishes, the line integral of the field

$$\int_{(x_0,y_0)}^{(x,y)} -\mathbf{E} \cdot d\mathbf{l}$$

is independent of the path joining the endpoints, and hence defines a potential function ϕ in a simply connected region. The potential is related to the field through

$$E = -\nabla\phi = -\left(\frac{\partial\phi}{\partial x}, \frac{\partial\phi}{\partial y}\right).$$

Since the electric field (in a region of constant permittivity ϵ_0) containing distributed charge of surface density ρ is governed by

$$\nabla \cdot \mathbf{E} = \frac{\rho}{\epsilon_0},$$

we obtain

$$\nabla^2 \phi = \nabla \cdot \nabla\phi = -\frac{\rho}{\epsilon_0},$$

$$\frac{\partial^2 \phi}{\partial x^2} + \frac{\partial^2 \phi}{\partial y^2} = -\frac{\rho}{\epsilon_0}$$

as the equation for the potential. This equation is referred to either as the potential equation or as *Poisson's equation*. The equation

$$\nabla^2 \phi = 0$$

holding in regions of zero charge density (source-free regions) is called *Laplace's equation*.

The boundary conditions for the above are derived, as usual, from the imposed physical constraints.

A typical condition is that the potential ϕ may be prescribed on the boundary of some region of interest, for example, a rectangle. The complete problem formulation (for, say, Laplace's equation) may take the form

$$\frac{\partial^2 \phi}{\partial x^2} + \frac{\partial^2 \phi}{\partial y^2} = 0, \ 0 < x < a, \ 0 < y < b, \tag{3.8}$$

$$\phi(0, y) = f_1(y),$$

$$\phi(a, y) = f_2(y),$$

$$\phi(x, 0) = g_1(x),$$

$$\phi(x, b) = g_2(x)$$

in the case where the potential is prescribed on the boundary of the region. This situation may be visualized as arising from an elaborate arrangement of conducting plates and batteries.

Potential and Laplace equations also arise in the analysis of steady state heat flow.

Consider a planar region of thermally conducting material with a surface source density given by f. Observe that in a steady state the total source contribution to the heat content of a region must be exactly balanced by the heat flux through the boundary of the region. (See Figure 3.6). This results in the balance condition

$$-\int_S \alpha \nabla u \cdot \mathbf{n} \, dS = \int \int_R f \, dA,$$

where u represents temperature, and $\mathbf{n}$ is the outward normal along the boundary curve S. Using the Divergence Theorem (visualize a solid of unit depth in the z-direction) to convert the boundary integral to an integral over the region, we obtain

$$\int \int_R \{-\alpha \nabla^2 u + f\} \, dA = 0.$$

Since the region R is essentially arbitrary (and assuming sufficient smoothness on the functions u and f) we conclude that

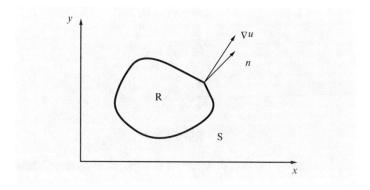

Fig. 3.6 Boundary heat flux

$$\nabla^2 u = -f(x, y)/\alpha,$$

so that the steady-state temperature distribution with heat sources present also satisfies Poisson's equation.

Boundary conditions for this case may be derived by arguments essentially identical to those of the heat equation case. Natural conditions include prescription of the values of the temperature along the boundary of some region. (This is essentially identical to the electrostatic case outlined above.) The case of an insulated boundary corresponds to a vanishing normal derivative on the surface in question. Heat transfer conditions analogous to those discussed above may also be imposed in a natural fashion (consistent with the condition that the problem posed makes physical sense).

The above boundary conditions may, of course, be imposed (on different portions of the boundary) simultaneously in one problem. The physical interpretation of the resulting system of equations should be clear.

Problems 3.4.

1. The above potential equations have been written in terms of a rectangular coordinate system. Using the usual formulas

$$x = r\cos(\theta),$$

$$y = r\sin(\theta),$$

   rewrite Laplace's equation in terms of polar coordinates.
2. Suppose that the boundary value problem

$$\frac{\partial^2 \phi}{\partial x^2} + \frac{\partial^2 \phi}{\partial y^2} = 0, \; 0 < x < 1, \; 0 < y < 2,$$

$$\phi(0, y) = 0,$$

$$\frac{\partial \phi}{\partial x}(1, y) = 0,$$

$$\frac{\partial \phi}{\partial y}(x, 2) = 0,$$

$$\phi(x, 0) = x^2$$

is given. Provide an explicit description of a physical system leading to this model.

3. Consider a two-dimensional flow of an incompressible fluid, and suppose that

$$\begin{bmatrix} u \\ v \end{bmatrix}$$

is a velocity field for the flow. $u(x, y, t)$ is the x-component of the velocity of the particles of fluid located at position (x, y) at time t.

Assuming that there are no sources or sinks of fluid in the region under consideration, show that conservation of mass leads to (the continuity equation)

$$\nabla \cdot \begin{bmatrix} u \\ v \end{bmatrix} = 0.$$

Hint: Consider the net flow through the boundaries of an arbitrary region in the flow. With no sources of fluid present in the flow field, this must vanish identically.

4. The velocity field of a fluid flow is called *irrotational* if the curl of the field vanishes. Using Problem 3 above, show that if the flow of a two-dimensional incompressible fluid is irrotational, then it is possible to define a potential function ϕ such that

$$\begin{bmatrix} u \\ v \end{bmatrix} = -\nabla \phi,$$

where (because of the continuity equation) ϕ satisfies

$$\nabla^2 \phi = 0.$$

(You may assume sufficient smoothness of the functions in question to justify required differentiations, and that the region is simply connected.)

5. Consider Problems 3 and 4, but consider this time that the fluid region contains sources and sinks distributed continuously with strength $f(x, y)$ at location

(x, y). Modify the continuity equation appropriately, and show that the velocity potential of an irrotational flow must now satisfy Poisson's equation

$$\frac{\partial^2 \phi}{\partial x^2} + \frac{\partial^2 \phi}{\partial y^2} = -f(x, y).$$

6. Consider the boundary value problem

$$\nabla^2 \phi = 0, \ x^2 + y^2 < 1,$$

with boundary conditions (in terms of polar coordinates)

$$\frac{\partial \phi}{\partial r}(1, \theta) = f(\theta).$$

a. Give an interpretation of the above as a steady-state heat flow problem.
b. Provide a physically based argument that one expects (on the basis of (a)) that the constraint that

$$\int_0^{2\pi} f(\theta) \, d\theta = 0$$

is an appropriate one for the boundary condition above.

3.5 Discrete Models of Boundary Value Problems

Discrete models of the partial differential equations introduced above arise in various contexts. Some such models arise in an attempt to solve partial differential equations on a numerical basis. Their analysis provides useful information for the choice of step sizes appropriate for such attempts, and some insights into the nature of difficulties associated with numerical treatment of such problems.

The above concerns are germane to attempts at digital computation of solutions. For some problems analog computation may provide a more cost effective means of obtaining the desired information. Physical models of the analogue provide the means for synthesis of such devices.

As a pedagogical device, well-constructed analogues mirror the essential properties of the boundary value problems from which they are derived. They therefore illuminate the solutions of the corresponding partial differential equations.

We consider first the diffusion equation

$$\frac{\partial u}{\partial t} = \kappa \frac{\partial^2 u}{\partial x^2}, \ 0 < x < 1.$$

(The scaling of the length to unity is a matter of notational convenience.)

We propose to replace the partial differential diffusion equation by a system of first-order ordinary differential equations (with respect to time) by use of the well-known Taylor series approximation for a second derivative

$$\frac{d^2 f(x)}{dx^2} \approx \frac{1}{h^2} [f(x+h) - 2f(x) + f(x-h)],$$

valid for sufficiently smooth functions of a single variable x.

Considering the diffusion model, we divide the spatial interval $[0, 1]$ into N-parts, take $h = \frac{1}{N}$, and rewrite the equation invoking the second derivative approximation. This gives

$$\frac{\partial}{\partial t} u(x, t) \approx \kappa N^2 (u(x + \frac{1}{N}, t) - 2u(x, t) + u(x - \frac{1}{N}, t)).$$

If we make the identifications

$$u(\frac{j}{N}, t) \leftrightarrow v_j(t),$$

then this leads directly to the system of equations

$$\frac{d}{dt} v_j(t) = \kappa N^2 (v_{j+1} - 2 v_j + v_{j-1})$$

as a plausible discrete model for the original diffusion model.

One question which arises in connection with the discrete model is the treatment of the (so far ignored) boundary conditions associated with the partial differential equation.

In the case of zero boundary conditions

$$u(0, t) = u(l, t) = 0,$$

the natural analog is to impose the constraints

$$v_0 = v_N = 0$$

on the discrete spatial model. The equation system is then written for j in the range $j = 1, 2, \ldots, N - 1$.

A physical realization of the discrete model is obtained by noting the resemblance of the governing equations to the equations of motion of an RC ladder network (Figure 3.7).

The relevant circuit equations are simply

$$C \frac{dv_j}{dt} = i_j = \frac{(v_{j+1} - v_j)}{R} - \frac{(v_j - v_{j-1})}{R}$$

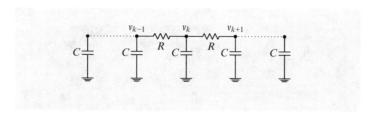

Fig. 3.7 RC ladder network

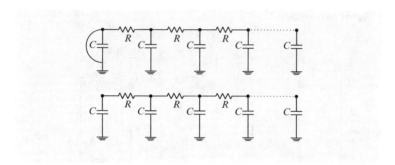

Fig. 3.8 RC circuit boundary conditions

or

$$\frac{dv_j}{dt} = \frac{1}{RC}[v_{j+1} - 2\,v_j + v_{j-1}].$$

The equations of the circuit and discrete model are identical provided that the constants satisfy

$$\frac{1}{RC} = \kappa N^2.$$

(Note that this correspondence is not actually required for the analogy, since the time scale of the differential equation is subject to arbitrary rescaling, which translates directly into a scaling of the constants.)

The boundary conditions of the diffusion equation are incorporated through the terminations of the ladder network. The cases of zero boundary condition and zero slope are illustrated in Figure 3.8.

It is a simple matter to see that the analog equations in the case of zero boundary conditions may be written in the form of the vector-matrix system

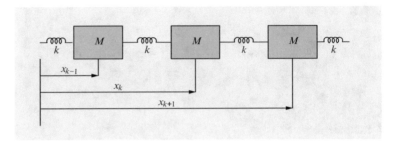

Fig. 3.9 Spring mass model

$$\frac{d}{dt}\begin{bmatrix} v_1 \\ v_2 \\ \vdots \\ v_{N-1} \end{bmatrix} = \frac{1}{RC}\begin{bmatrix} -2 & 1 & & & & \\ 1 & -2 & 1 & & & \\ & 1 & -2 & 1 & & \\ & & & \cdots & & \\ & & & 1 & -2 & 1 \\ & & & & 1 & -2 \end{bmatrix}\begin{bmatrix} v_1 \\ v_2 \\ \vdots \\ v_{N-1} \end{bmatrix}. \tag{3.9}$$

The coefficient matrix in the above evidently plays the role of the second derivative of the diffusion equation.

A discrete analog of the wave equation may be obtained from the physical example of a spring-mass string. (Figure 3.9).

Newton's law readily provides

$$\frac{M}{K}\frac{d^2 x_k}{dt^2} = [x_{k+1} - 2x_k + x_{k-1}],$$

while spatial discretization of

$$\frac{1}{c^2}\frac{\partial^2 u}{\partial t^2} = \frac{\partial^2 u}{\partial x^2}$$

leads to

$$\frac{1}{c^2}\frac{\partial^2 u}{\partial t^2}(x,t) = N^2\left(u(x+\frac{1}{N},t) - 2u(x,t) + u(x-\frac{1}{N},t)\right)$$

from which the analogies are evident. The reader may verify that the circuit arrangement of Figure 3.10

is also a wave equation analogue. The circuit is probably a more easily built analog model than the mechanism of Figure 3.9.

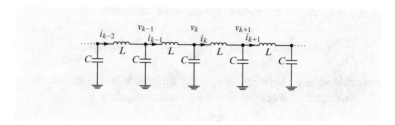

Fig. 3.10 LC ladder network

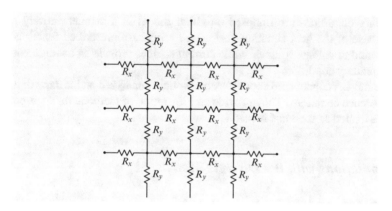

Fig. 3.11 Rectangular resistor network

Application of the second derivative approximation to Laplace's equation in a unit square (assuming N x-axis divisions, M y-axis divisions) gives

$$N^2\left[u(x + \frac{1}{N}, y) - 2\,u(x, y) + u(x - \frac{1}{N}, y)\right]$$
$$+ M^2\left[u(x, y + \frac{1}{M}) - 2\,u(x, y) + u(x, y - \frac{1}{M})\right] = 0.$$

Making the correspondence

$$u(\frac{k}{N}, \frac{j}{M}) \leftrightarrow v_{k,j},$$

we obtain the equations

$$N^2[v_{k+1,j} - 2\,v_{k,j} + v_{k-1,j}] + M^2[v_{k,j+1} - 2\,v_{k,j} + v_{k,j-1}] = 0.$$

This (somewhat imposing) set of linear algebraic equations may also be realized by a simple rectangular resistance network. The network is illustrated in Figure 3.11.

Conserving current at the node k, j leads to the relation

$$\frac{1}{R_x}[v_{k+1,j} - 2\,v_{k,j} + v_{k-1,j}] + \frac{1}{R_y}[v_{k,j+1} - 2\,v_{k,j} + v_{k,j-1}] = 0.$$

The correspondence between the partial difference equation and the resistive circuit is established by requiring that the parameters satisfy

$$\frac{R_x}{R_y} = \frac{M^2}{N^2}.$$

Boundary conditions are imposed on either model in a manner entirely parallel to the case of the heat equation analogue. In this case prescribed function values correspond to voltage sources, while current sources provide an analogue of a heat flux boundary condition.

The models constructed above may be explicitly analyzed on the basis of a theory of difference equations. They provide an illustration of methods for such problems and are treated as such in Chapter 7.

Computations with the Discrete Models

The discrete models represent potentially large systems of equations, but the form of the equations involves "sparse" systems of linear equations because of the localization of the interactions between elements.

Octave supports handling data as a *sparse matrix*, in order to conserve memory by storing only the nonzero entries. This can be used to numerically solve some of the discrete models in a relatively efficient way. We illustrate the procedure with the *RC* heat equation model.

The first problem is to construct the coefficient matrix for the governing equations. The spdiag Octave function constructs a sparse matrix by placing values along super- and sub-diagonal positions. The expectation is that this is a common structure in sparse problems.

Declaring the variables as global saves the complication of passing them as arguments in the function which computes the slope function for the differential equation.

OCTAVE
```
N=20;
e = repmat(1, N, 1);
d = repmat(-2, N, 1);

global A;
global RC;

A = spdiags([e d e], -1:1, N, N);

RC =.5;
```

Octave contains a variety of numerical methods for the solution of (vector) differential equation problems. They are constructed with a common argument format to facilitate experimentation with the methods.

The first argument is always the slope function for the governing differential equation. Since the governing equation of the *RC* model is a linear one, the slope calculation is a simple matrix multiplication.

OCTAVE

```
% rcheat
% script to solve   rc heat anaolg model

  function dydt = rcheat(t, y)
    global RC;
    global A;

    dydt = (1/RC)*A*y;
  endfunction
```

The procedure ode45 is a dual-order Runge-Kutta method that is a generally useful method, especially for problems that are not excessively "stiff." It has optional arguments for method tuning parameters, but the basic invocation requires a time interval and initial condition vector as second and third arguments.

The ode45 return values are a time vector, and a rectangular array whose columns are the solution vector samples. This makes it easy to plot the "solution surface" using the solution vector index as the "*x*-coordinate." The clear command removes the global variable definitions from the workspace.

OCTAVE

```
u0 =[repmat(0, N/4, 1); repmat(1, N/2, 1); repmat(0, N/4, 1)];

[t, u] =   ode45('rcheat', [0 5], u0);

surf((1:N)/N, t, u);

clear global A
clear global RC
```

The resulting surface is plotted in Figure 3.12.

Problems 3.5.

1. A boundary value problem not considered above is that associated with "heat flow in a laterally insulated thin ring." This may be visualized as the result of "bending a thin rod problem" into a circle and connecting the ends. The situation is illustrated in Figure 3.13.

    Deduce that an appropriate governing equation has the form

$$\frac{\partial u}{\partial t} = K \frac{\partial^2 u}{\partial \theta^2}.$$

    (The meaning of "thin laterally insulated" is that radial temperature gradients may be ignored.)

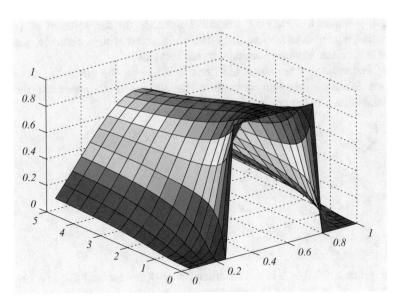

Fig. 3.12 Numerical solution of the RC analog system

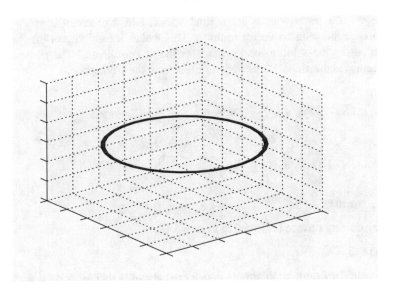

Fig. 3.13 Heat flow in a thin ring

2. Does calling the variable "θ" in the equation above make it an angle? Show that the boundary conditions

$$u(0, t) = u(2\pi, t),$$

$$\frac{\partial u}{\partial \theta}(0, t) = \frac{\partial u}{\partial \theta}(2\pi, t)$$

are required on the basis of the physical problem under consideration.

3. Construct an *RC*-network electrical analogue of the above boundary value problem.

4. Consider the *RC*-network analogue of the heat equation with zero slope (insulated ends) boundary conditions. Write out explicitly the equations of motion in vector-matrix notation analogous to equation (3.9) above.

5. Heat equations are encountered with "heat transfer" boundary conditions of the form

$$\alpha u - \beta \frac{\partial u}{\partial x} = 0.$$

Find a termination network for the *RC*-ladder appropriate for simulation of this boundary condition.

6. For the *LC*-ladder analogue of the wave equation, find the circuit terminations appropriate to the standard boundary conditions

$$u = 0$$

or

$$\frac{\partial u}{\partial x} = 0.$$

7. Use Octave to simulate and plot the solutions of the spring-mass analog of a wave equation.

   Hint: The easiest way to organize this problem is probably to let N be the number of masses, and generate matrices and arrays by stacking and concatenating N-dimensional pieces.

   OCTAVE
   ```
   u = [u1 ; u2];

   A = [A1 A2 ; A3 A4];

   pos = u(1:N, :);
   ```
   The position-time "surface" can easily be plotted.

8. Consider the discrete Laplace equation discussed above, and suppose that the grid division parameters M and N are equal.

   Show that the solution has the property that its value at each interior point of the grid is equal to the arithmetic average of the values at the four adjacent points. (A related averaging property also holds for Laplace's equation itself and forms the basis of certain numerical solution methods.)

9. Describe the physical arrangements associated with simulation of

   a. prescribed function values at the boundary,
   b. prescribed derivatives at the boundary for the case of the rectangular Laplace equation analogue. Show that Problem 9a corresponds to voltage sources, while Problem 9b may be identified with current sources.

10. What constraint does Kirchoff's current law (conservation of current) place on the solution of Problem 9a found above? Compare Problem 6, Section 3.4.

3.6 Separation of Variables

In this section we consider the solution of some of the boundary value problems developed in previous sections.

In the case of ordinary differential equations, elementary problems can often be solved by inspired guesswork. Solutions are sufficiently simple that this procedure causes no great amount of alarm.

The solutions of boundary value problems are rather more complicated. For this reason more systematic methods for solution are used. One such approach is the method of separation of variables, which we introduce through the example of the heat equation. The heat equation model which involves the smallest number of extraneous distractions is that associated with heat flow in a thin, laterally insulated ring. (See Problem 1, Section 3.5). The problem statement is

$$\frac{\partial u}{\partial t} = \kappa \frac{\partial^2 u}{\partial \theta^2}, \tag{3.10}$$

$$u(0, t) = u(2\pi, t),$$

$$\frac{\partial u}{\partial \theta}(0, t) = \frac{\partial u}{\partial \theta}(2\pi, t),$$

$$u(\theta, 0) = f(\theta), \ 0 < \theta < 2\pi.$$

The second and third conditions above impose continuity on the temperature and heat flux at the joint of the ring. The system (3.10) may be thought of as an initial value problem (for the equations of motion) for the temperature profile around the ring. The system is analogous to a system of first-order vector differential equations of the form

$$\frac{d\mathbf{u}}{dt} = \mathbf{A}\,\mathbf{u}, \tag{3.11}$$

$$\mathbf{u}(0) = \mathbf{f}.$$

Here $\mathbf{u}$ is a vector in some N-dimensional space, and $\mathbf{A}$ is an $N \times N$ matrix. The analogy may be made explicit by reference to the heat equation discrete analogue constructed in Section 3.5 if one desires, although the details of the equation are largely irrelevant to what follows.

The system of linear equations (3.11) may be solved by inspection in the case that the coefficient matrix $\mathbf{A}$ is a diagonal matrix. In the case that the matrix $\mathbf{A}$ may be diagonalized by a suitable change of basis, the solution also follows readily (this is the case in theory, although in practice a certain amount of pain is involved in the attempt to diagonalize $\mathbf{A}$).

Suppose that we make the explicit assumption that $\mathbf{A}$ in (3.11) is diagonalized by the non-singular matrix $\mathbf{P}$. Then from (3.11) we obtain

$$\frac{d}{dt}(\mathbf{P}^{-1}u) = (\mathbf{P}^{-1}\mathbf{A}\mathbf{P})(\mathbf{P}^{-1}\mathbf{u}),$$

$$\mathbf{P}^{-1}\mathbf{u}(0) = \mathbf{P}^{-1}\mathbf{f}.$$

Assuming that the eigenvalues of $\mathbf{A}$ are real and negative, and letting $\mathbf{z} = \mathbf{P}^{-1}\mathbf{u}$, the equation in new coordinates becomes

$$\frac{d}{dt}\mathbf{z} = \begin{bmatrix} -\lambda_1^2 & & & 0 \\ & -\lambda_2^2 & & \\ & & \ddots & \\ 0 & & & -\lambda_N^2 \end{bmatrix}\mathbf{z},$$

$$\mathbf{z}(0) = \mathbf{P}^{-1}\mathbf{f}.$$

Recall that under our assumption that $\mathbf{P}$ diagonalizes $\mathbf{A}$, the columns of $\mathbf{P}$ consist of the eigenvectors of $\mathbf{A}$, so that $\mathbf{z}$ is identified as the column vector of components of $\mathbf{u}$ with respect to the basis of eigenvectors of $\mathbf{A}$.

The explicit solution of the diagonalized system is

$$\mathbf{z} = \begin{bmatrix} e^{-\lambda_1^2 t} & & & 0 \\ & e^{-\lambda_2^2 t} & & \\ & & \ddots & \\ 0 & & & e^{-\lambda_N^2 t} \end{bmatrix}\mathbf{P}^{-1}\mathbf{f},$$

so that

$$\mathbf{u}(t) = \mathbf{P}\begin{bmatrix} e^{-\lambda_1^2 t} & & & 0 \\ & e^{-\lambda_2^2 t} & & \\ & & \ddots & \\ 0 & & & e^{-\lambda_N^2 t} \end{bmatrix}\mathbf{P}^{-1}\mathbf{f}.$$

If the columns of $\mathbf{P}$ are denoted by $\mathbf{X}_1, \mathbf{X}_2, \mathbf{X}_3 \ldots \mathbf{X}_N$ (recall these are the eigenvectors of the original $\mathbf{A}$), then "multiplying out" gives

$$\mathbf{u}(t) = \sum_{n=1}^{N} \mathbf{X}_n\, e^{-\lambda_n^2 t}\, b_n.$$

The numbers $\{b_n\}$ are the entries of $\mathbf{P}^{-1}\mathbf{f}$. As noted above, these are simply the components of the initial condition vector f with respect to the eigenvector basis. This is also evident from evaluating the solution when $t = 0$. Then

$$\mathbf{u}(0) = \mathbf{f} = \sum_{n=1}^{N} \mathbf{X}_n\, b_n.$$

The solution is expressed as a linear combination of the (vector) functions

$$\{\mathbf{X}_n\, e^{-\lambda_n^2 t}\}_{n=1}^{N}.$$

These are solutions of the differential equation

$$\frac{d\mathbf{u}}{dt} = \mathbf{A}\,\mathbf{u}, \tag{3.12}$$

and are distinguished among all solutions of the equation by the fact that the time dependence of the solution "factors out" of the vector expression. These special solutions may be arrived at by the following argument.

Suppose that we seek solutions of the system of equations in the special form

$$\mathbf{u}(t) = \mathbf{X}\, T(t),$$

where $\mathbf{X}$ is a vector to be determined, as is the scalar-valued function $T(t)$. The expression is a solution of the homogeneous system exactly when

$$T'(t)\mathbf{X} = \mathbf{A}\mathbf{X}T(t).$$

Assuming $T(t) \neq 0$, this is equivalent to

$$\frac{T'(t)}{T(t)}\mathbf{X} = \mathbf{A}\,\mathbf{X}. \tag{3.13}$$

The right side of this expression is evidently independent of time, so that in order for the equation to hold the quantity

$$\frac{T'(t)}{T(t)}$$

must also be a constant, independent of t. The name of this constant is a matter of taste and convenience. Assuming that the constant in question will turn out real and negative,[1] we call it $-\lambda^2$. Then

$$\frac{T'(t)}{T(t)} = -\lambda^2,$$

and $\mathbf{X}$ satisfies

$$\mathbf{A}\mathbf{X} = -\lambda^2\,\mathbf{X}. \qquad (3.14)$$

From this we conclude that $\mathbf{X}$ must be an eigenvector of $\mathbf{A}$ associated with eigenvalue $-\lambda^2$, and that the N candidates

$$\mathbf{u}_n(t) = \mathbf{X}_n\, e^{-\lambda_n^2 t}\ n = 1\ldots N$$

each satisfy the differential equation.

If we now consider the initial value problem (3.11), the individual solutions $\mathbf{u}_n(t)$ in general fail to satisfy the initial condition $\mathbf{u}(0) = \mathbf{f}$. However, since the differential equation (3.11) is a linear homogeneous one, the linear combination

$$\mathbf{u}(t) = \sum_{n=1}^{N} b_n\, \mathbf{X}_n\, e^{-\lambda_n^2 t}$$

is also a solution of (3.12). If $\{b_n\}_{n=1}^{N}$ are determined to meet the initial condition expansion, such a $\mathbf{u}$ also satisfies the initial condition required at the initial time $t = 0$.

In this way the standard solution of the differential equation initial value problem is recovered from the hypothesized solutions of the form $T(t)\,\mathbf{X}$.

We now consider the solution of the boundary value problem (3.10). The above digression on initial value problems for systems of linear ordinary differential equations provides the excuse for consideration of functions of the form

$$u(\theta, t) = X(\theta)\, T(t) \qquad (3.15)$$

as candidates for solutions of

$$\frac{\partial u}{\partial t} = \kappa\, \frac{\partial^2 u}{\partial \theta^2}$$

(subject to the periodic boundary conditions of(3.10).) The expression (3.15) is referred to as a separable solution, for obvious reasons. The separable expression (3.15) satisfies the partial differential equation exactly when

$$T'(t)X(\theta) = \kappa X''(\theta)T(t)$$

---

[1]In the boundary value problem motivating this discussion, there are physical and mathematical reasons why the constant is negative.

or

$$\frac{T'(t)}{\kappa T(t)} = \frac{X''(\theta)}{X(\theta)} \tag{3.16}$$

(at least wherever u of (3.15) is nonzero).

Since the left side of (3.16) is independent of θ, the right independent of t, each must be constant. A convenient choice of the name of the constant is, $-\lambda^2$.

In order that the separable solution (3.15) satisfy the boundary conditions

$$u(0, t) = u(2\pi, t),$$

$$\frac{\partial u}{\partial \theta}(0, t) = \frac{\partial u}{\partial \theta}(2\pi, t),$$

we require that X and $-\lambda^2$ (the two unknown quantities at this point) conspire that

$$X''(\theta) = -\lambda^2 X(\theta), \ 0 < \theta < 2\pi,$$

$$X(0) = X(2\pi),$$

$$X'(0) = X'(2\pi).$$

In the case that $\lambda^2 \neq 0$, the general solutions of the differential equation are of the form

$$X(\theta) = a\, e^{i\lambda\theta} + b\, e^{-i\lambda\theta}.$$

The boundary conditions require that

$$a + b = ae^{2\pi i\lambda} + be^{-2\pi i\lambda},$$

$$a - b = ae^{2\pi i\lambda} - be^{-2\pi i\lambda}.$$

This homogeneous system has a nontrivial solution exactly when

$$e^{(2\pi i\lambda)} = 1,$$

which implies that the exponent in the above is an integral multiple of $2\pi i$. That is,

$$\pm 2\pi i\lambda = 2\pi i n, \ n = 1, 2, \ldots.$$

For these values of λ, the corresponding time component equation

$$T'(t) = -\kappa\lambda^2\, T(t) = -\kappa\, n^2\, T(t)$$

has solution

$$T(t) = c_n e^{-\kappa n^2 t}.$$

This provides a list of separated solutions which may be organized as

$$u_n(\theta, t) = X_n(\theta) T_n(t)$$

$$= c_n e^{in\theta} e^{-\kappa n^2 t}.$$

This list of solutions arises from the case $\lambda^2 \neq 0$, since the form of solution used is only valid for that case. We must also consider the possibility that there are additional solutions meeting the boundary conditions arising from the situation where $\lambda^2 = 0$. [2]

If $\lambda^2 = 0$ in the separated X equation, the general form of the solution of the differential equation is a straight line,

$$X(\theta) = a + b\theta.$$

This satisfies the required periodicity conditions only for $b = 0$. Thus $X(\theta) = 1$ is a viable candidate from the $\lambda^2 = 0$ case, and the corresponding time component equation is

$$T'(t) = -\kappa 0^2 T(t) = 0,$$

with solution a constant:

$$T(t) = c_0.$$

This adds to the list above the solution

$$u_0(\theta, t) = c_0.$$

The whole collection of solutions obtained by separation of variables can be taken as[3]

$$\{c_n e^{in\theta} e^{-\kappa n^2 t}\}_{n=-\infty}^{\infty}.$$

The individual separated solutions satisfy the partial differential equation and the (periodic) boundary conditions but not, in general, the initial condition

$$u(\theta, 0) = f(\theta), \ 0 < x < 2\pi.$$

---

[2] The penalty associated with overlooking a possible λ^2-value in the separation process is usually an impasse arising several pages later in the calculation.

[3] There is a sleight of hand involved in the counting of solutions here. Because the T_n depends only on n^2, taking the X solution as $e^{in\theta}$ instead of the sum avoids counting the same form twice.

Because of the linearity of the governing equation and the fact that sums of the separated solutions still satisfy the spatial boundary conditions, we are led to the (formal) solution

$$u(\theta, t) = \sum_{n=-\infty}^{\infty} c_n e^{in\theta} e^{-\kappa n^2 t}.$$

To match the initial condition at $t = 0$, we require that

$$f(\theta) = u(\theta, 0) = \sum_{n=-\infty}^{\infty} c_n e^{in\theta}.$$

This uniquely determines the constants $\{c_n\}_{-\infty}^{\infty}$ of the expansion as the complex Fourier expansion coefficients for the initial temperature distribution function f. That is,

$$c_n = \frac{1}{2\pi} \int_0^{2\pi} f(\theta) e^{-in\theta} \, d\theta.$$

If this coefficient formula is incorporated in the solution formula, the resulting solution may be expressed as

$$u(\theta, t) = \sum_{n=-\infty}^{\infty} \left[\frac{1}{2\pi} \int_0^{2\pi} f(\theta) e^{-in\theta} \, d\theta \right] e^{in\theta} e^{-\kappa n^2 t}.$$

Example. Consider the problem of heat flow in a "thin laterally insulated" ring analyzed above. Suppose that the initial temperature profile is given by

$$f(\theta) = \theta, \ 0 < \theta < 2\pi.$$

Find an explicit solution in this case.

It is easy to see that

$$c_0 = \pi,$$

$$c_n = \frac{1}{-in}, \ n \neq 0.$$

The solution hence takes the form

$$u(\theta, t) = \pi + \sum_{\substack{n=-\infty \\ n \neq 0}}^{\infty} \frac{1}{-in} e^{in\theta} e^{-\kappa n^2 t}.$$

Plotting Solutions

The circular heat equation is in the form of a standard Fourier series, except for the fact that the expansion coefficient is time dependent. If it is desired to make plots of the solution, using a Cesaro sum will probably prove useful. To generate a surface plot of the solution (treating x, t as independent variables) we have to modify our series summation codes to deal with the extra variable.

There are actually two ways to handle this problem. The constraint on the code is the way the Octave `feval` operates, evaluating its first argument (name) at the remaining arguments. The choices are to write the partial sum code to use the number of arguments as a "switch variable," or to write separate routines for the different cases.

One can argue that execution speed suggests separate routines, so we adopt that choice.

OCTAVE

```
% partialsum for a t-dependent coefficient

function y = partialsum_1x_t(term, x, t, n)

    tempval = zeros(size (x));

    for kk = 1:n
       tempval = tempval + feval(term, x, t, kk);
    endfor

    y = tempval;

endfunction
```

The Cesaro sum procedure also has to be modified to accommodate the variable count.

OCTAVE

```
% Cesaro sum for t-dependent coefficients

function y = Cesaro_1x_t(term, x, t, N)

    tempval = zeros(size (x));

    for kk = 1:N
       tempval = tempval + partialsum_1x_t(term, x, t, kk);
    endfor

    y = tempval/N;

endfunction
```

Many partial differential equation models use material parameters, which naturally end up embedded in the solution. This puts another variable into the solution terms, and means a decision about how the variable is to be passed must be made.

Passing extra variables as arguments is a bad idea because of the `feval` complications. The parameters can actually be removed by redefining variables to absorb the constant,[4] but we adopt the convention to treat constants as global variables. The code for the term also reflects the convention that terms are counted from 1, and is written in real form to avoid roundoff issues.

OCTAVE
```
% term  for  heat  equation  on  a  circle
% straight  line  initial  condition

function  y  =  heatcircleterm (x,  t,  n)

    global  kappa;

    if  n  ==  1
      y  =  pi;
    else
      y  =  (-2/(n-1))*sin ((n-1)*x).*exp(-(n-1)*(n-1)*pi*pi*kappa*t);
    endif


endfunction
```

The solution is plotted as a parametric surface, and constants are handled with a strict *declare, use, discard* convention. If that seems tedious, the command sequence can be put in a script file and invoked as a unit (as long as the script .m file is on the user path.)

OCTAVE
```
    global  kappa

    [x  t]  =  meshgrid (0:.1  :2*pi,  0:.01  :  5);

    kappa  =  2;

    u  =  Cesaro_1x_t('heatcircleterm',  x,  t,  10);

    surf(x,  t,  u)

    clear  global  kappa
```
The resulting plot is shown in Figure 3.14.

The introduction of polar coordinates (r, θ) in Laplace's equation results in

$$\frac{\partial^2 u}{\partial r^2} + \frac{1}{r}\frac{\partial u}{\partial r} + \frac{1}{r^2}\frac{\partial^2 u}{\partial \theta^2} = 0.$$

This equation serves either as a model for the electrical potential in a circular region, or the steady state temperature distribution in a thin conducting disk.

---

[4]Let $\tau = \kappa t$, then declare τ pretentious and replace it by (a new) t.

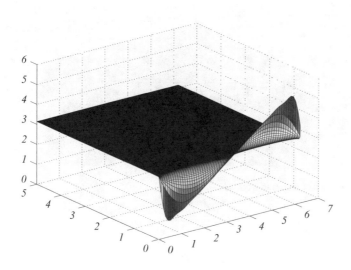

Fig. 3.14 Circular heat equation solution

Assuming that the circular region in question is of radius a, consider the boundary value problem

$$\frac{\partial^2 u}{\partial r^2} + \frac{1}{r}\frac{\partial u}{\partial r} + \frac{1}{r^2}\frac{\partial^2 u}{\partial \theta^2} = 0, \ 0 < r < a, \ 0 < \theta < 2\pi, \tag{3.17}$$

$$u(r, \theta) = u(r, \theta + 2\pi),$$

$$\frac{\partial}{\partial \theta} u(r, \theta) = \frac{\partial}{\partial \theta} u(r, \theta + 2\pi),$$

$$u(a, \theta) = f(\theta) \ 0 < \theta < 2\pi.$$

(See Figure 3.15).

We seek solutions of the first three equations in the separable form

$$u(r, \theta) = R(r)\,\Theta(\theta).$$

This leads to the requirement

$$\frac{R(r)}{r^2}\Theta^{''}(\theta) + \frac{R^{'}(r)}{r}\Theta(\theta) + R^{''}(r)\Theta(\theta) = 0,$$

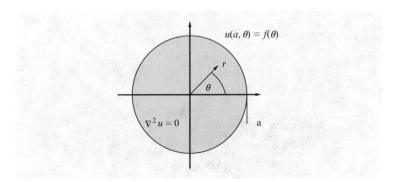

Fig. 3.15 Laplace's equation in a circle

or

$$-\frac{\Theta''(\theta)}{\Theta(\theta)} = \frac{r^2 R''(r)}{R(r)} + \frac{r R'(r)}{R(r)}. \tag{3.18}$$

Since the variables (r, θ) are separated in (3.18), each term in the equation must be a constant, which we call λ^2. The periodicity of the solution is imposed by requiring that $\Theta(\theta)$ satisfy

$$\Theta''(\theta) + \lambda^2 \Theta''(\theta) = 0,$$

$$\Theta(\theta) = \Theta(2\pi),$$

$$\Theta'(\theta) = \Theta'(2\pi).$$

As above, for $\lambda^2 \neq 0$, we obtain solutions (one for each nonzero integer n)

$$\Theta_n(\theta) = e^{in\theta}.$$

In case $\lambda^2 = 0$, solutions of the form

$$\Theta(\theta) = a + b\theta$$

arise, of which only

$$\Theta_0(\theta) = a$$

survives the imposition of the periodicity boundary conditions. The radial components of the separated solutions are obtained by returning to (3.18) with the information that $\lambda_n^2 = n^2$ are the appropriate separation constant values. This provides

$$r^2 R_n''(r) + r R_n'(r) - n^2 R_n(r) = 0,$$

recognizable as an equidimensional or Euler equation. Solutions (for $n \neq 0$) are of the form

$$r^p$$

where the exponent p satisfies the *indicial equation*

$$p(p-1) + p - n^2 = 0.$$

The general solution of the Euler equation hence in this case has the form

$$R_n(r) = A_n r^n + B_n r^{-n}$$

for some constants A_n, B_n.

Now if $n > 0$, and $B_n \neq 0$, then the solution is unbounded as $r \to 0$. Similarly, if $n < 0$ and $A_n \neq 0$, the same problem occurs. We conclude on a physical basis that this behavior must be excluded, so that of the general solutions we retain only

$$R_n(r) = \begin{cases} A_n r^n & n \geq 0, \\ B_n r^{-n} & n < 0. \end{cases}$$

This is more conveniently written as

$$R_n(r) = c_n r^{|n|},$$

which includes both cases of solution in a compact format.

The radial component associated with the case $\lambda^2 = 0$ satisfies

$$r^2 R_0''(r) + r R_0'(r) = 0,$$

with general solution

$$R_0(r) = c_0 + d_0 \ln(r).$$

Again on a physical basis we require that $d_0 = 0$. The collection of separated solutions may be written as

$$u_n(r, \theta) = c_n r^{|n|} e^{in\theta},$$

which includes the $\lambda^2 = 0$ case using the usual convention that $r^0 = 1$.

We superimpose the separated solutions in an attempt to meet the remaining condition of (3.17) at the boundary $r = a$. This gives (as a guess) the solution

$$u(r, \theta) = \sum_{-\infty}^{\infty} c_n r^{|n|} e^{in\theta}.$$

On the boundary $r = a$,

$$u(a, \theta) = f(\theta) = \sum_{-\infty}^{\infty} c_n a^{|n|} e^{in\theta},$$

from which we identify $\{c_n a^{|n|}\}$ as the Fourier expansion coefficient set for f^5,

$$c_n a^{|n|} = \frac{1}{2\pi} \int_0^{2\pi} f(\theta) e^{-in\theta} \, d\theta.$$

Incorporating this coefficient value into the solution expansion gives the single expression

$$u(r, \theta) = \sum_{-\infty}^{\infty} \left[\frac{1}{2\pi} \int_0^{2\pi} f(\theta) e^{-in\theta} \, d\theta \right] (\frac{r}{a})^{|n|} e^{in\theta}. \tag{3.19}$$

From this expression, it is evident that the solution consists of the Fourier expansion of the given boundary function f modified by the addition of the factors $(|r/a|)^n$.

Plotting Non-Rectangular Geometries

The most straightforward way to handle non-rectangular geometries when plotting with Octave is to make use of the facility for parametric surfaces. Parametric surface formalisms are familiar in the treatment of vector calculus and manifolds, and provide a solid foundation for understanding Octave plotting. In addition to the parametric surface approach, Octave routines often support "shorthand" versions accepting various vector arguments in place of the "uniform shaped array" arguments of the parametric equation view.

There is a valid argument to the effect that the parametric view is the "mathematically correct" one for the context, and our opinion is that consistently adopting it makes for more understandable and less likely surprising Octave code.

We have not specified the boundary value function for the Laplace equation example, so an explicit `laplace_eqn_term.m` function is not available. Modulo this missing function, a plot of the solution surface can be generated by the following "parametric surface" code. Think of the subscripts of the R, *Theta*, Z

---

[5]This relies on uniqueness of the Fourier coefficients of a function.

arrays as (samples) of the (u, v) parameter space. Then the rectangular coordinate calculations use componentwise multiplication to compute

$$Y(u, v) = \sin(Theta(u, v))\, R(u, v),$$
$$X(u, v) = \cos(Theta(u, v))\, R(u, v),$$

and the `Cesaro_2x` calculation is really done "in parallel" over the parametric coordinates,

$$Z(u, v) = Cesaro\_2x('laplace\_eqn\_term', Theta(u, v), R(u, v), 20).$$

OCTAVE

```
theta = linspace (0, 2*pi, 120);
r = linspace (0, 1, 10);

[R, Theta] = meshgrid (r, theta);


Y= sin (Theta).*R;
X= cos (Theta).*R;

Z = Cesaro_2x ('laplace_eqn_term', Theta, R, 20);

surf (X, Y, Z)
```

The fact that the "standard" Fourier series expansions appear in the above problems is a consequence of the circular geometry inherent in the physical problems. Such a problem for the wave equation is included in the problems for this section.

Boundary value problems with rectangular geometries give rise typically to the so-called half-range Fourier expansions and are treated in the following section.

Problems 3.6.

1. A three-dimensional analogue of a heat equation is given by the system [6]

$$\frac{d}{dt}\begin{bmatrix} v_1 \\ v_2 \\ v_3 \end{bmatrix} = \begin{bmatrix} -2 & 1 & 0 \\ 1 & -2 & 1 \\ 0 & 1 & -2 \end{bmatrix}\begin{bmatrix} v_1 \\ v_2 \\ v_3 \end{bmatrix}.$$

Find the solution of the above problem in two ways: first by diagonalizing the coefficient matrix of the system, and second by separation of variables methods.

---

[6]This is about the limit of what can be solved by hand using elementary means. There are "tricky" ways to diagonalize coefficient matrices of the given form. See Chapter 8.7.

2. a. Find the solution to the boundary value problem

$$\frac{\partial^2 u}{\partial r^2} + \frac{1}{r}\frac{\partial u}{\partial r} + \frac{1}{r^2}\frac{\partial^2 u}{\partial \theta^2} = 0,$$

$$\frac{\partial u}{\partial r}(1, \theta) = \pi - \theta, \ 0 < \theta < 2\pi.$$

  b. Give a description of a physical problem modeled by the above system.

  c. Suppose that

$$\frac{\partial u}{\partial r}(1, \theta) = f(\theta)$$

is imposed as a boundary condition for the above. What property of f is required for the problem posed to be solvable? Based on the physical meaning of the boundary condition, and the problem type, why is the required condition "obvious" (at least in hindsight)?

3. It is difficult to avoid the observation that pipe is conventionally manufactured with a circular cross section. This leads to an interest in the solutions of Laplace's equation in an annular region. See Figure 3.16.

    This is

$$\frac{\partial^2 u}{\partial r^2} + \frac{1}{r}\frac{\partial u}{\partial r} + \frac{1}{r^2}\frac{\partial^2 u}{\partial \theta^2} = 0, \ a < r < b, \ 0 < \theta < 2\pi,$$

$$u(a, \theta) = g(\theta), \ 0 < \theta < 2\pi,$$

$$u(b, \theta) = f(\theta), \ 0 < \theta < 2\pi.$$

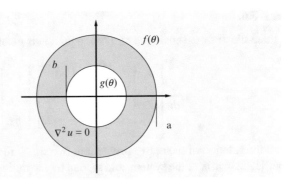

Fig. 3.16 Heat conduction in a pipe wall

a. Provide a description of a physical situation for which the above is appropriate.

b. Find a solution of the above boundary value problem.

4. The boundary value problems solved above were worked with the aid of the complex form of the usual Fourier series expansions. Show that by writing solutions of the appropriate separated equations in terms of trigonometric rather than complex exponential functions, a real form of the solutions is obtained.

5. Show that if the separation constant encountered solving the "heat equation in a ring" boundary value problem is called k (rather than $-\lambda^2$) the final answer remains the same. One really should do this problem once, in order to be convinced that calling the separation constant $-\lambda^2$ is an algebraic convenience born of experience (and a couple of theorems), and not a form of "cheating."

6. Consider the longitudinal vibrations of a "thin elastic ring," illustrated in Figure 3.17.

a. Show that an appropriate model is given by

$$\frac{1}{c^2}\frac{\partial^2 u}{\partial t^2} = \frac{\partial^2 u}{\partial \theta^2},$$

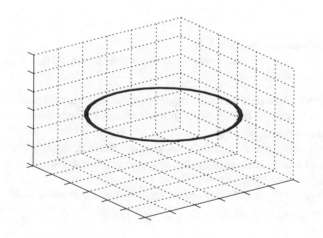

Fig. 3.17 Elastic ring model

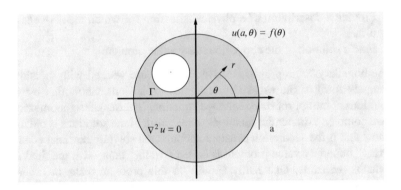

Fig. 3.18 Interior circular domain Γ

$$u(0, t) = u(2\pi, t),$$

$$\frac{\partial}{\partial \theta} u(0, t) = \frac{\partial}{\partial \theta} u(2\pi, t),$$

$$u(\theta, 0) = f(\theta),$$

$$\frac{\partial}{\partial t} u(\theta, 0) = g(\theta).$$

(See Problem 10, Section 3.3).

b. Find a solution of the above boundary value problem by the method of separation of variables.

7. Consider the solution of Laplace's equation in a circular region, with prescribed function values as the boundary condition imposed. In the interior of the region under consideration, construct a circle Γ (See Figure 3.18).

Using the solution of Laplace's equation in a circle derived above, show that the value of u at the center of the circle Γ is equal to the average of the values assumed on the circumference of Γ.

Hint: Consider the solution in a coordinate system appropriate to Γ. You may assume uniqueness of solutions to Laplace's equation in a circular region.

8. Using the Laplace equation solution

$$u(r, \theta) = \sum_{-\infty}^{\infty} \left[\frac{1}{2\pi} \int_0^{2\pi} f(\phi) e^{-in\phi} \, d\phi \right] (\frac{r}{a})^{|n|} e^{in\theta},$$

interchange the summation and integration and explicitly sum (the geometric series)

$$\sum_{-\infty}^{\infty} (\frac{r}{a})^{|n|} e^{in\theta}$$

to write the solution in the form

$$u(r, \theta) = \int_0^{2\pi} G(r, \theta - \phi) f(\phi) \, d\phi.$$

The function G appearing in the above form is called the Poisson kernel.

9. Find the explicit solution of Laplace's equation in a circle of radius 1, given the boundary condition

$$u(t, \theta) = \theta \, (2\pi - \theta).$$

10. Using the expansion coefficients computed in Problem 9, use Octave to plot the solution surface using a Cesaro sum of 30 terms.

3.7   Half-Range Expansions and Symmetries

The boundary value problems of the previous section arise from physical problems with an obvious "circular geometry," and lead to Fourier expansion problems of the standard mold.

The method of separation of variables may also be applied to boundary value problems with linear geometries, eventually leading to expansion problems referred to as half-range (or even "quarter-range") expansions. A prototype problem of this sort is the zero boundary diffusion problem

$$\frac{\partial u}{\partial t} = \kappa \frac{\partial^2 u}{\partial x^2}, \ 0 < x < L, \tag{3.20}$$

$$u(0, t) = u(L, t) = 0, \ t > 0,$$

$$u(x, 0) = f(x), \ 0 < x < L.$$

To apply the separation of variables technique to the above, we try a separable solution

$$u(x, t) = X(x)T(t).$$

Then (if this is a solution)

$$\frac{T'(t)}{\kappa T(t)} = \frac{X''(x)}{X(x)} = -\lambda^2$$

for some separation constant $-\lambda^2$.

In order that the separated solution should satisfy

$$u(0, t) = u(L, t) = 0$$

for all t, we are led to consider the spatial equation of the separated equation with homogeneous boundary conditions:

$$X''(x) + \lambda^2 X(x) = 0,$$
$$X(0) = X(L) = 0.$$

The general solution of this equation may be written as (for $\lambda^2 \neq 0$)

$$X(x) = A\cos(\lambda x) + B\sin(\lambda x).$$

The requirement that $X(0) = 0$ forces $A = 0$ in this solution. To satisfy the condition $X(L) = 0$ in a nontrivial fashion, the condition

$$\sin(\lambda L) = 0$$

must hold. From this requirement (and an intimate knowledge of the sine function) we conclude that

$$\lambda = \frac{n\pi}{L}, \ n = 1, 2, 3, 4, \ldots \tag{3.21}$$

are the relevant possible separation constants. (Negative n in the above provides only a multiple of the solution already obtained.)

Returning to the separation equation to solve the "T" equation, we find that the separable solutions are of the form

$$u_n(x, t) = b_n\, e^{-\kappa(\frac{n\pi}{L})^2 t}\, \sin(\frac{n\pi x}{L}).$$

We must also consider (briefly) the possibility that $\lambda^2 = 0$ is a separation constant. If this is the case, then the spatial solution is of the form

$$X(x) = C + Dx$$

and only the choice $C = D = 0$ provides a solution meeting the boundary conditions. Hence $\lambda^2 = 0$ provides nothing additional, and the list already obtained contains all nontrivial usable separated solutions.

In order to meet the boundary value problem initial condition constraint, we form linear combinations of the separable solutions, and try

$$\sum_{n=1}^{\infty} b_n e^{-\kappa(\frac{n\pi}{L})^2 t} \sin(\frac{n\pi x}{L})$$

as solution. To meet the initial condition at $t = 0$, we need

$$f(x) = \sum_{n=1}^{\infty} b_n \sin(\frac{n\pi x}{L}).$$

Recall that the standard form of the real Fourier series expansion involves both sine and cosine functions. The initial condition expansion encountered here gives slight pause since only the sine functions are present, and this is only "half" of the usual functions expected.

The standard expansion takes the form

$$f(x) = a_0 + \sum_{n=1}^{\infty} a_n \cos(n\,\omega_0\,x) + b_n \sin(n\,\omega_0\,x). \tag{3.22}$$

If the initial condition expansion is to be identified in these terms, we must identify the appropriate fundamental frequency, ω_0. Apparently we have

$$\omega_0 = \frac{\pi}{L} \tag{3.23}$$

and the period associated with the expansion must be

$$\frac{2\pi}{\frac{\pi}{L}} = 2L.$$

Note that this is *twice* the length of the interval over which f is prescribed in the original boundary value problem (3.20).

The correspondence between the initial condition expansion and (3.22) may be made by exploiting the above observation. We wish (3.22) to reduce to an expansion involving only $\sin(\frac{n\pi x}{L})$ terms over the interval $[0, L]$. However, we are free to define the function f as we please over the rest of the appropriate interval of length $2L$.

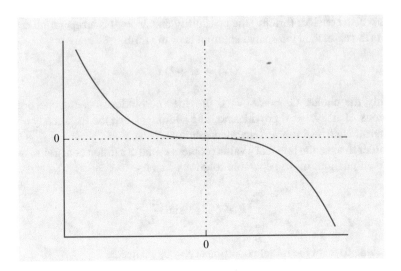

Fig. 3.19 Odd periodic extension

We define a new function $\tilde{f}$ to be an "odd $2L$-periodic extension" of f. This is illustrated in Figure 3.19.

$$\tilde{f}(x) = \begin{cases} f(x) & 0 < x < L, \\ -f(-x) & -L < x < 0, \\ \tilde{f}(x - 2L) & \text{otherwise.} \end{cases}$$

If the expansion (3.22) is computed for $\tilde{f}$ defined above, we obtain

$$b_n = \frac{2}{2L} \int_{-L}^{L} \tilde{f}(x) \sin(n\pi \frac{x}{L}) \, dx.$$

Since the integrand in this expression (as the product of two odd functions) is an even function, we can evaluate the integral over "the positive part of the domain" to obtain

$$b_n = \frac{2}{L} \int_{0}^{L} f(x) \sin(n\pi \frac{x}{L}) \, dx.$$

Since $\tilde{f}$ is an odd function, its average value vanishes, and $a_0 = 0$ in its Fourier expansion. The cosine expansion coefficients

$$a_n = \frac{2}{2L} \int_{-L}^{L} \tilde{f}(x) \cos(n\pi \frac{x}{L}) \, dx$$

are also identified as the average value of an odd function, and hence also vanish. We thus obtain the expansion

$$\tilde{f}(x) = \sum_{n=1}^{\infty} b_n \sin(n\pi \frac{x}{L}), \quad (-\infty < x < \infty).$$

Since f coincides with $\tilde{f}$ over the interval $[0, L]$,

$$f(x) = \sum_{n=1}^{\infty} b_n \sin(n\pi \frac{x}{L}), \quad 0 < x < L. \tag{3.24}$$

This expression is referred to as a *half-range sine series* expansion of f. The expansion provides f (in the usual sense of convergence of Fourier expansions) over the original physical interval $[0, L]$, and, of course $\tilde{f}$ otherwise.

To return to the solution of the boundary value problem (3.20), we see that the conclusion (3.24) ensures exactly the possibility of the required initial condition expansion. Appropriating the half-range coefficient formula we define $\{b_n\}$ through

$$b_n = \frac{2}{L} \int_0^L f(x) \sin(n\pi \frac{x}{L}) \, dx.$$

The complete solution of (3.20) is then given by

$$u(x, t) = \sum_{n=1}^{\infty} b_n e^{-\kappa(\frac{n\pi}{L})^2 t} \sin(n\pi \frac{x}{L}). \tag{3.25}$$

The introduction of $\tilde{f}$ above may appear initially to be a strictly mathematical artifice introduced for the purpose of justifying the half-range expansion. In fact, it is possible to deduce ((3.24), (3.25)) on the basis of entirely physical considerations, arguing as follows.

Consider again (3.20), and assume for convenience that the spatial length scale has been adjusted so that $L = \pi$. In conjunction with (3.20), consider again the thin ring heat equation of Section 3.6, and solve the boundary value problem

$$\frac{\partial u_{\text{ring}}}{\partial t} = \kappa \frac{\partial^2 u_{\text{ring}}}{\partial \theta^2},$$

$$u_{\text{ring}}(0, t) = u_{\text{ring}}(2\pi, t),$$

$$\frac{\partial}{\partial \theta} u_{\text{ring}}(0, t) = \frac{\partial}{\partial \theta} u_{\text{ring}}(2\pi, t),$$

$$u_{\text{ring}}(\theta, 0) = \tilde{f}(\theta).$$

(Recall the spatial scaling imposed above.)

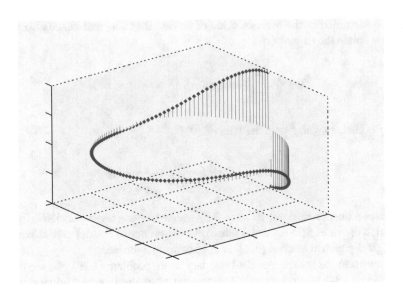

Fig. 3.20 Heat profile around a thin ring

This problem may be illustrated by constructing a cylinder through the ring, for the purpose of plotting solutions. The initial condition is illustrated in Figure 3.20. The temperature profile has odd symmetry with respect to the positions $\theta = 0$, $\theta = \pi$. This symmetry is retained as time evolves (this may be reasoned out or seen to follow from the form of the solution calculated in the previous section).

In particular, the temperature at $\theta = 0$ and $\theta = \pi$ remains at zero for all $t > 0$. The "first half" of the ring solution therefore provides a solution of (3.20) above. Conversely, we may easily verify that the solution (3.25) (with appropriate renaming and rescaling of the spatial variable) is a solution of the odd-symmetric ring problem.

The imposition of an insulated-end boundary condition with the heat equation leads to the need for half-range cosine series expansions. The boundary value problem is

$$\frac{\partial u}{\partial t} = \kappa \frac{\partial^2 u}{\partial x^2}, \tag{3.26}$$

$$\frac{\partial}{\partial x} u(0, t) = \frac{\partial}{\partial x} u(L, t) = 0,$$

$$u(x, 0) = f(x), \ 0 < x < L.$$

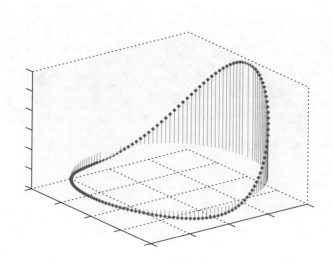

Fig. 3.21 Even extension ring problem

Separation of variables leads to the solution

$$u(x, t) = a_0 + \sum_{n=1}^{\infty} a_n e^{-\kappa(\frac{n\pi}{L})^2 t} \cos(n\pi \frac{x}{L}),\qquad(3.27)$$

$$a_0 = \frac{1}{L} \int_0^L f(x)\, dx,$$

$$a_n = \frac{2}{L} \int_0^L f(x) \cos(n\pi \frac{x}{L})\, dx.$$

(The a_0 term above arises from the case $\lambda^2 = 0$; see Problem 3 below).

Figure 3.21 illustrates the ring problem associated with the insulated-end boundary value problem.

The solution has even symmetry about the points $\theta = 0, \pi$. This even symmetry prevents the flow of heat across these junctions, and hence the temperature gradient vanishes there. This physical argument establishes the correspondence with (3.26).

Example. Symmetries of a different sort arise in the solutions of Laplace's equation in a rectangular region.

One model problem is that of the electric potential in a rectangle, with the potential prescribed on the edges (see Figure 3.22).

A frontal assault on this problem is possible, but the solution is more tedious than instructive. Rather than this, we consider first the (special case)

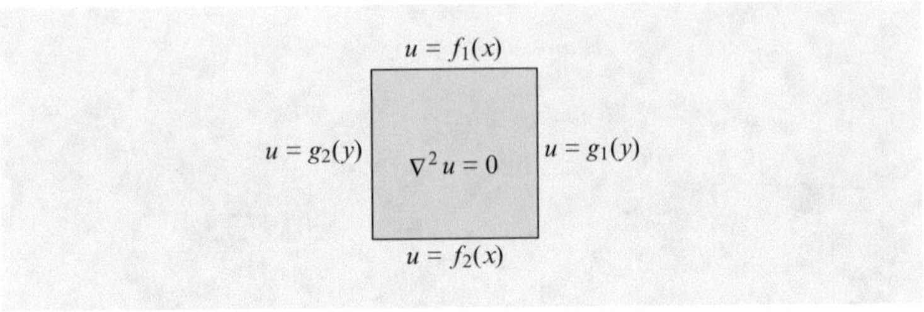

Fig. 3.22 Rectangular potential problem

$$\nabla^2 u = 0, \; 0 < x < a, \; 0 < y < b, \tag{3.28}$$
$$u(0, y) = 0,$$
$$u(x, b) = 0,$$
$$u(x, 0) = 0,$$
$$u(a, y) = f_2(y).$$

Attempting separation of variables, try

$$u(x, y) = X(x)\,Y(y)$$

to obtain

$$\frac{X''}{X} = -\frac{Y''}{Y}.$$

The choice of the form of the separation constant is suggested by the realization that solutions of the separated equations will give (on the face of it) trigonometric functions in one direction, and hyperbolic functions in the other, because of the differences of the sign in the equations. Since hyperbolic functions vanish at most once for real values of their argument, the boundary conditions (3.28) suggest choosing the sign of the separation constant to obtain trigonometric functions in the y-variable [7].

Electing to swim with the tide, we choose the separation constant "λ^2" and consider first the problem

$$Y'' + \lambda^2 Y = 0,$$
$$Y(0) = Y(b) = 0.$$

---

[7] Note that this is purely a cosmetic effect. If the "unfortunate" sign choice is made, the problem solution procedure will force the conclusion that the function arguments are purely imaginary, and the process will "self-correct."

This provides (as above, since this problem involves the same separated equation with the same boundary conditions)

$$\lambda_n^2 = (\frac{n\pi}{b})^2,$$

$$Y_n(y) = b_n \sin(n\pi \frac{y}{b}).$$

The corresponding homogeneous X equation

$$X_n'' - (\frac{n\pi}{b})^2 X_n = 0$$

has a general solution which can be written as

$$X_n(x) = \alpha_n \sinh(n\pi \frac{x}{b}) + \beta_n \cosh(n\pi \frac{x}{b}).$$

To meet the boundary condition $u(0, y) = 0$, we take $\beta_n = 0$ in the above, and write the separable solutions in the form

$$u_n(x, y) = b_n \sin(n\pi \frac{y}{b}) \sinh(n\pi \frac{x}{b}).$$

Since sums of such solutions still meet the homogeneous boundary conditions and satisfy the linear partial differential equation of (3.28), we postulate

$$u(x, y) = \sum_{n=1}^{\infty} b_n \sin(n\pi \frac{y}{b}) \sinh(n\pi \frac{x}{b})$$

as the solution of (3.28). The $\{b_n\}$ are evaluated from the remaining boundary condition

$$u(a, y) = \sum_{n=1}^{\infty} b_n \sin(n\pi \frac{y}{b}) \sinh(n\pi \frac{a}{b}) = f_2(y).$$

From this condition we deduce that the coefficients $\{b_n \sinh(n\pi \frac{a}{b})\}$ must be the half-range sine series expansion coefficients for the function f_2. Therefore we can immediately write

$$b_n \sinh(n\pi \frac{a}{b}) = \frac{2}{b} \int_0^b f_2(y) \sin(n\pi \frac{y}{b}) dy.$$

Incorporating this in the solution formula gives

$$u(x, y) = \sum_{n=1}^{\infty} \left[\frac{2}{b} \int_0^b f_2(y) \sin(n\pi \frac{y}{b}) dy \right] \sin(n\pi \frac{y}{b}) \frac{\sinh(n\pi \frac{x}{b})}{\sinh(n\pi \frac{a}{b})}. \qquad (3.29)$$

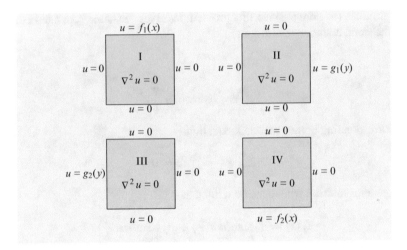

Fig. 3.23 Superimposed potential solutions

To solve the original problem of Figure 3.22 we observe that the solution (3.29) may be superimposed with the solutions of the problems II, III, and IV (Figure 3.23) to give a solution for Figure 3.22.

This works because the partial differential equation is linear and homogeneous, and each of the boundary functions of the "sub-problems" vanishes on three of the region sides. The solution for II is obviously

$$v(x, y) = \sum_{n=1}^{\infty} \left[\frac{2}{b} \int_0^b f_1(y) \sin(n\pi \frac{y}{b}) \, dy \right] \sin(n\pi \frac{y}{b}) \frac{\sinh(n\pi \frac{a-x}{b})}{\sinh(n\pi \frac{a}{b})}.$$

This is obtained either by the change of variable

$$x' = a - x$$

in connection with the problem above, or by the prescient choice of the functions $\sinh((a - x)\frac{n\pi}{b})$ and $\cosh((a - x)\frac{n\pi}{b})$ as solutions of the separated equations. Since Laplace's equation is invariant under an interchange of the x and y variables, solutions of III and IV are produced by applying such an interchange (with appropriate scale changes) to the above results. This gives

$$w(x, y) = \sum_{n=1}^{\infty} \left[\frac{2}{a} \int_0^a g_2(\zeta) \sin(n\pi \frac{\zeta}{a}) \, d\zeta \right] \sin(n\pi \frac{x}{a}) \frac{\sinh(n\pi \frac{y}{a})}{\sinh(n\pi \frac{b}{a})}$$

and

$$z(x, y) = \sum_{n=1}^{\infty} \left[\frac{2}{a} \int_0^a g_1(\zeta) \sin(n\pi \frac{\zeta}{a}) d\zeta \right] \sin(n\pi \frac{x}{a}) \frac{\sinh(n\pi \frac{b-y}{a})}{\sinh(n\pi \frac{b}{a})}.$$

The solution of the problem of Figure 3.22,

$$\nabla^2 \phi = 0 \quad 0 < x < a, \; 0 < y < b,$$
$$\phi(x, 0) = g_1(x),$$
$$\phi(x, b) = g_2(x),$$
$$\phi(0, y) = f_1(y),$$
$$\phi(a, y) = f_2(y),$$

is simply

$$\phi(x, y) = u(x, y) + v(x, y) + w(x, y) + z(x, y).$$

The reader may consider the prospect of obtaining this result by means of a direct attempt at the general version of the boundary value problem [8].

Example. Find the solution of

$$\nabla^2 \phi = 0, \; 0 < x < l, \; 0 < y < 2,$$
$$\phi(0, y) = \phi(x, 0) = \phi(x, 2) = 0,$$
$$\phi(l, y) = \frac{y}{2}.$$

The solution requires sine functions in the y-direction, hyperbolic sine functions in the x-direction;

$$\phi = \sum_{n=1}^{\infty} b_n \sin(n\pi \frac{y}{2}) \frac{\sinh(n\pi \frac{x}{2})}{\sinh(n\pi \frac{n\pi}{2})}.$$

The $\{b_n\}$ are the half-range expansion coefficients for the function

$$\phi(l, y) = \frac{y}{2}.$$

---

[8] In a direct attempt, there is no restriction of the sign of the separation constant, and all possibilities (positive, zero, negative) must be entertained at once. Obtaining a "neat" solution expression is then tedious at best.

By direct calculation

$$b_n = 2\frac{(-1)^{n+1}}{n\pi},$$

and

$$\phi(x, y) = \sum_{n=1}^{\infty} \left[2\frac{(-1)^{n+1}}{n\pi} \right] \sin(n\pi\frac{y}{2}) \frac{\sinh(n\pi\frac{x}{2})}{\sinh(n\pi\frac{n\pi}{2})}.$$

It should be obvious that an imposing list of boundary value problems may be created by the simple expedient of permuting the possible homogeneous boundary conditions of Section 3.4 with respect to Laplace's equation on a rectangle. Some of these permutations are included in the problems below.

Example. A straightforward boundary value problem for the wave equation is given by

$$\frac{1}{c^2}\frac{\partial^2 u}{\partial t^2} = \frac{\partial^2 u}{\partial x^2}, \tag{3.30}$$

$$u(x, 0) = f(x),$$

$$\frac{\partial u}{\partial t}(x, 0) = g(x),$$

$$\frac{\partial u}{\partial x}(0, t) = 0,$$

$$u(L, t) = 0.$$

This may be interpreted as describing the transverse motion of a string with one fixed and one free end, with given initial position and velocity.

We seek a separable solution

$$u(x, t) = X(x)T(t)$$

and obtain the separated systems

$$X'' + \lambda^2 X = 0, \tag{3.31}$$

$$X'(0) = 0,$$

$$X(L) = 0,$$

and

$$T'' + \lambda^2 c^2 T = 0, \tag{3.32}$$

with separation constant $-\lambda^2$. For $\lambda^2 \neq 0$ (the only relevant case), the general X solution is

$$X(x) = A \cos(\lambda x) + B \sin(\lambda x).$$

Use of the boundary conditions leads to

$$B = 0,$$

$$A \cos(\lambda L) = 0,$$

from which we find that

$$\lambda_n = \frac{(2n + 1)\pi}{2L}, \ n = 0, 1, 2, \ldots$$

lists the values of the separation constant required. The spatial solutions are therefore

$$X_n(x) = \cos((2n + 1)\pi \frac{x}{2L}).$$

The corresponding time dependence is determined from

$$T_n'' + \left(\frac{(2n + 1)\pi}{2L} \right)^2 c^2 T_n = 0,$$

with general time solution

$$T_n(t) = A_n \cos((2n + 1)\omega_0 t) + B_n \sin((2n + 1)\omega_0 t).$$

Here $\omega_0 = \frac{\pi c}{2L}$. The separated solutions are given by

$$u_n(x, t) = \cos((2n + 1)\pi \frac{x}{2L})[A_n \cos((2n + 1)\omega_0 t) + B_n \sin((2n + 1)\omega_0 t)].$$

These separated solutions have a simple geometrical interpretation (these and the solutions for the fixed-end case have musical consequences as well).

The separated solution may be rewritten (using trigonometric identities) as

$$u_n(x, t) = D_n \cos((2n + 1)\pi \frac{x}{2L}) \cos((2n + 1) \omega_0 t + \phi_n).$$

The interpretation of this solution is that it represents a standing wave. The shape of this standing wave is given by the spatial dependence

$$\cos(2n + 1)\frac{\pi x}{2L}.$$

Some of these so-called modes of vibration are illustrated in Figure 3.24.

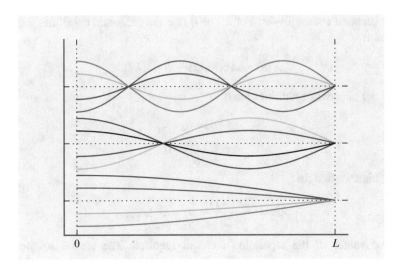

Fig. 3.24 Standing waves

These mode shapes are multiplied by a sinusoidal amplitude factor of frequency $(2n + 1)\omega_0$ in time. If ω_0 is regarded as the fundamental frequency associated with this phenomenon, then the amplitude factors from the higher modes consist of the odd harmonics of the fundamental frequency.

The model analyzed above also has application to the analysis of sound waves in a pipe. See the problems below.

The general solution of the problem (3.30) is obtained by constructing an arbitrary linear combination of the separated solutions,

$$u(x, t) = \sum_{n=0}^{\infty} \cos((2n + 1)\frac{\pi x}{2L}) (A_n \cos((2n + 1)\omega_0 t) + B_n \sin((2n + 1)\omega_0 t)).$$

(3.33)

From this we formally obtain

$$\frac{\partial}{\partial t} u(x, t) = \sum_{n=0}^{\infty} \cos((2n + 1)\frac{\pi x}{2L})$$

$$[-A_n (2n + 1) \omega_0 \sin((2n + 1)\omega_0 t) + B_n (2n + 1) \omega_0 \cos((2n + 1)\omega_0 t)].$$

The coefficients $\{A_n\}$ and $\{B_n\}$ are determined by requiring that these expressions reduce to the given initial position and velocity functions at $t = 0$.

$$f(x) = \sum_{n=0}^{\infty} \cos((2n + 1)\frac{\pi x}{2L})A_n,$$

$$g(x) = \sum_{n=0}^{\infty} \cos((2n + 1)\frac{\pi x}{2L})((2n + 1)\omega_0 B_n).$$

These are classified as *quarter-range expansions* of the given functions (see the problems below). The coefficients follow from

$$A_n = \frac{2}{L} \int_0^L f(x) \, \cos((2n+1)\frac{\pi x}{2L}) \, dx,$$

$$(2n+1)\omega_0 B_n = \frac{2}{L} \int_0^L g(x) \, \cos((2n+1)\frac{\pi x}{2L}) \, dx.$$

The final form of the solution may be obtained (if desired) by substitution of the coefficient formulas into expression (3.33).

Example. Casual observation of guitar players reveals that "hard rock" players tend to play "closer to the bridge" than, say, folk style artists. The problem is to account for this on the basis of a vibrating string model.

A model appropriate to the analysis is that of a vibrating string, fixed at both ends, and released from rest.

The problem is given by

$$\frac{1}{c^2} \frac{\partial^2 u}{\partial t^2} = \frac{\partial^2 u}{\partial x^2}, \tag{3.34}$$

$$u(0,t) = u(L,t) = 0,$$

$$\frac{\partial u}{\partial t}(x,0) = 0,$$

$$u(x,0) = f(x).$$

The initial string shape is that assumed by a string just prior to release. We model this by Figure 3.25.

From this figure,

$$f(x) = \begin{cases} \frac{x}{a} & 0 \le x \le a, \\ \frac{L-x}{L-a} & a \le x \le L. \end{cases}$$

The problem is solved by separation of variables. The relevant solutions of the separated spatial equation are

$$X_n(x) = \sin(n\pi \frac{x}{L}),$$

and the corresponding time dependence satisfies

$$T_n''(t) + (\frac{n\pi}{L})^2 T_n(t) = 0,$$

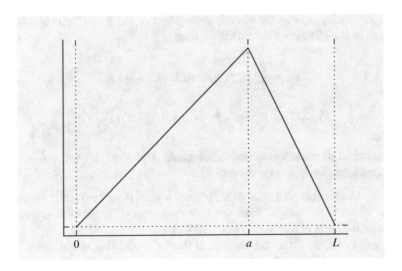

Fig. 3.25 Stretched guitar string

with solution

$$T_n(t) = A_n \cos(n\omega_0 t) + B_n \sin(n\omega_0 t).$$

(Here $\omega_0 = \frac{\pi c}{L}$.)

The general solution is

$$u(x, t) = \sum_{n=1}^{\infty} \sin(n \pi \frac{x}{L}) (A_n \cos(n\omega_0 t) + B_n \sin(n\omega_0 t)).$$

In view of the initial conditions we can see that

$$B_n = 0,$$

and matching the displacement

$$
\begin{aligned}
A_n &= \frac{2}{L} \int_0^L f(x) \sin(n\pi \frac{x}{L}) \, dx, \\
&= \frac{2}{L} \int_0^a \frac{x}{a} \sin(n\pi \frac{x}{L}) \, dx + \frac{2}{L} \int_a^L \frac{L-x}{L-a} \sin(n\pi \frac{x}{L}) \, dx, \\
&= \frac{2L^2}{\pi^2 a(L-a)} \frac{\sin(n\pi \frac{a}{L})}{n^2}.
\end{aligned}
$$

This solution can be plotted in Figure 3.26, and the oscillatory nature of the surface clearly appears.

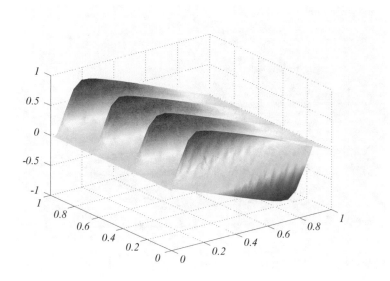

Fig. 3.26 Vibrating guitar string solution

On the basis of the solution calculation, we compare guitar styles.

The coefficient A_n gives the solution component associated with the nth mode, and hence the nth harmonic time component. If we assume that the mean-square value of the component measures the audible effect, then the difference in styles is embodied in the A_n. More than that, the essence of the difference is entirely in the factor $\sin(n\pi \frac{a}{L})$.

The relevant effect is illustrated by plotting A_n as a function of n for various $\frac{a}{L}$ This gives Figure 3.27.

It appears that smaller a corresponds to a greater proportion of higher harmonics and hence "explains" the observable effects. (e.g., more "twang" in rock guitar.)

We remark that the above analysis ignores various effects. One of these is pickup placement (which can be pursued on the basis of the mode shapes calculated above). Another physical effect is that higher modes are undoubtedly subject to damping effects not accounted for above. This should be recalled before consideration of Figure 3.27 for large (or even moderate) values of n.

In order to blunt the inevitable impression that we have left all of the more challenging boundary value problems for the exercises below, we now consider a wave equation with elastic constraints.

Example. The physical model consists of an elastic string with a fixed left-hand end and an elastically constrained right-hand end. See Figure 3.28. The parameter k is the elastic constant of the spring, while T is the usual string tension.

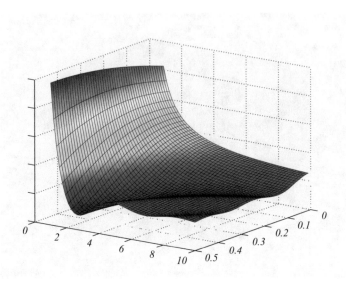

Fig. 3.27 Harmonic content

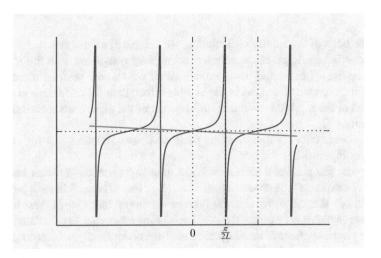

Fig. 3.28 Separation constant determination

The governing system of equations is

$$\frac{1}{c^2}\frac{\partial^2 u}{\partial t^2} = \frac{\partial^2 u}{\partial x^2},$$

$$u(x,0) = f(x),$$

$$\frac{\partial u}{\partial t}(x, 0) = g(x),$$

$$u(0, t) = 0,$$

$$-T\frac{\partial}{\partial x}u(L, t) = ku(L, t).$$

Turning the usual crank we assume

$$u(x, t) = X(x)\, T(t)$$

and produce the separated equations

$$\frac{1}{c^2}\frac{T''}{T} = \frac{X''}{X} = -\lambda^2$$

with boundary conditions

$$x(0) = 0,$$

$$-T\frac{dX}{dx}(L) = k\, X(L)$$

for the spatial component. This gives (for $\lambda^2 \neq 0$)

$$X(x) = A\cos(\lambda x) + B\sin(\lambda x)$$

as general solution. To meet the boundary conditions we require

$$A = 0$$

and

$$-T\lambda\cos(\lambda L) = k\sin(\lambda L). \tag{3.35}$$

The equation (3.35) determines the values of the separation constants. Since this is a transcendental equation, explicit solutions are not to be expected.

To come to terms with (3.35) we resort to graphical means. Rewrite the equation as

$$\tan(\lambda L) = -\frac{T}{k}\lambda$$

and plot both functions of this equation on a common set of axes. The curves obviously intersect at $\lambda = 0$. This must be regarded as a spurious solution, as the calculation of solutions leading to the equation are premised on the condition $\lambda \neq 0$.

As usual, the case $\lambda = 0$ is considered separately. The solution is again a straight line

$$X(x) = A + Bx,$$

and application of the boundary conditions leads to

$$A = 0, \ kBL = -TB.$$

Since k, L, and T are intrinsically positive parameters, there is no nonzero solution available. The case $\lambda = 0$ therefore provides no contribution.

The graphical determination of the constants is plotted in Figure 3.28. It is easy to see that the intersections for negative λ are just the negative of the corresponding positive solutions, since the solution has the form $\sin(\lambda x)$ and the separation constant appears in the form λ^2 in the T-equation, no additional solutions are obtained.

This provides the set of separated X-solutions

$$X_n(x) = \sin(\lambda_n x), \ n = 1, 2, 3, \ldots.$$

The corresponding T-equations are

$$T_n'' + \lambda_n^2 c^2 T = 0$$

with solutions

$$T_n(t) = a_n \cos(\lambda_n ct) + b_n \sin(\lambda_n ct).$$

The separated solutions take the form

$$u_n(x, t) = \sin(\lambda_n x) \, (a_n \cos(\lambda_n ct) + b_n \sin(\lambda_n ct)),$$

and in the usual fashion, we try

$$u(x, t) = \sum_{n=1}^{\infty} \sin(\lambda_n x) \, (a_n \cos(\lambda_n ct) + b_n \sin(\lambda_n ct)) \qquad (3.36)$$

for a solution. The initial conditions may be met by determining $\{a_n\}$ and $\{b_n\}$ so that

$$f(x) = \sum_{n=1}^{\infty} \sin(\lambda_n x) \, a_n,$$

$$g(x) = \sum_{n=1}^{\infty} \sin(\lambda_n x) \, (\lambda_n c \, b_n).$$

It is evident from Figure 3.28 that the solutions are approximated by $\lambda_n \approx \frac{n\pi}{L}$ for large odd values of n. If this were exactly the case for all values of n, then the required expansions reduce to standard (quarter-range) Fourier series problems.

Even though this is not the case, it is true that the functions in the expansions share with the standard sinusoids the crucial property of orthogonality:

$$\int_0^L \sin(\lambda_n x) \sin(\lambda_m x)\, dx = \begin{cases} 0 & m \neq n, \\ E_n^2 & m = n. \end{cases} \tag{3.37}$$

This property may be verified using the separated equations satisfied by the X functions. (See the problems below).

Given the orthogonality relations (3.37) the expansion coefficients follow in exactly the fashion leading to the original Fourier expansions of Chapter 2:

$$a_n = \frac{1}{E_n} \int_0^L f(x) \sin(\lambda_n x)\, dx,$$

$$\lambda_n c b_n = \frac{1}{E_n} \int_0^L g(x) \sin(\lambda_n x)\, dx.$$

The combination of these coefficients and (3.36) now provides the solution sought for the original problem (3.7).

The orthogonality of the separated solutions encountered above is, of course, not simply a happy turn of fate. It is a general property inherent in a wide class of boundary value problems, and having diverse other implications. The basis for this phenomenon is discussed in the following chapter.

Problems 3.7.

1. For the following boundary value problems, provide a description of a physical situation modeled by the given system of equations.

   a.

$$\frac{\partial u}{\partial t} = \kappa \frac{\partial^2 u}{\partial x^2}, \quad 0 < x < L,$$

$$u(0, t) = \frac{\partial}{\partial x} u(Lt) = 0,$$

$$u(x, 0) = x(x - L).$$

b.

$$\frac{\partial^2 u}{\partial x^2} + \frac{\partial^2 u}{\partial y^2} = 0, \ 0 < x < a, 0 < y < b,$$

$$u(x,0) = \frac{x}{a},$$

$$\frac{\partial}{\partial y} u(x,b) = 0,$$

$$u(0,y) = 0,$$

$$u(a,y) = 0.$$

c.

$$\frac{1}{c^2} \frac{\partial^2 u}{\partial t^2} = \frac{\partial^2 u}{\partial x^2}, \ 0 < x < 49,$$

$$\frac{\partial u}{\partial x}(0,t) = \frac{\partial u}{\partial x}(49,t) = 0,$$

$$\frac{\partial u}{\partial t}(x,0) = 0,$$

$$u(x,0) = \cos(\frac{2\pi}{49}x).$$

d.

$$\frac{\partial^2 u}{\partial r^2} + \frac{1}{r} \frac{\partial u}{\partial r} + \frac{1}{r^2} \frac{\partial^2 u}{\partial \theta^2} = 0, \ 0 < a < r < b, 0 < \theta < \frac{\pi}{4},$$

$$\frac{\partial u}{\partial \theta}(r,0) = \frac{\partial u}{\partial \theta}(r, \frac{\pi}{4}) = 0,$$

$$\frac{\partial}{\partial r} u(a,0) = 0,$$

$$u(b,\theta) = 0.$$

2. Find the solution of

$$\frac{\partial u}{\partial t} = 12 \frac{\partial^2 u}{\partial x^2}, \ 0 < x < 27,$$

$$\frac{\partial}{\partial x} u(0,t) = \frac{\partial}{\partial x} u(27t) = 0,$$

$$u(x,0) = x^2 (27 - x)^2.$$

3. Find the solution of the boundary value Problem 1c above.

4. Determine the solution of Problem 1b above.
5. Repeat 3 for the Problem 1a.
6. Repeat 4 for the Problem 1d.
7. Find the solution of the boundary value problem

$$\frac{\partial u}{\partial t} = 4\frac{\partial^2 u}{\partial x^2}$$

$$u(0, t) = 0,$$

$$u(l, t) = 100,$$

$$u(x, 0) = 0,\ 0 < x < l.$$

(See Problem 5, Section 3.2.)
8. The vibrating string equations above have been considered in the absence of gravity. To remedy this lack of reality, show that the governing equation in this case takes the form

$$\frac{\partial^2 u}{\partial t^2} = c^2 \frac{\partial^2 u}{\partial x^2} - g.$$

Find the solution of the boundary value problem

$$\frac{\partial^2 u}{\partial t^2} = c^2 \frac{\partial^2 u}{\partial x^2} - g,$$

$$u(0, t) = u(L, t) = 0,$$

$$\frac{\partial u}{\partial t}(x, 0) = 0,$$

$$u(x, 0) = f(x).$$

(See Problem 3, Section 3.5).
9. Find the solution of

$$\frac{1}{c^2}\frac{\partial^2 u}{\partial t^2} = \frac{\partial^2 u}{\partial x^2},\ 0 < x < L,$$

$$\frac{\partial}{\partial x}u(0, t) = \frac{\partial}{\partial x}u(L, t)\ = 0,$$

$$\frac{\partial u}{\partial t}(x, 0) = g(x),$$

$$u(x, 0) = f(x).$$

Assuming that f and g are smooth functions, what condition is required in order that the solution remain bounded for all t? Remark on the degree to which this is obvious from the physical interpretation of the problem.

10. Consider the boundary value problem

$$\frac{\partial u}{\partial t} = \kappa \frac{\partial^2 u}{\partial x^2}, \, 0 < x < L,$$

$$u(0, t) = \frac{\partial u}{\partial x}(L, t) = 0,$$

$$u(x, 0) = f(x).$$

a. Find a formal solution by means of separation of variables.
b. In connection with the initial condition expansion problem encountered in the system of equation (3.31), note that roughly one quarter of the usual functions are available. Find the "natural length" associated with the required expansion and the natural symmetries associated with the expansion.
c. Describe a boundary value problem for heat flow in a ring which provides the solution of (10a) using the symmetries arising in (10b).

11. Find a solution of the problem

$$\frac{\partial^2 u}{\partial x^2} + \frac{\partial^2 u}{\partial y^2} = 0, \, 0 < x < a, \, 0 < y < b,$$

$$\frac{\partial}{\partial y} u(x, 0) = f_1(x),$$

$$\frac{\partial}{\partial y} u(x, b) = f_2(x),$$

$$\frac{\partial}{\partial x} u(0, y) = g_1(y),$$

$$\frac{\partial}{\partial x} u(a, y) = g_2(y).$$

Use a method of superposition analogous to that in the text. For convenience, assume that the average values of the boundary functions above are all zero.

12. Find the solution of the boundary value problem

$$\frac{\partial^2 u}{\partial x^2} + \frac{\partial^2 u}{\partial y^2} = 0, \, 0 < x < 1, \, 0 < y < 2,$$

$$u(x, 0) = 0,$$

$$u(x, 2) = 0,$$

$$u(0, y) = 0,$$

$$(u + \frac{\partial u}{\partial x})(1, y) = e^y.$$

13. A model for heat conduction in the presence of insulation at one boundary and linear heat transfer at the other is

$$\frac{\partial u}{\partial t} = \kappa \frac{\partial^2 u}{\partial x^2}, \ 0 < x < l,$$

$$(u - \alpha \frac{\partial u}{\partial x})(0, t) = 0,$$

$$\frac{\partial}{\partial x} u(l, t) = 0,$$

$$u(x, 0) = f(x).$$

Find a solution of this boundary value problem.

14. The wave equation

$$\frac{1}{c^2} \frac{\partial^2 u}{\partial t^2} = \frac{\partial^2 u}{\partial x^2}$$

has been obtained as a model for the longitudinal vibrations of a beam (or perhaps spring) (Problem 10, Section 3.3), and for the transverse vibration of a (flexible) string. A different equation arises in the case of transverse vibrations of a simple beam.

The evident characteristic of a beam is that it resists bending. This resistance provides the restoring force in transverse vibration problems.

If a free-body element of a deformed beam is examined (see Figure 3.29), forces are distributed across the cut faces of the beam. These are resolved into components normal and transverse to the cut faces.

The transverse forces are idealized to a (net average) vertical shear. The normal forces associated with the axial compression and tension effects of the bending are replaced by net moments about the centerline of the beam. (See Figure 3.30).

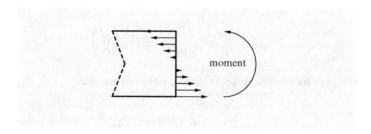

moment

Fig. 3.29 Free-body beam element

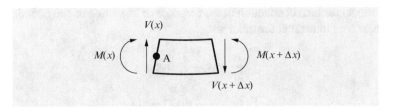

Fig. 3.30 Equivalent shear and bending moments

a. By taking moments about the point A of Figure 3.30, show that the bending moment and shear distributions are related by

$$\frac{\partial}{\partial x} M(x) = V(x).$$

b. The law governing the relationship of the bending moment and the beam displacement u is that

$$EI(x)\frac{\partial^2}{\partial x^2}u = M(x).$$

(E is a material constant and $I(x)$ is the moment of inertia of the cross section at location x. This relation is derived through the theory of elasticity.)
If ρ represents the mass per unit length of the beam, use the above relationships to derive the equation of motion

$$\rho(x)\frac{\partial^2 u}{\partial t^2} = -\frac{\partial^2}{\partial x^2}\left((EI(x))\frac{\partial^2 u}{\partial x^2}\right)$$

for the transverse vibration problem.

15. Consider the transverse vibrations of a free (flying, if you like) uniform beam. Arguing as in Problem 14 above, show that the governing equation is

$$\rho(x)\frac{\partial^2 u}{\partial t^2} = -\frac{\partial^2}{\partial x^2}\left((EI(x))\frac{\partial^2 u}{\partial x^2}\right)$$

and that the physically appropriate end conditions are that

$$\frac{\partial^2 u}{\partial x^2} = 0, \ (\text{at } x = 0, L),$$

$$\frac{\partial^3 u}{\partial x^3} = 0, \ (\text{at } x = 0, L).$$

a. Adjoin initial conditions to the above, and solve the resulting boundary value problem formally, assuming orthogonality of the separated X solutions.

b. For the construction of a well-tempered xylophone, the fundamental frequencies of the successive bars should be in the ratio of (semi-tones) $(2)^{\frac{1}{12}}$. Determine the lengths of the bars for a well-tempered xylophone whose lowest toned bar is of length 20 cm.

16. Consider Problem 14, and suppose that the ends of the beam in question are supported by frictionless pin joints. Show that the appropriate boundary conditions are

$$\frac{\partial^2 u}{\partial x^2} = u = 0, \text{ (at ends)}$$

and solve the resulting boundary value problem.

17. Analyze the motions of the classic diving board. You may assume that the board is firmly embedded in a rigid wall at one end. Argue first that the boundary conditions at the embedded end are

$$u = \frac{\partial u}{\partial x} = 0, \, x = 0.$$

18. The equations of compressible, inviscid, irrotational fluid flow (to mention just a few intimidating bits of terminology) can be made to yield a wave equation for sound waves. The physical variables of the problem are the velocity field

$$\mathbf{v}(x, y, z, t) = \begin{bmatrix} v_1(x, y, z, t) \\ v_2(x, y, z, t) \\ v_3(x, y, z, t) \end{bmatrix}$$

and the scalar density and pressure functions

$$\rho(x, y, z, t), \, p(x, y, z, t).$$

Conservation of mass (see Problem 3, Section 3.4) leads to the equation of continuity:

$$\frac{\partial \rho}{\partial t} + \nabla \cdot (\rho v) = 0.$$

Assuming that the only forces acting in the fluid arise from a normal stress from the pressure, Newton's law gives

$$\frac{\partial \mathbf{v}}{\partial t} + (\mathbf{v} \cdot \nabla)\mathbf{v} = -\frac{1}{\rho}\nabla p.$$

(The quadratic terms arise intuitively from the fact that $\mathbf{v}(x, y, z, t)$ is the velocity field, so that $\frac{\partial \mathbf{v}}{\partial t}$ accounts for only part of the particle acceleration. The equation may be derived from conservation of momentum considerations.)

In order to complete the system, a thermodynamic relation between pressure and density is introduced. This is the adiabatic law

$$p = k\rho^{\gamma}.$$

Assume that all velocities and deviations from an equilibrium density ρ_0 and pressure p_0 are small.

Introduce the irrotational assumption through the use of a velocity potential ϕ. Linearize the equations about the equilibrium point (neglect quadratic terms in small quantities) to obtain

$$\mathbf{v} = \nabla \phi,$$

$$\frac{\partial \phi}{\partial t} = \frac{\rho - \rho_0}{\rho_0}$$

and

$$\frac{\partial^2 \phi}{\partial t^2} = c^2 \nabla^2 \phi, \ c^2 = \frac{\gamma p_0}{\rho_0}$$

as the governing equations.

19. Consider sound waves in a one-dimensional pipe, open at one end. Show that the governing boundary value problem reduces to

$$\frac{1}{c^2} \frac{\partial^2 \phi}{\partial t^2} = \frac{\partial^2 \phi}{\partial x^2},$$

$$\frac{\partial \phi}{\partial x}(0, t) = 0,$$

$$\phi(L, t) = 0$$

(in addition to initial conditions).

(See Problem 18. Assume that the pressure at the open end is at the equilibrium value.)

20. Find the solution of the boundary value problem

$$\frac{\partial^2 u}{\partial r^2} + \frac{1}{r}\frac{\partial u}{\partial r} + \frac{1}{r^2}\frac{\partial^2 u}{\partial \theta^2} = 0, \, 0 < \theta < \frac{\pi}{4}, \, 0 < a < r < b,$$

$$\frac{\partial u}{\partial \theta}(r, 0) = \frac{\partial u}{\partial \theta}(r, \frac{\pi}{4}) = 0,$$

$$\frac{\partial}{\partial r}u(a, \theta) = 0,$$

$$u(b, \theta) = \theta.$$

21. Suppose that $K(x)$ is defined by

$$K(x) = \begin{cases} 1 & 0 < x < 1, \\ 4 & 1 < x < 2. \end{cases}$$

a. Find allowable values of λ and solutions X of the differential equation

$$K(x)X'' + \lambda^2 X = 0, \, 0 < x < 2,$$

$$X(0) = X(2) = 0.$$

(Seek solutions with continuous first derivative.)

b. If the set of solutions obtained above is denoted by $\{X_n(x)\}_{n=1}^{\infty}$, show that the solutions are orthogonal in the sense that

$$\int_0^2 X_n(x)X_m(x)\frac{1}{K(x)}\, dx = 0, \, m \neq n.$$

c. Use the results of (21a) and (21b) above to find a solution of the boundary value problem

$$\frac{\partial u}{\partial t} = K(x)\frac{\partial^2 u}{\partial x^2},$$

$$u(0, t) = u(2, t) = 0,$$

$$u(x, 0) = f(x), \, 0 < x < 2.$$

Give a physical interpretation of the problem above.

22. Show that an appropriate model for the vibrations of an elastic string with elastically constrained ends is

$$\frac{1}{c^2}\frac{\partial^2 u}{\partial t^2} = \frac{\partial^2 u}{\partial x^2},$$

$$(T\frac{\partial}{\partial x}u - ku)(0, t) = 0,$$

$$(-T\frac{\partial}{\partial x}u - ku)(L, t) = 0,$$

$$\frac{\partial}{\partial t}u(x, 0) = g(x),$$

$$u(x, 0) = f(x).$$

Find a solution of the above boundary value problem.

23. The separated X solutions for the one-elastic-end wave equation problem analyzed above were given as

$$X_n(x) = \sin(\lambda_n x),$$

solutions of

$$X_n'' + \lambda_n^2 X_n = 0$$

subject to the boundary conditions

$$X_n(0) = 0,$$

$$-T\frac{dX_n}{dx}(L) = k X_n(L).$$

Show that these solutions satisfy the conditions

$$\int_0^L X_n(x) X_m(x)\, dx = 0, \ n \neq m.$$

(This can be done either by liberal use of trigonometric identities, or more elegantly by using integration by parts and the governing differential equations.)

24. Something like a world-beating boundary value problem for Laplace's equation on a rectangle can be obtained by prescribing heat transfer boundary conditions (with free parameters) on all four sides. Specific solutions (in principle) could then be obtained by adjusting parameters and functions to suit.
The explicit description of such a problem is

$$\frac{\partial^2 u}{\partial x^2} + \frac{\partial^2 u}{\partial y^2} = 0,$$

$$\left(\alpha u + \beta \frac{\partial}{\partial x} u\right)(0, y) = f_1(y),$$

$$\left(\gamma u + \delta \frac{\partial}{\partial x} u\right)(a, y) = f_2(y),$$

$$\left(\epsilon u + \phi \frac{\partial}{\partial y} u\right)(x, 0) = g_1(x),$$

$$\left(\kappa u + \lambda \frac{\partial}{\partial y} u\right)(x, b) = g_2(x).$$

Invoking the usual superposition and symmetry arguments, it suffices to solve the above problem for the case where

$$f_1(y) = g_1(x) = g_2(x) = 0.$$

Find (in principle) a solution of this latter problem.

3.8   Some Matters of Detail

In the previous sections we have found by separation of variables methods certain expressions which are candidates for solutions of various boundary value problems.

These expressions strictly should be regarded as candidates for solutions since we have made no effort to verify that they in fact "solve" the original problems posed. The main reason for this is that it is first necessary to *define* the notion of solution, and the sense in which a solution is to satisfy the indicated equation and boundary conditions.

Coupled with these considerations are issues of existence and uniqueness of solutions, and of the continuous dependence of the solution on the given problem data. On the basis of the physical nature of the problems from which the boundary value problems arose, one expects (or more properly, hopes) that these issues can be resolved in the affirmative sense. In fact, this is so strongly felt that problems for which these properties hold are conventionally called *well posed*. Problems for which one or more of those properties fail have been labeled by the pejorative term *ill-posed*.

There is some tendency to (psychologically at least) raise the above expectation to a certainty, and to assume that some equation system is well posed on the basis of its origin in some underlying physical problem.

This tendency should be regarded as a tactical error for at least two reasons (beyond the support of the existence and uniqueness theorem industry). In the first place, there exist physical problems that are inherently ill-posed. This revelation may be one of the most illuminating aspects of the analysis of such problems. To rule it out is to miss the central feature of the problem in such a case. A second point rests on the relationship between a given physical phenomenon and a purported mathematical model of the phenomenon. The translation from phenomenon to model is a dynamical process involving refinement, analysis, comparison, and so forth. The model at any stage is not the phenomenon itself, and the demonstration that the problem is well posed should be regarded as simply one of the (possibly) desirable characteristics of the model. (On a more mundane level, it is not unknown that someone should "get the equations wrong.")

The problem of existence of solutions of partial differential equations in general is a technically complex one. For the case of the standard boundary problems considered above, the problem largely disappears, as "solutions" evidently not only exist but are available in closed form.

The heat equation and wave equation considered above are related in that each has an interpretation as an *equation of evolution*.

In the case of the heat equation, the solution formulas obtained above transform the initial function f at time $t = 0$ into a second function at a later time t. In the case of the zero-boundary condition problem, the transformation $f(x) \mapsto u(x, t) = S_t f(x)$ is defined by

$$u(x, t) = S_t f(x) = \sum_{n=1}^{\infty} b_n e^{-\kappa(\frac{n\pi}{L})^2 t} \sin(\frac{n\pi x}{L}) \tag{3.38}$$

where

$$b_n = \frac{2}{L} \int_0^L f(x) \sin(\frac{n\pi x}{L}) dx$$

should be thought of as a linear transformation mapping the initial function f into the (profile shape function at time t) u according to the above formula.

In view of Parseval's formula the solution energy is

$$\int_0^L |u(x, t)|^2 dx = \frac{1}{2} \sum_{n=1}^{\infty} |b_n|^2 e^{-2\kappa(\frac{n\pi}{L})^2 t},$$

so that S_t evidently transforms square-integrable functions (initial conditions) into functions (temperature profiles) square-integrable at each time $t > 0$. S_t hence maps the vector space $L^2[0, L]$ into itself for $t > 0$.

Another consequence of Parseval's relation is that it is very likely that, unless f is suitably restricted, the integral

$$\int_0^L |u(x,t)|^2 \, dx$$

will actually *diverge* for $t < 0$. This indicates that the heat equation is ill-posed running in the direction of negative time. (Arbitrarily small initial conditions amplify to arbitrarily large levels for any fixed $t < 0$, and the solution is hence not continuous with respect to the initial condition for $t < 0$.)

The essential idea behind a linear time-invariant evolution model is that the solution obtained at time $(t + \tau)$ should be the same as that obtained from evolving for a time interval τ from an initial condition resulting from a t-interval transfer from an initial condition. This is much more compactly expressed as (the *semi-group property*)

$$(S_{t+\tau} f) = S_t(S_\tau f).$$

That this is true for S_t defined above follows from the observation that

$$\{b_n e^{-\kappa(\frac{n\pi}{L})^2 t}\}$$

are self-evidently the half-range expansion coefficients of the function defined by $u(x,t) = S_t f(x)$.

The transformation from these initial coefficients through an interval τ is obtained by multiplying the half-range expansion coefficients by $\{e^{-\kappa(\frac{n\pi}{L})^2 \tau}\}$. Hence

$$S_\tau(S_t f)(x) = \sum_{n=1}^\infty e^{-\kappa(\frac{n\pi}{L})^2 \tau} \left(b_n e^{-\kappa(\frac{n\pi}{L})^2 t}\right) \sin(\frac{n\pi x}{L})$$

$$= \sum_{n=1}^\infty b_n e^{-\kappa(\frac{n\pi}{L})^2 t + \tau} \sin(\frac{n\pi x}{L})$$

$$= (S_{t+\tau} f)(x),$$

which verifies the semi-group property for this particular problem.

These considerations give a natural sense in which to consider the problem of satisfaction of initial conditions. We may consider (the squared distance of the solution from the initial condition)

$$\frac{1}{L} \int_0^L |u(x,t) - f(x)|^2 \, dx,$$

and ask whether the above quantity approaches zero as $t \to 0^+$. If so, we may say that the solution satisfies the initial condition in the mean-square sense.

Using Parseval's Theorem, we calculate explicitly

$$\frac{1}{L}\int_0^L |u(x,t) - f(x)|^2\, dx = \frac{1}{2}\sum_{n=1}^{\infty} |b_n|^2 \left(e^{-\kappa(\frac{n\pi}{L})^2 t} - 1\right)^2.$$

To show that this is less than, say, ϵ for t sufficiently small, the standard tactic is to waste half of ϵ on the tail of the series above.

Since by Parseval's Theorem the series

$$\sum_{n=1}^{\infty} |b_n|^2$$

is convergent, there exists $N(\frac{\epsilon}{2})$ such that

$$\sum_{n=N+1}^{\infty} |b_n|^2 < \frac{\epsilon}{2}.$$

Since $|e^{-at} - 1|^2 < 1$ for all $at > 0$, we have

$$\sum_{n=N+1}^{\infty} |b_n|^2 \left(e^{-\kappa(\frac{n\pi}{L})^2 t} - 1\right)^2 < \frac{\epsilon}{2},$$

independent of $t > 0$. Now choose $\delta(t)$ so that for $0 < t < \delta$,

$$\sum_{n=1}^{N} |b_n|^2 \left(e^{-\kappa(\frac{n\pi}{L})^2 t} - 1\right)^2 < \frac{\epsilon}{2}.$$

This is possible since this finite series is a function continuous in t, with a limit of 0 at $t = 0$) We conclude that for any $\epsilon > 0$ there exists a δ such that for $0 < t < \delta$ we have

$$\frac{1}{L}\int_0^L |u(x,t) - f(x)|^2\, dx < \epsilon.$$

Since ϵ is arbitrary, this shows that the solution converges to the initial condition in the mean-square sense as $t \to 0^+$.

A small variation of this argument (using linearity of S_t) shows that in the mean-square sense the solution is a continuous function of the initial data. This provides one answer to the continuous dependence issue raised above.

Since the solution (3.38) involves Fourier series in an essential way, mean-square considerations are natural for these problems. One may, of course, also consider pointwise convergence of the associated series solutions.

Since when $t = 0$ the expression (3.38) reduces to

$$u(x, 0) = \sum_{n=1}^{\infty} b_n \sin(\frac{n \pi x}{L})$$ (3.39)

it is evident from the discussions of Chapter 2 that some restrictions must be placed on $\{b_n\}$ before pointwise convergence of (3.39) can be expected.

In fact, the considerations of Chapter 2 may be applied intact to the problem (3.39). If the odd periodic extension of f is sufficiently smooth, then (3.39) converges pointwise. A similar appropriation of result may be made for the problem of uniform convergence.

In the solution (3.38), the presence of the $e^{-\kappa(\frac{n\pi}{L})^2 t}$ factors in the series expansion has an enormous "taming effect" on the Fourier expansion coefficients $\{b_n\}$. Since

$$\sum_{n=1}^{\infty} \left| e^{-\kappa(\frac{n\pi}{L})^2 t} \sin(\frac{n \pi x}{L}) \right|^2$$

is (for $t > 0$) uniformly convergent in x (the terms are less than, say, $\frac{M}{n^p}$ for any $p > 0$) Parseval's Theorem implies that

$$\sum_{n=1}^{\infty} b_n e^{-\kappa(\frac{n\pi}{L})^2 t} \sin(\frac{n \pi x}{L})$$

is convergent (uniformly in x).

It is also easy to see that the expression

$$\sum_{n=1}^{\infty} (\frac{n\pi}{L})^2 b_n e^{-\kappa(\frac{n\pi}{L})^2 t} \sin(\frac{n \pi x}{L})$$

is also uniformly convergent with respect to both x and t as long as t is bounded away from zero. The consequence of this is that the solution expression

$$u(x, t) = \sum_{n=1}^{\infty} b_n e^{-\kappa(\frac{n\pi}{L})^2 t} \sin(\frac{n \pi x}{L})$$ (3.40)

may be differentiated termwise once with respect to t, and twice with respect to x as long as $t > 0$. Hence

$$\frac{\partial u}{\partial t} = \kappa \frac{\partial^2 u}{\partial x^2}$$

holds pointwise in (x, t) for $t > 0$. The solution (3.40) hence satisfies the heat equation in this sense, assuming only that the initial function f is square-integrable.

We summarize these conclusions in the form of a theorem.

Theorem. *Consider the boundary value problem*

$$\frac{\partial u}{\partial t} = \kappa \frac{\partial^2 u}{\partial x^2}, \; 0 < x < L, \; t > 0,$$

$$u(0, t) = u(L, t) = 0,$$

$$u(x, 0) = f(x).$$

Then, assuming that

$$\int_0^L |f(x)|^2 \, dx < \infty,$$

the function

$$(S_t f)(x) = \sum_{n=1}^{\infty} (\frac{n\pi}{L})^2 \, b_n \, e^{-\kappa(\frac{n\pi}{L})^2 t} \, \sin(\frac{n\pi x}{L}),$$

$$b_n = \frac{2}{L} \int_0^L f(x) \, \sin(\frac{n\pi x}{L}) \, dx$$

satisfies the conditions

$$S_\tau(S_t f)(x) = (S_{\tau + t} f)(x), \; t, \tau > 0,$$

$$\lim_{t \to 0+} \int_0^L |u(x, t) - f(x)|^2 \, dx = 0,$$

$$\frac{\partial u}{\partial t} = \kappa \frac{\partial^2 u}{\partial x^2} \, t > 0, \; 0 < x < L,$$

$$u(0, t) = u(L, t) = 0,$$

where both the solution series and the indicated partial derivatives of the governing partial differential equation are uniformly convergent for t > 0.

Further, if the odd periodic extension f of the initial condition is sufficiently smooth,[9] the solution series converges pointwise to f even when t = 0.

It should be noted that the above allows, in particular, discontinuous initial conditions (which need not even satisfy the formal boundary conditions $u(0, 0) = u(L, 0) = 0$ in a pointwise sense). The solution can therefore serve as a model for a heated bar instantaneously brought into contact with ice blocks. The inherent physical ambiguity about the boundary conditions appropriate for such a problem is in a real sense resolved by the solution (see Problem 7).

---

[9]That is, f is absolutely continuous with square-integrable derivative over [0, 2L].

The formal solution of the wave equation obtained above has some properties in common with the heat equation, but differs in some respects. The differences arise chiefly because the solution formulas involve none of the smoothing tendencies exploited so heavily above.

For concreteness, we consider the wave equation problem

$$\frac{1}{c^2}\frac{\partial^2 u}{\partial t^2} = \frac{\partial^2 u}{\partial x^2},$$

$$u(0, t) = u(L, t) = 0,$$

$$u(x, 0) = f(x),$$

$$\frac{\partial u}{\partial t}(x, 0) = g(x).$$

The formal solution obtained in the previous section is given by

$$u(x, t) = \sum_{n=1}^{\infty} (a_n \cos(n\omega_0 t) + b_n \sin(n\omega_0 t)) \sin(\frac{n\pi x}{L}) \tag{3.41}$$

where

$$a_n = \frac{2}{L} \int_0^L f(x) \sin(\frac{n\pi x}{L}) \, dx,$$

$$n \omega_0 b_n = \frac{2}{L} \int_0^L g(x) \sin(\frac{n\pi x}{L}) \, dx.$$

Since the problem requires the specification of both a position and a velocity function, it is appropriate to consider in addition to (3.41) an expression representing the velocity distribution at time t. Formal differentiation gives

$$v(x, t) = \sum_{n=1}^{\infty} (a_n n\omega_0 \sin(n\omega_0 t) + b_n n\omega_0 \cos(n\omega_0 t)) \sin(\frac{n\pi x}{L}). \tag{3.42}$$

If the initial displacement f is restricted so that the odd $2L$-periodic extension of f is a "good" function, while g is square-integrable,

$$\frac{1}{L} \int_0^L |f'(x)|^2 \, dx < \infty,$$

$$\frac{1}{L} \int_0^L |g(x)|^2 \, dx < \infty,$$

then these conditions imply (by Parseval's Theorem)

$$\sum_{n=1}^{\infty} |a_n|^2 (n\omega_0)^2 < \infty, \quad \sum_{n=1}^{\infty} |b_n|^2 < \infty.$$

These relations serve to guarantee that (3.41) and (3.42) represent mean-square convergent Fourier series.

While the restrictions may be regarded as having been forced by the requirement that (3.42) make mathematical sense, it actually has a strong physical basis. The original derivation of the vibrating string model (or consideration of a discrete spring-mass analogue) leads one to consider (see Problems 3 and 4 below)

$$\frac{1}{L} \int_0^L T (f'(x))^2 + \rho (g(x))^2 \, dx$$

as the mechanical energy associated with the string. The restriction on the initial conditions is simply the natural restriction to initial conditions of finite energy. Using (3.41), (3.42) it is easy to see that the solution formulas also have finite energy in this sense (and in fact, the same energy as the initial conditions).

Using these considerations (3.41) and (3.42) may be used to define a (time-varying) transformation of initial functions analogous to (3.38) of the heat equation case. Define $\tilde{S}_t$ by

$$\begin{bmatrix} u(x,t) \\ v(x,t) \end{bmatrix} = \tilde{S}_t \begin{bmatrix} f \\ g \end{bmatrix}.$$

From the defining formulas (3.41) and (3.42) it follows that for f, g satisfying the finite energy condition,

$$\tilde{S}_\tau \cdot \tilde{S}_t \begin{bmatrix} f \\ g \end{bmatrix} = \tilde{S}_{t+\tau} \begin{bmatrix} f \\ g \end{bmatrix}$$

for all values of t, τ. The fact that all real values of t, τ are allowable in this expression is a consequence of the time reversibility of the wave equation, and an analytical expression of the intuitive interpretation of the wave equation as a time invariant initial value problem (or evolution equation).

The question of the satisfaction of boundary and initial conditions for the wave equation is resolved in a manner parallel to the heat equation case. The expression (3.41) evidently satisfies the boundary conditions exactly, since all of the sine functions vanish at the ends of the interval.

The initial conditions are satisfied in the sense that

$$\int_0^L |u(x,t) - f(x)|^2 \, dx,$$

$$\int_0^L |v(x,t) - g(x)|^2 \, dx$$

both approach zero as $t \to 0$ as long as f, g are of finite energy. It is also the case that the initial displacement condition is satisfied pointwise under this condition. In view of (3.42), pointwise convergence of the velocity expansion evidently requires a further degree of differentiability of the initial functions f, g.

The solution (3.41) may be interpreted as a "shuffling of energy" among the system modes. It involves no smoothing effects of the sort present in the heat equation case. As a consequence, the question of the sense in which (3.41) satisfies the original partial differential equation is more delicate than in the heat equation case.

The finite energy restriction ensures that (3.42) is convergent in mean-square only. Since the governing equation involves a second derivative with respect to time, we contemplate the formal series

$$\sum_{n=1}^{\infty} (n\omega_0)^2 \left\{ a_n \cos(n\omega_0 t) + \frac{b_n}{n\omega_0} \sin(n\omega_0 t) \right\} \sin\left(\frac{n\pi x}{L}\right).$$

This has an interpretation as a mean-square convergent series provided that

$$\sum_{n=1}^{\infty} (n\omega_0)^4 |a_n|^2 < \infty,$$

$$\sum_{n=1}^{\infty} (n\omega_0)^2 |b_n|^2 < \infty.$$

If these restrictions are imposed, then (3.41) can be interpreted as satisfying the wave equation in the mean-square sense. If one wishes to obtain the more classical satisfaction in the pointwise sense, the initial functions must be even smoother in order to guarantee that the differentiated series are pointwise convergent.

In view of the welter of restrictions, sub-cases and caveats associated with the above, we avoid construing the above as a theorem.

The above discussion touches on aspects of the issues raised at the beginning of this section with the notable exception of uniqueness of solutions. One approach to this for the problems discussed above is through the theory of semi-groups and evolution equations. This is well beyond the level of our presentation. Another approach is the use of integral methods to obtain uniqueness for solutions with sufficient differentiability to justify manipulations of the governing equations. Such a result for Laplace's equation is given below; some versions for the equations above are given in the problems.

The solutions of Laplace's equation derived in the previous section have a smoothing effect inherent in the formulas which is similar to that noted for the heat equation. Laplace's equation in a circle (with prescribed boundary function) has a solution

$$u(r, \theta) = \sum_{n=-\infty}^{\infty} c_n \, r^{|n|} \, e^{in\theta}, \ r < 1, \qquad (3.43)$$

$$c_n = \frac{1}{2\pi} \int_0^{2\pi} f(\theta) \, e^{in\theta} \, d\theta$$

in terms of the boundary function f. In view of the "taming effect" of the factors $r^{|n|}$ on the Fourier coefficients $\{c_n\}$, the series in (3.43) is pointwise convergent for $r < 1$ as long as

$$\int_0^{2\pi} |f(\theta)|^2 \, d\theta < \infty.$$

The series, even better, is uniformly convergent, even after being differentiated twice (or more) with respect to the variables (r, θ) (or (x, y), if one makes the associated rectangular-polar conversion).

A consequence of this is that $u(\cdot, \cdot)$ defined by (3.43) satisfies

$$\nabla^2 u = 0, \ 0 < r < 1, \ 0 < \theta < 2\pi,$$

pointwise as long as $r < 1$.

A limit argument entirely similar to that employed in the case of the heat equation shows that

$$\lim_{r \to 1^-} \frac{1}{2\pi} \int_0^{2\pi} |u(r, \theta) - f(\theta)|^2 \, d\theta = 0.$$

The solution (3.43) therefore satisfies the expected boundary condition in the mean-square sense.

If the given boundary value function f is sufficiently restricted so that its Fourier series converges pointwise, then the solution (3.43) (which reduces to the Fourier series when $r = 1$) obviously converges to f pointwise on the boundary.

Laplace's equation models arise from the consideration of the vector-calculus formulations of conservation laws. In view of this connection between Laplace's equation and these theorems, it is not surprising that these appear in the context of uniqueness theorems for Laplace's equation.

We describe a version of such a result for the three-dimensional Laplace equation. The underlying physical model is that of steady state heat flow in a uniform body of finite extent, with prescribed temperature on the surface. See Figure 3.31.

We denote by V the interior of the body and by Σ the surface. We assume Σ sufficiently regular to justify an application of Gauss's Theorem.

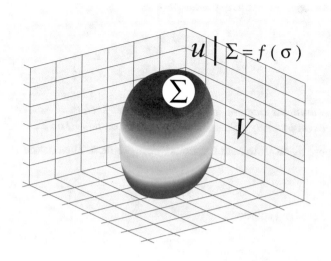

Fig. 3.31 Heated uniform body V

If ϕ and ψ are twice continuously differentiable scalar fields, then a standard vector identity is

$$\nabla(\phi\nabla\psi) = \phi\nabla^2\psi + \nabla\phi \cdot \nabla\psi.$$

Integrating this identity over the volume and applying Gauss's Theorem to evaluate the divergence term gives

$$\iiint_V \{\phi\nabla^2\psi + \nabla\phi \cdot \nabla\psi\}\, dV = \iiint_V \nabla \cdot (\phi\nabla\psi)\, dV$$

$$= \iint_\Sigma (\phi\nabla\psi) \cdot \mathbf{n}\, dS$$

$$= \iint_\Sigma \phi\frac{\partial\psi}{\partial n}\, dS.$$

In case $\phi = \psi$ the expression reduces to

$$\iiint_V \{\phi\nabla^2\phi + \nabla\phi \cdot \nabla\phi\}\, dV = \iint_\Sigma \phi\frac{\partial\phi}{\partial n}\, dS.$$

This last identity can provide a uniqueness result for Laplace's equation. The result provides uniqueness in a class of solutions sufficiently smooth to justify application of the vector calculus results leading to the above identity.

Theorem. *Consider the boundary value problem*

$$\nabla^2 u = 0 \ \text{in} \ V,$$

$$u\Big|_{\Sigma} = f,$$

where f is a continuous function on the boundary of V. Then there exists at most one solution twice continuously differentiable in V, and continuous on Σ.

Proof. Suppose u and v are both solutions of the problem. Then define

$$\phi = u - v.$$

Since u and v are both solutions,

$$\nabla^2 \phi = 0,$$

and

$$\phi\Big|_{\Sigma} = 0.$$

Apply the integral identity to find

$$\int\int\int_V \{0 + \nabla\phi \cdot \nabla\phi\} \, dV = \int\int_\Sigma 0 \, dS = 0,$$

and hence that $\nabla\phi = 0$, and so ϕ is constant in V. Since $\phi\Big|_{\Sigma} = 0$, this constant vanishes, and

$$\phi = u - v = 0.$$

The solution is hence unique.

The above result is informative in the event that such solutions may be shown to exist. This is the case, but since such demonstrations cannot be considered light reading, we refer the interested reader to other references.

It should be clear that the above result can be "cut down" to the two-dimensional case by use of Green's Theorem in the plane. Combining this with the discussion above of Laplace's equation in a circle provides an existence and uniqueness package for that problem.

Problems 3.8.

1. Consider the function transformations $\{S_t\}$ associated with the wave equation

$$S_t \begin{bmatrix} f \\ g \end{bmatrix} = \begin{bmatrix} u(x,t) \\ v(x,t) \end{bmatrix}$$

where u, v are given in (3.41) and (3.42) above. Show that as long as the initial conditions have finite energy, s_t satisfies

$$S_{t+\tau} = S_t \cdot S_\tau.$$

2. For the wave equation considered above, prove that for finite energy initial conditions,

$$u(x,t) \to f(x),$$
$$v(x,t) \to g(x)$$

in the mean-square sense as $t \to 0$.

3. Consider the spring-mass analogue of the wave equation (Figure 3.9) with $x_0 = x_M = 0$ as boundary conditions. Define the total mechanical energy

$$E = \sum_{j=1}^{M} \frac{1}{2} k (x_j - x_{j-1})^2 + \frac{1}{2} M \left(\frac{dx_j}{dt} \right)^2$$

and show (by differentiation, to cite one way) that E is independent of time.

4. a. By analogy with Problem 3 above, argue that

$$E = \frac{1}{2} \int_0^L T \left(\frac{\partial u}{\partial x} \right)^2 + \rho \left(\frac{\partial u}{\partial t} \right)^2 dx$$

   is the world's only candidate for the energy in a vibrating string.

   b. Using equations (3.41) and (3.42) above and Parseval's Theorem, show that the energy in the string considered above is finite provided that f and g are finite energy initial conditions.

   c. Show that the string energy is in fact constant, independent of time, and given by

$$\frac{1}{2} \int_0^L T \left(\frac{\partial f}{\partial x} \right)^2 + \rho (g(x))^2 \, dx.$$

5. Provide the formal proof that the solution for Laplace's equation in a circle

$$u(r, \theta) = \sum_{n=-\infty}^{\infty} c_n r^{|n|} e^{in\theta}$$

converges in mean-square to the boundary function f as $r \to 1^-$.

6. For $|r| < 1$, the solution formula of (3.43) may be given an interpretation as an inner product by use of the product form of Parseval's Theorem. Express the solution in the form

$$u(r, \theta) = \frac{1}{2\pi} \int_0^{2\pi} f(\phi) G(r, \theta, \phi) \, d\phi$$

by explicitly finding the function whose Fourier coefficients are

$$\{r^{|n|} e^{in\theta}\}$$

and expressing the inner product as an integral. This is an example of a "Green's function" solution of the boundary value problem.

7. The solution of Laplace's equation in a unit square with one prescribed nonzero boundary value function is

$$u(x, y) = \sum_{n=1}^{\infty} b_n \sin(n\pi y) \frac{\sinh(n\pi x)}{\sinh(n\pi)},$$

$$b_n = 2 \int_0^1 f(y) \sin(n\pi y) \, dy.$$

8. a. Show that the above is pointwise convergent as long as $0 < x < 1$, and f is square-integrable.

   b. Show that the twice differentiated series is uniformly convergent for $0 \le x \le 1 - \delta$, $\delta > 0$, and hence that u above satisfies

$$\frac{\partial^2 u}{\partial x^2} + \frac{\partial^2 u}{\partial y^2} = 0,$$

   on the inside of $0 < x < 1, 0 < y < 1$.

9. For the solution u above, show that $u(x, \cdot)$ approaches f in mean-square as $x \to 1^-$.

10. Prove an analogue of the three-dimensional uniqueness theorem for the Laplace equation valid for the two-dimensional unit circle problem.

11. Prove a continuous-dependence result for the heat equation (standard version). To be precise, show that for fixed $t > 0$, and given $\epsilon > 0$, there exists a $\delta > 0$ such that (uniformly in x)

$$|u_f(x, t) - u_g(x, t)| < \epsilon$$

provided that

$$\int_0^L |f(x) - g(x)|^2 \, dx < \delta.$$

Here u_f and u_g represent the solutions for initial conditions f and g of the standard zero boundary condition heat equation.

12. The energy conservation result of Problem 4 above is derived by use of Parseval's relation on an explicit form of a solution of the governing equation.

a. Use the wave equation

$$\frac{1}{c^2} \frac{\partial^2 u}{\partial t^2} = \frac{\partial^2 u}{\partial x^2},$$

$$u(0, t) = u(L, t) = 0$$

to show that (assuming that differentiation under the integral sign and some integration by parts can be justified)

$$\frac{\partial}{\partial t} E = \frac{1}{2} \frac{\partial}{\partial t} \int_0^L T \left(\frac{\partial u}{\partial x} \right)^2 + \rho \left(\frac{\partial u}{\partial t} \right)^2 dx = 0,$$

so that E is constant.

b. Use the result of Problem 12a to prove that solutions of the wave equation which have continuous second partial derivatives are unique.

13. Show that a solution of the heat equation

$$\frac{\partial u}{\partial t} = \kappa \frac{\partial^2 u}{\partial x^2},$$

$$u(0, t) = u(L, t) = 0,$$

which has one continuous partial derivative with respect to time, and two continuous partial derivatives with respect to space has the property that

$$\frac{\partial}{\partial t} \int_0^L |u(x, t)|^2 \, dx \le 0.$$

Use this to conclude that solutions with this degree of smoothness are unique.

Further Reading

R. Aris, *Vectors, Tensors, and the Basic Equations of Fluid Mechanics* (Prentice-Hall, Englewood Cliffs, 1962)

W.E. Boyce, R.C. DiPrima, *Elementary Differential Equations and Boundary Value Problems*, 3rd edn. (Wiley, New York, 1977)

H.S. Carslaw, J.C. Jaeger, *Conduction of Heat in Solids*, 2nd edn. (Oxford University Press, London, 1986)

R.V. Churchill, *Fourier Series and Boundary Value Problems*, 2nd edn. (McGraw-Hill, New York, 1963)

J.H. Davis, *Differential Equations with Maple* (Birkhäuser Boston, Boston, 2000)

E.C. Jordan, K.G. Balmain, *Electromagnetic Waves and Radiating Systems*, 2nd edn. (Prentice-Hall, Englewood Cliffs, 1968)

H. Lewy, An example of a smooth linear partial differential equation without solutions. Ann. Math. **66**, 155–158 (1957)

C.C. Lin, L.A. Segel, *Mathematics Applied to Deterministic Problems in the Natural Sciences* (MacMillan, New York, 1974)

S.V. Marshall, G.G. Skitek, *Electromagnetic Concepts and Applications* (Prentice-Hall, Englewood Cliffs, 1990)

L.M. Milne-Thompson, *Theoretical Hydrodynamics*, 4th edn. (MacMillan, London, 1962)

J.A. Sommerfeld, *Mechanics of Deformable Bodies* (Academic, New York, 1964)

J.A. Sommerfeld, *Partial Differential Equations in Physics* (Academic, New York, 1964)

Chapter 4
Sturm–Liouville Theory and Boundary Value Problems

4.1 Further Boundary Value Problems

The boundary value problems of the previous chapter are distinguished by the fact that it is possible to derive more or less explicit solutions for the cases considered. A partial exception to this is the problem of the wave equation with an elastic constraint, for which the solution breaks the standard Fourier series mold.

The explicit analytical tractability encountered in the problems of the previous chapter is more the exception than the rule among the general run of boundary value problems. While explicit calculations (in terms of elementary functions) are not possible in the general case, it is true that the essence of the standard techniques employed above carries over intact to boundary value problems of fairly general type.

In the rest of this section we discuss some examples of these more general boundary value problems. The following sections consider the theoretical basis for the success of the separation of variables methods. The chapter concludes with some solution methods for the more general problems discussed below.

One obvious limitation of the models considered so far is that they were all based on an assumption of uniformity of material property (conductivity, specific heat, density, permittivity, etc.) in the region under consideration. If this hypothesis is dropped, the resulting model equations are of course modified.

A heat equation model for the "three-dimensional, variable medium" case may be produced as follows.

We consider heat conduction in a nonuniform medium, characterized by density, specific heat, and conductivity functions $\rho(\cdot,\cdot,\cdot)$, $c(\cdot,\cdot,\cdot)$ and $\alpha(\cdot,\cdot,\cdot)$. Retaining Newton's law of cooling (heat flows downhill) as the governing physical law, we extract a (smooth) volume of material V from the bulk of the solid under consideration. See Figure 4.1.

© Springer International Publishing Switzerland 2016
J.H. Davis, *Methods of Applied Mathematics with a Software Overview*,
Applied and Numerical Harmonic Analysis, DOI 10.1007/978-3-319-43370-7_4

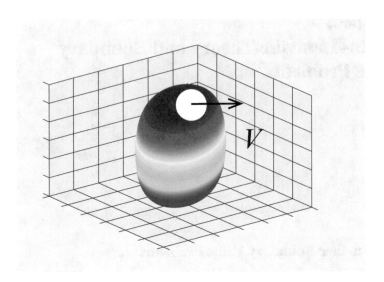

Fig. 4.1 Newton's law of cooling

Equating rate of change of heat content in V to the net heat flux across the boundary of V, ∂V, we obtain

$$\frac{\partial}{\partial t}\int\int\int_V \rho(x,y,z)c(x,y,z)u(x,y,z,t)\,dV = \int\int_S \alpha(x,y,z)\,(\nabla u)\cdot \mathbf{n}\,dS.$$

Assuming suitable differentiability, we convert the surface integral by use of the divergence theorem (Gauss's Theorem) to obtain

$$\int\int\int_V \left\{\frac{\partial}{\partial t}(\rho c u) - \nabla\cdot(\alpha\,\nabla u)\right\}\,dV = 0.$$

Since V is arbitrary, the governing equation is

$$\frac{\partial}{\partial t}(\rho c u) - \nabla\cdot(\alpha\,\nabla u) = 0,$$

or

$$\frac{\partial}{\partial t}(u) = \frac{1}{\rho c(x,y,x)}\nabla\cdot(\alpha\,\nabla u).$$

Even in the one-dimensional case this differs from the previous result in that

$$\frac{1}{\rho c(x)}\frac{\partial}{\partial x}\left(\alpha(x)\frac{\partial}{\partial x}u\right)$$

evidently consists of a combination of first and second derivatives with spatially variable coefficients.

The one-dimensional boundary conditions are easily modified to the three-dimensional case. If the region containing the body is denoted by Γ, with (smooth) boundary Σ, then a complete specification of the boundary value problem is

$$\frac{\partial}{\partial t}(u) = \frac{1}{\rho c(x, y, x)} \nabla \cdot (\alpha \nabla u) \text{ in } \Gamma, \tag{4.1}$$

$$a(\sigma)u(\sigma) + b(\sigma)\frac{\partial u}{\partial n}(\sigma) = 0 \text{ on } \Sigma,$$

$$u(x, y, z, 0) = u_0(x, y, z).$$

In case $a\,b \neq 0$ in the above, the interpretation is a linear heat transfer boundary condition. If one or the other of the coefficients vanishes, the interpretation is of an insulated or ice-bound boundary region.

A variable coefficient analog of Laplace's equation is obtained from the above by seeking steady state (time independent) solutions of (4.1), in the presence of heat sources on the boundary. The model reduces to

$$\nabla \cdot (\nabla u) \text{ in } \Gamma,$$

with boundary conditions

$$a(\sigma)u(\sigma) + b(\sigma)\frac{\partial u}{\partial n}(\sigma) = f(\sigma) \text{ on } \Sigma.$$

A wave equation involving two spatial coordinates arises from an analysis of the vibration of an elastic membrane (more prosaically referred to as a drum).

The derivation may be approached through the construction of a discrete analogue. Consider a rectangular array of point masses, interconnected by (ideal linear) springs. See Figure 4.2.

This network is stretched (to tension the springs) and the boundaries are attached to a rigid framework in Figure 4.2. The masses are naturally labeled by integer

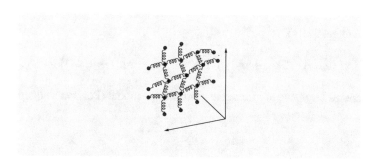

Fig. 4.2 Spring mass network

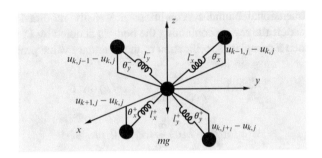

Fig. 4.3 Force balance in the elastic network

coordinates (k, j), and the condition for equilibrium may be obtained by extracting the "typical" mass from the array (Figure 4.3). The vertical force balance is (assuming uniform spring constants)

$$k \left(l_x^+ \sin(\theta_x^+) - l_x^- \sin(\theta_x^-) + l_y^+ \sin(\theta_y^+) - l_y^- \sin(\theta_y^-) \right) = mg.$$

By elementary trigonometry, this reduces to (u_{kj} denotes vertical displacement)

$$k \left((u_{k+1j} - u_{kj}) + (u_{k-1j} - u_{kj}) + (u_{kj+1} - u_{kj}) - (u_{kj-1} - u_{kj}) \right) = mg$$

or

$$\left((u_{k+1j} - 2\,u_{kj} + u_{k-1j}) + (u_{kj+1} - 2\,u_{kj} - u_{kj-1}) \right) = +\frac{mg}{k}.$$

This result is of course just a discrete form of Poisson's equation.

To make the transformation to an elastic membrane, we consider subdividing the net constructed above, in such a way that the total mass involved (as well as the spring constant) remain fixed throughout. If there are N particles in such a net, with M total mass, then the only change required in the above is to replace m by M/N (and allow (k, j) to range appropriately). The equation then takes the form

$$\left((u_{k+1j} - 2\,u_{kj} + u_{k-1j}) + (u_{kj+1} - 2\,u_{kj} - u_{kj-1}) \right) = +\frac{M}{N^2}\frac{g}{k}$$

or (more suggestively)

$$\left(N^2\,(u_{k+1j} - 2\,u_{kj} + u_{k-1j}) + N^2\,(u_{kj+1} - 2\,u_{kj} - u_{kj-1}) \right) = Mg.$$

As N becomes large, we are forced to consider the model (Poisson's equation)

$$\frac{\partial^2 u}{\partial x^2} + \frac{\partial^2 u}{\partial y^2} = \frac{M}{k}g$$

as a limiting form of the discrete model, and to refer to the evanescent elastic net described by the resulting partial differential equation as an elastic membrane.

The above is a derivation of an equilibrium equation. A wave equation is obtained by considering the equations of motion of the net constructed above. The simplest route to the desired equation is to declare Figure 4.2 to be in motion, and recall that an "inertial force" must be incorporated in the force balance equation for such a situation. The governing equation is hence

$$k\left(u_{k+1j} - 2u_{kj} + u_{k-1j}\right) + k\left(u_{kj+1} - 2u_{kj} + u_{kj-1}\right) = mg + m\frac{d^2}{dt^2}u_{kj}.$$

Proceeding as above, we obtain

$$\frac{\partial^2 u}{\partial t^2} = \frac{k}{M}\nabla^2 u + g$$

as the equation of evolution for a vibrating elastic membrane. The partial differential equation of course must be supplemented by boundary and initial conditions to give a complete problem statement.

This equation may also be derived on the basis of an assumption about the stress distribution in such a membrane. This approach is described in the problems below.

If the method of separation of variables is applied to the heat conduction problem (4.1) above, we assume

$$u = XT$$

and obtain (assuming uniform material, so that α, ρ, and c are independent of the time and spatial coordinates)

$$\frac{T'}{T}(\frac{\alpha}{\rho c}) + \frac{X''}{X} = -\lambda^2.$$

If the problem under consideration is, say, heat conduction in a pipe, then the introduction of cylindrical coordinates is appropriate (so that the physical boundary conditions translate readily to conditions in terms of the separated solutions). The Laplacian in terms of rectangular polar coordinates is given by

$$\nabla^2 X = \frac{\partial^2 X}{\partial r^2} + \frac{1}{r}\frac{\partial X}{\partial r} + \frac{1}{r^2}\frac{\partial^2 X}{\partial \theta^2} + \frac{\partial^2 X}{\partial z^2}.$$

For problems in which symmetry in the θ and z directions is appropriate, the separated equation involves only the radial coordinate and takes the form ($X(r, \theta, z) = R(r)$),

$$\frac{d^2 R}{dr^2} + \frac{1}{r}\frac{dR}{dr} + \lambda^2 R(r) = 0. \tag{4.2}$$

Appropriate boundary conditions for constant temperature boundary conditions are

$$R(a) = R(b) = 0.$$

(This assumes the removal of a steady state solution associated with different temperatures on the interior and exterior circumferences of the "pipe" under consideration.)

The equation (4.2) is an instance of *Bessel's equation*; more precisely, it is an instance of Bessel's equation of order zero.

Bessel's equation is a constant feature of boundary value problems with a cylindrical geometry. The cases of certain wave and Laplace's equation models are treated in the problems below.

To complete the simple problem of radial heat conduction in a pipe, we consider the Bessel equation (4.2).

If the pattern of the simple boundary value problems of the previous chapter is to be repeated, we hope to discover that this equation has solutions only for certain values of the separation constant, say for

$$\{\lambda_n^2\}_{n=1}^\infty$$

(assuming only a countable number of such λ_n^2 appear) with corresponding solutions $R_n(\cdot)$:

$$R_n'' + \frac{1}{r} R_n' + \lambda_n^2 R_n = 0, \tag{4.3}$$

$$R_n(a) = R_n(b) = 0.$$

Corresponding to such radial solutions (assuming they exist) would be the corresponding time dependence

$$T_n'(t) = -\left(\frac{\alpha}{\rho c}\right) \lambda_n^2 T_n$$

$$= -\kappa \lambda_n^2 T_n,$$

so that the typical separated solution would take the form

$$u_n(r, t) = b_n R_n(r) e^{-\kappa \lambda_n^2 t}.$$

This leads to the guess

$$u(r, t) = \sum_{n=1}^\infty b_n R_n(r) e^{-\kappa \lambda_n^2 t}.$$

as solution. If the radial temperature distribution at the initial time $t = 0$ is prescribed as say f, we evidently must determine $\{b_n\}$ from

$$f(r) = u(r, 0) = \sum_{n=1}^{\infty} b_n R_n(r).$$

This requirement is evidently an expansion problem analogous to the Fourier expansions of the previous chapter. The functions involved, however, are generally not trigonometric in nature, so that the possibility of such expansion does not follow directly from our previous results about Fourier series. In addition to this final issue, there is also the problem of the existence of the separation constants and solutions (hoped for) in the separated spatial equation above.

Since we have taken the effort to outline this scenario, it must be the case that these problems have a satisfactory resolution. In fact, the problems raised can be treated in fairly wide generality, which includes the Bessel equation of order zero as a special case. The general classification of these issues is the theory of Sturm–Liouville problems. We treat this topic in the following sections.

Problems 4.1.

1. Suppose that a drum (of conventional shape) is modeled by a circular elastic membrane

$$\frac{1}{c^2} \frac{\partial^2 u}{\partial t^2} = \frac{\partial^2 u}{\partial r^2} + \frac{1}{r} \frac{\partial u}{\partial r} + \frac{1}{r^2} \frac{\partial^2 u}{\partial \theta^2},$$

$$u(b, \theta) = 0, \text{ for } 0 < \theta < 2\pi,$$

$$\frac{\partial u}{\partial r}(0, \theta) = 0 \text{ for } 0 < \theta < 2\pi.$$

Assuming that the solution sought is independent of the angular variable θ, show that separation of variables leads to consideration of Bessel's equation of order zero.

2. Suppose that an elastic membrane is defined as an ideal object which has, first, a mass (surface) density of ρ, and, second, a force characteristic describable as a "two-dimensional tension." That is, an (infinitesimal) element of the membrane is subject to a net force from the remaining membrane equivalent to a tangential force of T per unit length of the perimeter, acting normal to the perimeter. (See Figure 4.4)

   Construct a force balance for such an element and hence deduce (for small displacements) the governing equation

$$\frac{\partial^2 u}{\partial t^2} = c^2 \nabla^2 u$$

in the absence of a gravitational force.

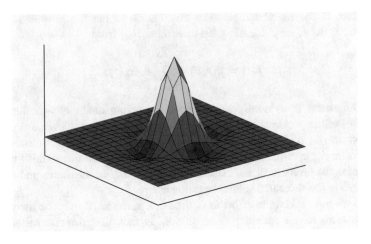

Fig. 4.4 Elastic membrane

3. Show that with a change of variables, the zero-order Bessel equation

$$R'' + \frac{1}{r}R' + \lambda^2 R = 0,$$

$$R(a) = R(b) = 0$$

can be converted to

$$\frac{d^2X}{dx^2} + \frac{1}{x}\frac{dX}{dx} + X = 0,$$

$$x(\lambda a) = x(\lambda b) = 0.$$

(This saves "dragging the λ" through a solution of Bessel's equation.)

4.2   Self-adjoint Eigenvalue Problems

From the discussions of the previous section, it is clear that second-order partial differential expressions appear regularly in the governing differential equations of the classical boundary value problems. If the method of separation of variables succeeds for such a problem, the appearance of a second-order ordinary differential equation of the form

$$X''(x) + P_1(x)X'(x) + [P_2(x) + \lambda^2 P_3(x)]X(x) = 0 \qquad (4.4)$$

is an inevitable consequence. The typical associated boundary conditions are of the form

$$\alpha X(a) + \beta X'(a) = 0, \tag{4.5}$$

$$\gamma X(b) + \delta X'(b) = 0.$$

What is required of (4.4) and (4.5) is often the information that solutions exist only for certain values of the separation parameter λ^2, and further that the associated solutions are applicable in certain expansion problems (arising from initial or boundary conditions in the original boundary value problem).

In the case of the elementary boundary value problems of the previous chapter, these questions are resolved in an atmosphere of comfortable familiarity, due largely to the exalted position conventionally accorded to the trigonometric functions. For more complicated (realistic, interesting, challenging,...) problems, the separated equation (4.4) has variable coefficients, and solutions in terms of elementary functions are not available. For this reason, discussion of (4.4) must proceed on its own terms.

The properties of (4.4) may be understood and explained on the basis of a deep connection with certain problems of linear algebra.

The relevant topic is the analysis of self-adjoint matrices. Recall that the adjoint of a linear mapping L in a finite-dimensional inner product space is defined to be the mapping L^* such that

$$(L\mathbf{u}, \mathbf{v}) = (\mathbf{u}, L^*\mathbf{v})$$

for all vectors $\mathbf{u}, \mathbf{v}$ in the inner product space. In terms of matrices computed with respect to an orthonormal basis, the matrix of L^* is simply the complex conjugate transpose of that of L.

Of special interest are those linear mappings which are their own adjoint, so that

$$L = L^*$$

or

$$(L\mathbf{u}, \mathbf{v}) = (\mathbf{u}, L^*\mathbf{v}) = (\mathbf{u}, L\mathbf{v})$$

for all vectors $\mathbf{u}, \mathbf{v}$. Such mappings turn out to have remarkable properties. We review these properties in the case of finite-dimensional vector spaces (for which the results can be stated in terms of matrices) below.

Theorem. *The eigenvalues of a self-adjoint matrix are real.*

Proof. Let λ be such an eigenvalue of the self-adjoint matrix L, and $\mathbf{u}$, $\|\mathbf{u}\| = 1$, a corresponding eigenvector. Then

$$(L\mathbf{u}, \mathbf{u}) = (\lambda\mathbf{u}, \mathbf{u}) = \lambda(\mathbf{u}, \mathbf{u}) = \lambda,$$

while also

$$(L\mathbf{u}, \mathbf{u}) = (\mathbf{u}, L^*\mathbf{u}) = (\mathbf{u}, L\mathbf{u}) = (\mathbf{u}, \lambda\mathbf{u}) = \overline{\lambda}(\mathbf{u}, \mathbf{u}) = \overline{\lambda}.$$

Hence $\lambda = \overline{\lambda}$, and λ is real.

Theorem. *The eigenvectors u, v associated with distinct eigenvalues λ, μ are orthogonal.*

Proof.

$$\lambda(\mathbf{u}, \mathbf{v}) = (\lambda\,\mathbf{u}, \mathbf{v}) = (L\mathbf{u}, \mathbf{v}) = (\mathbf{u}, L\mathbf{v}) = (\mathbf{u}, \mu\,\mathbf{v}) = \overline{\mu}(\mathbf{u}, \mathbf{v}) = \mu\,(\mathbf{u}, \mathbf{v})$$

so that

$$(\lambda - \mu)\,(\mathbf{u}, \mathbf{v}) = 0.$$

Since $\lambda \neq \mu$, $(\mathbf{u}, \mathbf{v}) = 0$, and so $\mathbf{u}$ and $\mathbf{v}$ are orthogonal.

Theorem. *There exists an orthonormal basis for the inner product space V consisting of eigenvectors of the self-adjoint matrix L. If $\mathbf{u}\mathbf{u}^*$ denotes the linear mapping defined by (for each eigenvector $\mathbf{u}$)*

$$\mathbf{u}\mathbf{u}^*(\mathbf{x}) = \mathbf{u}\,(\mathbf{x}, \mathbf{u}),$$

then each $\mathbf{x} \in U$ has a unique expansion

$$\mathbf{x} = \left(\sum_{i=i}^{\dim U} \mathbf{u}_i\,\mathbf{u}_i^* \right)(\mathbf{x})$$

$$= \sum_{i=i}^{\dim U} \mathbf{u}_i\,(\mathbf{x}, \mathbf{u}_i),$$

while L is representable (in the diagonalized form)

$$L = \left(\sum_{i=i}^{\dim U} \lambda_i\,\mathbf{u}_i\,\mathbf{u}_i^* \right)$$

with $\{\lambda_i\}$ a nonincreasing sequence.

Proof Outline. We outline a proof which has some use in the case of eigenvalue problems for differential equations. Consider the quadratic form maximization problem

$$P_0 : \underset{\|\mathbf{u}\|=1}{\mathrm{Max}}\,(L\mathbf{u}, \mathbf{u}).$$

The indicated maximum actually exists and is attained for some vector $\mathbf{u}_1$, $\|\mathbf{u}\| = 1$. If the maximum value is λ_1, then $\mathbf{u}_1$ is actually an eigenvector of L associated with λ_1;

$$L\mathbf{u}_1 = \lambda_1\,\mathbf{u}_1.$$

We now form the matrix

$$L1 = L - \lambda_1\,\mathbf{u}_1\,\mathbf{u}_1^*.$$

If $L1 \neq 0$, we consider the maximization problem

$$P_1: \quad \operatorname*{Max}_{\|\mathbf{u}\|=1,\,(\mathbf{u},\mathbf{u}_1)=0} (L_1\mathbf{u},\mathbf{u}).$$

Again the indicated maximizing vector exists, and is an eigenvector of L_1, associated with an eigenvalue named λ_2. However,

$$L1\,\mathbf{u}_2 = L\mathbf{u}_2 - \lambda_1\mathbf{u}_1(\mathbf{u}_1^*\mathbf{u}_2)$$

$$= L\mathbf{u}_2 - 0$$

$$= \lambda_2\,\mathbf{u}_2,$$

(since $\mathbf{u}_2$ is an eigenvector of L_1) so that $\mathbf{u}_2$ is also an eigenvector of L itself. Moreover, by construction of problem P_1 the maximum value satisfies the inequality

$$\Lambda_2 \le \lambda_1.$$

This procedure continues by induction until dim U orthogonal vectors are obtained, at which time the linear mapping

$$L - \lambda_1\,\mathbf{u}_1\,\mathbf{u}_1^* - \cdots - \lambda_{\dim U}\,\mathbf{u}_{\dim U}\,\mathbf{u}_{\dim U}^*$$

vanishes identically. The claimed result is then established.

These results as stated apply to linear mappings in a finite-dimensional vector space, while vector spaces of functions are involved in the analysis of boundary value problems. However, the intuition that something related to the self-adjoint matrix results must be involved can be generated by reconsideration of the standard (zero boundary) heat equation problem. When variables were separated in this problem, we were led to the equation

$$X'' = -\lambda^2 X \tag{4.6}$$

with the boundary conditions $x(0) = x(1) = 0$. We found

1. an infinite sequence of real numbers $\lambda_n^2 = (\frac{n\pi}{l})^2$ and functions

$$X_n(x) = \sin(\frac{n\pi x}{l})$$

satisfying the equation and boundary conditions

$$X_n''(x) = -\lambda_n^2 X_n(x) = -(\frac{n\pi}{l})^2 X_n(x)$$

$$X_n(0) = X_n(l) = 0;$$

2. that the eigenfunctions X_n and X_m associated with different values of λ^2 (i.e., $\lambda_n^2 = (\frac{n\pi}{l})^2$ and $\lambda_m^2 = (\frac{m\pi}{l})^2$) could be chosen to be real and were orthogonal:

$$\frac{2}{l} \int_0^l X_n(x) X_m(x)\, dx = \begin{cases} 0 \; m \neq n, \\ 1 \; m = n, \end{cases}$$

and finally,

3. that an arbitrary square-integrable function $f \in L^2[0, l]$ could be expanded in a mean-square convergent series

$$f(x) = \sum_{n=1}^{\infty} (f, X_n) X_n(X) = \sum_{n=1}^{\infty} b_n \sin(\frac{n\pi x}{l}).$$

We now claim that the properties (4.4) and (4.5) above can be seen to be inevitable through the device of casting the equation (4.6) as an eigenvalue problem, and duplicating the largely formal manipulations employed in the matrix theorems above. The property of completeness of the eigenfunctions, relying as it does on the Riesz–Fisher theorem, involves some hard analysis and is not available on the cheap.

In order to carry this out, we must discover the appropriate analogue of the self-adjoint matrix L above. If the expression

$$X'' = -\lambda^2 X$$

is to represent an eigenvalue equation (with eigenvalue $-\lambda^2$), then evidently L in some sense corresponds to the second derivative expression $\frac{d^2}{dx^2}$. Since the eigenfunctions obtained are evidently orthogonal with respect to the standard $L^2[0, l]$ inner product

$$(f_1(\cdot), f_2(\cdot)) = \frac{2}{l} \int_0^l f_1(x) \bar{f}_2(x)\, dx,$$

the required vector space V must be identified with (the Hilbert Space) $L^2[0, l]$.

The explicit definition of the appropriate linear mapping requires some care. In the first place, it must involve the boundary conditions of (4.6), since these enter in the eigenvalue determination in an essential way. The second point is that the answer cannot be "simply $\frac{d^2}{dx^2}$" since an arbitrary $L^2[0, l]$ function need not have two derivatives. Even if two derivatives exist in some appropriate sense, the second derivative may fail to be square-integrable, which would inhibit the formation of inner products with a clear conscience. These problems are solved essentially by refusing to apply $\frac{d^2}{dx^2}$ to any functions for which these problems arise.

The device which accomplishes this is to define a subset of $L^2[0, l]$ by

$$\mathcal{D} = \left\{ f \mid f \in L^2[0, l], \ (f, f' \text{ absolutely continuous }), \right.$$

$$\left. \frac{2}{l} \int_0^l |f''(x)|^2 \, dx < \infty, \text{ and satisfying } f(0) = f(l) = 0 \right\}.$$

The linear mapping L is defined by

$$Lf = \frac{d^2}{dx^2} f, \qquad f \in \mathcal{D}.$$

$\mathcal{D}$ is actually a subspace of the inner product (Hilbert) space $L^2[0, l]$. This follows since linear combinations of elements of functions in $\mathcal{D}$ still satisfy the homogeneous boundary conditions, and retain the required differentiability and integrability properties.

The question of what L does to the rest of $L^2[0, l]$ does not arise. L is simply undefined outside of $\mathcal{D}$ (usually called the domain of L).

Definition. The linear mapping L, defined on a domain $\mathcal{D}$, a subspace of an inner product space V, is called *formally self-adjoint*[1] if the condition

$$(L\mathbf{u}, \mathbf{v}) = (\mathbf{u}, L\mathbf{v})$$

holds for all $u, v \in \mathcal{D}$.

Example. The linear mapping L defined above is formally self-adjoint. To verify this, take u, v in $\mathcal{D}$, and form

$$(L\mathbf{u}, \mathbf{v}) = \frac{2}{l} \int_0^l \frac{d^2 u}{dx^2} \overline{v}(x) \, dx$$

(well defined since $u \in \mathcal{D}$ guarantees square-integrability of the second derivative).

---

[1]This definition suffices to obtain the desired results and avoids the technicalities inherent in the rigorous definition of "adjoints" for mappings of the sort constructed above.

Integrating by parts, we obtain

$$
\begin{aligned}
(Lu, v) &= \frac{2}{l} \left\{ \bar{v}(x) \frac{du}{dx} \Big|_0^l - \int_0^l \frac{du}{dx} \frac{d\bar{v}}{dx} \, dx \right\} \\
&= \frac{2}{l} \left\{ 0 \frac{du}{dx}(l) - 0 \frac{du}{dx}(0) - \int_0^l \frac{du}{dx} \frac{d\bar{v}}{dx} \, dx \right\} \\
&= \frac{2}{l} \left\{ -u \frac{d\bar{v}}{dx} \Big|_0^l + \int_0^l u \frac{d^2\bar{v}}{dx^2} \, dx \right\} \\
&= \frac{2}{l} \left\{ -0 \frac{d\bar{v}}{dx}(l) + 0 \frac{d\bar{v}}{dx}(0) + \int_0^l u \frac{d^2\bar{v}}{dx^2} \, dx \right\} \\
&= \frac{2}{l} \left\{ \int_0^l u \frac{d^2\bar{v}}{dx^2} \, dx \right\} \\
&= (u, Lv)
\end{aligned}
$$

so that L is in fact formally self-adjoint.

Example. Separation of variables in the heat equation with insulated boundaries leads to the equation and boundary conditions

$$
X'' + \lambda^2 X = 0, \qquad 0 < x < l,
$$

$$
X'(0) = X'(1) = 0.
$$

In connection with this problem we construct a linear mapping $\hat{L}$ with domain

$$
\mathcal{D} = \left\{ f \, \middle| \, f \in L^2[0, l], \, (f, f' \text{ absolutely continuous }) \right.
$$

$$
\left. \frac{2}{l} \int_0^l |f''(x)|^2 \, dx < \infty, \text{ and satisfying } f'(0) = f'(l) = 0 \right\}.
$$

$\hat{L}$ is defined by

$$
\hat{L} f = \frac{d^2}{dx^2} f \text{ for } f \in \mathcal{D}.
$$

Taking the same inner product as the previous example, two quick integrations by parts show that

$$
(\hat{L} u, v) = (u, \hat{L} v).
$$

This mapping is also formally self-adjoint.

The notion of formal self-adjointness is sufficiently strong to recover statements analogous to the linear algebra theorems above.

Theorem. *Suppose that the formally self-adjoint linear mapping L defined on a domain $\mathcal{D}$ of an inner product space V has an eigenvector* **u** *(in $\mathcal{D}$) associated with eigenvalue λ. Then λ is real.*

Proof. Take $\|\mathbf{u}\| = 1$, and form

$$\lambda = \lambda(\mathbf{u}, \mathbf{u}) = (\lambda\mathbf{u}, \mathbf{u}) = (L\mathbf{u}, \mathbf{u}) = (\mathbf{u}, L\mathbf{u}) = (\mathbf{u}, \lambda\mathbf{u}) = \overline{\lambda}(\mathbf{u}, \mathbf{u}) = \overline{\lambda}$$

so that $\lambda = \overline{\lambda}$, and λ is real.

Theorem. *Let L be as in the previous theorem, and let* **u** *be an eigenvector of L associated with λ,* **v** *an eigenvector associated with μ. Then if $\lambda \neq \mu$,* **u** *is orthogonal to* **v**.

Proof. Just as in the matrix case,

$$\lambda(\mathbf{u}, \mathbf{v}) = (\lambda\mathbf{u}, \mathbf{v}) = (L\mathbf{u}, \mathbf{v}) = (\mathbf{u}, L\mathbf{v}) = (\mathbf{u}, \mu\mathbf{v}) = \overline{\mu}(\mathbf{u}, \mathbf{v}) = \mu(\mathbf{u}, \mathbf{v}).$$

Hence $(\lambda - \mu)(\mathbf{u}, \mathbf{v}) = 0$, and hence (since $\lambda - \mu \neq 0$) we have $(\mathbf{u}, \mathbf{v}) = 0$, making **u** and **v** orthogonal.

We remark that the above results assume that eigenvalues and eigenvectors exist for the L in question. The proof that such is the case requires something analogous to the eigenvalue existence argument of the matrix case above and is a nontrivial problem. There exist formally self-adjoint linear mappings which fail to have eigenvectors (examples arise naturally in quantum mechanics, for example). The problem of existence of eigenvalues and eigenvectors for the boundary value eigenvalue problems motivating this discussion is treated (gently) in the following section.

We recall that some simplifications in computation were obtained in previous sections through the judicious use of $-\lambda^2$ as the name of a separation constant. This can be understood on the basis of manipulations in the style of this section.

Example. Show that the eigenvalues associated with the zero boundary heat equation are strictly negative (so that only a masochist would call the separation constant other than $-\lambda^2$).

To show this (on the basis of no explicit computation) consider the linear mapping L constructed above for this problem. Let u, $\|u\| = 1$ be an eigenvector associated with eigenvalue μ. Then

$$Lu = \mu u$$

and (because u is a unit vector)

$$(Lu, u) = (\mu u, u) = \mu.$$

But (for this case)

$$(Lu, u) = \frac{2}{l} \int_0^l \frac{d^2u}{dx^2} \bar{u}(x)\, dx$$

$$= -\frac{2}{l} \int_0^l \|\frac{du}{dx}\|^2\, dx$$

(take $u = v$ in the example above, or recreate the required integration by parts). Hence

$$\mu = -\frac{2}{l} \int_0^l \|\frac{du}{dx}\|^2\, dx \le 0,$$

and μ is non-positive. In fact, $\mu \ne 0$. For if $\mu = 0$, then $\frac{du}{dx} = 0$, and since u satisfies the boundary conditions $u(0) = u(l) = 0$, this forces $u = 0$ in contradiction to $\|u\| = 1$. Hence $\mu < 0$, and the eigenvalues are strictly negative.

Problems 4.2.

1. Fill in the details required to complete the proof of the matrix theorem. Treat the constrained maximization problem by, for example, Lagrange multiplier methods.
2. Show that the linear mapping $\hat{L}$ associated with the insulated-end heat equation is formally self-adjoint. Explicitly note the points of the argument where the definition of the domain must be invoked to justify the existence of an integral or the vanishing of some term.
3. Consider the boundary value problem for heat conduction in a thin ring. Explicitly construct (including a precise domain description) a formally self-adjoint mapping $\hat{L}$ associated with this problem; verify formal self-adjointness for your construct.
4. Show that the eigenvalues associated with the doubly insulated-end heat equation are non-positive. Show also that the eigenvalues associated with the heat equation in the case of one zero and one insulated end are strictly negative.
5. Suppose that a function of two variables is defined by

$$K(t, s) = \sum_{i=1}^{N} f_i(t)g_i(s),$$

where $\{f_i\}$ and $\{g_i\}$ are (at least) square-integrable on the interval $[0, 1]$. Define a linear mapping $L : L^2[0, 1] \mapsto L^2[0, 1]$ by

$$(Lx)(t) = \int_0^1 K(t, s)\, x(s)\, ds.$$

What condition is required so that L is formally self-adjoint (with respect to the integral / product inner product)?

Show that L in fact has eigenvalues, and describe how to reduce the computation of these to the problem of eigenvalue computation for a matrix derivable from the kernel function K above.

6. Suppose that a formally self-adjoint linear mapping is constructed as in (5) above. Show that K may be expressed in the form

$$K(t,s) \sum_{i=1}^{N} \lambda_i \phi_i(t) \overline{\phi}_i(s)$$

where $\{\phi_i(t)\}$ is an orthonormal set of vectors in $L^2[0,1]$.

Hint: The tools of linear algebra are sufficient to deduce this from results of this section.

7. Suppose that $\{u_i\}_{i=1}^{N}$ are a set of orthonormal functions in $L^2[0,l]$. Show that the N^2 functions $\{v_{ij}(\cdot,\cdot)\}_{i,j=1}^{N}$ which are defined as

$$v_{ij}(t,s) = u_i(t)\, u_j(s)$$

are orthonormal in the inner product space $L^2[(0,l) \times (0,l)]$, with inner product

$$(f,g) = \int_0^1 \int_0^1 f(t,s)\overline{g}(t,s)\, dt ds.$$

Given a function of two variables $K(\cdot,\cdot)$ such that (it is smooth, or minimally)

$$\int_0^1 \int_0^1 |K(t,s)|^2 dt ds < \infty,$$

find a set of constants $\{\alpha_{ij}\}_{i,j=1}^{N}$ such that

$$\int_0^1 \int_0^1 \left[\left| K(t,s) - \sum_{i=1}^{N} \sum_{j=1}^{N} \alpha_{ij} v_{ij}(t,s) \right| \right]^2 dt ds$$

is as small as possible.

Problems 5, 6, and 7 find application in the computation of eigenvalues of differential equations and "integral operators" of the type in Problem 5 above.

8. A linear mapping L in a Hilbert space is called bounded if there exists a constant M such that

$$\|L\mathbf{u}\| < M\|\mathbf{u}\|.$$

Show (by producing examples of u exceeding any purported size bound M) that the mapping L associated with the zero boundary heat equation is not bounded.

9. Define a function $k(\cdot)$ such that

$$k(x) = \begin{cases} 0, & 0 \le x < \frac{1}{2}, \\ 1, & \frac{1}{2} \le x \le 1, \end{cases}$$

and a linear mapping L in $L^2[0, 1]$ by

$$(Lf)(x) = k(x)f(x).$$

Show that both 0 and 1 are eigenvalues of L, and produce at least 17 eigenvectors associated with each eigenvalue.

10. Let $k(x) = x$, $0 \le x \le 1$, and define L in $L^2[0, 1]$ by

$$(Lf)(x) = k(x)f(x) = xf(x).$$

Show that L is formally self-adjoint, but that L has no eigenvalues.

4.3 Sturm–Liouville Problems

A Sturm–Liouville problem may be described as a standard form, formally self-adjoint eigenvalue problem for a second-order ordinary differential equation. The standard form of such problems is given by

$$\frac{d}{dx}\left(p(x)\frac{dX}{dx}\right) + (q(x) + \lambda^2 r(x))X = 0, \ a < x < b, \qquad (4.7)$$

$$\alpha X(a) + \beta X'(a) = 0,$$

$$\gamma X(b) + \delta X'(b) = 0.$$

This is close to the typical form of the equations obtained in a separation of variables exercise. As noted in previous sections, this is

$$X'' + P_1(x)X' + P_2(x)X + \lambda^2 P_3(x)X = 0$$

(with boundary conditions as in (4.7)). This form is (superficially) different from (4.7), which is specifically constructed to facilitate procedures of integration by parts.

In fact the typical form equation can be made equivalent to the standard by use of the integrating factor device familiar from the theory of ordinary differential equations. It is evidently required to interpret the expression $X'' + P_1(x)X'$ as a total derivative (in order to identify it with $(pX')'$ of (4.7)). To do this we introduce an integrating factor

$$I(x) = \exp\left[\int_\alpha^x P_1(\eta)d\eta\right]$$

(the lower limit of the integral in this expression is essentially arbitrary, subject only to the constraint that the integral is convergent so that the integrating factor $I(\cdot)$ is well defined). This gives

$$I(x) X'' + I(x) P_1(x) X' + [(P_2(x) + \lambda^2 P_3(x)] I(x) X = 0$$

or

$$\frac{d}{dx}[I(x) X'] + (P_2(x) I(x) + \lambda^2 P_3(x) I(x)) X = 0.$$

This is exactly in the form of the standard differential equation of (4.7), provided that we identify the coefficients

$$p(x) = I(x),$$

$$q(x) = P_2(x) I(x),$$

$$r(x) = P_3(x) I(x).$$

Example. Bessel's equation of order n is given by

$$X'' + \frac{1}{x} X' + (\lambda^2 - \frac{n^2}{x^2}) X = 0.$$

To put this in standard form, choose

$$I(x) = \exp\left[\int_1^x \frac{1}{\eta} d\eta\right] = x.$$

Then

$$p(x) = x,$$

$$q(x) = -\frac{n^2}{x^2} x = -\frac{n^2}{x},$$

$$r(x) = 1 \cdot x = x,$$

and the standard form is

$$\frac{d}{dx}(x X') + (\lambda^2 - \frac{n^2}{x}) X = 0.$$

In the treatment of Sturm–Liouville problems, it is necessary to make some assumptions regarding the behavior of the coefficient functions $p(\cdot)$, $q(\cdot)$, and $r(\cdot)$ in (4.7).

We assume that $p(\cdot)$, $q(\cdot)$, and $r(\cdot)$ are real valued, $p(\cdot)$ and $r(\cdot)$ are continuous, and that $p(\cdot)$ is continuously differentiable. (These are much more restrictive assumptions than are required to obtain the desired results).

We also assume that the functions $p(\cdot)$ and $r(\cdot)$ vanish nowhere on the interval $[a, b]$.

The nonvanishing of $p(\cdot)$ is connected with the non-singularity of the governing differential equation (4.7). If $p(\cdot)$ vanishes somewhere, then (since $p(\cdot)$ is the coefficient of the highest-order term in the differential equation) special methods are required to analyze solutions of the equation (standard existence theorems apply in the contrary case). The assumption that $r(\cdot)$ has one sign allows definition of an inner product associated with the equation.

The case in which $p(\cdot)$ or $r(\cdot)$ vanishes is referred to as a "singular Sturm–Liouville equation." Bessel's equation is in fact singular at the origin. The singular case receives special treatment.

It turns out that the function $r(\cdot)$ serves as a weight function in the definition of an associated inner product. For this reason, it is conventional to assume that r (which is of single sign) is in fact positive on the interval in question. If this is not initially the case, the original equation (4.7) can be multiplied by (-1) to give an equation in which $r(x)$ is positive.

It is also conventional to arrange the equation so that the coefficient p is positive, in order to make unambiguous statements about eigenvalue locations. This is accomplished again by multiplying the equation by (-1), but this time absorbing the $-$ sign into the definition of the eigenvalue parameter λ^2. We write (if necessary)

$$\frac{d}{dx}\left((-p(x))\frac{dX}{dx}\right) + ((-q(x)\,X) + (-\lambda^2)\,r(x)\,X = 0,$$

or

$$\frac{d}{dx}\left((p_1(x))\frac{dX}{dx}\right) + ((q_1(x)\,X) + (\lambda^2)\,r_1(x)\,X = 0.$$

With these conventions of notation established, we are now prepared to treat the Sturm–Liouville problem in the style to which we have become accustomed in the previous sections. To make (4.7) look like an eigenvalue problem, we rewrite it in the form

$$\frac{1}{r(x)}\frac{d}{dx}\left(p(x)\frac{dX}{dx}\right) + \frac{q(x)}{r(x)}X(x) = -\lambda^2 X,$$

$$\alpha x(a) + \beta X'(a) = 0,$$

$$\gamma X(b) + \delta X'(b) = 0.$$

(We also assume $\alpha, \beta, \gamma, \delta$ all real, and $\alpha^2 + \beta^2 \neq 0$, $\gamma^2 + \delta^2 \neq 0$, so that some constraint appears at each boundary).

We must associate with this problem a linear mapping L with domain $\mathcal{D}$, in some inner product space V. Intuitively, $\mathcal{D}$ consists of functions satisfying the given boundary conditions, and sufficiently smooth that application of the indicated differentiations in the differential equation results in a function in V.

The vector space V is to be determined so that (the chosen L will be formally self-adjoint)

$$\left(\frac{1}{r}\{(pu')' + qu\}, v\right) = \left(u, \frac{1}{r}\{(pv')' + qv\}\right).$$

In view of this requirement we are led to define the inner product for V by

$$(u(\cdot), v(\cdot)) = \int_a^b u(x)\overline{v}(x)r(x)dx.$$

(Since $r(\cdot)$ is strictly positive, this does define an inner product. The resulting space may be interpreted as a Hilbert space in a manner analogous to the case $r(\cdot) = 1$).

The linear mapping L is defined with domain

$$\mathcal{D} = \left\{ f(\cdot) : \int_a^b |f(x)|^2 r(x)dx < \infty, (f(\cdot), f'(\cdot) \text{ absolutely continuous}),\right.$$

$$\int_a^b |f''(x)|^2 r(x)dx < \infty, \text{ and satisfying}$$

$$\left. \alpha f(a) + \beta f'(a) = 0, \gamma f(b) + \delta f'(b) = 0 \right\}.$$

L is defined by

$$(Lf)(x) = \frac{1}{r(x)} \left\{ \frac{d}{dx}\left(p(x)\frac{df}{dx}\right) + q(x)f(x) \right\}$$

for all $f(\cdot) \in \mathcal{D}$.

Now (with all of the work buried in the construction of L and $\mathcal{D}$), it is a simple matter of two integrations by parts to verify that L is formally self-adjoint. Take $u(\cdot), v(\cdot)$ in $\mathcal{D}$ and form

$$(Lu, v) = \int_a^b \frac{1}{r(x)}\{(pu')' + qu\}(x)\overline{v}(x)r(x)dx$$

$$= p(b)u'(b)\overline{(v)}(b) - p(a)u'(a)\overline{(v)}(a)$$

$$- p(b)\overline{(v)}'(b)u(b) + p(a)\overline{v}'(a)u(a)$$

$$+ \int_a^b u(x)\overline{\left[\frac{1}{r(x)}\{(pv')' + qv\}(x)\right]}r(x)dx.$$

Since functions in $\mathcal{D}$ satisfy the boundary conditions, we have

$$\alpha u(a) + \beta u'(a) = 0,$$

$$\gamma u(b) + \delta u'(b) = 0,$$

$$\alpha v(a) + \beta v'(a) = 0,$$

$$\gamma v(b) + \delta v'(b) = 0.$$

These equations are used to eliminate the "boundary terms" from the integrations by parts. Since $\alpha^2 + \beta^2 \neq 0$, $\gamma^2 + \delta^2 \neq 0$, to be specific, assume $\beta \neq 0$, $\delta \neq 0$. Then

$$u'(a) = -\frac{\alpha}{\beta} u(a),$$

$$u'(b) = -\frac{\gamma}{\delta} u(b),$$

and substitution of these above gives

$$(Lu, v) = \int_a^b u(x) \overline{\left[\frac{1}{r(x)} \{(pv')' + qv\}(x) \right] } r(x)\,dx,$$

$$= (u, Lv),$$

so that L is formally self-adjoint as predicted (in case $\beta \neq 0, \delta \neq 0$). Pursuit of the remaining permutations of nonvanishing coefficients shows that in fact

$$(Lu, v) = (u, Lv), \qquad u, v \in \mathcal{D}$$

as long as $\alpha^2 + \beta^2 \neq 0$, $\gamma^2 + \delta^2 \neq 0$, verifying formal self-adjointness for the general case.

From this computation we immediately conclude the following.

Theorem. *The eigenvalues of the Sturm–Liouville linear mapping L above (assuming they exist) are all real.*

Theorem. *The eigenvectors corresponding to distinct eigenvalues of the Sturm–Liouville mapping L (assuming they exist) are mutually orthogonal (with respect to the inner product which made L formally self-adjoint). That is, if*

$$Lu = \lambda u, \qquad u \in \mathcal{D},$$

$$Lv = \mu v, \qquad v \in \mathcal{D},$$

and $\lambda \neq \mu$, then

$$\int_a^b u(x)\bar{v}(x)\, r(x)\, dx = 0.$$

The proofs of these results are identical (literally) to those of the previous section.

Example. It is crucial to note that the inner product with respect to which the eigenvectors are orthogonal is that arising from the construction of L and the inner product space V.

That is, it is the "integral inner product" with the weighting factor $r(\cdot)$ arising from the standard form of the Sturm–Liouville equation which is appropriate.

Example. The Bessel equation of order zero is given by

$$X'' + \frac{1}{x}X' + \lambda^2 X = 0, \qquad a \leq x \leq b,$$

with boundary conditions of the usual sort, say for example, $X(a) = X(b) = 0$. The standard form is

$$(xX')' + \lambda^2 x X = 0,$$

from which we identify

$$r(x) = x,$$
$$p(x) = x.$$

The theorem then assures that if

$$(xu')' - \lambda x u = 0,$$
$$u(a) = u(b) = 0$$

and

$$(xv')' - \lambda x v = 0,$$
$$v(a) = v(b) = 0$$

then

$$\int_a^b u(x)\,\bar{v}(x)\, x\, dx = 0$$

provided that $\lambda \neq \mu$.

Although we are retaining the possibility of complex-valued solutions of the Sturm–Liouville problem, in fact the solution may be chosen to be real. To see this, note that since the coefficients of the differential equation and boundary conditions are real-valued, the complex conjugate of a solution to the problem is also a solution. That is,

$$(p\overline{X}')' + q\overline{X} + \lambda^2 r\overline{X} = 0,$$

$$\alpha\overline{X}(a) + \beta\overline{X}(a) = 0,$$

$$\gamma\overline{X}(b) + \delta\overline{X}(b) = 0.$$

The linearity and homogeneity of the governing equations then ensure that the (real valued) function

$$\tilde{X}(x) = X + \overline{X}$$

is also a solution. This combination is the desired real-valued solution.

The results above are the Sturm–Liouville version of the self-adjointness results of the previous section. They are essentially immediate consequences of the algebraic properties of the relevant definitions.

The Sturm–Liouville analogue of the trigonometric eigenfunction expansion result, in contrast, requires some substantial investment in analysis. A rigorous proof carries with it a proof of the completeness of the resulting eigenfunctions (such results are proved in [9]). Since we have declared our willingness to accept this result for the Fourier series expansion in previous sections, we are content to do the same for the analogous Sturm–Liouville Theorem.

One of the properties of the eigenvalue problems encountered in the previous sections is that the eigenvalues obtained are (at least predominantly) negative. This is a general property of Sturm–Liouville systems, which is easily seen in the case of special boundary conditions.

Consider the standard equation

$$(p\,u')' + q\,u = \lambda r\,u, \qquad a < x < b, \tag{4.8}$$

and assume that λ is an eigenvalue associated with the unit norm

$$\int_a^b r(x)|u(x)|^2 dx = 1$$

eigenvector $u(\cdot)$. Then multiplying the standard equation by $\overline{u}(\cdot)$, and integrating over $[a, b]$ gives

$$\int_a^b (p\,u')'\overline{u}dx + \int_a^b q\,|u|^2 dx = \lambda \int_a^b r\,|u|^2 dx.$$

Integrating the first term by parts,

$$(p\,u')\overline{(u)}|_a^b - \int_a^b p\,|u'|^2 dx + \int_a^b \frac{q}{r} r|u|^2 dx = \lambda.$$

Assuming u (or u') vanishes on the boundaries, the boundary term above vanishes and we have

$$\lambda = \int_a^b \frac{q}{r} r |u|^2 dx - \int_a^b p |u'|^2 dx.$$

Since $p(\cdot)$ is assumed positive, we obtain

$$\lambda \le \int_a^b \frac{q}{r} r |u|^2 dx \le \max(\frac{q}{r}) \int_a^b r |u|^2 dx = \max(\frac{q}{r}).$$

This elementary exercise shows that all eigenvalues are bounded above by some fixed constant. The derivation also illustrates our previous remark that assuming $p(\cdot) > 0$ amounts to establishing a sign convention for the eigenvalues.

The discussion above is based on a premise of the existence of eigenvalues. There are a variety of arguments leading to such a conclusion. An argument having close connections to the matrix case discussed earlier is based on the conversion of the original differential equation into an equivalent integral equation. This process is in general an engaging exercise in ordinary differential equations. We are content to illustrate to procedure for the "standard" prototype example, and briefly outline the general result.

Example. The problem is to show that the eigenvalue problem

$$X'' = \lambda X, \ X(0) = X(1) = 0$$

can be posed in the form of (an integral equation)

$$X(x) = -\lambda \int_0^1 K(x, \xi) x(\xi) \, d\xi, \ 0 < x < 1.$$

To do this, we integrate the governing equation once to obtain first

$$X'(x) - X'(0) = \lambda \int_0^x X(\xi) d\xi.$$

Integrate a second time to obtain

$$X(1) - X(x) = X'(0) \int_x^1 1 \, d\eta + \lambda \int_x^1 \left(\int_0^\eta X(\xi) d\xi \right) d\eta$$

$$-X(x) = X'(0)(1 - x) + \lambda \int_x^1 \left(\int_0^\eta X(\xi) d\xi \right) d\eta.$$

To meet the boundary condition $X(0) = 0$, we determine $X'(0)$ from

$$0 = X'(0) + \lambda \int_0^1 \left(\int_0^\eta X(\xi) d\xi \right) d\eta.$$

In order to obtain the required form, we interchange the order of integration of the last term in the integral equation. The relevant diagram is given in Figure 4.5.

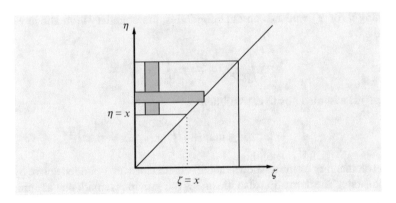

Fig. 4.5 Integration region

Performing the integration with respect to η first, two cases arise. This gives

$$\int_x^1 \left(\int_0^\eta X(\xi)d\xi \right) d\eta = \int_0^x \left(\int_x^1 X(\xi)d\eta \right) d\xi + \int_x^1 \left(\int_\xi^1 X(\xi)d\eta \right) d\xi$$

$$= \int_0^x (1-x)X(\xi)d\xi + \int_x^1 (1-\xi)X(\xi)d\xi.$$

Identifying the integral term in the boundary condition expression as an extreme case of the above,

$$X'(0) = -\lambda \int_0^1 (1-\xi)X(\xi)d\xi.$$

Using these results in the integrated form gives in turn

$$-X(x) = \lambda \int_0^1 -(1-x)(1-\xi)X(\xi)d\xi$$

$$+ \lambda \int_0^x (1-x)X(\xi)d\xi + \lambda \int_x^1 (1-\xi)X(\xi)d\xi$$

$$= \lambda \int_0^x \{-(1-x)(1-\xi) + (1-x)\}X(\xi)d\xi$$

$$+ \lambda \int_x^1 \{-(1-x)(1-\xi) + (1-\xi)\}X(\xi)d\xi$$

$$= \lambda \int_0^1 K(x,\xi)X(\xi)d\xi$$

where

$$K(x, \xi) = \begin{cases} \xi(1 - x), & \xi \leq x, \\ x(1 - \xi), & \xi \geq x. \end{cases}$$

Evidently λ is an eigenvalue of of the differential equation exactly in case

$$(KX)(x) = \int_0^1 K(x, \xi)X(\xi)d\xi = -\frac{1}{\lambda}X(x).$$

If we regard the left side of this expression as defining a linear mapping of $L^2[0, 1]$ into itself, then the equation is exactly an eigenvalue problem for the mapping K (with eigenvalue "$-1/\lambda$"). K evidently has as eigenvalues the negative reciprocals of those of the differential equation problem. Hence (since we explicitly found the differential equation eigenvalues and vectors earlier),

$$KX = \mu x$$

has solutions

$$\mu_n = \frac{1}{(n\pi)^2},$$

$$X_n = \sin n\pi x,$$

as one may verify directly.

A similar problem conversion may be carried out in the case of the general problem

$$(pX')' + qX = \lambda r X$$

(although the general procedure may require adding a sufficiently large multiple of X to both sides of the differential equation, and a redefinition of λ to assure that the eigenvalues are strictly negative). It is possible to show (under these conditions) that the Sturm–Liouville differential equation eigenvalue problem is equivalent to an integral equation

$$X(x) = -\lambda \int_a^b K(x, \xi) X(\xi) d\xi,$$

for a suitable "kernel function" $K(\cdot, \cdot)$.

The "kernel" function $K(\cdot, \cdot)$ may be regarded as the continuous analogue of a matrix. The analogy is extremely close, as the linear mapping K defined through the kernel has many properties in common with matrices. Rewriting the equation as

$$KX = -\frac{1}{\lambda}X = \mu X,$$

it should come as no surprise that K is formally self-adjoint with respect to the inner product with weight function $r(\cdot)$.

One route to eigenvalues in the matrix case is through the use of quadratic forms. The mapping K may be used to define a form in "infinitely many variables." Define

$$(Kf, f) = \int_a^b \bar{f}(x) \int_a^b K(x, \xi) f(\xi) \, d\xi \, dx,$$

and consider the maximization problem P_0:

$$\max (Kf, f) \text{ subject to } ||f||^2 = \int_a^b |f(x)|^2 r(x) dx = 1.$$

Beyond being formally self-adjoint, K (with the "eigenvalue shifting" mentioned above) is also positive definite and bounded in the sense that (for some M)

$$0 \le \int_a^b \bar{f}(x) \int_a^b K(x, \xi) f(\xi) \, d\xi \, dx \le M \int_a^b |f(x)|^2 \, r(x) \, dx.$$

It is possible to show next that a solution of this maximization problem exists. If μ_1 denotes the maximum value of this quadratic form, then μ_1 is an eigenvalue of K with eigenfunction $\phi(\cdot)$:

$$\int_a^b K(x, \xi) \phi_1(\xi) d\xi = \mu_1 \cdot \phi_1(x)$$

with

$$\int_a^b |\phi_1(x)|^2 r(x) dx = 1.$$

This establishes the existence of an eigenvalue for K; the other eigenvalues are derived by a procedure analogous to that of the matrix case. That is, we define a (smaller) kernel function by

$$K1(x, \xi) = K(x, \xi) - \mu_1 \phi_1(x) \bar{\phi}_1(\xi)$$

and consider maximizing the "quadratic form"

$$P_1 : \ max(K_1 f, f) \text{ subject to } ||f||^2 = 1, \ (f, \phi_1) = 0.$$

K_1 is also a positive form, and its maximum value is no larger than that associated with K. (In fact, in this particular case it is smaller. The argument for this relies on uniqueness of the corresponding differential equation solution.) Since $K_1(\cdot, \cdot)$ is a kernel of the same sort as $K(\cdot, \cdot)$ it is possible to deduce again existence of an eigenvalue and eigenvector for K_1 (and in turn for K, as in the matrix case). Continuing inductively, we deduce in this fashion the existence of an infinite number of positive eigenvalues

$$\mu_1 > \mu_2 > \mu_3 > \mu_4 > \cdots$$

and corresponding orthonormal eigenvectors $\{\phi_n(\cdot)\}$. With some further effort, it may be shown that

1. $\mu_n \to 0$,
2. $\{\phi_n(\cdot)\}$ are complete,
3. $K(x, \xi) = \sum_{n=1}^{\infty} \mu_n \phi_n(x) \bar{\phi}_n(\xi)$.

Referring these results back to the original Sturm–Liouville differential equation (with appropriate formally self-adjoint boundary conditions) we then conclude the existence of $\{\lambda_n\}(\lambda_n = -\frac{1}{\mu_n})$,

$$\lambda_1 > \lambda_2 > \lambda_3 > \lambda_4 > \cdots, \ \lambda_n \to -\infty,$$

and a corresponding complete orthonormal (with respect to the weighted inner product) set of eigenfunctions $\{X_n(\cdot)\}$ $(X_n(\cdot) = \phi_n(\cdot)$ above).

The above outline (or even a fully detailed version of the above program) gives little feeling for the detailed nature of the eigenfunctions and eigenvalues.

The "low eigenvectors" can be approximately calculated by various means. One method is to simply approximate the integral equation introduced above by a matrix equation. The approximate calculation is then a problem in numerical linear algebra. One approach to this has been outlined in the problems of the previous section. For certain classes of equations, solutions may be constructed on the basis of power series. This is a classical method in the theory of ordinary differential equations, leading directly to the standard power series representations for the so-called special functions. These methods also provide the technical basis for the analysis of singular Sturm–Liouville problems, since the results provide a detailed description of the solutions in the neighborhood of a singular point. This use is illustrated in the treatment of Bessel functions in Section 4.5. Power series solutions are discussed in some detail in the following section.

The Sturm–Liouville equation

$$(p X')' + q X + \lambda^2 r X = 0 \tag{4.9}$$

may also be considered for the case of large "λ" (corresponding to the "high" eigenvectors). It is possible to show that the solutions of (4.9) are in fact related to the familiar trigonometric functions.

There are of course certain cases of the equation whose solutions are exactly expressible in trigonometric terms. One such case is that in which (4.9) results from an "unfortunate" change of variables in connection with the "simple" equation

$$\frac{d^2 X}{d\xi^2} + \lambda^2 X = 0.$$

Suppose that the (x, ξ) variables are related by the twice continuously differentiable monotone function $\phi(\cdot)$:

$$x = \phi(\xi), \ \xi = \phi^{-1}(x) = f(x). \tag{4.10}$$

Then the chain rule provides

$$\frac{dg}{d\xi} = \frac{dg}{dx}\frac{dx}{d\xi} = g'(\phi(\xi))\phi'(\xi) = g'(x)\frac{1}{f'(x)},$$

$$\frac{d^2g}{d\xi^2} = g''(x)\frac{1}{[f'(x)]^2} + g'(x)\frac{-f''(x)}{[f'(x)]^3}.$$

The original differential equation expressed in terms of the variable "x" takes the form

$$\frac{1}{[f'(x)]^2}X''(x) - \frac{f''(x)}{[f'(x)]^3}X'(x) + \lambda^2 X(x) = 0. \tag{4.11}$$

Identifying the relevant terms in this equation and the Sturm–Liouville standard form, we see that the corresponding terms are

$$\frac{1}{[f'(x)]^2} = \frac{p(x)}{r(x)}.$$

This identifies the change of variables in terms of the Sturm–Liouville coefficients $p(\cdot)$ and $r(\cdot)$:

$$\frac{d\xi}{dx} = f'(x) = \sqrt{\frac{r(x)}{p(x)}}, \quad \xi = \int_a^x \sqrt{\frac{r(\eta)}{p(\eta)}}\, d\eta. \tag{4.12}$$

In order that the change of variables provide exactly the solution of (4.9), it is of course required that the first derivative terms match as well. On the other hand, it is possible to suggest that the change of variable is relevant to (4.9) even if this is not the case.

Recall that, intuitively speaking, differentiation has an inherent tendency to increase the "size" of a function. In fact, one may see that, for large λ, (4.11) essentially represents a balance between the λ^2 term and the second derivative. The first derivative term is of magnitude "only" λ, and on a relative basis is negligible. It is tempting to suspect the same sort of behavior in (4.9) for large λ^2. If (4.9) and (4.11) both exhibit a balance between the second derivative and the large λ^2 terms, then the fact that only the lower-order terms differ in the two equations makes the change of variables (4.12) of interest for (4.9) as well as (the obviously contrived) (4.11).

With these considerations in mind, we seek a solution of (4.9) in the form

$$X(x) = e^{i\lambda\xi(x)} a(x), \tag{4.13}$$

where the function $\xi(\cdot)$ is given by (4.12). The function $a(\cdot)$ may be called the "amplitude function" associated with (4.13). In the case of (4.11), obviously $a(\cdot)$

is a constant; in the general case the hope is that $a(\cdot)$ is "slowly varying" (so that the rapid oscillatory behavior expected in (4.13) is confined to the exponential term).

Since the exponential term in (4.13) never vanishes, it is always possible to express solutions in the indicated form. The problem, of course, is to characterize the resulting $a(\cdot)$. To find $a(\cdot)$, we substitute in the equation (4.9)

$$X' = i\lambda \xi' e^{i\lambda \xi(x)} a(x) + e^{i\lambda \xi(x)} a'(x),$$

and

$$(pX')' + qX + \lambda^2 r X = p'\,[i\lambda \xi' e^{i\lambda \xi(x)} a(x) + e^{i\lambda \xi(x)} a'(x)]$$
$$+ p[i\lambda \xi'' e^{i\lambda \xi(x)} a(x) + (i\lambda)^2 (\xi')^2 e^{i\lambda \xi(x)} a(x)$$
$$+ i\lambda \xi' e^{i\lambda \xi(x)} a'(x) + i\lambda \xi' e^{i\lambda \xi(x)} a'(x) + e^{i\lambda \xi(x)} a''(x)]$$
$$+ qa(x) e^{i\lambda \xi(x)} + \lambda^2 r e^{i\lambda \xi(x)} a(x).$$

Since $(\xi')^2 = \frac{r}{p}$, the λ^2 terms in the above cancel, and, in order that (4.9) hold, we must determine $a(\cdot)$ from

$$p\,a'' + a'\,[p' + 2pi\lambda \xi'] + a\,[i\lambda p'\,\xi' + i\lambda p\,\xi'' + q] = 0.$$

This equation is an example of a singularly perturbed differential equation. Such equations are widely studied, and numerous asymptotic results are available. The dominant behavior of the equation may be discovered by rewriting it in the form

$$i\lambda\Big[\frac{p'}{p} + \frac{\xi''}{\xi'} + 2\frac{a'}{a}\Big] = -\frac{\xi\,a'' + a'\,p' + a\,q}{a\,p\,\xi'};$$

since

$$\frac{d}{dx}\Big(\frac{a'}{a}\Big) = \frac{a''}{a} - \frac{a'^2}{a^2}$$

the equation may also be written as

$$i\lambda\left[\frac{p'}{p} + \frac{\xi''}{\xi'} + 2\frac{a'}{a}\right] = -\frac{1}{p\xi'}\left\{\left[p\Big(\frac{a'}{a}\Big)' + p\Big(\frac{a'}{a}\Big)^2\right] + p'\frac{a'}{a} + q\right\}.$$

Arguing that $\frac{a'}{a}$ and $(\frac{a'}{a})'$ are expected to be "small," while λ is "large," this is really of the form

$$\frac{p'}{p} + \frac{\xi''}{\xi'} + 2\frac{a'}{a} \approx 0.$$

Replacing the approximation with an equality and integrating gives

$$\ln(p\xi'a^2) = const, \tag{4.14}$$

$$a(x) = const\frac{1}{\sqrt{\xi'p}} = const\frac{1}{\sqrt{\sqrt{r(x)p(x)}}}$$

$$= const[r(x)p(x)]^{-1/4}.$$

From this result, we see that, provided $r(\cdot)$ and $p(\cdot)$ are smooth, the approximate $a(\cdot)$ is slowly varying.

The result is at least consistent with the expectation leading to the expression. Formal justification of (4.14) may be obtained either through the use of asymptotic methods, or by verifying (4.14) as an approximate solution of an integral equation version of the governing equation.

Assuming the validity of the approximation obtained this way (known as the WKB approximation), two linearly independent (near) solutions of (4.9) are obtained as

$$\phi_1(x) \approx c_1 \, (r(x)p(x))^{-1/4} e^{i\lambda\xi(x)},$$

$$\phi_2(x) \approx c_2 \, (r(x)p(x))^{-1/4} e^{-i\lambda\xi(x)}.$$

The large eigenvalues are approximated by applying the Sturm–Liouville boundary conditions to the approximate general solution formed from these near solutions.

Example. Bessel's equation of order n is

$$X'' + \frac{1}{x}X' + (\lambda^2 - \frac{n^2}{x^2})X = 0.$$

We have previously identified

$$p(x) = x,$$

$$q(x) = -n^2/x,$$

$$r(x) = x,$$

hence for this example

$$\xi(x) = \int_a^x \sqrt{1}d\eta = x - a,$$

and

$$(rp)^{-1/4} = \frac{1}{\sqrt{x}}.$$

The approximate solutions are hence

$$\phi_1(x) = c_1 \frac{1}{\sqrt{x}} e^{i\lambda(x-a)},$$

$$\phi_2(x) = c_2 \frac{1}{\sqrt{x}} e^{-i\lambda(x-a)}.$$

If the boundary conditions $X(a) = X(b) = 0$ are imposed, from

$$X(x) = \frac{1}{\sqrt{x}} \{ c_1 e^{i\lambda(x-a)} + c_2 e^{-i\lambda(x-a)} \}$$

we obtain

$$\frac{1}{\sqrt{a}}(c_1 + c_2) = 0, \ X(x) = c_3 \frac{1}{\sqrt{x}} \sin \lambda(x-a),$$

and finally

$$\frac{1}{\sqrt{b}} \sin \lambda(x-a) = 0.$$

The approximate large eigenvalues and eigenvectors are hence

$$\lambda_n^2 \approx \left(\frac{n\pi}{b-a} \right)^2,$$

$$X_n(x) \approx b_n \frac{1}{\sqrt{x}} \sin \frac{n\pi(x-a)}{b-a}.$$

For the sake of reference, we collect the "Sturm–Liouville facts" discussed above in the following.

Theorem (Sturm–Liouville). *Consider the Sturm–Liouville problem (in standard form)*

$$(p X')' + q X + \lambda^2 r X = 0, {}^2 \, a < x < b,$$

$$\alpha X(a) + \beta X'(a) = 0,$$

$$\gamma X(b) + \delta X'(b) = 0$$

and suppose

1. $p(\cdot), q(\cdot), r(\cdot), \alpha, \beta, \gamma, \delta$ all are real valued;
2. $p(\cdot), q(\cdot), r(\cdot) > 0$;
3. $p(\cdot), q(\cdot), r(\cdot)$ are continuously differentiable (for convenience).

---

[2]Here it is convenient to write the parameter as λ^2 as used above, because it is possible for a finite number of the eigenvalues to be negative if there is no assumption on q.

Then we conclude:

A. *There exist an infinite number of eigenvalues $\{\lambda_n^2\}$, and a fixed real number M such that*

$$M < \lambda_1^2 < \lambda_2^2 < \lambda_3^2 < \lambda_4^2 < \cdots ,$$

$$\lambda_n^2 \to \infty \text{ as } n \to \infty \text{ (see below)}.$$

B. *The eigenfunctions $X(\cdot)$,*

$$(p X')' + q X + \lambda^2 r X = 0,$$

may be chosen real, are pairwise orthogonal with respect to the weight function $r(\cdot)$ (and may be normalized), so that

$$\int_a^b X_n(x) \, X_m(x) \, r(x) \, dx = \delta_{mn} = \begin{cases} 1 & m = n, \\ 0 & m \neq n. \end{cases}$$

C. *The eigenfunctions are complete in the vector (Hilbert) space of functions square-integrable with weight $r(\cdot)$. If*

$$\int_a^b |f(x)|^2 r(x) dx < \infty,$$

then

$$f(x) = \sum_{n=1}^{\infty} b_n X_n(x)$$

with $b_n = \int_a^b f(x) X_n(x) r(x) dx$.
This equality holds (initially) in the sense of convergence in the associated Hilbert space (pointwise with restricted $f(\cdot)$).

D. *The large eigenvalues are approximately determined by application of the indicated boundary conditions to the approximate general solution of*

$$X_n(x) \approx (rp)^{-1/4} \{c_1 \, e^{i\lambda_n \xi(x)} + c_2 \, e^{-i\lambda_n \xi(x)}\}$$

where

$$\xi(x) = \int_a^x \sqrt{\frac{r(\eta)}{p(\eta)}}.$$

Problems 4.3.

1. Reduce the following second-order equations to the standard Sturm–Liouville form. In each case, find the appropriate weight function and write out the integral form of the associated orthogonality condition.

   (a) $x^2 X'' + X' + \lambda^2 X = 0$, $1/2 < x < 1$.

   (b) $(1 + x^2)X'' + xX' + \lambda^2(1 + x^4)X = 0$, $0 < x < 1$.

   (c) $e^{-x}X'' + 2e^{-x}X' + e^x \lambda^2 X = 0$, $0 < x < 2$.

   (d) $(1 - x^2)X'' + xX' + \lambda^2 X = 0$, $0 < x < 1$.

2. Consider the eigenvalue problem associated with the transverse vibrations of a beam. This is

$$\frac{d^2}{dx^2}\left[EI(x)\frac{d^2 X}{dx^2} \right] + \lambda^2 X = 0.$$

   Show that, for the physically natural boundary conditions associated with this problem (see Problems 14, 14b, and 15 of Section 3.7) there is an associated formally self-adjoint differential equation, and hence that orthogonal eigenvectors are expected.

3. For the problem of the vibrating beam (with boundary conditions of Section 3.7) above, show that the eigenvalues are positive (with the possible exception of the "free-end" case.)

4. Consider the problem

$$\rho(x) X'' + \lambda^2 X = 0,$$

$$X(0) = X(1) = 0,$$

   with eigenfunctions $\{X_n(x)\}$. Show that the functions $\{\frac{X_n'(x)}{\lambda_n}\}$ are eigenvectors of a Sturm–Liouville system with the same eigenvalues as above. Show further that $\{\frac{X_n'(x)}{\lambda_n}\}$ is in fact orthonormal with respect to the associated inner product, provided that $\{X_n(x)\}$ are normalized with respect to the problem. (You may assume $\rho(\cdot)$ continuously differentiable and positive.)

5. Consider the elastically restrained string example of Section 3.7 above. Show (in two lines of argument) that the separated X-solutions for the problem are orthogonal.

6. Let $K(\cdot, \cdot)$ be defined as

$$K(x, \xi) = \begin{cases} \xi(1 - x) & \xi \leq x, \\ x(1 - \xi) & \xi \geq x. \end{cases}$$

   Let the linear mapping K be defined in $L_2[0, 1]$ by

$$(Kf)(x) = \int_0^1 K(x, \xi)f(\xi)d\xi.$$

Show that K satisfies (for all $u, v \in L_2$)

$$(Ku, v) = (u, Kv),$$

and so that K is formally self-adjoint.

7. Define K as in Problem 6, and show by direct calculation that K has eigenvalues $\{\frac{1}{(n\pi)^2}\}$ with corresponding eigenfunctions $\{\sin n\pi x\}$.

8. Suppose that $f(\cdot)$ is a twice continuously differentiable function. Show that (with K as in Problem 6 above)

$$\frac{d^2}{dx^2}(Kf)(x) = -f(x),$$

$$K\frac{d^2f}{dx^2}(x) = -f(x),$$

so that $(-K)$ acts as an inverse of the second derivative mapping, at least for sufficiently smooth $f(\cdot)$.

9. Consider the Sturm–Liouville equation

$$\frac{d}{dx}(p(x)u') + q(x)u = 0$$

(λ is "buried" in the $q(\cdot)$). Define

$$X(x) = u(x),$$
$$Y(x) = p(x)\, u'(x),$$

and introduce polar coordinates by

$$X(x) = r(x)\, \cos\theta(x),$$
$$Y(x) = r(x)\, \sin\theta(x).$$

Show that $r(\cdot)$, $\theta(\cdot)$ satisfy

$$r' = \left(\frac{1}{p(x)} - q(x)\right) r \sin\theta\, \cos\theta,$$

$$\theta' = -q(x)\cos^2\theta - \frac{1}{p(x)}\sin^2\theta.$$

(This transformation is a useful device in the pursuit of general properties of Sturm–Liouville systems.)

10. Show that by patient quadrature it is possible to reduce the (slightly degenerate) Sturm–Liouville equation

$$(p(x)X')' = \lambda^2\, r(x)X, \ 0 < x < 1,$$

$$X(0) = X(1) = 0$$

to the form

$$X(x) = -\lambda \int_0^1 K(x, \xi) X(\xi) d\xi.$$

Show that the mapping K defined by the integral above is formally self-adjoint with respect to (the entirely predictable) integral inner product with weight function $r(\cdot)$.

11. Construct the WKB-method/ large λ approximate solutions for the following Sturm–Liouville equations:

(a) $xX'' + X' + xe^{-x}\lambda^2 X = 0.$

(b) $\frac{d}{dx}((1 - x^2)X') + (\lambda^2 - \frac{n^2}{1-x^2})X = 0$, (Legendre's equation).

(c) $\frac{d}{dx}((1 - x^2)X') + \lambda^2 X = 0.$

12. Suppose that the WKB approximation is applied to

$$(pX')' + qX' + \lambda^2 r(x) X = 0, \ a < x < b,$$

with boundary conditions

$$\alpha_0 X(a) + \beta_0 X'(a) = 0,$$

$$\gamma_0 X(b) + \delta_0 X'(b) = 0.$$

Find the coefficients associated with a constant coefficient Sturm–Liouville problem whose large eigenvalues approximate those of the general problem above.

13. Use the result of Problem 12 together with some facility in classical standard boundary value problems (Problem 24, Section 3.7 for example) to deduce that the estimate

$$\frac{\lambda_n^2}{n^2} \approx \text{constant}$$

holds for Sturm–Liouville eigenvalue locations in the case of large values of the eigenvalue.

4.4   Power Series and Singular Sturm–Liouville Problems

The asymptotic WKB results of the previous section show that for the standard (that is, non-singular) Sturm–Liouville problems the eigenfunctions corresponding to large eigenvalues may be approximated in terms of familiar functions. This leaves open the question of calculation of these functions for small values of the eigenvalue.

A classical approach to this problem is through the method of power series solutions of the governing differential equations. This leads to power series expansions for the special functions (conventionally defined as solutions of the differential equations resulting from separation of variables applied to Laplace's equation in various coordinate systems).

For the discussion of power series solutions, it is convenient to use a different standard form of the governing differential equation. We consider

$$P(x)X'' + Q(x)X' + R(x)X = 0, \tag{4.15}$$

or

$$X'' + \frac{Q(x)}{P(x)}X' + \frac{R(x)}{P(x)}X = 0,$$

$$X'' + P_1(x)X' + P_2(x)X = 0.$$

(There is no need to identify λ^2 in the above; it is treated as a parameter absorbed into the coefficient functions.)

Definition. If $P_1(\cdot)$ and $P_2(\cdot)$ above are analytic functions in some neighborhood of the point x_0, then x_0 is called an ordinary point of the differential equation (4.15).

A standard result concerning ordinary points is the following.

Theorem (Ordinary point solutions). *If x_0 is an ordinary point of (4.15), then the solutions of (4.15) are expressible in the form*

$$X(x) = \sum_{n=0}^{\infty} a_n(x - x_0)^n = a_0 y_0(x) + a_1 y_1(x)$$

where y_0, y_1 are linearly independent, analytic functions with a radius of convergence at least as large as the minimum of the radii of convergence associated with $P_1(\cdot)$, $P_2(\cdot)$ at x_0.

This result is proved in many differential equations texts, essentially by substituting the supposed form and estimating the behavior of the formal series so obtained.

Definition. If both of the functions $(x-x_0) P_1(x)$ and $(x-x_0)^2 P_2(x)$ define functions analytic in a neighborhood of x_0, then x_0 is called a *regular singular point* of the differential equation (4.15).

This definition essentially requires that the singularity of $P_1(\cdot)$ at x_0 is at worst a first-order pole, and the singularity of $P_2(\cdot)$ at worst a second-order pole. A point x_0 at which the singularities are worse than this is called an irregular singular point. We treat only the regular case below.

Example. 1. For the equation

$$X'' + \lambda^2 X = 0$$

every point x_0 is an ordinary point.
2. Bessel's equation of order n is

$$x^2 X'' + xX' + (x^2 - n^2)X = 0,$$

so that $P_1(x) = \frac{1}{x}$, $P_2(x) = (1 - \frac{n^2}{x^2})$ for this case. The point $x_0 = 0$ is thus a regular singular point for Bessel's equation.
3. Legendre's equation is given by

$$(1 - x^2)X'' - 2xX' + \lambda(\lambda + 1)X = 0.$$

Here

$$P_1(x) = \frac{-2x}{1 - x^2},$$

$$P_2(x) = \frac{\lambda(\lambda + 1)}{1 - x^2}.$$

The points $x_0 = 1$, $x_0 = -1$ are both regular singular points of this equation. The point $x_0 = 0$ is an ordinary point.

We illustrate the procedure for the ordinary point case with the following example.

Example. Consider

$$X'' + \lambda^2 X = 0,$$

and (taking the base point $x_0 = 0$) seek a solution of the form

$$X(x) = \sum_{n=0}^{\infty} c_n x^n.$$

Since the above theorem guarantees that the solution is analytic in any neighborhood of the origin, we differentiate termwise and use the results in the differential equation

$$X'(x) = \sum_{n=1}^{\infty} n c_n x^{n-1},$$

$$X''(x) = \sum_{n=2}^{\infty} n(n-1) c_n x^{n-2},$$

and

$$\sum_{n=2}^{\infty} n(n-1)c_n x^{n-2} + \sum_{j=0}^{\infty} \lambda^2 c_j x^j = 0.$$

To express this in terms of a single series, change the summation index of the first term above using

$$n - 2 = j, \ n - 1 = j + 1.$$

Hence

$$\sum_{j=0}^{\infty} c_{j+2}(j+2)(j+1)x^j + \lambda^2 c_j x^j = 0,$$

and the coefficients are determined by

$$c_{j+2} = \frac{-\lambda^2}{(j+1)(j+2)} c_j.$$

The calculations

$$c_2 = -\frac{1}{1 \cdot 2} \lambda^2 c_0$$

$$c_4 = \frac{1}{1 \cdot 2 \cdot 3 \cdot 4} \lambda^4 c_0$$

$$c_6 = -\frac{1}{1 \cdot 2 \cdot 3 \cdot 4 \cdot 5 \cdot 6} \lambda^6 c_0$$

$$c_3 = -\frac{1}{2 \cdot 3} \lambda^2 c_1$$

$$c_5 = \frac{1}{2 \cdot 3 \cdot 4 \cdot 5} \lambda^4 c_1$$

$$c_7 = -\frac{1}{2 \cdot 3 \cdot 4 \cdot 5 \cdot 6 \cdot 7} \lambda^6 c_1$$

identify the solution (in familiar terms) as (for $\lambda \neq 0$)

$$X(x) = c_0 \cos \lambda x + \frac{c_1}{\lambda} \sin \lambda x.$$

We may remark at this point that there is a tendency (born of familiarity) to regard the trigonometric functions as "known," and somehow more tractable than other special functions arising in boundary value problems. With some justice one may claim that these functions are known to the extent of the power series calculated above. To the extent that other special functions may also be calculated in such terms, they are also known to the same degree. It is also true that analogues of addition theorems, differentiation and integration rules, and so on exist for most special functions (although we do not treat such topics). Such results may be found in classical references such as [10], [11], and a presentation of a formal framework underlying such topics is contained in the reference [4].

In the case of a singular point problem, it is not in general the case that analytic solutions of the differential equation (4.15) exist. Some insight into this problem is provided by the example of the Euler (or equidimensional) equation.

Example. The recollection that

$$\frac{d}{dx} x^r = r x^{r-1}$$

leads one to try solutions of the form

$$X(x) = x^r$$

in the (Euler) equation

$$x^2 X'' + \alpha x X' + \beta X = 0.$$

This provides a solution as long as the exponent satisfies the *indicial equation*

$$r(r-1) + \alpha r + \beta = 0.$$

Various cases arise depending on the character of the solutions of the indicial equation.

a. If these roots are real and distinct, then x^{r_1}, x^{r_2} provide two linearly independent solutions.
b. If the roots are distinct but complex valued, then there are also two solutions, formally x^{r_1}, x^{r_2}. If $r_1 = a + ib$, then

$$|x|^{r_1} = e^{(a+ib)\ln|x|} = e^{a\ln|x|} e^{ib\ln|x|}$$
$$= |x|^a(\cos b \ln|x| + i \sin b \ln|x|)$$

is a more accurate and familiar description of the solution appropriate for real x.
c. If the indicial equation has a repeated root, then the above procedure yields only a single solution. A second solution is obtained by noting that the equation

$$x^2 (x^r)'' + \alpha x (x^r)' + \beta x^r = (r - r_1)^2 x^r$$

amounts to an algebraic identity in the variables (x, r). Differentiation with respect to r gives (the identity)

$$x^2(\frac{\partial}{\partial r} x^r)'' + \alpha x(\frac{\partial}{\partial r} x^r)' + \beta(\frac{\partial}{\partial r} x^r) = \frac{\partial}{\partial r}[(r - r_1)^2 x^r].$$

Evaluating this identity at $r = r_1$ gives the conclusion

$$x^2 \left(\frac{\partial}{\partial r} x^r \Big|_{x=r_1} \right)'' + \alpha x \left(\frac{\partial}{\partial r} x^r \Big|_{x=r_1} \right)' + \beta \left(\frac{\partial}{\partial r} x^r \Big|_{x=r_1} \right) = 0,$$

so that

$$\frac{\partial}{\partial r} x^r \Big|_{x=r_1} = x^r \ln x \Big|_{x=r_1} = x^{r_1} \ln x$$

is (at least for $x > 0$) the missing second solution.

This example illustrates that some type of singularity is typically to be expected of solutions at a singular point.

We now discuss solutions in the regular singular point case. A key fact which simplifies some of the formulas involved is provided by a representation of the result of multiplying two power series. (This is in fact closely related to the convolution theorem of discrete-transform theory and is discussed again in this context in Chapter 7.) The result in question is that

$$\left(\sum_{n=0}^{\infty} a_n x^n \right) \cdot \left(\sum_{k=0}^{\infty} a_k x^k \right) = \sum_{n=0}^{\infty} \sum_{k=0}^{\infty} a_n b_k x^{n+k}$$

$$= \sum_{l=0}^{\infty} x^l \left(\sum_{j=0}^{l} a_j b_{l-j} \right).$$

The sequence $\{c_l\}$ generated according to

$$c_l = \sum_{j=0}^{l} a_j b_{l-j}$$

is referred to as the convolution of the sequences $\{a_n\}_{n=0}^{\infty}$ and $\{b_k\}_{k=0}^{\infty}$. In the context of analytic function expansions, the central result concerning this is that $\sum_{l=0}^{\infty} c_l x^l$ has radius of convergence at least as large as that of the common radius of convergence of the original two series multiplied together.

It is convenient to assume (that is, translate the origin so) that the singular point occurs at $x = 0$. We then write (4.15) in the form

$$x^2 X'' + x [x P_1(x)] X' + [x^2 P_2(x)] X = 0 \qquad (4.16)$$

in order to exploit analogies with the Euler equation case. Since $x_0 = 0$ is a regular singular point,

$$p(x) = x P_1(x) = \sum_{n=0}^{\infty} p_n x^n$$

and

$$q(x) = x^2 P_2(x) = \sum_{n=0}^{\infty} q_n x^n$$

are by definition both analytic in a neighborhood of the origin. We seek solutions of the form

$$X(x) = x^r y(x)$$

where

$$y(x) = \sum_{n=0}^{\infty} a_n x^n$$

is analytic on some neighborhood of the origin. If this assumed form satisfies the differential equation, then substitution provides

$$\sum_{n=0}^{\infty} a_n(n+r)(n+r-1)x^{n+r} + \left(\sum_{n=0}^{\infty} p_n x^n\right)\left(\sum_{n=0}^{\infty} a_n(n+r)x^{n+r}\right)$$

$$+ \left(\sum_{n=0}^{\infty} q_n x^n\right)\left(\sum_{n=0}^{\infty} a_n x^{n+r}\right) = 0.$$

Exploiting the power series convolution formulas above we obtain

$$\sum_{n=0}^{\infty} \left[a_n(n+r)(n+r-1) + \sum_{j=0}^{n} a_j(j+r)p_{n-j} + \sum_{j=0}^{n} a_j q_{n-j} \right] x^{n+r} = 0.$$

We attempt to determine $\{a_n\}$ from this expansion by equating each coefficient of x to zero. This gives first (from $n=0$)

$$a_0 r(r-1) + a_0 r p_0 + a_0 q_0 = 0.$$

Since to avoid degeneracy we require $a_0 \neq 0$, this gives the *indicial equation*

$$F(r) = r(r-1) + r p_0 + q_0 = 0. \tag{4.17}$$

This relation determines allowable values of the parameter r. As in the Euler equation, various distinctions arise from the nature of the solutions of the indicial equation (4.17).

The coefficient of x_{n+r} in the expansion may be identified as

$$a_n(n+r)(n+r-1) + \sum_{j=0}^{n} a_j(j+r)p_{n-j} + \sum_{j=0}^{n} a_j q_{n-j}.$$

Equating this to zero, we see that a_n (which appears in all the terms) must satisfy

$$a_n F(n + r) = - \sum_{j=0}^{n-1} a_j \left[(j + r) p_{n-j} + q_{n-j} \right]$$

where (see (4.17))

$$F(n + r) = (n + r)(n + r - 1) + (n + r) p_0 + q_0.$$

The above relation determines a_n (recursively) in terms of $a_0, a_1, \ldots, a_{n-1}$ as long as $F(n + r) \neq 0$. This leads to the following procedure.

Frobenius Method I

- From the differential equation (4.16) construct the indicial equation

$$F(r) = 0$$

and find the roots r_1, r_2:

$$F(r) = (r - r_1)(r - r_2).$$

- Select a_0 arbitrarily, one of the roots r_1, and start computing

$$a_n = \frac{-1}{F(n + r_1)} \sum_{j=0}^{n-1} a_j \left[(j + r_1) p_{n-j} + q_{n-j} \right].$$

As long as $F(n + r_1) \neq 0$, that is, $n + r_1 \neq r_2$, or $r_2 - r_1 \neq n$ for some integer, this sequentially produces the a_n. A critical case is therefore seen to arise whenever the roots of the indicial equation differ by an integer. In this latter case the method produces a candidate solution only for the root of the larger real part. If the roots are distinct and do not differ by an integer, two solutions are produced in this way. These have the form:

$$y_1(x) = x^{r_1} \sum_{n=0}^{\infty} a_n(r_1) x^n,$$

$$y_2(x) = x^{r_2} \sum_{n=0}^{\infty} a_n(r_2) x^n.$$

If $r_1 = r_2$, or $r_2 - r_1 = N$, N some positive integer, we obtain only the second solution of this set.

The cases not covered by the above method (repeated or "integer-differing" roots) are handled by methods similar to those used for the repeated root Euler equation. The first troublesome case for this more complicated problem is that of a repeated root in the indicial equation.

Frobenius Method II

In the above method we seek solutions of the form

$$X(x, r) = x^r \sum_{n=0}^{\infty} a_n x^n. \tag{4.18}$$

The result of the computations carried out above is that the consequence of substituting the above form in the differential equation (4.16) is the formula

$$LX(x, r) = x^2 X'' + xp(x)X' + q(x)X$$

$$= x^r a_0 F(r) + \sum_{n=1}^{\infty} \left[a_n F(n + r) + \sum_{j=0}^{n-1} a_j((j + r)p_{n-j} + q_{n-j}) \right] x^{n+r}$$

If (as above) we choose to define $a_n(\cdot)$ as a function of the variable r by

$$a_n(r) = -\frac{1}{F(n + r)} \sum_{j=0}^{n-1} a_j ((j + r) p_{n-j} + q_{n-j}),$$

then the above provides the algebraic identity analogous to the Euler equation case:

$$LX(x, r) = a_0 x^r F(r).$$

If the indicial equation has a repeated root, then

$$F(r) = (r - r_1)^2,$$

and the above is

$$LX(x, r) = a_0 x^r (r - r_1)^2.$$

To obtain a second solution of the equation, we formally differentiate the above with respect to r, and put $r = r_1$. Then (assuming $x > 0$)

$$L(\frac{\partial}{\partial r} X(x, r)) = a_0 x^r \ln x (r - r_1)^2 + a_0 x^r (2) (r - r_1),$$

and

$$L\left(\frac{\partial}{\partial r}X(x,r)\bigg|_{r=r_1}\right) = 0,$$

this gives the second solution

$$y_2(x) = \frac{\partial}{\partial r}\left(x^r \sum_{n=0}^{\infty} a_n(r)x^n\right)\bigg|_{r=r_1}$$

$$= x^{r_1}\ln(x)\sum_{n=0}^{\infty} a_n(r_1)x^n + x^{r_1}\sum_{n=0}^{\infty} a_n'(r)x^n$$

$$= \ln(x)y_1(x) + \sum_{n=0}^{\infty} b_n(r_1)x^n.$$

(As a practical matter, the second solution may be found either by differentiation of $a_n(r)$ (assuming it is available in closed form), or by substituting this last form in (4.16) to determine $\{b_n(r_1)\}$.)

The remaining troublesome case is that in which the roots differ by some integer N. If $r_2 = r_1 + N$ is the "larger" root, we have one solution

$$y_2(x) = x^{r_2}\sum_{n=0}^{\infty} a_n(r_2)x^n.$$

The second solution may be produced by an elaboration of the considerations immediately above.

Frobenius Method III

Proceeding as above, we obtain the identity

$$LX(x,r) = x^r a_0 F(r)$$

provided that the coefficient sequence $\{a_n(r)\}$ is defined (as usual) by

$$a_n(r) = -\frac{1}{F(n+r)}\sum_{j=0}^{n-1} a_j((j+r)p_{n-j} + q_{n-j}). \tag{4.19}$$

This fails to produce the hoped-for solution when $r = r_1$ because

$$F(N+r_1) = F(r_2) = 0,$$

leaving $a_N(r_1)$ undefined (at best). A way around this problem is provided by a redefinition of a_0 to include a factor of $r - r_1$. Due to the linearity of the recursion, all a_j then contain a factor of $r - r_1$; this effectively cancels the singularity in (4.19) which otherwise occurs when $r = r_1$. This tactic is equivalent to the multiplication of the assumed solution (4.18) by the factor $r - r_1$. In these terms,

$$L\left((r - r_1)X(x, r)\right) = x^r (r - r_1) F(r) a_0 \qquad (4.20)$$

$$= x^r (r - r_1)^2 (r - r_2) a_0,$$

and this readily produces

$$\frac{\partial}{\partial r}[(r - r_1)X(x, r)]\Big|_{r=r_1}$$

as the (candidate for) missing second solution.

Using

$$(r - r_1) X(x, r) = (r - r_1) x^r \sum_{n=0}^{\infty} a_n(r) x^n,$$

we compute (for $x > 0$)

$$\frac{\partial}{\partial r}[(r - r_1) X(x, r)]\Big|_{r=r_1} = x^r \ln(x) \sum_{n=0}^{\infty}(r - r_1) a_n(r) x^n \Big|_{r=r_1}$$

$$+ x^r \sum_{n=0}^{\infty} \frac{\partial}{\partial r}[(r - r_1) a_n(r)] x^n \Big|_{r=r_1} .$$

This is one form of the second solution. In fact, the first term of the expression above can be identified in familiar terms.

In the indicated summation, the first N terms vanish by virtue of the factor $r - r_1$.

$$x^r \ln(x) \sum_{n=0}^{\infty}(r - r_1) a_n(r) x^n \Big|_{r=r_1} = x^{r_1} \ln(x) \sum_{n=N}^{\infty} a_n(r) (r - r_1) \Big|_{r=r_1} x^n$$

$$= x^{r_1} \ln(x) \sum_{j=0}^{\infty} a_{N+j}(r) (r - r_1) \Big|_{r=r_1} x^{N+j}.$$

Since $F(r) = (r - r_1)(r - r_2)$, we have

$$F(r + N) = (r + N - r_1)(r + N - r_2)$$

$$= (r + N - r_1)(r - r_1)$$

and

$$a_N(r) = \frac{1}{(r + N - r_1)(r - r_1)} \sum_{j=0}^{n-1} a_j(r) \left[(j + r) p_{n-j} + q_{n-j} \right],$$

$$(r - r_1) a_N(r) \Big|_{r=r_1} = \frac{1}{N} \sum_{j=0}^{n-1} a_j(r_1) \left[(j + r_1) p_{n-j} + q_{n-j} \right] = \tilde{a}_0.$$

If we now compute

$$(r - r_1) a_{N+1}(r) \Big|_{r=r_1}$$

$$= \frac{1}{F(N+1+r_1)} \sum_{j=0}^{\infty} (r - r_1) a_j(r) \left[(j + r) p_{N+1-j} + q_{N+1-j} \right] \Big|_{r=r_1},$$

we obtain

$$(r - r_1) a_{N+1}(r) \Big|_{r=r_1} = \frac{1}{F(N + r_1 + 1)} \tilde{a}_0 \left[(N + 1 + r_1) p_1 + q_1 \right]$$

$$= \frac{1}{F(r_2 + 1)} \tilde{a}_0 \left[(1 + r_2) p_1 + q_1 \right]$$

$$= \tilde{a}_1(r_2).$$

In short, the above coefficient is exactly the "1" coefficient in an "r_2-solution" computation starting with the (scale factor) coefficient $\tilde{a}_0$. From the recursive form of the coefficient calculations we deduce that this pattern persists. We therefore obtain

$$x^{r_1} \ln(x) \sum_{j=0}^{\infty} a_{N+j}(r) (r - r_1) \Big|_{r=r_1} x^{N+j} = \ln(x) x^{r_1+N} \sum_{j=0}^{\infty} \tilde{a}_j(r_2) x^j$$

$$= \alpha \, \ln(x) y_2(x).$$

The form of the second solution is therefore

$$\alpha \ln(x) y_2(x) + x^{r_1} \sum_{n=0}^{\infty} \frac{\partial}{\partial r} \left[(r - r_1) a_n(r) \right] \Big|_{r=r_1} x^n.$$

Synopsis

The solution forms determined above are formal solutions, in the sense that we have made no effort to establish the convergence of the infinite series computed for the general case. Such general results are available in texts on ordinary differential equations. In the case of particular examples of the above (e.g., Bessel's equation) convergence of the series often can be handled on an individual basis. We illustrate the above results with the Bessel equation example in the following section.

We recall that in the discussion of the Sturm–Liouville equation

$$(p X')' + q X' + \lambda^2 r X = 0, \, a < x < b$$

the problem is called non-singular, provided that neither $p(\cdot)$ nor $r(\cdot)$ vanishes on the interval $[a, b]$ in question. The general results concerning eigenvalues and completeness of eigenfunctions were based on this premise. For the case of non-singular Sturm–Liouville equations for which the equation coefficient functions are analytic, the "ordinary point" methods of this section suffice in principle for eigenfunction calculation.

On the other hand, the definition of the regular singular point case for the differential equation (4.15) above ensures that the corresponding Sturm–Liouville problem is singular, since the coefficient $p(\cdot)$ evidently vanishes in this case at $x = 0$ (assuming that $x = 0$ belongs to the range of interest). We have noted earlier that Bessel's equation on the interval $(0, 1)$ is singular. This circumstance prevents the direct application of our Sturm–Liouville results to the problem of heat conduction in a solid cylinder.

The treatment of singular Sturm–Liouville equations in the general case is a subject of some complexity. The basic motivation is, on the other hand, easy to describe.

Suppose that

$$(p X')' + q X' + \lambda^2 r X = 0,$$

(with some boundary conditions) is singular on the interval $[a_0, b]$ by virtue of the fact that $p(a_0) = 0$, (while otherwise $p(\cdot), r(\cdot) > 0$). In order to treat a boundary value problem in which this arises in the "classic" fashion, we require eigenfunctions (a complete set) and knowledge of eigenvalues.

To analyze the singular problem, we note that, for $\epsilon > 0$, the problem

$$(p X')' + q X' + \lambda^2 r X = 0$$

on the interval $[a_0 + \epsilon, b]$ is entirely standard. That is, there exist eigenvalues $\{\lambda_n^2(\epsilon)\}$, and eigenfunctions $\{X_n(x; \epsilon)\}$ complete in the expected sense (for appropriate boundary conditions). The "singular" $[a_0, b]$ problem is now treated by analyzing the above results as $\epsilon \to 0^+$. Careful analysis of the limiting forms may then yield the desired orthogonality and completeness results.

The relevance of the power series solutions discussed above to this procedure should be clear. The typical solutions in the singular case may show singular behavior in the neighborhood of the point a_0 (in the sense that they become unbounded). The ability to estimate this behavior is crucial to the success of the limiting procedure outlined above. The power series Frobenius solutions above contain exactly the explicit information required.

This procedure is illustrated in the following section in the case of Bessel's equation.

Problems 4.4.

1. For the equations below, identify the singular points. Classify the singular points as either regular or irregular in each case.

   a. $(1 - x^2)X'' - xX' + \lambda^2 X = 0$, (Tchebycheff's equation).
   b. $e^{1/x}X'' + xX' + X = 0$.
   c. $(\sin x)^3 X'' + (\cos x)^2 X' + X = 0$.
   d. $(\sin x)^2 X'' + (\cos x)^2 X' + X = 0$.

2. For each of the regular singular point cases of Problem 1, reduce the equation to the standard form required for application of the Frobenius method.

3. Find the solution of the equation

$$x^2 X'' + x X' + (x^2 - \frac{1}{4})0$$

   (this is Bessel's equation of order $\frac{1}{2}$) corresponding to the larger root of the indicial equation.

4. Find the power series solution of Legendre's equation

$$(1 - x^2)X'' - 2xX' + m(m+1)X = 0$$

   in the neighborhood of $x = 0$. (Note this is an ordinary point problem.)
   Show that the series obtained diverge for $|x| > 1$, except in the case for which m is an integer, for which polynomial solutions are obtained.

4.5 Cylindrical Problems and Bessel's Equation

In Section 4.1 above we have produced the equation

$$\frac{1}{c^2}\frac{\partial^2 u}{\partial t^2} = \frac{\partial^2 u}{\partial r^2} + \frac{1}{r}\frac{\partial u}{\partial r} + \frac{1}{r^2}\frac{\partial^2 u}{\partial \theta^2}$$

as a description of the vibrations of a circular elastic membrane (a drum). In connection with this equation, we may consider both "annular" and the usual models. These are illustrated in Figure 4.6.

In the case of an annular region, the appropriate boundary conditions are that

$$u(a, t) = u(b, t) = 0.$$

For the case of the circular region, one condition is evidently that

$$u(b, t) = 0.$$

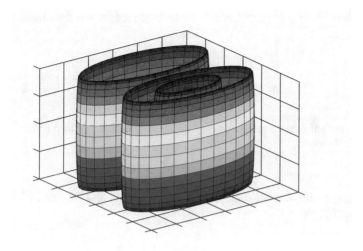

Fig. 4.6 "Drum" geometries

Since the separated radial equation is a second-order equation (in fact, as seen above, a Bessel equation) we evidently require an additional condition. This additional condition is intimately connected with the fact that the Bessel equation for this problem is a singular Sturm–Liouville problem, so that the appropriate limiting procedure outlined in the previous section should be invoked in this case. This procedure in essence produces the "missing" boundary condition as a consequence of the analysis.

An entirely similar situation arises in the case of the boundary value model for the conduction of heat in a cylindrical bar. The governing equation is

$$\frac{1}{\kappa}\frac{\partial u}{\partial t} = \frac{\partial^2 u}{\partial r^2} + \frac{1}{r}\frac{\partial u}{\partial r} + \frac{1}{r^2}\frac{\partial^2 u}{\partial \theta^2} + \frac{\partial^2 u}{\partial z^2},$$

and conduction in either a solid rod or an annular cylinder (a pipe) may be readily considered. See Figure 4.6.

The separated radial equation is again an equation of the Bessel type, and boundary condition considerations identical to those above again arise.

If solutions of the above boundary value problems are sought with only a radial spatial dependence, then the separated radial equation is Bessel's equation of order zero. We illustrate the power series methods of the previous section with this example.

For the case of an annular region, we have

$$R''(r) + \frac{1}{r}R'(r) + \lambda^2 R(r) = 0,$$

$$R(a) = R(b) = 0.$$

Introducing the change of variables $x = \lambda r$ produces the standard Bessel equation of order zero,

$$X''(x) + \frac{1}{x}X'(x) + X(x) = 0,$$

$$X(\lambda a) = X(\lambda b) = 0.$$

To seek Frobenius-style solutions, we rewrite this in the standard form of a Bessel equation of order 0,

$$x^2 X'' + x X' + x^2 X = 0. \tag{4.21}$$

The origin is a regular singular point of this equation, and we readily identify the coefficient functions

$$p(x) = \sum_{n=0}^{\infty} p_n x^n = 1 = p_0,$$

$$q(x) = \sum_{n=0}^{\infty} q_n x^n = x^2 = q_2 x^2$$

for use in the solution procedure.

The indicial equation associated with this model is

$$F(r) = r(r-1) + r p_0 + q_0$$

$$= r(r-1) + r \cdot 1 + 0$$

$$= r^2 = 0.$$

The indicial equation therefore has a repeated root $r = 0$.

From the results of the previous section, there is one solution of the form

$$y(x) = x^0 \sum_{n=0}^{\infty} a_n(0) x^n,$$

where $\{a_n(r)\}$ is determined recursively from

$$a_n(r) = \frac{-1}{F(n+r)} \sum_{j=0}^{n-1} a_j(r) [(j+r)p_{n-j} + q_{n-j}]$$

$$= \frac{-1}{F(n+r)} a_{n-2}(r) q_2$$

$$= \frac{-1}{F(n+r)} a_{n-2}(r)$$

(since q_2 is the only nonvanishing coefficient in the above sum). The solution is determined by selecting $a_0(0)$ arbitrarily (say $a_0(0) = 1$), and using the coefficient recursion as it stands to determine a_n for even n in terms of a_0.

$$a_n(0) = -\frac{1}{(n+0)^2} a_{n-2}(0),$$

so that (with $a_0 = 1$)

$$a_{2m} = \frac{(-1)^m}{(2m)^2(2m-2)^2(2m-4)^2 \cdots 2^2} = \frac{(-1)^m}{2^{2m}(m!)^2}.$$

To determine the odd coefficients, we require first $a_1(0)$. But

$$a_1(r) = -\frac{1}{F(1+r)} a_0[rp_1 + q_1],$$

$$a_1(0) = 0,$$

so that all odd coefficients vanish.

This first solution of Bessel's equation of order zero is denoted by $J_0(\cdot)$. We have just explicitly calculated

$$J_0(x) = \sum_{m=0}^{\infty} \frac{(-1)^m}{2^{2m}(m!)^2} x^{2m}. \tag{4.22}$$

This series is readily seen by the ratio test to have an infinite radius of convergence.

A second solution of (4.21) is given by (from our results for the repeated root case)

$$y_2(x) = \ln(x)J_0(x) + \sum_{n=1}^{\infty} a'_n(0)x^n.$$

Proceeding as above, we readily calculate the required function

$$a_{2m}(r) = \frac{(-1)^m}{(2m+r)^2(2m-2+r)^2(2m-4+r)^2 \cdots (2+r)^2},$$

$$a_{2m+1}(r) = 0.$$

Writing

$$a_{2m}(r) = \prod_{i=1}^{m} \frac{1}{(2i+r)^2},$$

we evaluate the required derivative by logarithmic differentiation. Since

$$\ln \alpha_{2m}(r) = -2 \sum_{i=1}^{m} \ln(2i + r),$$

we have

$$\frac{d}{dr} \ln \alpha_{2m}(r) = \frac{1}{\alpha_{2m}(r)} \alpha'_{2m}(r)$$

$$= -2 \sum_{i=1}^{m} \frac{1}{(2i + r)}.$$

This gives

$$\alpha'_{2m}(r) = -2 \prod_{j=1}^{m} \frac{1}{(2j + r)^2} \sum_{i=1}^{m} \frac{1}{(2i + r)},$$

and finally

$$\alpha'_{2m}(r) = (-1)^{m+1} \prod_{j=1}^{m} \frac{1}{(2j)^2} \sum_{i=1}^{m} \frac{1}{i}$$

$$= \frac{(-1)^{m+1}}{2^{2m}(m!)^2} H_m,$$

where $H_m = 1 + 1/2 + 1/3 + \cdots + 1/m$ is the partial sum of the harmonic series. This gives a second solution in the form

$$y_2(x) = \ln(x) J_0(x) + \sum_{i=1}^{m} \frac{(-1)^{m+1} H_m x^{2m}}{2^{2m}(m!)^2}.$$

The solution J_0 is called a Bessel function of the first kind and order zero. In preference to this form, the combination

$$Y_0(x) = \frac{2}{\pi} [y_2(x) + (\gamma - \ln 2) J_0(x)] \tag{4.23}$$

(called a Bessel function of the second kind and order zero) is often used as a standard form of a second solution of (4.21). The parameter γ is the so-called "Euler-Mascheroni" constant,

$$\gamma = \lim_{m \to \infty} (H_m - \ln(m)) = 0.5772 \cdots.$$

From our derivation the sudden appearance of $Y_0(\cdot)$ may appear a bit strained. However, the constants in the formula are a convenience in some problems.

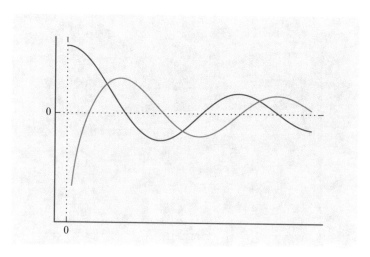

Fig. 4.7 Bessel functions

The general appearance of the solutions derived in this way is illustrated in Figure 4.7.

With the solutions of the zero-order Bessel equation in hand, the solution of the boundary value problem

$$\frac{1}{r}\frac{\partial u}{\partial r} + \frac{\partial^2 u}{\partial r^2} = 0, \ a < r < b,$$

$$u(a, t) = u(b, t) = 0$$

may be completed. (Recall that physically this is a model for radial heat conduction in a pipe.)

The separated radial equation

$$R''(r) + \frac{1}{r}R'(r) + \lambda^2 R(r) = 0,$$

$$R(a) = R(b) = 0$$

has the general solution

$$R(r) = c_1 J_0(\lambda r) + c_2 Y_0(\lambda r).$$

The boundary conditions require that

$$\begin{bmatrix} J_0(\lambda a) & Y_0(\lambda a) \\ J_0(\lambda b) & Y_0(\lambda b) \end{bmatrix} \begin{bmatrix} c_1 \\ c_2 \end{bmatrix} = \begin{bmatrix} 0 \\ 0 \end{bmatrix}$$

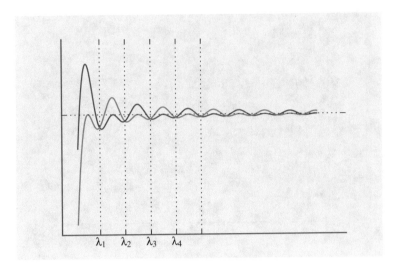

Fig. 4.8 Intersecting curves determine eigenvalues

possess a nontrivial solution, so that the determinant condition

$$J_0(\lambda a)Y_0(\lambda b) - J_0(\lambda b)Y_0(\lambda a) = 0$$

must hold. This equation determines the eigenvalues (separation constants) associated with the problem. This may be visualized by defining

$$g_1(\lambda) = J_0(\lambda a)Y_0(\lambda b),$$

$$g_2(\lambda) = J_0(\lambda b)Y_0(\lambda a)$$

and constructing the graphs of Figure 4.8. The intersection points were calculated with Octave as described for the related J_0 problem below.

From the general theory of Sturm–Liouville systems, we know that

1. There exists a countably infinite set of eigenvalues

$$\lambda_1^2 < \lambda_2^2 < \lambda_3^2 < \cdots$$

   satisfying

$$J_0(\lambda_n a)Y_0(\lambda_n b) - J_0(\lambda_n b)Y_0(\lambda_n a) = 0,$$

   together with
2. a set of eigenfunctions which may be expressed in the form

$$\tilde{X}_n(r) = J_0(\lambda_n r)Y_0(\lambda_n a) - J_0(\lambda_n a)Y_0(\lambda_n r)$$

which are orthogonal with respect to the weight function $r(\cdot)$ associated with the Sturm–Liouville equation.

$$\int_a^b \tilde{X}_n(r)\,\tilde{X}_m(r)\,r\,dr = \begin{cases} c_n^2 & m = n, \\ 0 & m \neq n. \end{cases}$$

The functions

$$X_n(\cdot) = \frac{\tilde{X}_n(\cdot)}{c_n}$$

are therefore orthonormal, and complete in the associated Hilbert space.

The solution of the original boundary value problem is completed by solving the separated "T" equation

$$T_n'(t) = -\kappa \lambda_n^2\, T_n(t)$$

to give

$$T_n(t) = e^{-\kappa \lambda_n^2 t}\, T_n(0),$$

and consequently separated solutions of the form

$$u_n(r, t) = X_n(r)e^{-K\lambda_n^2 t}T_n(0),$$

The general solution of the heat equation is then obtained by superposition,

$$u(r, t) = \sum_{n=0}^{\infty} B_n X_n(r)\, e^{-\kappa \lambda_n^2 t}.$$

If the associated initial condition is

$$u(r, 0) = f(r),$$

we obtain from the general solution the expansion problem

$$f(r) = \sum_{n=0}^{\infty} B_n X_n(r).$$

This identifies $\{B_n\}$ *uniquely* as the coefficients for an expansion of $f(\cdot)$ with respect to the complete orthonormal set $\{X_n(\cdot)\}$. Hence $\{B_n\}$ is computable from the inner products

$$B_n = \int_a^b f(r) X_n(r)\, r\, dr.$$

The above is appropriate for $f(\cdot)$ restricted by the requirement that

$$\int_a^b |f(r)|^2 \, r \, dr < \infty.$$

The series expansion then is convergent in the mean-square sense (as discussed in some generality in Chapter 2 above).

If instead of the annular problems considered above one considers the solid cylinder (or circular drum) case, then the separated radial equation is still the Bessel equation of zero order,

$$R''(r) + \frac{1}{r} R'(r) + \lambda^2 R(r) = 0.$$

The solution valid in the region $0 < r < b$ is

$$R(r) = c_1 J_0(\lambda r) + c_2 Y_0(\lambda r),$$

and the (most evident) boundary condition is that

$$R(b) = c_1 J_0(\lambda b) + c_2 Y_0(\lambda b) = 0$$

as in the previous case. From the representations

$$J_0 = \sum_{m=0}^{\infty} \frac{(-1)^m x^{2m}}{2^{2m} (m!)^2}$$

$$Y_0(x) = \frac{2}{\pi} \left[(\ln \frac{x}{2} + \gamma) J_0(x) + \sum_{m=0}^{\infty} \frac{(-1)^{m+1} H_m x^{2m}}{2^{2m} (m!)^2} \right],$$

we see that

$$J_0(0) = 1,$$

$$J_0'(0) = 0,$$

while

$$\lim_{x \to 0^+} Y_0(x) = -\infty.$$

In view of this it appears evident (on a physical basis) that an appropriate second boundary condition for the solid cylinder case is simply

$$R'(0) = 0,$$

since this will serve to eliminate the offending Y_0 term. In a similar fashion, the imposition of the requirement "$R(0)$ bounded" leads to the same formal conclusion. The difficulty is that such procedures lend no credibility to the required orthogonal expansion procedure of the associated boundary value problem.

In order to justify the validity of these expansion procedures, the standard approach is to apply a limiting procedure, as described in the previous section. Careful analysis of the limiting procedure (see reference [9] for example) then identifies the relevant expansion formulas and the nature of the "missing boundary conditions." In place of a general discussion, we illustrate the result of this procedure for the case of the zero-order Bessel equation.

On a physical basis, we expect, as noted above, that the appropriate "missing" boundary condition is

$$R'(0) = 0.$$

We therefore consider the problem

$$(r R')' + \lambda^2 r R = 0$$

with boundary conditions

$$R(b) = 0,$$

$$R'(a) = 0.$$

The last of these conditions has a physical interpretation (if one desires such) as arising from an annular heat conduction problem with a "solid insulating core." Applying these boundary conditions to the general solution

$$R = c_1 J_0(\lambda r) + c_2 Y_0(\lambda r)$$

gives

$$\lambda[c_1 J_0'(\lambda a) + c_2 Y_0'(\lambda a)] = 0.$$

In view of the expected limiting solution, we assign $c_1 = 1$, and solve this for c_2. This gives

$$c_2(a) = -\frac{J_0'(\lambda a)}{Y_0'(\lambda a)},$$

where

$$J_0'(\lambda a) = \sum_{m=0}^{\infty} \frac{(-1)^m (\lambda a)^{2m-1}(2m)}{2^{2m}(m!)^2},$$

$$Y_0'(\lambda a) = \frac{2}{\pi}\left[\frac{1}{\lambda a}J_0(\lambda a) + \ln(\frac{\lambda a \gamma}{2})J_0'(\lambda a)\right.$$

$$\left. + \sum_{m=1}^{\infty} \frac{(-1)^{m+1}(\lambda a)^{2m-1}H(2m)}{2^{2m}(m!)^2}\right].$$

We conclude from the expansions that $c_2(a)$ is well defined for small values of a, and that in fact,

$$\lim_{a \to 0^+} c_2(a) = 0.$$

The solution

$$R(r) = J_0(\lambda r) - \frac{J_0'(\lambda a)}{Y_0'(\lambda a)} Y_0(\lambda r)$$

approaches the limiting $J_0(\lambda r)$ (uniformly and in mean-square over $[a, b]$, since

$$\frac{J_0(\lambda a)}{Y_0'(\lambda a)} \approx a^2$$

as $a \to 0^+$).

The limiting positions of the eigenvalues are determined by

$$J_0(\lambda b) = 0, \tag{4.24}$$

the limiting form as $a \to 0^+$ of

$$J_0(\lambda b) + c_2(a) Y_0(\lambda b) = 0.$$

This determines λ_n as ($1/b$ times) the nth zero of $J_0(\cdot)$. The limiting form of the orthogonality condition is simply

$$\int_a^b J_0(\lambda_n r) J_0(\lambda_m r) \, r \, dr = \begin{cases} c_n^2 & m = n, \\ 0 & m \neq n. \end{cases}$$

Assuming[3] the completeness of these limiting eigenfunctions, the solution of the "kettle drum" problem may be found.

We pose (considering only symmetric solutions)

$$\frac{1}{c^2} \frac{\partial^2 u}{\partial t^2} = \frac{1}{r} \frac{\partial u}{\partial r} + \frac{1}{r^2} \frac{\partial^2 u}{\partial r^2}, \quad 0 < r < b,$$

$$\frac{\partial u}{\partial t}(r, 0) = 0,$$

$$u(r, 0) = u_0(r),$$

$$u(b, t) = 0 \, (u'(0, t) = 0),$$

---

[3]Since the problem is singular, the Sturm–Liouville results of Section 4.3 do not apply directly, and completeness remains to be proven. Proofs of completeness for some singular problems appear in reference [9].

and find separated solutions

$$X_n(r) = J_0(\lambda_n r)/c_n,$$

$$T_n''(t) = -\lambda_n^2 T_n(t),$$

$$T_n(t) = A_n \cos \lambda_n ct + B_n \sin \lambda_n ct,$$

$$u_n(r, t) = X_n(r)[A_n \cos \lambda_n ct + B_n \sin \lambda_n ct]$$

and general solution (for angular-symmetric solutions)

$$u(r, t) = \sum_{n=1}^{\infty} X_n(r)[A_n \cos \lambda_n ct + B_n \sin \lambda_n ct],$$

$$u_0(r) = \sum_{n=1}^{\infty} A_n X_n(r),$$

$$A_n = \int_0^b u_0(r) \left[\frac{J_0(\lambda_n r)}{c_n} \right] r dr,$$

$$B_n = 0.$$

From the observation (on the basis of numerical computation, for example) that

$$\lambda_n \neq \frac{n\omega_0}{c}$$

(the Bessel function zeroes are not integer multiples of a fixed number) we conclude that a drum is unlikely to be mistaken for a stringed instrument.

In separation of variables problems involving polar coordinates for which an angular dependence of the solutions is allowed, the Bessel equation of order n arises. This is

$$x^2 X'' + x X' + (x^2 - n^2) X = 0. \tag{4.25}$$

The origin is a regular singular point for this equation, and the Frobenius method may be applied to determine solutions of this problem.

The indicial equation for Bessel's equation of order n (4.25) is easily seen to be

$$F(r) = r^2 - n^2 = 0.$$

Since (see Section 4.6) n is an integer in the problem under consideration, this is a case in which the roots differ by an integer.

Applying the standard methods described above to this problem, we find that solutions may be expressed in the form

$$J_n(x) = \sum_{m=0}^{\infty} \frac{(-1)^m (\frac{x}{2})^{n+2m}}{m!(m+n)!},$$
(4.26)

$$Y_n(x) = \frac{2}{\pi}(\gamma + \ln \frac{x}{2})J_n(x) - \frac{1}{\pi} \sum_{m=0}^{n-1} \frac{(n-m-1)!}{m!}(\frac{x}{2})^{2m-n}$$

$$- \frac{1}{\pi} \sum_{m=0}^{\infty} \frac{(-1)^m (\frac{x}{2})^{n+2m} H_{m+n}}{m!\,(m+n)!}.$$

These computations are left as an exercise.

Finding Bessel Function Zeroes

The eigenvalues determined from zeroes of the Bessel function J_0 exist on the basis of general Sturm–Liouville theory, but if any computations are to be made using these, numerical values are needed. Since Octave includes both Bessel function routines and standard zero finding routines, it is fairly easy to compute a few dozen of the zeros.

The zero finding routine is named fzero. The simplest invocation of fzero uses the function "name" (more accurately, it is using the address) and a value close to the desired zero as arguments. The assumption is that the function takes only a single variable, so that if we intend to apply fzero to the Bessel function we have to work around the fact that Octave computes Bessel functions of all orders with a single call with x and n (the order) as arguments. We need a wrapper function.

OCTAVE
```
% j0(x)
% zero order Bessel function of first kind wrapper

function y = j0(x)

y = besselj(0, x);
endfunction
```

As an aid for finding the zeros, first generate a plot.

OCTAVE
```
fs = 1000;
x = 0: 1/fs :100;
y = j0(x);
plot(x, y)
```

The plot in Figure 4.9 suggests that the zeroes in the interval $(0, 100)$ should be relatively easy to handle numerically, and this turns out to be the case.

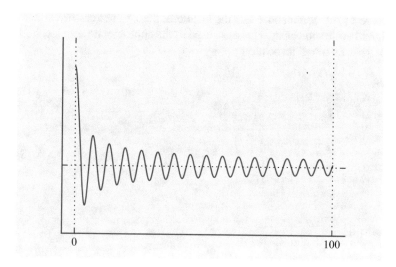

Fig. 4.9 J_0 on the interval $[0, 100]$

Looking at the y data one can see that (to the sampling resolution of .001) the zeroes can be located from the zero-crossings of the data. To avoid a lot of typing we construct a "vector" zero finder that invokes `fzero` at a vector of candidates, and returns the corresponding located zeroes.

OCTAVE

```
% getfzeros(fun, candidates)
% variant of fzero for computing multiple zeros

function answers = getfzeros(fun, candidates)

  answers = 0.0 * candidates;

  for k = 1:length(candidates)
    answers(k) = fzero(fun, candidates(k), []);
  endfor

endfunction
```

The calculation is based on the assumption that the zeroes can be located near locations where two successive function values have an opposite sign. Of course, this is wrong if one of the sample values happens to land exactly on one of the zeroes. In the face of numerical roundoff this is an event of very low probability. A "double zero" also can cause trouble, although that is unlikely in a Sturm–Liouville eigenvalue problem. In any event, the calculated results can be checked against the graph of Figure 4.9.

The code functions by multiplying y by a copy of itself shifted "up" by one sample slot, and recording the vector indices where the product is negative using the Octave `find` command. From the locations, the x values can be extracted from the original evaluation array. These are used as the approximate zero location vector for the `getfzeros` invocation.

OCTAVE

```
fs = 1000;
x = 0: 1/fs :100;
y = j0(x);

changes = find(y' .*   [y(2:end)';0] < 0);

candidates = x(changes);

answers = getfzeros('j0', candidates);

[candidates' answers']

ans =
```

2.4040	2.4048
5.5200	5.5201
8.6530	8.6537
11.7910	11.7915
14.9300	14.9309
18.0710	18.0711
21.2110	21.2116
24.3520	24.3525
27.4930	27.4935
30.6340	30.6346
33.7750	33.7758
36.9170	36.9171
40.0580	40.0584
43.1990	43.1998
46.3410	46.3412
49.4820	49.4826
52.6240	52.6241
55.7650	55.7655
58.9060	58.9070
62.0480	62.0485
65.1890	65.1900
68.3310	68.3315
71.4720	71.4730
74.6140	74.6145
77.7560	77.7560
80.8970	80.8976
84.0390	84.0391
87.1800	87.1806
90.3220	90.3222
93.4630	93.4637
96.6050	96.6053
99.7460	99.7468

The count of answers checks with what is indicated on the plot, so unless the plot is grossly misleading, the first 32 zeroes of J_0 have been located.

Problems 4.5.

1. Derive the recurrence relation

$$\frac{2n}{x} J_n(x) = J_{n+1}(x) + J_{n-1}(x)$$

for the Bessel functions $J(\cdot)$.

2. Bessel's equation of order n is given by

$$x^2 X'' + x X' + (x^2 - n^2) X = 0$$

Use Frobenius' methods to derive the solutions $J_n(x)$, $Y_n(x)$ of (4.26) above.

3. Use the power series representation for $J_n(\cdot)$ to derive the differentiation formula

$$J_{n-1}(x) - J_{n+1}(x) = 2\frac{d}{dx} J_n(x),$$

and (using 1)

$$x\frac{d}{dx} J_n(x) = n J_n(x) - x J_{n+1}(x).$$

4. Derive the integration formula

$$\int x J_n^2(x) \, dx = \frac{x^2}{2} [J_n^2(x) - J_{n-1}(x) J_{n+1}(x)]$$

(required to determine normalization constants for Bessel function expansions).

5. Consider the problem of radial heat conduction in a solid cylinder with an insulated perimeter. Show that an appropriate model is

$$\frac{\partial u}{\partial t} = \kappa \left(\frac{1}{r} \frac{\partial u}{\partial r} + \frac{\partial^2 u}{\partial r^2} \right),$$

$$\frac{\partial u}{\partial r}(0, t) = \frac{\partial u}{\partial r}(b, t) = 0,$$

$$u(r, 0) = f_0(r).$$

Find the solution for this model.

6. Use Octave to plot J_0, J_1, J_2, J_3 all on the same graph.
7. Find the first 20 zeroes of J_1 using Octave.
8. Use Octave to evaluate $\{j_{m,n}\}_{m,n=1}^{20}$, where $j_{m,n}$ is the nth zero of J_m.
9. The later zeroes of J_0 seem to be about 3.14 apart. Explain that.

4.6 Multidimensional Problems and Forced Systems

Although boundary value problems involving more than two independent variables
have been previously mentioned, we have so far produced explicit solutions only for
two-variable cases.

In this section we consider some higher-dimensional problems for the standard
models. The solutions of these are conceptually identical to the two-variable case,
although some greater tolerance for bookkeeping is required. The results of these
problems also may be employed to shed some light on the idealizations ("laterally
insulated bars," etc.) employed in the two-variable problem formulations.

This section also contains some remarks of a general nature, and an introductory
treatment of boundary value problems with forcing terms.

Consider the heat equation

$$\frac{1}{\kappa}\frac{\partial u}{\partial t} = \frac{\partial^2 u}{\partial x^2} + \frac{\partial^2 u}{\partial y^2} + \frac{\partial^2 u}{\partial z^2} \tag{4.27}$$

describing heat conduction in a rectangular solid (see Figure 4.10).

We assume that the faces $x = 0$, $x = a$ are held at zero temperature, while
the remaining four sides are insulated. The appropriate boundary conditions are
therefore

$$u(0, y, z, t) = u(a, y, z, t) = 0,$$

$$\frac{\partial u}{\partial y}(x, 0, z, t) = \frac{\partial u}{\partial y}(x, b, z, t) = 0,$$

$$\frac{\partial u}{\partial z}(x, y, 0, t) = \frac{\partial u}{\partial z}(x, y, c, t) = 0.$$

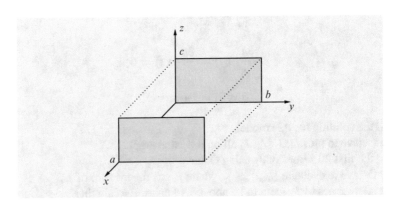

Fig. 4.10 Rectangular solid

If we seek separable solutions of this problem, we form

$$u = X(x)\, Y(y)\, Z(z)\, T(t)$$

and in the usual fashion find that

$$\frac{X''}{X} + \frac{Y''}{Y} + \frac{Z''}{Z} = -\lambda^2 = \text{constant.} \tag{4.28}$$

This equation may be further separated to provide

$$-\frac{X''}{X} = \frac{Y''}{Y} + \frac{Z''}{Z} + \lambda^2 = k_x^2.$$

In order to meet the boundary conditions in the x-direction, we require

$$X'' + k_x^2 X = 0,$$

$$x(0) = X(a) = 0.$$

These give a countable number of solutions

$$X_l(x) = \sin\frac{l\pi x}{a},\ l = 1, 2, 3, \ldots$$

and identify (the "x-separation constants")

$$k_x^2 = (\frac{l\pi}{a})^2.$$

Returning to (4.28) and separating the z dependence

$$\frac{Z''}{Z} - (\frac{l\pi}{a})^2 = -[\lambda^2 + \frac{Y''}{Y}] = -k_z^2.$$

We require

$$Z'' + (k_z^2 - (\frac{l\pi}{a})^2) Z = 0,$$

$$Z'(0) = Z'(c) = 0.$$

This provides another countable set of solutions (for each l)

$$k_z^2 - (\frac{l\pi}{a})^2 = (\frac{n\pi}{c})^2,\ n = 0, 1, 2, \ldots,$$

$$z_0(z) = 1,$$

$$z_n(z) = \cos\frac{n\pi z}{c}.$$

From equation (4.28), Y is determined from

$$Y'' + \left(\lambda^2 - [(\frac{l\pi}{a})^2 + \frac{n\pi}{c})^2] \right) Y = 0$$

with boundary conditions

$$Y'(0) = Y'(b) = 0.$$

This system has solutions

$$Y_m(y) = \cos \frac{m\pi y}{b},$$

$$Y_0(y) = 1$$

and identifies

$$\lambda_{lmn}^2 - [(\frac{l\pi}{a})^2 + \frac{n\pi}{c})^2] = (\frac{m\pi}{b})^2,$$

$$\lambda_{lmn}^2 = [(\frac{l\pi}{a})^2 + \frac{m\pi}{b})^2] + (\frac{n\pi}{c})^2.$$

The separable solutions are therefore

$$u_{lmn}(x, y, z, t)$$

$$= B_{lmn} \sin \frac{l\pi x}{a} \cos \frac{m\pi y}{b} \cos \frac{n\pi z}{c} e^{-\kappa[(\frac{l\pi}{a})^2 + (\frac{m\pi}{b})^2 + (\frac{n\pi}{c})^2]t}.$$

The corresponding general solution takes the form of a superposition

$$u(x, y, z, t)$$

$$= \sum_{l=1}^{\infty} \sum_{m=0}^{\infty} \sum_{n=0}^{\infty} B_{lmn} \sin \frac{l\pi x}{a} \cos \frac{m\pi y}{b} \cos \frac{n\pi z}{c} e^{-\kappa[(\frac{l\pi}{a})^2 + (\frac{m\pi}{b})^2 + (\frac{n\pi}{c})^2]t}.$$

Associated with the model (4.27) is an initial temperature distribution

$$u(x, y, z, 0) = u_0(x, y, z).$$

Using this with the general solution, we find the expansion problem

$$u_0(x, y, z) = \sum_{l=1}^{\infty} \sum_{m=0}^{\infty} \sum_{n=0}^{\infty} B_{lmn} \sin \frac{l\pi x}{a} \cos \frac{m\pi y}{b} \cos \frac{n\pi z}{c}.$$

This is an example of a *multiple Fourier series*.

The inner product theory of such expansions is (fortunately) identical to that for the single variable case. The functions

$$X_{lmn} = \frac{\sqrt{8}}{\sqrt{abc}} \sin \frac{l\pi x}{a} \cos \frac{m\pi y}{b} \cos \frac{n\pi z}{c}$$

(allowing modification of the scale factor in case $m, n = 0$) are readily seen to be orthogonal (in fact orthonormal) with respect to the inner product defined by

$$(f(\cdot, \cdot, \cdot), g(\cdot, \cdot, \cdot)) = \int_0^a \int_0^b \int_0^c f(x, y, z)\overline{g}(x, y, z) \, dx \, dy \, dz.$$

That is,

$$\int_0^a \int_0^b \int_0^c X_{ijk} X_{lmn} \, dx \, dy \, dz = \begin{cases} 1 & \text{if } i = l, j = m, k = n, \\ 0 & \text{otherwise.} \end{cases}$$

In fact, the collection $\{X_{lmn}\}$ is complete in the (Hilbert) space of functions square-integrable over the volume of the block in question.

These facts identify the initial condition series as an orthonormal expansion of the sort discussed in generality in Chapter 2. The coefficients $\{B_{lmn}\}$ are therefore given by (the inner products) $(m, n \neq 0)$

$$B_{lmn} = \frac{8}{abc} \int_0^a \int_0^b \int_0^c u_0(x, y, z) \sin \frac{l\pi x}{a} \cos \frac{m\pi y}{b} \cos \frac{n\pi z}{c} \, dx \, dy \, dz.$$

A consequence of this view is that initial temperatures square-integrable over the rectangle are an appropriate class of initial conditions (since this is the (Hilbert) space associated with the expansion problem. The initial condition expansion therefore holds initially in mean-square sense. With suitable complications, the convergence may be discussed in a fashion parallel to that given in Section 3.8 for the one variable case. It is useful to discuss the general solution in the light of the simple bar models of Chapter 3.

If the initial condition specified is dependent only on the x-variable, then it is easy to see that

$$B_{lmn} = 0 \text{ for } m \neq 0, n \neq 0,$$

while

$$B_{l00} = \frac{2}{abc} \int_0^a \int_0^b \int_0^c u_0(x) \sin \frac{l\pi x}{a} \, dx \, dy \, dz$$

$$= \frac{2}{a} \int_0^a u_0(x) \sin \frac{l\pi x}{a} \, dx.$$

The general three-dimensional solution reduces to

$$u(x, y, z, t) = \sum_{l=1}^{\infty} \left(\frac{2}{a} \int_0^a u_0(\xi) \sin \frac{l\pi \xi}{a} \, d\xi \right) \sin \frac{l\pi x}{a} \, e^{-\kappa (\frac{l\pi}{a})^2 t},$$

which is identical to the one-variable solution of the previous chapter. This result is the source of the "laterally insulated bar" terminology of the previous chapter.

A more subtle connection between these problems arises from the case in which one length scale is much larger than the other two. Say for concreteness that $a \gg b$, $a \gg c$. Then write

$$\lambda_{lmn}^2 = (\frac{l\pi}{a})^2 + \frac{a^2}{b^2}(\frac{m\pi}{a})^2 + \frac{a^2}{c^2}(\frac{n\pi}{a})^2.$$

This shows that the higher harmonics are relatively much more heavily damped in the x, y directions. For fixed t, the solution is reasonably approximated by

$$\sum_{l=1}^{N} B_{l00} \, e^{-\kappa (\frac{l\pi}{a})^2 t} \sin \frac{l\pi x}{a}$$

(assuming that the power spectrum of the initial distribution is concentrated in the lower frequency range). These considerations identify the one-variable model as relevant for "thin insulated bars" even in absence of initial uniformity in the transverse directions.

Plotting Heat Equation Solutions

Generally speaking, displaying multidimensional data sets is a problem. For the heat equations with three spatial coordinates there are four independent variables for the transient version of the problem, and some thought must be taken to visualize the results.

Octave is particularly strong in its support for data visualization. Data generated from measurements or external programs can be imported, as long as it is in a standard format.[4] The basic methods for visualization involve taking "slices" through a dataset, and representing data values by such visual attributes as color, transparency, and lighting.

The simplest version of the issue arises with a problem where there are three independent variables and a single "observed quantity" (dependent variable). As an instance of this, we can consider a transient heat equation with two spatial dimensions. The model is

---

[4]The HDF format developed by the National Center for Supercomputing Applications (NCSA) is supported for both import and export by Octave.

$$\frac{\partial u}{\partial t} = \frac{\partial^2 u}{\partial x^2} + \frac{\partial^2 u}{\partial y^2},$$

$$u(x, 0, t) = u(x, 1, t) = u(0, y, t) = u(1, y, t) = 0,$$

$$u(x, y, 0) = f(x, y), \ (x, y) \in (0, 1) \times (0, 1).$$

The spatial domain is a square, and the physical interpretation is of an initially heated block (uniform in the z direction) cooled with temperature set to 0 on all sides. This problem may be solved easily following the previous example, and the solution will be

$$u(x, y, t) = \sum_{n=1}^{\infty} \sum_{n=1}^{\infty} b_{nm} \sin(n \pi x) \sin(n \pi y) e^{-(n^2 + m^2) \pi^2 t}.$$

If the initial temperature distribution is taken as constant, this is a physically realistic case. For this case the expansion coefficients were computed earlier for the square wave example. The formula for the term (the normal mode terminology might also be used) is

OCTAVE

```
function y = heat_2x_t_term(x, y, t, n, m)

y = (2/(n*pi))*(1 - (-1)^n).* sin (n*pi*x)...
    .*((2/(m*pi))*(1 - (-1)^m).* sin (m*pi*y))...
    .*(exp(-n*n*pi*pi*t - m*m*pi*pi*t));

endfunction
```

To get a result without a lot of overshoot at the edges of the domain, we employ a two-dimensional Cesaro sum to evaluate the solution series. The code is the expected[5]

OCTAVE

```
function y = Cesaro_2x_t(term, x, y, t, N, M)

    tempval = zeros(size (x));

    for kn = 1:N
      for km = 1:M
        tempval = tempval + partialsum_2x_t(term, x, y, t, kn, km);
      endfor
    endfor

    y = tempval/(N*M);

endfunction
```

---

[5]One might also evaluate the partial sums along the diagonal $M = N$ and average only N rather than MN partial sums. The execution time is much lower, but the overshoot is evident in the plots.

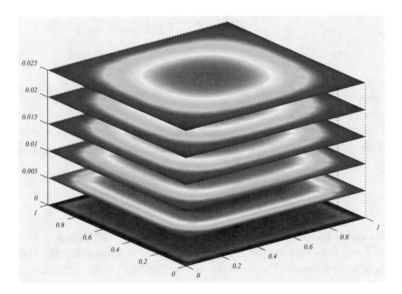

Fig. 4.11 Time slices through a heat equation solution

To generate a plot, we first construct a three-dimensional grid. Then (a good time to take in a movie) calculate a two-dimensional Cesaro sum out to terms $M = 50$, $N = 50$. To generate plots, extract slices through the temperature values at sequential time samples.

The default slices have embedded grid lines and colored boxes. A smooth looking plot is generated by removing the grid lines and having the color (representing temperature) appear interpolated from the evaluated sample values.

The default slices appear opaque. This is fine, but a different appearance results from changing the mapping between temperature and color, and assigning transparency to the slices according to their color (that is, temperature) values.

OCTAVE

```
[X, Y, T] = meshgrid(0: .05: 1, 0: .05: 1, 0: .005: .1);

u = Cesaro_2x_t('heat_2x_t_term', X, Y, T, 50, 50);

h = slice(X, Y, T, u, [], [], [0 .005 .010 .015 .020 .025]);

set(h, 'EdgeColor', 'none', 'FaceColor', 'interp');
```

The resulting plot is shown in Figure 4.11.

In the previous section the problem of vibration of a circular membrane was treated in the case of radial dependence alone. The general problem is described by

$$\frac{1}{c^2}\frac{\partial^2 u}{\partial t^2} = \frac{\partial^2 u}{\partial r^2} + \frac{1}{r}\frac{\partial u}{\partial r} + \frac{1}{r^2}\frac{\partial^2 u}{\partial \theta^2}, \ 0 < r < b \tag{4.29}$$

with boundary conditions

$$u(b, \theta, t) = 0,$$

$$u(r, \theta + 2\pi, t) = u(r, \theta, t),$$

$$\frac{\partial u}{\partial r}(0, \theta, t) = 0,$$

and initial conditions

$$u(r, \theta, 0) = u_0(r, \theta), \ 0 < r < b,$$

$$\frac{\partial u}{\partial t}(r, \theta, 0) = 0, \ 0 < \theta < 2\pi.$$

Separation of variables in the form

$$u = T(t) R(r) \Theta(\theta)$$

provides

$$\Theta_m(\theta) = e^{im\theta}, \ m = 0, \pm 1, \pm 2,$$

$$\frac{1}{c^2} T''(t) = -\lambda^2 T(t),$$

and

$$r^2 R'' + r R' + \{\lambda^2 r^2 - m^2\} R = 0, \tag{4.30}$$

$$R(b) = 0, R'(0) = 0.$$

(This second boundary condition should be understood in the sense of the singular problem analysis of the previous section, strictly speaking.)

The equation (4.30) is a Bessel equation of order m. The indicial equation is

$$r(r - 1) + r - m^2 = 0,$$

$$r^2 = m^2.$$

This means that the roots of the indicial equation differ by an integer. The first solution (corresponding to $m \geq 0$) may be calculated as

$$J_m(\lambda r) = (\lambda r)^m \left[1 + \sum_{n=1}^{\infty} \frac{(-1)^n (\frac{\lambda r}{2})^{2n}}{n!(n+m)!} \right].$$

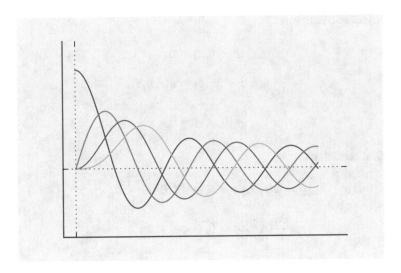

Fig. 4.12 Zeros of Bessel functions

The second solution has the form

$$a\, J_m(\lambda r)\ln \lambda r + (\lambda r)^{-m}\left[1 + \sum_{n=1}^{\infty}[(r+m)a_n(r)]'\bigg|_{r=-m}(\lambda r)^n\right],$$

and is singular at $r = 0$. Since $J_m'(0) = 0$, we arrive at

$$R(r) = J_m(\lambda r)$$

as the relevant solution of the radial equation (4.30).

The separation constants are determined from

$$J_m(\lambda b) = 0.$$

The λ-values are therefore determined in terms of the zeroes of the mth Bessel function of the first kind. See Figure 4.12.

If $j_{m,n}$ denotes the nth zero of the mth Bessel function of the first kind, so that

$$J_m(j_{m,n}) = 0,$$

we have

$$\lambda_{m,n} = \frac{j_{m,n}}{b}.$$

The results we expect to obtain from an analysis of this singular Sturm–Liouville problem are that (for fixed $m > 0$)

1. $j_{m,n} \to \infty$ as $n \to \infty$,

2. $\int_0^b J_m(\frac{j_{m,n}r}{b}) J_m(\frac{j_{m,k}r}{b}) r\,dr = \begin{cases} 0 & n \neq k, \\ c_{mn}^2 & n = k. \end{cases}$,

3. the set $\{X_n\}$,

$$X_n(r) = J_m(\frac{j_{m,n}r}{b}),$$

is complete with respect to the Hilbert space associated with (2).

With the λ_{mn} so determined, the general solution of the separated "T equation" is

$$T_{mn}(t) = A_{mn} \cos(\lambda_{mn} c\,t) + B_{mn} \sin(\lambda_{mn} c\,t).$$

The general problem solution takes the form

$$u(r, \theta, t) = \sum_{m=-\infty}^{\infty} \sum_{n=0}^{\infty} A_{mn} \cos((\lambda_{mn} c\,t) e^{im\theta} J_{|m|}(\frac{j_{m,n}r}{b}).$$

(In this the sinusoidal time dependence is dropped because of the zero initial velocity. The $|m|$ arises from the convention that $m > 0$ arising from the Bessel equation solution.)

Assuming the (inevitable) completeness of the set $\{X_{mn}\}$,

$$X_{mn}(r, \theta) = e^{im\theta} J_{|m|}(\frac{j_{m,n}r}{b}),$$

we compute the $\{A_{mn}\}$ through

$$A_{mn} = \frac{1}{c_{mn}^2} \int_0^{2\pi} \int_0^b u_0(r, \theta) e^{-im\theta} J_{|m|}(\frac{j_{m,n}r}{b}) r\,dr\,d\theta.$$

This completes the solution of (4.29).

Another common class of boundary value problems arises from models with a spherical geometry. For such models, the introduction of spherical-polar coordinates (Figure 4.13) is appropriate.

Laplace's equation in this system takes the form

$$\frac{1}{\rho^2} \frac{\partial^2 u}{\partial \rho^2} + \frac{2}{\rho^2} \frac{\partial u}{\partial \rho} + \frac{1}{\rho^2 \sin(\phi)} \frac{\partial u}{\partial \phi}\left(\sin(\phi) \frac{\partial u}{\partial \phi}\right) + \frac{1}{\rho^2 \sin(\phi)^2} \frac{\partial^2 u}{\partial \theta^2} = 0.$$

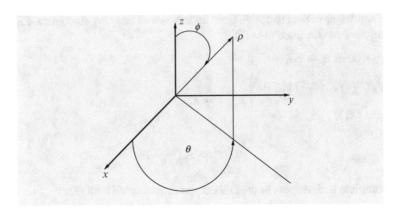

Fig. 4.13 Spherical polar coordinates

To apply separation of variables methods to this instance, we assume the form

$$u(\rho, \theta, \phi) = R(\rho)\,\Theta(\theta)\,\Phi(\phi).$$

Separating the variables after substitution we obtain

$$\frac{\sin^2 \phi}{R}\frac{\partial}{\partial \rho}\left(\rho^2 R'\right) + \frac{\sin \phi}{\Phi}\frac{\partial}{\partial \phi}\left(\sin \phi\, \Phi'\right) = -\frac{\Theta''}{\Theta} = n^2.$$

Since the geometry requires that

$$\Theta(\theta + 2\pi) = \Theta(\theta),$$

we conclude that

$$\Theta(\theta) = e^{in\theta},\ n = 0, \pm 1, \pm 2, \ldots$$

are the appropriate θ components of the separated solutions. Knowing that the "n" in the separated equation is integer-valued, we further separate the equation to obtain

$$\frac{1}{R}\frac{\partial}{\partial \rho}\left(\rho^2 R'\right) = -\frac{1}{\Phi \sin \phi}\frac{\partial}{\partial \phi}\left(\sin \phi\, \Phi'\right) + \frac{n^2}{\sin^2 \phi}.$$

It is convenient (see the problems below) to write the separation constant in this equation in the form $m(m + 1)$ to obtain

$$\frac{d}{d\rho}\left(\rho^2 R'\right) - m(m + 1)R = 0, \tag{4.31}$$

$$\frac{d}{d\phi}\left(\sin \phi \Phi'\right) + \left[n(m + 1)\sin \phi - \frac{n^2}{\sin^2 \phi}\right]\Phi = 0.$$

The radial component equation is an Euler equation with solution

$$R(\rho) = A\,\rho^m + B\,\rho^{-(m+1)}.$$

The polar-angle equation is cast in more familiar form by making the change of variables

$$x = \cos\phi,$$

$$X(x) = \phi(\cos^{-1}x)$$

to obtain

$$\frac{d}{dx}\left[(1-x^2)\frac{dX}{dx}\right] + \left[m(m+1) - \frac{n^2}{1-x^2}\right]X = 0,\ -1 < x < 1,$$

which is known as an *associated Legendre equation*.

A special case arises when $n = 0$. Here the so-called *Legendre equation*

$$\frac{d}{dx}\left[(1-x^2)\frac{dX}{dx}\right] + m(m+1)X = 0,\ -1 < x < 1$$

is obtained. This is an equation of Sturm–Liouville type, singular at (both end points of the domain) $x = \pm1$.

The Legendre equation may be analyzed on the basis of the power series methods of Section 4.4 (see the problems below). The details of such analysis show that the equation has a polynomial solution P_m for each integral value of m. Further pursuit of detail shows that (for integral m) the second linearly independent solution has a logarithmic singularity at $x = \pm1$, and is therefore excluded (compare Section 4.4 and the discussion of Y_0). If m is nonintegral, then both solutions of the Legendre equation are singular. This restricts useful separated solutions to nonnegative integral values of m.

The solutions of the associated Legendre equation could in principle be discussed in a parallel fashion. In practice, it is simpler to verify that

$$P_m^n(x) = (-1)^n\,(1-x^2)^{\frac{n}{2}}\,\frac{d^n}{dx^n}P_m(x)\ (n > 0) \tag{4.32}$$

($P_m(x)$ is the polynomial solution of Legendre's equation (see the problems below)) is a solution of the associated equation. Considerations of the kind alluded to above rule out use of the second linearly independent solution.

This procedure provides

$$U_{mn}(\rho,\theta,\phi) = e^{in\theta}\left(A_m\,\rho^m + B_m\,\rho^{-(m+1)}\right)P_m^{|n|}(\cos\phi)$$

as separated solutions. In the case of physical problems involving a solid sphere, we take $B_m = 0$ to eliminate the singularity at the origin, giving the solutions

$$e^{in\theta}\,\rho^m\,P_m^{|n|}(\cos\phi),$$

known as *spherical harmonics*. If the problem under consideration is, say, steady state heat conduction in a sphere with prescribed temperature on the surface, then the formal solution is

$$u(\rho, \theta, \phi) = \sum_{n=-\infty}^{\infty} \sum_{m=0}^{\infty} B_{mn} e^{in\theta} \, \rho^m \, P_m^{|n|}(\cos \phi).$$

The expansion coefficients are determined from the requirement that

$$u(1, \theta, \phi) = f(\theta, \phi) = \sum_{n=-\infty}^{\infty} \sum_{m=0}^{\infty} B_{mn} e^{in\theta} \, P_m^{|n|}(\cos \phi)$$

using the expected orthogonality relations.

In the examples above, the separated spatial solutions have turned out orthogonal. One explanation of this is that it is a consequence of the fact that we have continued separation to the level of separate Sturm–Liouville problems for each spatial variable. From this point of view, the overall orthogonality is simply a consequence of the separate orthogonality associated with the individual Sturm–Liouville solutions.

This line of argument rests on the supposition that the separated equation (for the heat and wave equation cases)

$$\nabla^2 X = -\lambda^2 X$$

is capable of complete separation in the above sense. There are a number of coordinate systems for which this separation is possible. The reference [5] provides a list of such.

However, in order to ensure the success of the separation method as employed above, more is required than just separability of the equation. In order to obtain boundary conditions for the separated equations, the boundary conditions for the original problem must be specified on level coordinate surfaces for the variables utilized. The reader may illustrate this by attempting to solve a circular membrane problem using rectangular coordinates.

In fact, it is easy to construct physically plausible boundary value problems (heat conduction in an automobile engine?) for which separation of variables (down to the single coordinate level) fails to be a useful method.

It would be useful (for, say, numerical analysis) to discover that the broad outlines of the solution methods used above persist in the absence of complete separability of the model equation. For both the diffusion equation

$$\frac{\partial u}{\partial t} = \nabla^2 u \tag{4.33}$$

and wave equation

$$\frac{\partial^2 u}{\partial t^2} = \nabla^2 u, \tag{4.34}$$

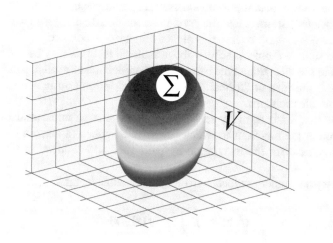

Fig. 4.14 Generic volume

the assumption of time/space separable solutions of the form

$$u = T(t) X(\mathbf{x})$$

(using $\mathbf{x}$ to denote the coordinate vector of a generic spatial coordinate system) leads to the separated equations

$$\frac{T'}{T} = \frac{\nabla^2 X}{X} = -\lambda^2,$$

$$\frac{T''}{T} = \frac{\nabla^2 X}{X} = -\lambda^2.$$

A typical problem is posed in a volume V with surface Σ as illustrated in Figure 4.14.

Associated with the problem are boundary conditions expressible in the form (assuming suitable parameterization of the surface Σ)

$$\alpha(s)\, u(s) + \beta(s)\frac{\partial u}{\partial n}(s) = 0,\ s \in \Sigma$$

(with α, β real valued, not both vanishing). Thus in either case we are led to (the eigenvalue problem for ∇^2)

$$\nabla^2 X + \lambda^2 X = 0,\ \mathbf{x} \in V,$$

$$\alpha(s)\, X(s) + \beta(s)\frac{\partial X}{\partial n}(s) = 0,\ s \in \Sigma.$$

If it can be shown that this problem has properties analogous to those of the (one-dimensional) Sturm–Liouville problems considered earlier, then we expect our general methods to carry over to this higher-dimensional context.

The two sample problems solved above indicate that this may be the case. The problem is to see this without the necessity for production of explicit solutions. The key to this is the idea of "formal self-adjointness" introduced in connection with Sturm–Liouville problems.

To show that ∇^2 is formally self-adjoint, we require an inner product space, a specification of the functions to which ∇^2 is to be applied (a domain for ∇^2), and some replacement for the integration by parts method employed in the previous case.

An appropriate inner product is given by the volume integral

$$(f, g) = \int \int \int_V f(x)\overline{g}(x) \, dV.$$

A vector space of functions $\mathcal{D}$ (a domain for ∇^2) is defined including the properties (at least required for formal calculations) that for $u \in \mathcal{D}$

1. u is square-integrable in V,
2. $\alpha(s) u(s) + \beta(s) \frac{\partial u}{\partial n}(s)$ is well defined and vanishes on Σ, and
3. the Green's Theorem (4.35) below is justified.

We recall (see Section 3.8) the Green's Theorem identity (valid for sufficiently smooth functions and regions)

$$\int \int \int_V \left(v\nabla^2 u - u\nabla^2 v \right) dV = \int \int_\Sigma \left(v\frac{\partial u}{\partial n} - u\frac{\partial v}{\partial n} \right) dS. \qquad (4.35)$$

This is the operational equivalent of two integrations by parts, and is the obvious candidate for the major tool of a calculation of formal self-adjointness. With these preliminaries, it is easy to obtain the following (formal) result.

Theorem. *Consider the partial differential expression ∇^2, restricted to a subspace of functions $\mathcal{D}$, defined on a compact volume V with smooth boundary. Then L, defined by*

$$Lu = \nabla^2 u, \ u \in \mathcal{D},$$

is formally self-adjoint with respect to the natural inner product

$$(f, g) = \int \int \int_V f(x)\overline{g}(x) \, dV.$$

Proof. It is required to show that

$$(Lu, v) = (u, Lv), \ u, v \in \mathcal{D}.$$

But

$$(Lu, v) = \iiint_V \bar{v}\nabla^2 u \, dV$$

$$= \iiint_V u\nabla^2\bar{v} \, dV + \iint_\Sigma \left(\bar{v}\frac{\partial u}{\partial n} - u\frac{\partial \bar{v}}{\partial n}\right) dS.$$

The natural boundary condition requirement is that

$$\alpha^2(s) + \beta^2(s) \neq 0, \; s \in \Sigma,$$

and this condition is used to eliminate the boundary term in the inner product calculation. The general case requires some tedious patchwork, so assume for concreteness that $\beta(s) \neq 0$ on Σ. Then the boundary integral takes the form

$$\iint_\Sigma \left(\bar{v}\frac{\partial u}{\partial n} - u\frac{\partial \bar{v}}{\partial n}\right) dS = \iint_\Sigma \left(\bar{v}\frac{\alpha}{\beta}u - u\frac{\overline{\alpha}}{\beta}v\right)$$

$$= \iint_\Sigma \frac{\alpha}{\beta}(\bar{v}u - u\bar{v}) \, dS = 0.$$

This verifies the special case, and should make the general result credible. We conclude, therefore, that ∇^2 is indeed formally self-adjoint.

With this result we conclude that

1. eigenvalues of ∇^2 are real (if they exist), and
2. eigenvectors corresponding to different eigenvalues (assuming existence) are orthogonal.

These conclusions are sufficient to describe the formal solution process for the heat equation (4.33) or the wave equation (4.34). Assuming that there exist $\{\lambda_n^2\}$, a countable set of eigenvalues with eigenvectors $\{X_n(\mathbf{x})\}$,

$$\nabla^2 X_n(\mathbf{x}) = -\lambda_n^2 X_n(\mathbf{x})$$

we obtain a formal solution of (4.33) in the form

$$u(\mathbf{x}, t) = \sum_n B_n X_n(\mathbf{x})e^{-\lambda_n^2 t},$$

and of (4.34) in the form

$$u(\mathbf{x}, t) = \sum_n \{A_n \cos(\lambda_n t) + B_n \sin(\lambda_n t)\} X_n(\mathbf{x}).$$

The unknown constants in the above are determined from the initial conditions (in the case of (4.33))

$$u(\mathbf{x}, 0) = \sum_n B_n X_n(\mathbf{x}).$$

As long as $\{X_n(\mathbf{x})\}$ is a complete normalized set of spatial functions, this uniquely determines the $\{B_n\}$ as

$$B_n = \int \int \int_v u(\mathbf{x}, 0) X_n(\mathbf{x}) \, dV$$

(assuming that $\{X_n\}$ has been normalized, so that $\{X_n\}$ represents an orthonormal set).

We conclude, then, that the general problem (4.33) leads to a solution of the form

$$u(\mathbf{x}, t) = \sum_n \left(\int \int \int_v u(\mathbf{x}, 0) X_n(\mathbf{x}) \, dV \right) X_n(\mathbf{x}) e^{-\lambda_n^2 t}.$$

This is of course the same as that for the original simple problems, allowing some liberty for the counting associated with writing the general summation as $\sum_n$.

To complete the parallels of this discussion with the Sturm–Liouville case, we mention that completeness results may be established for such problems, and that variational characterizations of the eigenvalue problem exist. Such results establish the validity (in a rigorous sense) of the formal procedure outlined above.

The boundary value problems treated so far are all homogeneous problems, in that the governing equations involve no forcing functions. Such terms arise in the source term for Poisson's equation

$$\nabla^2 u = -f(x) \tag{4.36}$$

for which f may be considered as a forcing term added to Laplace's equation.

In the context of heat equation models, internal heat generation processes (from say, chemical or radioactive sources) may produce a governing equation of the form

$$\frac{\partial u}{\partial t} = \nabla^2 u + f(x, t).$$

Similarly, wave equations with driving forces of the form

$$\frac{\partial^2 u}{\partial t^2} = \nabla^2 u + f(x, t)$$

may arise.

Associated with these equations are, of course, boundary conditions. In the homogeneous case, these may take the usual form

$$\alpha(s) u(s) + \beta(s) \frac{\partial u}{\partial n}(s) = 0.$$

In problems where forcing functions are present, these may take the more general form

$$\alpha(s) u(s) + \beta(s) \frac{\partial u}{\partial n}(s) = g(s).$$

In both problem cases, g above may depend on both the boundary location and time, and has evident physical interpretations.

One method for treating such problems (of some general utility) is the method of Laplace transforms. The eigenfunction methods of the previous sections may also be adapted to provide formal solutions for such problems. We consider first Poisson equation (4.36). In the first place, a reduction may be made to the case of the homogeneous boundary conditions. This is accomplished by first finding v satisfying (Laplace's equation)

$$\nabla^2 v = 0,$$

$$\alpha v + \beta \frac{\partial v}{\partial n} = g,$$

and defining $\tilde{u} = u - v$. The function $\tilde{u}$ then (formally) satisfies (4.36) with homogeneous boundary conditions.

Assume (as expected from the discussion above) that for the eigenvalue problem

$$\nabla^2 X = -\lambda^2 X,$$

$$\alpha X + \beta \frac{\partial X}{\partial n} = 0,$$

there exist eigenvalues $\{\lambda_n^2\}$ and a complete set of eigenfunctions (orthonormalized) $\{X_n\}$

$$\nabla^2 X_n = -\lambda_n^2 X_n,$$

$$\alpha X_n + \beta \frac{\partial X_n}{\partial n} = 0.$$

We seek a solution of (4.36) in the form

$$u(x) = \sum_n B_n X_n(x),$$

and expand the forcing function f in the form

$$f(\mathbf{x}) = \sum_n c_n X_n(\mathbf{x}),$$

$$c_n = \int \int \int_V f(\zeta) X_n(\zeta) \, dV.$$

The solution procedure is entirely analogous to that employed for periodic solutions of ordinary differential equations: we substitute the series expansions in (4.36), and equate expansion coefficients. The result is

$$\nabla^2 \left(\sum_n B_n X_{(\mathbf{x})} \right) = - \sum_n c_n X_n(\mathbf{x}),$$

$$\sum_n -\lambda_n^2 B_n X_n(\mathbf{x}) = - \sum_n c_n X_n(\mathbf{x}),$$

or

$$\lambda_n^2 B_n = c_n,$$

$$B_n = \frac{c_n}{\lambda_n^2}$$

(as long as $\lambda_n^2 \neq 0$). This gives the solution of (4.36) in the form

$$u(\mathbf{x}) = v(\mathbf{x}) + \frac{\sum_n \left(\int \int \int_V f(\zeta) X_n(\zeta) \, dV \right) X_n(\mathbf{x})}{\lambda_n^2}.$$

"Conceptual solutions" in the above style may also be derived for the inhomogeneous heat and wave equations.

The system

$$\frac{\partial u}{\partial t} - \nabla^2 u = f(\mathbf{x}, t),$$

$$\alpha(s)u(s) + \beta(s)\frac{\partial u}{\partial n}(s) = 0$$

may be solved by a similar method. Expand $f(\cdot, t)$ (for each fixed t) in an orthonormal expansion

$$f(\mathbf{x}, t) = \sum_n c_n(t) X_n(\mathbf{x})$$

with

$$c_n(t) = \int \int \int_V f(\zeta, t) X_n(\zeta) \, dV,$$

and seek a solution in the form

$$u(\mathbf{x}, t) = \sum_n B_n(t) X_n(\mathbf{x}).$$

This gives (upon substitution in (4.33))

$$\sum_n \frac{\partial B_n}{\partial t} X_n = \sum_n -\lambda_n^2 B_n X_n + \sum_n c_n X_n,$$

or

$$\frac{\partial B_n}{\partial t} = -\lambda_n^2 B_n + c_n$$

with solution

$$B_n(t) = B_n(0) e^{-\lambda_n^2 t} B_n(0) + \int_0^t e^{-\lambda_n^2 (t-\tau)} c_n(\tau) \, d\tau.$$

The solution is then

$$u(\mathbf{x}, t) = \sum_n \left\{ e^{-\lambda_n^2 t} B_n(0) + \int_0^t e^{-\lambda_n^2 (t-\tau)} c_n(\tau) \, d\tau \right\} X_n(\mathbf{x}).$$

$\{B_n(0)\}$ is (as usual) the initial condition expansion sequence. The result may be written in different form by incorporating this and the initial condition expansion formulas within the solution expression.

The case of inhomogeneous boundary conditions may be reduced to the case just treated. To do this, first we solve Laplace's equation

$$\nabla^2 v = 0,$$

$$\alpha(s)v(s) + \beta(s)\frac{\partial v}{\partial n}(s) = g(s, t),$$

treating t as a parameter (constant as far as the spatial derivatives of this auxiliary solution are concerned. This produces a solution $v(\mathbf{x}, t)$ formally satisfying

$$\nabla^2 v(\mathbf{x}, t) = 0, \, \mathbf{x} \in V,$$

$$\alpha(s)v(s, t) + \beta(s)\frac{\partial v}{\partial n}(s, t) = g(s, t), \, s \in \Sigma.$$

The solution of the inhomogeneous problem

$$\frac{\partial u}{\partial t} - \nabla^2 u = f(\mathbf{x}, t),$$

$$\alpha(s)u(s, t) + \beta(s)\frac{\partial u}{\partial n}(s, t) = g(s, t)$$

is then sought in the form

$$u = \tilde{u} + v(\mathbf{x}, t).$$

Evidently $\tilde{u}$ meets the homogeneous boundary condition

$$\alpha(s)\tilde{u}(s, t) + \beta(s)\frac{\partial \tilde{u}}{\partial n}(s, t) = 0$$

and $\tilde{u}$ satisfies the equation

$$\frac{\partial \tilde{u}}{\partial t} = \nabla^2 \tilde{u} + f(\mathbf{x}, t) + \frac{\partial}{\partial t} v(x, t).$$

This is a forced heat equation problem with homogeneous boundary conditions, and the formal solution follows as above by series methods.

Example. As an example of the procedures advocated above, we consider the problem

$$\frac{\partial u}{\partial t} = \frac{\partial^2 u}{\partial x^2} + f(x, t), \ 0 < x < 1,$$

$$u(0, t) = g_0(t), \ u(1, t) = g_1(t),$$

$$u(x, 0) = u_0(x), \ 0 < x < 1$$

as the most familiar model incorporating the complications in question.
  In this case the eigenvalue problem reduces to the standard

$$\frac{d^2}{dx^2} X(x) = -\lambda^2 X(x),$$

$$x(0) = x(1) = 0,$$

with eigenfunctions and eigenvalues

$$X_n(x) = \sqrt{2} \sin n\pi x,$$

$$\lambda_n^2 = (n\pi)^2.$$

The auxiliary "Laplace equation" solution $v(x, t)$ is determined by

$$\frac{d^2 v}{dx^2} = 0,$$

$$v(0, t) = g_0(t), \ v(l, t) = g_1(t),$$

so that

$$v(x, t) = g_0(t) + x(g_1(t) - g_0(t)).$$

The function u is determined by solving the forced heat equation

$$\frac{\partial \tilde{u}}{\partial t} = \nabla^2 \tilde{u} + f(\mathbf{x}, t) - \frac{\partial}{\partial t}(g_0(t) + x(g_1(t) - g_0(t)))$$

with homogeneous boundary conditions $u(0) = u(1) = 0$.
  Following the eigenfunction expansion method, we find a solution in the form

$$\tilde{u}(x, t) = \sum_{n=1}^{\infty} B_n(t) \sin n\pi x,$$

where

$$B_n(t) = B_n(0) \, e^{-(n\pi)^2 t} + \int_0^t e^{-(n\pi)^2(t-\tau)} c_n(\tau) \, d\tau.$$

The appropriate $c_n(\tau)$ is given by

$$c_n(\tau) = \int_0^1 \sqrt{2} \sin n\pi\zeta \left\{ f(\zeta, t) - \frac{\partial}{\partial t}(g_0(t) + \zeta \, (g_1(t) - g_0(t))) \right\} d\zeta$$

$$= \sqrt{2} \int_0^1 \sin n\pi\zeta f(\zeta, t) \, d\zeta - g_0'(\tau) \frac{1 - (-1)^n}{n\pi} \sqrt{2}$$

$$- \left\{ g_1'(\tau) - g_0'(\tau) \right\} \frac{(-1)^{n+1} \sqrt{2}}{n\pi}.$$

Hence

$$B_n(t) = B_n(0) e^{-(n\pi)^2 t} + \int_0^t e^{-(n\pi)^2(t-\tau)} \left\{ \sqrt{2} \int_0^1 \sin n\pi\zeta f(\zeta, \tau) \, d\zeta \right.$$

$$\left. - g_0'(\tau) \frac{\sqrt{2}}{n\pi}(1 - 2(-1)^n) - g_1'(\tau) \frac{\sqrt{2}}{n\pi}(-1)^{n+1} \right\} d\tau.$$

$B_n(0)$ is determined from

$$B_n(0) = \sqrt{2} \int_0^1 \sin n\pi\zeta \, \tilde{u}(\zeta, 0) \, d\zeta$$

$$= \sqrt{2} \int_0^1 [u(\zeta, 0) - \{g_0(0) + \zeta(g_1(0) - g_0(0))\}] \sin n\pi\zeta \, d\zeta.$$

Finally, u is recovered as

$$u(x, t) = \tilde{u}(x, t) + v(x, t).$$

We remark that the method evidently requires differentiability of g_0, g_1 in order to make sense of some of the above formal calculations.

With obvious modifications the same general method may be adopted to the case of inhomogeneous wave equations. Details are left as an exercise.

Problems 4.6.

1. Consider heat conduction in a cylindrical solid, with the surfaces $r = b$, $z = 0$, $z = c$ all insulated. Show that the governing equation is

$$\frac{1}{\kappa} \frac{\partial u}{\partial t} = \frac{\partial^2 u}{\partial r^2} + \frac{1}{r} \frac{\partial u}{\partial r} + \frac{1}{r^2} \frac{\partial^2 u}{\partial \theta^2} + \frac{\partial^2 u}{\partial z^2},$$

with boundary and initial conditions

$$\frac{\partial u}{\partial r}(b, \theta, z, t) = 0,$$

$$\frac{\partial u}{\partial z}(r, \theta, 0, t) = \frac{\partial u}{\partial z}(r, \theta, c, t) = 0,$$

$$u(r, \theta, z, 0) = u_0(r, \theta, z),$$

$$u(r, \theta, z, t) \text{ bounded.}$$

Find a solution of this problem.

2. Show that for Problem 1, if either the initial condition is suitably chosen, or the cylinder is "slim" then the solution reduces to a familiar one-dimensional form in an appropriate sense.
3. Show that as $c \to 0$ in the solution of Problem 1, one obtains a solution of a "thin insulated disk" heat conduction problem. (This is in an approximate sense for arbitrary initial conditions, and exact for initial conditions independent of z.)
4. Using rectangular polar coordinates, formulate and solve the heat conduction problem for a laterally insulated ring. See Figure 4.15.

Show that the "thin ring" solution approaches that for a circular heat equation model (in an appropriate sense).

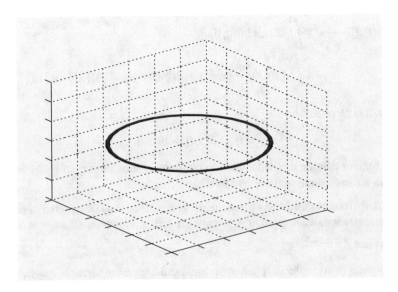

Fig. 4.15 Laterally insulated thin ring

5. Use Octave to generate a view of an $x = 0$ slice through the two-dimension transient heat equation solution discussed above. That should display the time variation of the temperature in the region.

6. Repeat Problem 5 for a model where insulation is applied to two opposite sides of the block, leaving the other two at a temperature $u = 0$.

7. The Cesaro sum used for the two-dimensional solution averages the two-dimensional Fourier partial sums in both dimensions. A much faster computation would average partial sums, while taking the same number of terms in both the x, y expansions. This would use

$$\frac{1}{N} \sum_{n=1}^{N} partial\_sum\_2x\_t(term, x, y, t, n, n),$$

and effectively try to evaluate a two-dimensional limit along the diagonal of the limit variables.

   Try this in Octave, and verify on the heat equation example that some Gibb's phenomenon artifacts seem to appear in the result (compared to the two-dimensional averaging).

8. Establish a version of Parseval's Theorem applicable to the case of three-dimensional Fourier series expansions arising from the partially insulated block heat conduction problem of the text.

9. We showed how to evaluate the zeroes of the Bessel function J_0 using Octave. Use Octave to generate an array whose entries are $\{j_{mn}\}$, the nth zero of the Bessel function J_m.

10. With the result of Problem 9 you should be able to evaluate and display a solution of a heat equation in a circular region. The messiest part of this problem is calculating the expansion coefficients with respect to the Bessel function basis. Completion of this problem probably deserves an Octave merit badge as a reward.

11. Use the indicated change of variables to obtain the associated Legendre equation

$$\frac{d}{dx}\left[(1 - x^2)\frac{dX}{dx}\right] + \left[m(m + 1) - \frac{n^2}{1 - x^2}\right] X = 0, \ 0 < x < 1$$

   from the separated equation (4.31).

12. Use the Frobenius method to show that Legendre's equation

$$\frac{d}{dx}\left[(1 - x^2)\frac{dX}{dx}\right] + [m(m + 1)] X = 0$$

   has a polynomial solution for nonnegative integer values of m.

13. Directly verify that the expression in equation (4.32) is a solution of the associated Legendre equation.

14. Use the vector identity

$$\int\int\int_V v\,\nabla^2 u + \nabla u \cdot \nabla v \, dV = \int\int_\Sigma v\frac{\partial u}{\partial n}\,dS$$

to show that the eigenvalues of ∇^2 under the boundary conditions of vanishing normal derivative are non-positive.

15. Use the ideas of problem (14) and the "universal heat equation" solution of the text to show that the temperature of an insulated body tends as time $t \to \infty$ to the average of the initial temperature.

    Hint: Under insulated boundary conditions ∇^2 has an obvious eigenvector and eigenvalue.

16. Verify that under the standard boundary conditions on ∇^2, the eigenvalues are real and eigenvectors orthogonal (assuming that these exist.)

17. Heat conduction in an inhomogeneous medium leads to the model equation

$$\frac{\partial u}{\partial t}(x,t) = \frac{1}{\rho(x)c(x)}\,\nabla\cdot(\alpha(x)\,\nabla u)\,,\ x\in V.$$

Associate with this problem (on a compact region V) a formally self-adjoint eigenvalue problem, and describe the resulting formal solution and expansion problems.

18. Carry out the formal solution procedure for the "general wave equation"

$$\frac{1}{c^2}\frac{\partial^2 u}{\partial t^2} = \nabla^2 u,\ \mathbf{x}\in V,$$

$$\alpha(s)u(s) + \beta(s)\frac{\partial u}{\partial n}(s) = 0,\ s\in\Sigma,$$

$$u(\mathbf{x},0) = f(x),$$

$$\frac{\partial u}{\partial t}(x,0) = g(x).$$

19. Consider the "general solution" of Poisson's equation in a circular domain formally derived in the text. Suppose that the boundary condition is that

$$\frac{\partial u}{\partial n} = 0.$$

Show that $\lambda = 0$ is an eigenvalue, and that the formal solution procedure in this case requires that $c_0 = 0$. Interpret this condition physically.

20. Find the solution of the heat equation problem

$$\frac{\partial u}{\partial t} = \frac{\partial^2 u}{\partial x^2},\ 0 < x < 1,$$

$$u(0, t) = 0,$$

$$u(l, t) = \sin(\omega_0 t),$$

$$u(x, 0) = 0.$$

21. Outline in detail the solution procedure for the "general linear inhomogeneous wave equation"

$$\frac{1}{c^2} \frac{\partial^2 u}{\partial t^2} = \nabla^2 u, +f(\mathbf{x}, t) \ \mathbf{x} \in V,$$

$$\alpha(s)u(s) + \beta(s)\frac{\partial u}{\partial n}(s) = g(s, t), \ s \in \Sigma,$$

$$u(\mathbf{x}, 0) = h(x),$$

$$\frac{\partial u}{\partial t}(x, 0) = j(x).$$

(You may assume existence of an appropriate complete set of eigenvectors).

22. Find the solution of the following concrete case of Problem 21.

$$\frac{1}{c^2} \frac{\partial^2 u}{\partial t^2} = \frac{\partial^2 u}{\partial x^2} u, \ 0 < x < 1,$$

$$u(x, 0) = 0,$$

$$\frac{\partial u}{\partial t}(x, 0) = 0,$$

$$.u(l, t) = 0,$$

$$u(0, t) = \sin(\omega_0 t).$$

23. Provide a physical interpretation for the time variable boundary condition

$$\alpha(s) u(s) + \beta(s)\frac{\partial u}{\partial n}(s) = g(s, t), \ s \in \Sigma,$$

in the case of a single spatial variable.

24. In various chemical processes, pipes carry material of differing temperatures. At times of transition the material of the pipe may be subject to thermal stress from these changes. One model for such a problem is that of an insulated pipe section, subject to a time varying internal temperature.

Assuming angular uniformity, show that an appropriate model is

$$\frac{\partial u}{\partial t} = \kappa \left(\frac{\partial^2 u}{\partial r^2} + \frac{1}{r}\frac{\partial u}{\partial r} + \frac{\partial^2 u}{\partial z^2} \right), \ a < r < b, 0 < z < c,$$

with boundary conditions

$$u(a, z, t) = g(z, t),$$

$$\frac{\partial u}{\partial r}(b, z, t) = 0,$$

in the radial direction, and, assuming the pipe ends at fixed temperatures,

$$u(r, 0, t) = T_0,$$

$$u(r, c, t) = T_1.$$

Assuming that

$$u(r, z, 0) = T_0 + \frac{z}{c}(T_1 - T_0),$$

find a solution of this problem.

4.7   Finite Differences and Numerical Methods

While the theory associated with partial differential equations having a formally self-adjoint spatial derivative component has an impressive scope, there are problems of interest for which it is inadequate.

One such class of problems is, of course, the nonlinear case. Nonlinear problems for which an adequate analytical treatment exists stand as landmarks. For the bulk of such problems, numerical methods of approximation provide the only available avenue of analysis. There are also linear problems for which this is appropriate. Many of these involve complicated physical geometries far removed from the possibility of exact solution, even though in principle the usual eigenfunction expansions may exist. In such cases, direct numerical calculations may represent the most computationally economical means of obtaining the desired results.

Numerical methods for partial differential equations can be broadly divided into methods based on local finite-difference approximations to the governing equations, and methods based on construction of more global approximations to the unknown solution functions. In this section we introduce finite difference methods, while methods loosely based on approximation techniques are introduced in the following section.

To specify a function of one or more real variables completely requires specification of an "infinite amount of information."[6] To represent a function computationally, only a finite number of values (with finite precision) can be carried as data.

---

[6]E.g., the value at an infinite number of points, or an infinite number of Fourier coefficients.

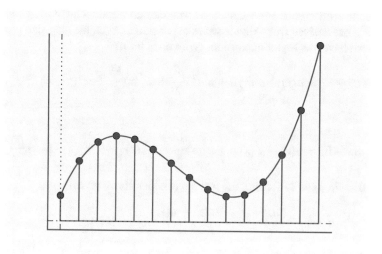

Fig. 4.16 Sampled solution values

In the case of numerical methods based on finite differences, the interpretation is that this data represents the function values at a set of node or sample points in the function domain. In the case of a single independent variable, this is illustrated in Figure 4.16.

The function in question is represented by the sample values

$$\{u_j = u(x_j)\}_{j=0}^{N}$$

and is calculated between sample points by an interpolation of the sample values.

The essence of the finite difference approach is to approximate the derivatives of an unknown function in terms of the function sample values, and to substitute these approximations into the governing differential equations. This converts the original partial (or ordinary) differential equation into a (typically very large) system of simultaneous algebraic equations which are then solved for the unknown sample values.

It may be noted from this description that there are two sources of error in calculations made with the finite differences approach. The first is a *discretization or truncation error* arising from the replacement of the derivatives by their finite difference approximation, while the second is a *round off error* introduced from the necessarily inexact solution of the resulting algebraic equations. In most applications the truncation error should be most significant. However, roundoff errors typically grow with the size of system being solved, and so the strategy of driving down the truncation error estimates by increasing the number of sample points is not one that can be carried to extremes.

The finite difference approximation formulas and the associated truncation error estimates are derived from Taylor series expansions. Given the step size parameter $h > 0$, we have the Taylor expansions (with error term)

$$u(x + h) = u(x) + u'(x)h + u''(x)\frac{h^2}{2} + u'''(x)\frac{h^3}{6} + u^{iv}(x + \theta_1 h)\frac{h^4}{24},$$

and

$$u(x - h) = u(x) - u'(x)h + u''(x)\frac{h^2}{2} - u'''(x)\frac{h^3}{6} + u^{iv}(x + \theta_2 h)\frac{h^4}{24},$$

where $0 < \theta_1, \theta_2 < 1$. Adding the two expressions above gives

$$u''(x) = \frac{(u(x+h) - 2u(x) + u(x-h))}{h^2} + M\frac{h^2}{12}, \tag{4.37}$$

where M is of order max $|u^{iv}(x)|$. Writing out the expansion of $u(x_h)$ using the third-order Taylor error formula provides

$$u'(x) = \frac{(u(x+h) - u(x-h))}{2h} + M_1\frac{h^2}{6}, \tag{4.38}$$

where M_1 is of order max $|u'''(x)|$. The formulas (4.37) and (4.38) are called *central difference approximations*, and provide the finite difference approximations

$$u''(x) \approx \frac{(u(x+h) - 2u(x) + u(x-h))}{h^2},$$

$$u'(x) \approx \frac{(u(x+h) - u(x-h))}{2h}$$

with a truncation error of order h^2 in the variable increment. The forward difference approximation

$$u'(x) \approx \frac{u(x+h) - u(x)}{h}$$

by contrast has a truncation error of order equal to the step size h.

As a sample problem for the finite difference method, we consider the basic diffusion equation

$$\frac{\partial u}{\partial t} = \frac{\partial^2 u}{\partial x^2}, \quad 0 < x < 1,$$

$$u(0, t) = u(1, t) = 0,$$

$$u(x, 0) = \sin(\pi x).$$

In the construction of a finite difference approximation to a given problem, the choice of finite difference formulas used determines (among other things) the ease

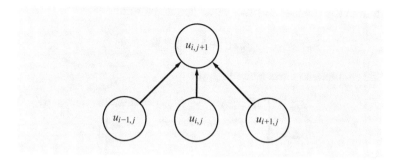

Fig. 4.17 Dependency diagram

with which the resulting algebraic system may be solved. If one chooses a forward difference in the time coordinate, and a centered difference in the spatial coordinate, the resulting finite difference equation takes the form

$$\frac{u_{i,j+1} - u_{i,j}}{k} = \frac{u_{i-1,j} - 2\,u_{i,j} + u_{i+1,j}}{h^2}.$$

Here we assume that the sample points are equally spaced in the time direction with interval k, and in the spatial direction with step size h. The finite difference equation may be rewritten in the form

$$u_{i,j+1} = \left(r\,u_{i-1,j} - (2r - 1)\,u_{i,j} + r\,u_{i+1,j}\right), \qquad (4.39)$$

where i, j range over $j \geq 0, 1 \leq i \leq N - 1$, with the step parameters $r = \frac{k}{h^2}$, $h = \frac{1}{N}$.

This equation determines the interior spatial samples at time $j + 1$ in terms of the interior and boundary values at time j. This dependence is illustrated in Figure 4.17.

Note that the iteration (4.39) will generate sample values for an arbitrary combination of step sizes h and k. We expect that h and k must be chosen "small" in order that the replacement of derivatives by finite differences should be justifiable. However, there is also a constraint on the relative sizes of h and k that must be taken into account.

This constraint arises from consideration of the propagation of roundoff error through the course of a computation based on (4.39). The update algorithm (4.39) expresses the sample point values at time $j + 1$ as a linear combination of those at time j. If these sample values are written in vector form (by stacking the spatial samples) with the definition

$$\mathbf{u}_j = \begin{bmatrix} u_{1,j} \\ u_{2,j} \\ u_{3,j} \\ \vdots \\ u_{N-1,j} \end{bmatrix},$$

the system of equations (4.39) may be written in the form

$$\mathbf{u}_{j+1} = \mathbf{A}\,\mathbf{u}_j,\ j \geq 0,$$

where the coefficient matrix takes the form

$$\mathbf{A} = \begin{bmatrix} (l-2r) & r & \cdots & 0 \\ r & (l-2r) & r & \cdots \\ \vdots & \ddots & \ddots & \vdots \\ 0 & \cdots & r & (l-2r) \end{bmatrix}.$$

Because the difference equation is linear, an error in **u** at any time is propagated forward according to the same equation. The effect of an initial error **e** after j steps is therefore the expression

$$\mathbf{E} = \mathbf{A}^j\,\mathbf{e}. \tag{4.40}$$

If this initial error effect is to disappear as the computation progresses, it is necessary that $\mathbf{A}^j \to \mathbf{0}$ as $j \to \infty$. In order for this to occur, it is necessary and sufficient that the eigenvalues of A should all be less than 1 in magnitude.

The special form of the coefficient matrix **A** makes it possible (see Section 8.4) to calculate eigenvalues in closed form. The result of this calculation is that the eigenvalues are given by

$$\lambda_k(\mathbf{A}) = 1 - 4\,r\sin^2\frac{k\pi}{N},\ k = 1, 2, \ldots, N-1.$$

In order that the effect of roundoff should be damped out, it is required that r be chosen so that

$$\left| 1 - 4\,r\sin^2\frac{k\pi}{N} \right| < 1,\ k = l, 2, \ldots, N-1.$$

The constraint on the relative sizes of the time and spatial steps is therefore that h and k should be chosen to ensure that

$$\frac{k}{h^2} = r < 1/2. \tag{4.41}$$

If this constraint is satisfied (so that the roundoff effects are damped) the numerical method (4.39) is said to be *numerically stable*. The effect of the condition (4.41) can be seen dramatically in sample computations. See the problems below.

The constraint (4.41) requires that one scale the time axis as the square of the scaling of the spatial axis in order to maintain stability as the step sizes are decreased. There are other finite difference methods which do not have this stringent requirement. An example is the *Crank–Nicholson* scheme defined by (averaging the present and past second derivatives)

$$\frac{u_{i,j+1} - u_{i,j}}{k} = \left\{ \frac{u_{i+1,j+1} - 2u_{i,j+1} + u_{i-1,j+1}}{2\,h^2} + \frac{u_{i+1,j} - 2u_{i,j} + u_{i-1,j}}{2\,h^2} \right\} \tag{4.42}$$

This is an example of an *implicit method*, so described because the matrix form of the vector iteration is

$$\mathbf{B}\,\mathbf{u}_{j+1} = \mathbf{C}\,\mathbf{u}_j, \tag{4.43}$$

defining the "updated" sample values implicitly, and requiring the solution of the linear system (4.43) in order to proceed from one time step to the next. It can be shown that (4.42) is numerically stable for all positive values of $r = k/h$ This may be regarded as the trade-off received from adoption of the implicit scheme (4.43).

The implicit scheme (4.43) requires the solution of a system of dimension equal to the number of spatial sample points at each time step. Larger scale numerical linear algebra problems arise from the application of finite differences to Laplace's equation

$$\nabla^2 \phi = \frac{\partial^2 \phi}{\partial x^2} + \frac{\partial^2 \phi}{\partial y^2} = 0 \text{ in } \Omega,$$

$$\phi \Big|_{\partial \Omega} = f.$$

If central differences are used with this equation, using step sizes of h and k results in the numerical problem

$$\frac{\phi_{i-1,j} - 2\,\phi_{i,j} + \phi_{i+1,j}}{h^2} + \frac{\phi_{i,j-1} - 2\,\phi_{i,j} + \phi_{i,j+1}}{k^2} = 0,$$

for i,j values corresponding to the interior of Ω together with (the boundary conditions)

$$\phi_{i,j} = f, \ i,j \in \partial \Omega.$$

A diagram illustrating the dependencies similar to Figure 4.17 can be given for this algorithm also.

The above equations represent a system determining (all of) the interior sample values in terms of the boundary values of f. The size of the resulting system in the case of most real problems demands that special methods tailored to the form of the equations be used to solve the resulting set of equations. We refer the reader to the numerical analysis literature for further discussion of these problems.

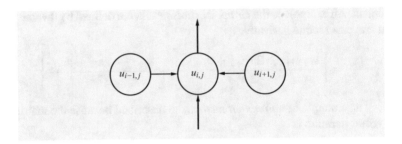

Fig. 4.18 Method of lines diagram

A numerical method related closely to finite difference methods (and also to the "analogue models" of Section 4.4) is the so-called *method of lines*. In our discussion above, all of the derivatives in the given partial differential equation have been replaced with finite differences. It is also possible to retain the derivatives with respect to one of the variables, and convert the given partial differential equation into a system of ordinary differential equations. In the case of the diffusion equation the resulting equations take the form

$$\frac{d}{dt} u_i = \frac{(u_{i-1} - 2\,u_i + u_{i+1})}{h^2},$$

which represents a system of equations for the evolution over time of the sample values of the original system. The diagram analogous to Figure 4.16 in the case of this system is given in Figure 4.18. This illustrates the continuous evolution of the sample points over time and is the source of the terminology "method of lines."

Software

The method of lines in the past has been used to construct systems solvable by analog computer. More recently, the method has been used to convert partial differential equations into a problem solvable by use of some of the sophisticated numerical ordinary differential equation packages available. In effect, this relies on the package to construct the time discretization required for the solution, while the spatial discretization converts the original partial differential equation to a system of approximating ordinary differential equations.

Octave contains various numerical methods for the solution of ordinary differential equations. The result illustrated earlier in Figure 3.12 can be taken as an example of using Octave to solve a partial differential heat equation by the method of lines.

Octave also supports solving partial differential equations with the finite element method, and we introduce that capability below.

Problems 4.7.

1. Assuming that the function u is four times continuously differentiable, show that

$$u''(x) = \frac{(u(x+h) - 2u(x) + u(x-h))}{h^2} + M\frac{h^2}{12},$$

where $|M| \leq \sup |u^{iv}(\zeta)|$, $\zeta \in [x-h, x+h]$.

2. Show that (assuming u three times continuously differentiable)

$$u'(x) = \frac{(u(x+h) - u(x-h))}{2h} + M_1\frac{h^2}{6},$$

where $|M_1| \leq \sup |u'''(\zeta)|$, $\zeta \in [x-h, x+h]$.

3. For the basic heat equation model

$$\frac{\partial u}{\partial t} = \kappa \frac{\partial^2 u}{\partial x^2},$$

$$u(0, t) = u(1, t) = 0,$$

$$u(x, 0) = f(x),$$

show that

$$\frac{\partial}{\partial t} \int_0^1 u^2(x, t)\, dx < 0.$$

4. For the discretized heat equation

$$u_{i,j+1} - u_{i,j} = \frac{k}{h^2}\left(u_{i-1,j} - 2u_{i,j} + u_{i+1,j}\right)$$

define (the energy at time j)

$$E_j = \sum_{i=0}^{N} h(u_{i,j})^2$$

and compute

$$E_{j+1} - E_j.$$

What do you hope for, in view the properties of the true solution?

5. Use the Jordan canonical form for matrices to show that

$$\mathbf{A}^j \mathbf{e} \to \mathbf{0}$$

as $j \to \infty$ if and only if all eigenvalues of $\mathbf{A}$ are less than 1 in magnitude.

6. Write a program to compute the iteration (4.39)

$$u_{i,j+1} = \left(r\, u_{i-1,j} - (2r - 1)\, u_{i,j} + r\, u_{i+1,j}\right),\, j > 0,\, 1 < i < N - 1$$

with

$$u_{0,j} = u_{N,j} = 0,$$

and

$$u_{i,0} = \sin\left(\frac{i\pi}{N}\right),\, i = 0, .., N$$

for values of $r = .3, .4, .45, .55, .6$, each over 10 time steps. Plot your results together with the exact solution of the original equation.

7. Determine the coefficient matrices **B** and **C** of the update formula (4.43) associated with the *Crank–Nicholson iteration* (4.42).

8. Write the Crank–Nicholson iteration (4.43) as (by inverting the coefficient matrix)

$$u_{j+1} = \mathbf{A}\, u_j.$$

Use the fact that the eigenvalues of

$$\mathbf{D} = \begin{bmatrix} -2 & 1 & \cdots & \cdots & 0 \\ 1 & -2 & 1 & \cdots & 0 \\ 0 & 1 & -2 & 1 & \cdots \\ \vdots & \ddots & \ddots & \cdots & 0 \\ 0 & \cdots & \cdots & 1 & -2 \end{bmatrix}$$

(**D** is $N - 1 \times N - 1$) are given by

$$\lambda_k(\mathbf{D}) = -\sin^2\left(\frac{k\pi}{2N}\right),\, k = 1, ..., N - 1,$$

to show that

$$\mathbf{A}^j\, \mathbf{e} \to \mathbf{0}$$

for the Crank–Nicholson scheme, independent of the value of $r = k/h$.

9. Show that applying the "method of lines" (using finite differences in the spatial direction) to the wave equation

$$\frac{\partial^2 u}{\partial t^2} = \kappa\, \frac{\partial^2 u}{\partial x^2},$$

$$u(0, t) = u(1, t) = 0,$$

$$u(x, 0) = f(x),$$

$$\frac{\partial u}{\partial t}(x, 0) = g(x)$$

results in a system of equations identical to that of the spring-mass wave equation analogue of Section 3.5.

10. What equations result if one attempts "separation of variables" in order to solve the discrete variable problem

$$u_{i,j+1} = r\, u_{i-1,j} - (2r - 1)\, u_{i,j} + r\, u_{i+1,j}?$$

(Solution methods for such equations are discussed in Chapter 8).

4.8   Variational Models and Finite Element Methods

The finite difference methods introduced in the previous section have had a relatively long history of application. They are based essentially on local approximations to the governing partial differential equations, and have error characteristics largely explainable by arguments based on the classical Taylor series.

There are also numerical methods for partial differential equations whose mode of approximation has a flavor more global in character. The Rayleigh–Ritz method had its origin in Rayleigh's work on sound waves. The finite element method is of more recent origin, with the major impetus having arisen from the needs of calculations in structural mechanics. Finite element methods have since found wide application to a variety of problems, and constitute an area under active development.

The sense in which these methods seek an approximate solution on a more global basis than finite difference methods can be seen from consideration of the framework in which the problem is formulated. To this point, we have characterized functions specifying quantities of physical interest as solutions of partial differential equations obtained "from (differential) first principles." However, there are alternative "first principles" that may be used to characterize such quantities. A variational formulation of the physical problem forms the starting point for many finite element methods. This different starting point leads to the difference in the method.

The general topic of variational formulations of physical problems is one with a wide literature, and strong connections to the classical (and modern) theory of the calculus of variations. For our purposes it suffices to introduce the ideas involved with a specific case.

A suitable example is provided by the problem of equilibrium of a two-dimensional membrane subject to edge displacements. The problem from a differential point of view leads to the boundary value problem for Laplace's equation

$$\nabla^2 u = 0 \text{ in } \Sigma, \qquad\qquad (4.44)$$

$$u\Big|_{\partial\Sigma} = f,$$

where f is the boundary displacement. Our immediate goal is to describe the variational formulation for this problem, and its connection with the conventional formulation (4.44) above.

A variational formulation of the membrane equation may be found on the basis of the spring-mass discussion in Section 4.1 which ultimately leads to the equation (4.44). Recall that in this problem we seek a description of a state of mechanical equilibrium, and that, in mechanics, attaining a state of equilibrium in a system with conservative interactions corresponds to minimizing the potential energy of the system. In the absence of gravity (the situation which corresponds to equation (4.44) above) the potential energy of the system is given by the sums of the squares of the spring displacements (times one half the spring constant). For small displacements from the horizontal, the corresponding form of the potential energy associated with the limiting partial differential equation model takes the form

$$E = \int\int_\Sigma \frac{1}{2}\left\{ \left(\frac{\partial u}{\partial x}\right)^2 + \left(\frac{\partial u}{\partial y}\right)^2 \right\} dA \qquad\qquad (4.45)$$

$$= \int\int_\Sigma |\nabla u|^2 \, dA.$$

A variational formulation of the original problem is the assertion that at equilibrium the potential E above is minimized, subject to the constraint that

$$u\Big|_{\partial\Sigma} = f.$$

This minimization problem is an example of a problem in the calculus of variations. The variable with respect to which the minimization takes place is a function (usefully regarded as an unknown vector in an appropriate vector space). The conventional equations (4.44) arise intuitively from the process of setting the "derivative" of the objective functional E given by (4.45) to zero in order to solve the minimization problem.

The problem of showing that a minimizing function u for the problem

$$\min_u \int\int_\Sigma |\nabla u|^2 \, dA, \text{ subject to } u\Big|_{\partial\Sigma} = f \qquad\qquad (4.46)$$

actually exists in a suitably chosen class of functions is a difficult one. A treatment of such problems is given in [7].

Assuming that such a solution exists with a suitable degree of differentiability, it is possible to derive the necessary condition that such a solution must satisfy.

We assume that there exists a function u minimizing (4.45) and satisfying the boundary constraint. This means that nearby functions still satisfying the constraint that the boundary values coincide with f must correspond to larger values of the potential functional E. To construct a family of nearby functions meeting this constraint we define an *admissible variation function* v to be one such that

$$v\bigg|_{\partial \Sigma} = 0,$$

and

$$\int \int_{\Sigma} |\nabla v|^2 \, dA < \infty.$$

This guarantees that (for any ϵ) the function

$$u + \epsilon \, v$$

has $\int \int_{\sigma} |\nabla (u + \epsilon v)|^2 \, dA < \infty$ with $(u + \epsilon v)\bigg|_{\partial \Sigma} = f.$

We may now evaluate the objective functional E for the family of functions of the form

$$u + \epsilon v.$$

With u the optimum value, and v a (temporarily fixed) admissible variation, this defines a function

$$\phi(\epsilon) = \int \int_{\Sigma} |\nabla (u + \epsilon v)|^2 \, dA$$

of the parameter ϵ. If u is the optimum value, then ϕ must have a local minimum value at $\epsilon = 0$ for each admissible variation v, so that we must have as a necessary condition

$$\frac{d}{d\epsilon} \phi(\epsilon)\bigg|_{\epsilon=0} = 0$$

for each admissible v. The introduction of the parameterized family of functions meeting the problem boundary conditions ("admissible") is the basic device in the calculus of variations. The effect is to bring the original problem (4.46) within range of the standard calculus method.

For the particular case under consideration we have the explicit form for $\phi(\epsilon)$ given as

$$\phi(\epsilon) = \int \int_{\Sigma} |\nabla u|^2 + 2\epsilon \, \nabla u \cdot \nabla v + \epsilon^2 |\nabla v|^2 \, dA$$

so that

$$\frac{d}{d\epsilon}\phi(\epsilon)\Big|_{\epsilon=0} = 2 \int\int_{\Sigma} \nabla u \cdot \nabla v \, dA.$$

The necessary condition for the optimality of u is therefore that

$$\int\int_{\Sigma} \nabla u \cdot \nabla v \, dA = 0$$

for all admissible variation functions v.

To see what this condition implies for the optimal function u, recall the Green's Theorem identity

$$\int\int_{\Sigma} \left(v\nabla^2 u + \nabla u \cdot \nabla v\right) dA = -\int_{\partial\Sigma} v\frac{\partial u}{\partial n} \, dS$$

(assuming that the optimum is a sufficiently well-behaved function). Since v is an admissible variation, $v\Big|_{\partial\Sigma} = 0$, and so the identity implies that for the optimal u and an admissible v,

$$\int\int_{\Sigma} \left(v\nabla^2 u\right) dA = -\int\int_{\Sigma} \left(\nabla u \cdot \nabla v\right) dA.$$

The necessary condition then is that

$$\int\int_{\Sigma} \left(v\nabla^2 u\right) dA = 0$$

for all admissible variations v. Since v is arbitrary (save vanishing at the boundary), we conclude from this condition that the optimal u must satisfy the conditions[7]

$$\nabla^2 u = 0 \text{ in } \Sigma$$

as well as

$$u\Big|_{\partial\Sigma} = f.$$

This argument shows the connection between (the variational formulation) (4.46) and the conventional differential model (4.44) for the elastic membrane problem. The equilibrium position u may be characterized as minimizing the potential energy

---

[7] If $\nabla^2 u \neq 0$ in some region of Σ, one can construct a smooth v vanishing outside of this region, and such that $\int\int v\nabla^2 u \, dA \neq 0$. Therefore $\nabla^2 u$ must vanish in Σ.

E in (4.45), subject to the boundary constraint. The minimization of (4.46) by use of the "calculus of variations" shows the usual model equations as the necessary conditions for the minimization.

A numerical method can be based on finding an approximate solution to the minimization problem (4.46), rather than attempting an approximate solution of the differential equation (4.44) as is done with finite difference methods. Since a computational solution is to be attempted, only a finite number of variables can be handled. The idea then is to carry out an approximate solution of the problem by minimizing the energy functional E given by (4.46) over a class of candidate functions which is finite dimensional.

The resulting approximate minimizer is taken as the numerical solution to the problem. The power of these methods comes from the fact that it is often possible to adapt the approximating functions to the problem at hand, and obtain a relatively accurate solution with the use of comparatively few parameters in the computation.

The general form of the computations to be made can be seen by writing the approximating function in the form

$$u_N(\mathbf{x}, \alpha_1, \ldots, \alpha_N)$$

where $\{\alpha_i\}_{i=1}^N$ are the parameters, often sample values of the unknown function. The approximate minimization is then (for our sample problem)

$$\min_{\alpha_1,\ldots,\alpha_N} \int\int_\Sigma |\nabla u_N(\mathbf{x}, \alpha_1, \ldots, \alpha_N)|^2 \, dA. \tag{4.47}$$

The constraint that the approximating function must meet the boundary conditions (at least approximately) is taken care of by the construction process for u_N. The minimization problem leads to the system of equations

$$\frac{\partial}{\partial \alpha_i} \int\int_\Sigma |\nabla u_N(\mathbf{x}, \alpha_1, \ldots, \alpha_N)|^2 \, dA = 0 \tag{4.48}$$

for the unknown parameters $\alpha_1, \ldots, \alpha_N$. In the common case, ∇u_n is linear in the parameters $\alpha_1, \ldots, \alpha_N$ and (4.48) therefore represents a system of linear equations for the unknowns. The detailed form of the equations (4.48) is highly dependent on the nature of the approximating functions chosen.

The main characteristic which distinguishes the *finite element method* from others based on the variational problem formulation (such as the Rayleigh-Ritz method) is the choice of the form of the functions used in the approximation process. For the finite element method, the physical problem domain is subdivided into *finite elements*, for example, by triangulation of the physical domain. The approximation is obtained by use of polynomial interpolation over each finite element. The overall approximation is generated by "splicing together" the approximations on the finite element subdomains (see Figure 4.19)

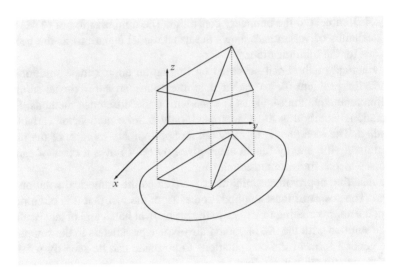

Fig. 4.19   Triangular element decomposition

The *finite element* terminology arises from the possibility of interpreting the corresponding solution elements as idealized structural members in certain structural mechanics applications.

Mathematically, the basic finite elements are defined through a process of piecewise polynomial interpolation. There exists a hierarchy of finite elements which have been investigated. The cases considered vary with the dimension of the underlying problem domain (one, two, or more independent variables) as well as with the degree of differentiability required of the resulting overall interpolation.

This system is potentially a very large scale linear algebra problem. However, the nature of the finite elements results in a *sparse* coefficient matrix $\mathbf{M}$ (a given u-value is associated with at most six different finite elements), which is an aid in the numerical solution in a typical application case,

The mathematical basis behind variational methods is discussed in the references [7] and [3]. The [7] reference contains an introduction to finite elements of basic types. The fact is that the finite elements that are used in practice vary with the problem under consideration.

The range of available choices is discussed in the book documenting the FEniCS project [8]. This project provides free software for finite element calculations, and supports both C++ libraries and a python interface to the libraries.

The FEniCS Project allows symbolic description of the variational problem to be solved. The physical domain is described by a mesh: this is generated by an external program in general, or put together from built-in examples of relative simple geometry. The variational form is described and processed by the library codes to generate a solution.

The basic library is written in C++, and it is possible to write applications entirely in C++ using the API defined by the library. At the time the library was constructed, it was planned to provide a python interface, so the library could be used with a python scripting interface. As a result there is not a large speed penalty associated with using python. There is also an Octave package that uses the C++ library to give an Octave extension. The octave user interface is basically the same as that encountered with using the C++ library directly, while the python FEniCS extension has a higher, more abstract feel to the API.

An example problem in python is given below. It is a small variant on one of the demo scripts provided by the fenics code package.

4.9 Computational Finite Element Methods

The finite element method is a very powerful method for generating numerical solutions to a variety of partial differential equation problems. For problems with physical geometries outside of the standard "classical shapes," finite element methods represent a well-understood practical method to generate some representative solutions.

Of course, applying the finite element method to a particular problem requires defining elements adapted to the physical domain, and managing the "bookkeeping" associated with refinements of the finite element mesh as the solution process proceeds. For something like the problem of heat conduction in an exhaust manifold casting, construction and management of a finite element mesh description is a nontrivial problem.

If one considers this issue with an intent to handle general three-dimensional shapes, it becomes evident that the first issue is the mathematical (and computational) description of the physical shape of the problem domain. This is also the initial issue with computer aided design (CAD) programs, so it is natural to utilize the graphical "front end" of a CAD program as the "input device" for finite element calculations.

There are a large number of commercial software programs in this area. In particular there are "add on" programs for popular CAD environments like Autocad [TM], as well as other tools. Consider also the development of open source CAD library code hosted at *Open Cascade* [6]. There are now some public standards for graphical CAD formats,[8] so easy to obtain and use finite element tools should become more available over time. There is a free CAD program [2] based on the Open Cascade libraries. This has some support for working with meshes, and may prove useful for generating mesh models for use in fenics.

---

[8]Look for *IGES*, CAD, FEM, and similar things on the world wide web.

Python `fenics`

PYTHON
```
"""
FEniCS tutorial demo program: Poisson equation with Dirichlet
conditions. Simplest example of computation and visualization
with FEniCS.

Modified slightly for a 3d domain

-Laplace(u) = f on the unit square.
u = u0 on the boundary.
f = concentrated internal heat source
"""

from dolfin import *
```
\# Create mesh and define function space
```
mesh = UnitCubeMesh(16, 16, 16)
V = FunctionSpace(mesh, 'Lagrange', 1)
```
\# Define boundary conditions
```
u0 = Constant(0.0)

def u0_boundary(x, on_boundary):
    return on_boundary

bc = DirichletBC(V, u0, u0_boundary)

f = Expression("10*exp(-(pow(x[0] - 0.25, 2)\
+ pow(x[1] - 0.5, 2)  + pow(x[2] - 0.25, 2))) / (0.02)")
```
\# Define variational problem
```
u = TrialFunction(V)
v = TestFunction(V)

a = inner(nabla_grad(u), nabla_grad(v))*dx
L = f*v*dx
```
\# Compute solution
```
u = Function(V)
solve(a == L, u, bc)
```
\# Plot solution and mesh
```
plot(u)
plot(mesh)
```
\# Dump solution to file in VTK format
```
file = File('poisson-cube.pvd')
file << u
```
\# Hold plot
```
interactive()
```

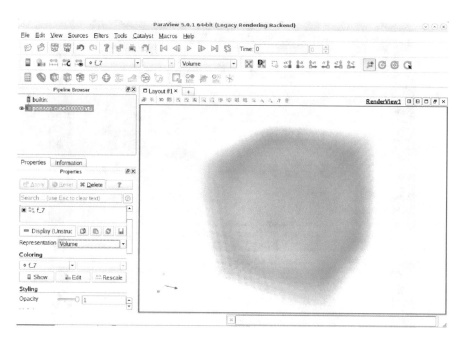

Fig. 4.20   FEniCS vtk output viewed in Paraview

The data is saved in a standard "vtk" format, for which there are several viewers (Figure 4.20). The difficulty is in interpreting the data from a multi-dimensional calculation. This is particularly acute in problems involving fluid transport of various properties carried along with the flow. "VTK" is a format for storing such data, and there are pieces of software for viewing such data. One is "paraview" used below on page 281. A much more powerful visualization tool is Mayavi2, which requires scripting in python to manage the data view, and a corresponding investment in learning the viewing API (Figure 4.21). The website for further information is illustrated on page 282.

Problems 4.9.

1. Find a differential equation for the function (of one variable) u minimizing

$$\int_0^1 \left(\frac{du}{dx}\right)^2 dx$$

   subject to $u(0) = 1, u(1) = 0$.
2. Find the energy expression corresponding to equation   (4.45) for the case of equilibrium of a membrane subject to the force of gravity.
3. Show that the variation of the energy derived in Problem 2 leads to the partial differential equation

Fig. 4.21 Website for Mayavi2 information

$$\nabla^2 u = -\frac{\rho g}{\kappa},$$

where ρ is the mass density, and g the acceleration due to gravity.

4. What differential equation is associated with variation of the functional (with respect to u)

$$\int_0^1 \left\{ \left(p(x) \frac{du}{dx} \right)^2 - r(x) \left(u(x) \right)^2 \right\} \, dx$$

subject to the constraints $u(0) = 0, u(1) = 1$?

5. Show that for $x_j \neq x_{j-1}$, the matrix

$$\begin{bmatrix} 1 & x_{j-1} & (x_{j-1})^2 & (x_{j-1})^3 \\ 1 & x_j & (x_j)^2 & (x_j)^3 \\ 0 & 1 & 2x_{j-1} & 3(x_{j-1})^2 \\ 0 & 1 & 2x_j & 3(x_j)^2 \end{bmatrix}$$

is invertible.

6. Suppose that it is desired to find the approximate solution of

$$\min_{u} \int_0^1 \left(e^x \frac{du}{dx} \right)^2 dx$$

subject to $u(0) = 0, u(1) = 1$. Find the form of the approximate minimization problem obtained by approximating u above by a basic linear finite element approximation, using four elements.

7. Suppose that it is required to solve

$$\nabla^2 u = 0 \text{ in } \Sigma,$$

$$u |_{\partial \Sigma} = f$$

where Σ is the unit circle, by means of finite element methods. FEniCS supports built-in meshes for a circle, so look through the demo programs to develop a solution to the unit circle problem.

8. Plot a finite element mesh for a triangular domain using the Octave `gmsh`.

9. Try to use the Octave fenics facilities to solve the problem of steady state heat conduction in a triangular domain. Look at the documentation for the python-fenics package.

10. With the FEniCS python facilities approximate the lowest eigenvector for ∇^2 on a semi-circular domain and zero boundary conditions. You'll have to look through the FEniCS book [8] for information on computing eigenvalues.

11. Explicitly solve the eigenvalue problem of Problem 10 and compare the exact and computed solutions.

12. Prepare a report on finite element mesh refinement, as described in [1]. Discover what FEniCS is using for mesh refinement.

References

1. R.E. Banks, *PLTMG: A Software Package for Solving Elliptic Partial Differential Equations* (Society for Industrial and Applied Mathematics, Philadelphia, 1990)
2. FreeCAD Developers, *Freecad* (2016). http://www.freecadweb.org/. Accessed 25 May 2016
3. J.L. Lions, *Optimal Control of Systems Governed by Partial Differential Equations* (Springer, New York, 1970)
4. W. Miller, *Lie Theory and Special Functions* (Academic, New York, 1968)
5. P.M. Morse, H. Feshbach, *Methods of Theoretical Physics* (McGraw-Hill, New York, 1953)
6. OPEN CASCADE SAS, Open cascade (2016). http://www.opencascade.com/. Accessed 31 May 2016
7. J.T. Oden, J.N. Reddy, *Mathematical Theory of Finite Elements* (Wiley, New York, 1976). Republished 2011, Dover Publications
8. The FEniCS Project, Automated solution of differential equations by the finite element method (2016). http://fenicsproject.org/book/index.html#book. Accessed 25 May 2016
9. E.C. Titchmarsh, *Eigenfunction Expansions Associated With Second Order Differential Equations* (Oxford University Press, London, 1946)

10. C.N. Watson, *A Treatise on the Theory of Bessel Functions*, 2nd edn. (Cambridge University Press, Cambridge, 1944)
11. E.T. Whittaker, C.N. Watson, *A Course of Modern Analysis*, 4th edn. (Cambridge University Press, Cambridge, 1927)

Further Reading

A.K. Aziz (ed.), *Symposium on the Mathematical Foundations of the Finite Element Method with Applications to Partial Differential Equations, University of Maryland, Baltimore* (Academic, New York, 1972)

C.M. Bender, S.A. Orzag, *Advanced Mathematical Methods for Scientists and Engineers* (McGraw-Hill, New York, 1978)

E.A. Coddington, N. Levinson, *Theory of Ordinary Differential Equations* (McGraw-Hill, New York, 1955)

R. Courant, D. Hilbert, *Methods of Mathematical Physics*, vol. 1 (Interscience, New York, 1953)

R. Courant, D. Hilbert, *Methods of Mathematical Physics*, vol. 2 (Interscience, New York, 1962)

K.E. Gustafson, *Introduction to Partial Differential Equations and Hilbert Space Methods* (Wiley, New York, 1980)

C.C. Lin, L.A. Segel, *Mathematics Applied to Deterministic Problems in the Natural Sciences* (MacMillan, New York, 1974)

J.E. Marsden, A.J. Tromba, *Vector Calculus* (W.H. Freeman, San Francisco, 1976)

L.M. Milne-Thompson, *Theoretical Hydrodynamics*, 4th edn. (MacMillan, London, 1962)

D.H. Norrie, C. de Vries, *An Introduction to Finite Element Analysis* (Academic, New York, 1978)

C.D. Smith, *Numerical Solution of Partial Differential Equations* (Oxford University Press, London, 1965)

G.D. Smith, *Numerical Solution of Partial Differential Equations* (Oxford University Press, London, 1965)

J.A. Sommerfeld, *Mechanics of Deformable Bodies* (Academic, New York, 1964)

J.A. Sommerfeld, *Partial Differential Equations in Physics* (Academic, New York, 1964)

W.C. Strang, C.J. Fix, *An Analysis of the Finite Element Method* (Prentice-Hall, Englewood Cliffs, 1973)

Chapter 5
Functions of a Complex Variable

5.1 Complex Variables and Analytic Functions

In earlier chapters, complex-valued functions appeared in connection with Fourier series expansions. In this context, while the function assumes complex values, the argument of the function is real-valued. There is a highly developed theory of (complex-valued) functions of a complex-valued argument. This theory contains some remarkably powerful results which are applicable to a variety of problems.

In this chapter, we provide an account of some of these basic results. These are applied to certain boundary value problems in this chapter, and to various transform methods in the following chapters.

Since a complex variable

$$z = x + iy$$

has a real and an imaginary part, the domain of definition of a function f of a complex variable z may be identified with (a subset of) the plane R^2. This is illustrated in Figure 5.1.

Through this identification it is usual to visualize a complex variable as a point in the plane. The representation of complex quantities in this manner is called the *Argand diagram*, and the plane of Figure 5.1 is referred to as the *complex plane*.

This representation is based initially on the use of rectangular coordinates in the plane. Points may also be written in *polar form* with the use of polar coordinates. The relationships are the usual rectangular polar conversion formulas. Given

$$z = x + iy,$$

the quantity

$$r = \sqrt{x^2 + y^2}$$

© Springer International Publishing Switzerland 2016
J.H. Davis, *Methods of Applied Mathematics with a Software Overview*,
Applied and Numerical Harmonic Analysis, DOI 10.1007/978-3-319-43370-7_5

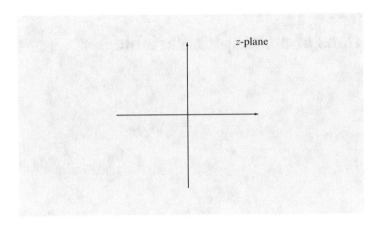

Fig. 5.1 Complex plane

is referred to as the *modulus* of the complex number z and is denoted by

$$|z| = \sqrt{x^2 + y^2}.$$

The angle θ in the representation

$$z = x + iy = r e^{i\theta} = r\cos\theta + ir\sin\theta$$

is referred to as the *argument* of the complex number z and is denoted by

$$\theta = \arg(z).$$

With this notation, the polar form may be written as

$$z = |z| e^{i\,\arg(z)}.$$

To say that f is a complex-valued function of a complex argument means that

$$f(x + iy) = u(x, y) + iv(x, y)$$

where u and v are the real and imaginary parts of f respectively. From the point of view of real-variable functions, the complex function f is equivalent to the pair of functions (u, v) and may be thought of as defining a function from the plane R^2 to R^2 by sending

$$\begin{bmatrix} x \\ y \end{bmatrix} \mapsto \begin{bmatrix} u(x, y) \\ v(x, y) \end{bmatrix}.$$

It might be thought that this observation would render a theory of functions of a complex variable unnecessary, in view of the wealth of the available results of multivariable calculus. However, it turns out that the economy of the notions and power of the results of complex variable theory make it useful in its own right.

To develop a complex variable calculus, we first consider the problem of defining differentiation for such functions. We introduce the natural definition

$$f'(z_0) = \lim_{h \to 0} \frac{f(z_0 + h) - f(z_0)}{h}$$

as the candidate for the derivative of the function $f(\cdot)$ a complex variable z at the point z_0. While this is formally identical to the corresponding real-variable definition, a crucial difference arises from the fact that the variables are complex-valued, and that the indicated limit is required to exist independent of the manner in which the "complex increment" h tends to zero. On the other hand, the formal identity of the definition results in the carry over of most real variable differentiation formulas to the complex domain.

The function f is said to be *differentiable* at the point z_0, provided that the limit exists.

This terminology describes behavior at a point, while it is convenient in the theory to have a term for functions that are differentiable at each point of some (open) domain [1] D. It is often said that such a function is *analytic* in D. However the term analytic is occasionally used for functions defined and differentiable at all but a finite number of points of D, and it is sometimes used to indicate that the function is defined and differentiable at each point of some (unspecified) domain.

Because of this potential ambiguity in the word "analytic," many books say that a function defined and differentiable at each point of the (open) domain D is "*regular in D*" or "*holomorphic in D.*"[2]

In this book, we use the terms "analytic" and "regular" with reference to a specified domain, and "analytic" (alone) to indicate regularity with respect to some domain.

If a function is defined and differentiable at each point of some open domain Γ (containing a point a) with the exception of the point a, the function is said to have an isolated singular point, or singularity, at the point $z = a$.

Example. Let $f(z) = z^2$. Then

$$\frac{f(z_0 + h) - f(z_0)}{h} = 2z_0 + h$$

and $f'(z) = 2z$ as might be expected, so that $f(z) = z^2$ is analytic for all z, $|z| < \infty$.

---

[1] The term "domain" has a technical meaning. See page 304.

[2] The consequences of assuming that a function is analytic are surprisingly large. The real variable case gives little clue how "well behaved" analytic functions are.

Example. Let $f(z) = e^z = e^x (\cos y + i \sin y)$. Then it may be checked that for all z, $|z| < \infty$,

$$\lim_{\alpha, \beta \to 0} \frac{e^{x+\alpha} (\cos(y + \beta) + i \sin(y + \beta)) - e^x (\cos y + i \sin y)}{\alpha + i\beta}$$

exists, with the value e^z. The function e^z is therefore everywhere analytic, for $|z| < \infty$.

Example. Let

$$f(z) = R(z) = \frac{p(z)}{q(z)}$$

where p, q are relatively prime polynomials in z. Then it is easy to see that $R(z)$ is differentiable at each z_0 such that $q(z_0) \neq 0$, and that the derivative formula coincides with what is expected from real-variable calculus. At each zero of q, R has an isolated singularity.

The fact that at a point of differentiability of a function f the limit

$$\lim_{h \to 0} \frac{f(z_0 + h) - f(z_0)}{h}$$

must exist independent of the manner in which $h \to 0$ implies certain relationships between the partial derivatives of the component functions u and v of an analytic function f. To derive these, we compute the derivative of f using both real and imaginary variable increments. Then we have (using a real increment α)

$$\begin{aligned}
f'(z_0) &= \lim_{\alpha \to 0} \frac{f(z_0 + \alpha) - f(z_0)}{\alpha} \\
&= \lim_{\alpha \to 0} \frac{u(x_0 + \alpha, y_0) - u(x_0, y_0)}{\alpha} + i \frac{v(x_0 + \alpha, y_0) - v(x_0, y_0)}{\alpha} \\
&= \frac{\partial u}{\partial x}(x_0, y_0) + i \frac{\partial v}{\partial x}(x_0, y_0),
\end{aligned}$$

while also (using the purely imaginary increment $i\beta$)

$$\begin{aligned}
f'(z_0) &= \lim_{\alpha \to 0} \frac{u(x_0, y_0 + \beta) - u(x_0, y_0)}{i\beta} + i \frac{v(x_0, y_0 + \beta) - v(x_0, y_0)}{i\beta} \\
&= -i \frac{\partial u}{\partial y}(x_0, y_0) + \frac{\partial v}{\partial y}(x_0, y_0).
\end{aligned}$$

Since we are assuming that $f'(z_0)$ exists, equating the real and imaginary parts of the two alternative expressions leads to the conclusion that

$$\frac{\partial u}{\partial x} = \frac{\partial v}{\partial y}, \tag{5.1}$$

$$\frac{\partial v}{\partial x} = -\frac{\partial u}{\partial y} \tag{5.2}$$

at a point of differentiability of the function f. These relations are called the *Cauchy–Riemann equations* for the real and imaginary parts of the analytic function f.

One of the central results of complex function theory is that a function f which is analytic in some domain D actually has (complex) derivatives of all orders in D. In particular, second order partial derivatives of the component u, v functions exist and are continuous in such a domain. Assuming this information, the partial differentiation of (5.1) may be justified in a region of analyticity of f to give the results that

$$\frac{\partial^2 u}{\partial y \partial x} = \frac{\partial^2 v}{\partial y^2}$$

$$\frac{\partial^2 u}{\partial x \partial y} = -\frac{\partial^2 v}{\partial x^2}$$

$$\frac{\partial^2 v}{\partial x \partial y} = \frac{\partial^2 u}{\partial x^2}$$

$$\frac{\partial^2 v}{\partial y \partial x} = -\frac{\partial^2 u}{\partial y^2}$$

in the region. Equating the (continuous) respective mixed partial derivatives shows that

$$\frac{\partial^2 u}{\partial x^2} + \frac{\partial^2 u}{\partial y^2} = 0$$

and

$$\frac{\partial^2 v}{\partial x^2} + \frac{\partial^2 v}{\partial y^2} = 0.$$

This means that both the real and imaginary parts of an analytic function are solutions of Laplace's equation in the region of analyticity. This observation is the basis of the application of complex variable methods to boundary value problems. This application is pursued in Sections 5.5 and 5.6 below.

Visualizing Complex Functions

A complex function of a complex argument is basically a two-dimensional vector-valued function whose domain is (a subset of) a plane. This means that complex functions may be visualized by treating them as a pair of "surfaces" defined by the real and imaginary parts (or magnitude and phase) of the function.

Octave supports complex-valued variables, so the definition of a complex function is indistinguishable from that of a real valued one. A rational function example is

OCTAVE

```
% f_of(z)
% rational function

function y = f_of(z)

y = (1./(z -1.001)) + 1./(z.^2 + 2*z +1.001) ;

endfunction
```

The odd numbers are a simple way to prevent Octave from dividing by zero when a simple coarse grid is used for plotting.

The function has a first order singularity near $z = 1$ and a double singularity close to $z = -1$. This means that attempts to make plots will encounter very large values, which "take over" the plot. This effect can be controlled by defining a function to "clip" the values on their way to the plot.

OCTAVE

```
% clip(x, limits)
% works for real valued things only, probably

function y = clip(x, limits)

  y = min(max(x, limits(1)), limits(2));

endfunction
```

We treat surface plots in the parametric form.

OCTAVE

```
% Plot of rational function

[X Y] = meshgrid(-2: .1 :2);

Z = X + i*Y;

W = f_of(Z)

u = real(f_of(Z)); v = imag(f_of(Z));

surf(X, Y, abs(W), gradient(abs(W)));
```

```
surf(X, Y, clip(abs(W), [-1, 5]), gradient(abs(W)));
surf(X, Y, clip(abs(W), [-1, 5]));

h = surf(X, Y, clip(abs(W), [-1, 5]));

set(h, 'EdgeColor', 'none', 'FaceColor', 'interp');

adjust_plot

print -depsc 'absratfun.epsc'
```

This results in the plot shown in Figure 5.2. The effect of the singular points can be clearly seen.

An interesting plot is obtained by constructing overlaid contour plots of the real and imaginary parts. The commands are

OCTAVE
```
[X Y] = meshgrid(-3: .05: 3);
 Z = X + i*Y;

W = f_of(Z);
u = real(W);
v = imag(W);

contour(X, Y, clip(u, [-4 4]), 30);
hold on
contour(X, Y, clip(v, [-4 4]), 30);
```

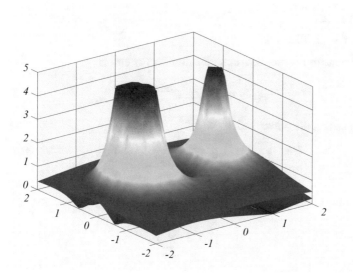

Fig. 5.2 Absolute value of a rational function

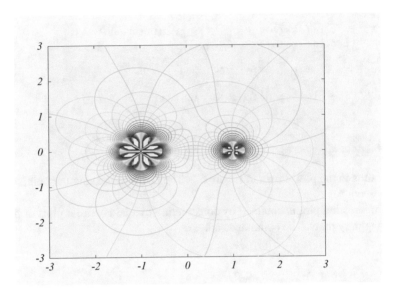

Fig. 5.3 Real and imaginary part level curves

The resulting plot in Figure 5.3 illustrates the orthogonality of the level curves. This property is a consequence of the Cauchy-Riemann conditions.

Problems 5.1.

1. Show that the modulus of a complex number satisfies the relations

   a. $|a\,b| = |a|\,|b|$,
   b. $|\frac{1}{a}| = \frac{1}{|a|}$,
   c. $|a + b| \le |a| + |b|$.

2. Show that for two nonzero complex numbers z_1 and z_2,

$$z_1\, z_2 = |z_1|\,|z_2|\, e^{i\,(\arg(z_1) + \arg(z_2))}.$$

3. Suppose that the power series

$$\sum_{n=0}^{\infty} a_n\, z^n$$

   has radius of convergence R. Show that

$$f(z) = \sum_{n=0}^{\infty} a_n\, z^n$$

   defines f as an analytic function for $|z| < R$.

4. Show that the function $a(\cdot)^3$ is not an analytic function.

$$a(z) = \bar{z}$$

5. Is $f(z) = |z|^2$ analytic?
6. Prove that, assuming f is analytic in a domain D, $|f|^2$ is not analytic unless f is constant in D.
7. Prove that the level curves of the real and imaginary parts of an analytic function are orthogonal to each other.
8. Suppose that an analytic function f is written in terms of polar variables as

$$f(re^{i\theta}) = u(r, \theta) + i\,v(r, \theta).$$

Express the Cauchy–Riemann conditions in terms of the polar variable functions r and θ.

9. Suppose that f is analytic on some domain D, and that g is analytic on some domain containing $f(D)$. Prove that $g \circ f : g \circ f(z) = g(f(z))$ is analytic on D, and that the derivative may be calculated using the chain rule.
10. Assume $e^z = 1 + z + \frac{z^2}{2!} + \frac{z^3}{3!} + \cdots$ and that

$$\cos z = \frac{e^{iz} + e^{-iz}}{2},$$

$$\sin z = \frac{e^{iz} - e^{-iz}}{2i}.$$

Compute

a. $\frac{d}{dz}\sin(z)$,
b. $\frac{d}{dz}\cos(z)$,
c. $\frac{d}{dz}\tan(z)$.

11. Prove that the quotient rule and product rule of differentiation apply to the quotient and product of analytic functions.
12. Suppose that f is analytic in some domain D and that $f(z) \neq 0$ for $z \in D$. Show that

$$\log |f(z)|$$

satisfies Laplace's equation in D.

$^3\bar{z}$ is the complex conjugate of $z = x + iy$, defined by $\bar{z} = x - iy$.

5.2   Domains of Definition of Complex Functions

The previous section introduced complex variable-versions of several "usual" functions. The attempt to produce complex versions of other common functions (e.g., logarithm, fractional root) leads to a problem of apparent "multi-valuedness." Since (by definition) a function must be single-valued, some extra care is required in the handling of such functions.

A problem involving domain specifications is encountered in the definition of inverse trigonometric functions in elementary calculus. There the problem is handled by restricting the domain of a function in such a way that the multi-valuedness does not appear. An analogous domain restricting convention appears in complex function theory in the guise of branch cuts. In the case of complex functions the problem may also often be circumvented through the introduction of an expanded domain of definition for the function in question. This expanded domain has a convenient geometrical visualization and is known as a Riemann surface.

To introduce the idea of an alternate domain description, we consider the Riemann sphere description of the conventional complex variable z-plane domain. The geometrical construct involved is illustrated in Figure 5.4. The idea is to associate points on the sphere (in a smooth, one-to-one fashion) with those on the plane. A function originally regarded as having a domain of definition in the plane may be provided with a domain on the Riemann sphere by defining the mapping from the plane to the sphere to be the point of intersection of a line from the sphere's north pole down to the point on the plane. To define the function on the sphere, the value at a spherical coordinate is assigned to be the value assumed by the original function at the corresponding point in the plane.

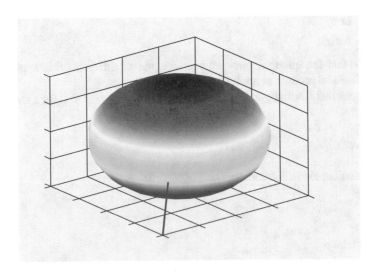

Fig. 5.4 Riemann sphere and mapping

This visualization of the z-plane is useful in that it places the point $z = \infty$ on the same footing as the other points in the (finite) plane. This point of view is useful in the discussion of functions transforming the z-plane into itself, where it is convenient to consider the mapping of certain points "to ∞." An example is the mapping

$$g(z) = \frac{1}{z},$$

which may be viewed as an interchange of the top and bottom hemispheres of the Riemann sphere. Similar mappings appear in the treatment of fluid flow and heat flow problems by complex variable methods (see Section 5.6 below), and it is convenient to treat them on a uniform basis.

In consideration of rational functions of z, the introduction of the Riemann sphere is more of a conceptual convenience than a necessity. Cases in which domain redefinitions are required occur in the treatment of the logarithm and power law functions.

A simple example is $f(z) = \sqrt{z}$. Intuition suggests that (at least for $z \neq 0$) f is analytic, with f' defined by $f'(z) = 1/2\sqrt{z}$. On the other hand, consider the "natural" definition of $\sqrt{z}$ as (in terms of polar representation)

$$f(z) = r^{\frac{1}{2}} e^{i\frac{\theta}{2}} \tag{5.3}$$

and consider the evaluation of $f(z)$ along a path encircling the origin (Figure 5.5). As one traverse of the path is made, we see that the value of the function changes $i\theta$ continuously from $r_0^{\frac{1}{2}} e^{i\frac{\theta_0}{2}}$ to the value

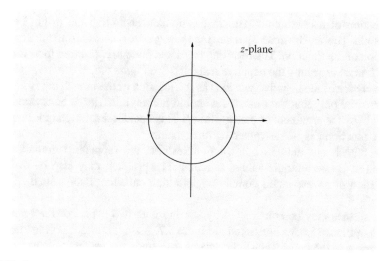

z-plane

Fig. 5.5 Complex plane

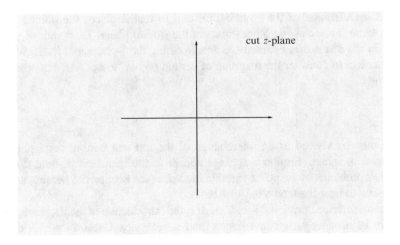

Fig. 5.6 Complex plane with cut

$$r_0^{\frac{1}{2}} e^{i \frac{\theta_0 + 2\pi}{2}} e^{i\pi} = r_0^{\frac{1}{2}} e^{i \frac{\theta_0}{2}} e^{i\pi}.$$

Although in this case the function value returns to a value different from the original, it is easy to see that a closed path that fails to encircle the origin provides no net change of the function along the traverse. This apparent ambiguity in the function value indicates that the specification (5.3) fails to adequately define a single-valued analytic function f. In addition to the formula (5.3) it is necessary to prescribe the domain of definition of the function in such a way that the resulting function is single-valued on its domain.

One way to accomplish this is to restrict the domain in such a way that the construction of a continuous path in the domain encircling the origin becomes impossible. This can be done by *removing the negative real axis* from the domain of definition of the function. The resulting restricted domain is referred to as the plane with a *branch cut* along the negative real axis. See Figure 5.6.

The choice of the negative real axis as the location of the branch cut is arbitrary. It is essential only that the branch cut should present a barrier to encirclements of the origin by the argument of the function f. This domain restriction in turn ensures that the function f is single-valued on its domain.

The branch cut domain construction restricts the (apparent) domain of the function to create a single-valued function. The problem may also be overcome by creation of an expanded domain of definition called a "two-sheeted Riemann surface."

The surface may be constructed by starting with two copies of the branch cut domain of Figure 5.6, as illustrated in Figure 5.7. A single surface is constructed by superimposing the two branch cut domains, and smoothly pasting together the "cut edges" of the two "sheets." The "pasting" is done in such a way that the upper edge of the sheet I branch cut is attached to the lower edge of the sheet II branch cut; at the same time the upper edge of the sheet II branch cut is attached to the lower edge of the sheet I branch cut.

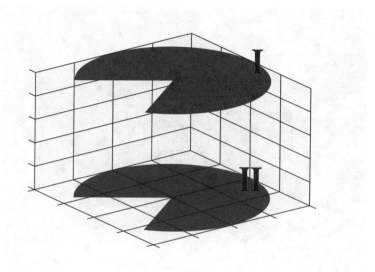

Fig. 5.7 Plane copies

This operation produces a two-dimensional surface which may be roughly visualized as two parallel rubber sheets, connected along the single line representing the negative real axis. See Figure 5.8. This three-dimensional visualization is of course imperfect, as the surface must be regarded as crossing through itself (in accordance with the attachment scheme specified above) along the branch cut.

The function f is defined with this surface as domain as follows. Points on the surface are described with "double wrapped" polar coordinates: points of the form

$$r\,e^{i\theta},\, r > 0,\, 0 \le \theta \le \pi$$

are regarded as belonging to the first and second quadrant of sheet I. For

$$\pi \le \theta \le 3\pi$$

the points are regarded as belonging to the sheet II; finally the points

$$3\pi \le \theta \le 4\pi$$

make up the third and fourth quadrants of sheet I. A point traveling in a circle about the origin (on sheet I) therefore moves through the branch cut crossover point, around the origin on sheet II, back onto sheet I through the branch cut, and finally returns along sheet I to its origin. Note that it is necessary to increase the angular argument by 4π in order to complete a closed path about the origin. An increase of argument by 2π serves to transfer the given point to the corresponding location on the opposite sheet.

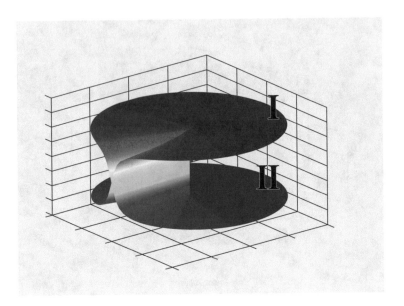

Fig. 5.8 Pasted cut planes

This discussion so far serves to describe the domain as a set; it remains to define the function on this domain. The definition is simply that for

$$z = r e^{i\theta}, \ 0 \le \theta \le 4\pi, \ r \ne 0,$$

we take

$$\sqrt{z} = r^{\frac{1}{2}} e^{i\frac{\theta}{2}}.$$

The double-sheeted nature of the domain ensures that f so defined is single-valued. One can also verify that

$$f'(z) = \frac{1}{\sqrt{z}}, \ z \ne 0,$$

so that f so defined is analytic for $z \ne 0$.

There are a number of functions whose "natural domain" can be regarded as a Riemann surface closely related to the example above. One such function is the *natural logarithm*.

The natural requirement for such a definition is that the function should act as an inverse of the exponential function e^z (defined by Euler's formula, or by the obvious Taylor series). However, since

$$e^{z+2n\pi i} = e^z, \ n = 0, \pm 1, \pm 2, \ldots,$$

the value of the inverse function is evidently ambiguous if the original z-plane is regarded as the domain. The solution to this difficulty is to construct the domain so that each of the values $z + 2n\pi i, \ n = 0, \pm 1 \ldots$ corresponds to a different point in

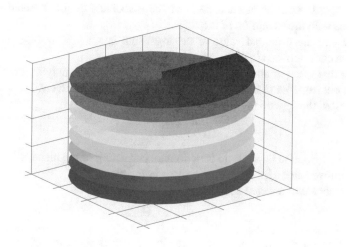

Fig. 5.9 Riemann surface for a logarithm

the domain. A domain for this example is constructed by pasting an infinite number of copies of the cut plane Figure 5.6 into a domain conceptually resembling a spiral staircase. The construction is essentially that of the $\sqrt{z}$ domain, without the need for returning to the original sheet. See Figure 5.9.

Associated with this domain are labels (r, θ, N) corresponding to conventional polar coordinates in the plane, as well as the "sheet number" N. For such an argument, the logarithm is defined by

$$\log z = \ln r + i\theta + 2N\pi i.$$

A version of the logarithm function may also be defined by adopting the branch-cut domain of Figure 5.6 as the domain of definition. This function is known as the principal value of the logarithm, and is defined for z in the cut plane, i.e., z of the form

$$z = r e^{i\theta}, -\pi \leq \theta \leq \pi$$

by

$$\text{Log}(z) = \ln r + i\theta.$$

Both of these logarithms are analytic on their respective domains of definition, with (expected) derivatives $1/z$.

Problems 5.2.

1. Show that the function $\sqrt{z}$ is analytic in the domain consisting of the plane with a branch cut removed along the negative real axis.

   Hint: For smooth functions, the Cauchy Riemann relations are necessary and sufficient for analyticity.

2. Show $\sqrt{z}$ is analytic at each point of the associated double-sheeted Riemann surface with the exception of the origin.

3. Show that the principal value of the logarithm is analytic in the plane with a branch cut removed along the negative real axis.

4. Show that $\sqrt{z+a}$ and $\sqrt{z-a}$ may each be defined as analytic functions in a plane cut from $-a$ to infinity, and from a to infinity respectively.

5. Show that the function

$$f(z) = \sqrt{z^2 - a^2}$$

may be defined to be analytic on a domain consisting of the plane, with a branch cut removed along the line from $-a$ to a in the plane.

6. Construct a two-sheeted Riemann surface for the function

$$f(z) = \sqrt{z^2 - a^2}.$$

7. The Riemann surface "pictures" were plotted with the Octave parametric surface capability. Can you generate a "picture" of the surface of

$$f(z) = \sqrt{z^2 - a^2}$$

using Octave? Appendix B has more information on surface plots, and a "better" picture of the square root Riemann surface that might be helpful.

8. Describe a Riemann surface for the function

$$f(z) = z^{\frac{1}{n}},$$

where n is an integer, $n > 0$.

9. Use Octave to generate a visualization of the Riemann surface for the function

$$f(z) = z^{\frac{1}{n}},$$

for $n = 3, 4, 5$.

5.3 Integrals and Cauchy's Theorem

In Section 5.1 above we introduced complex differentiation as the natural analogue of the real-value case definition. Complex integration in turn may be readily defined in terms of the familiar line integrals of vector calculus. What are perhaps the most striking and useful results of complex function theory arise from the interplay between complex integration and analytic functions. The central fact in this circle of results is Cauchy's Theorem described below.

These topics can be pursued at various levels of generality and rigor. The approach taken below is to present the main results in a form consistent with the usual vector calculus results, and usable for most applications.

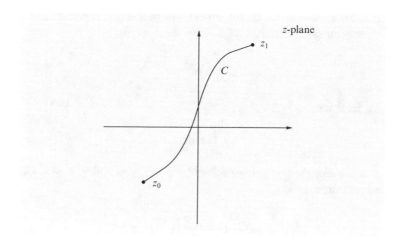

Fig. 5.10 Path in the complex plane

Given a curve C in the complex plane from z_0 to z_1 (see Figure 5.10), and a complex-valued function f, we are required to define the formal expression

$$\int_C f(z)\,dz.$$

Given a parameterization of the curve C,

$$z(t) = x(t) + i\,y(t),\ 0 \le t \le 1,$$

$$z(0) = z_0,$$

$$z(1) = z_1,$$

the above expression may be interpreted in terms of conventional line integrals. The appropriate formalism is simply

$$\int_C f(z)\,dz = \int_C (u(x,y) + i\,v(x,y))\,(dx + i\,dy)$$

$$= \int_C (u\,dx - v\,dy) + i \int_C (v\,dx + u\,dy).$$

Assuming that the curve C is a piecewise smooth rectifiable path in the plane, the above integrals may be evaluated in terms of the parameterization to give

$$\int_0^1 \left(u(x(t), y(t))\,\frac{dx}{dt} - v(x(t), y(t))\,\frac{dy}{dt} \right) dt$$

$$+i \int_0^1 \left(v(x(t), y(t))\,\frac{dx}{dt} + u(x(t), y(t))\,\frac{dy}{dt} \right) dt.$$

We assume the above-mentioned smoothness restriction on the path C, and take the above expression as the definition of the contour integral

$$\int_C f(z)\,dz.$$

Example. Evaluate

$$\int_C z^2 \, dz,$$

where C is the straight line parameterized by $x = t, y = t, 0 \leq t \leq 1$. Using the indicated parameterization,

$$\int_C z^2 \, dz = \int_0^1 (t + i\,t)^2 \, (1 + i) \, dt$$

$$= (1 + i)^3 \int_0^1 t^2 \, dt = \frac{(1 + i)^3}{3}.$$

Example. Evaluate the complex integral

$$\int_C z^2 \, dz,$$

where C' consists of the horizontal line segment $x = t, y = 0, 0 \leq t \leq 1$, followed by the vertical line segment $x = 1, y = t, 0 \leq t \leq 1$. We have

$$\int_C z^2 \, dz = \int_0^1 (t + 0)^2 \, (1 + 0) \, dt + \int_0^1 (1 + it)^2 \, (0 + i) \, dt$$

$$= 1/3 + i \int_0^1 (1 + 2\,it + (it)^2) \, dt$$

$$= \frac{(1 + i)^3}{3}.$$

The above examples produce a result consistent with a formal application of the familiar "anti-derivative" rule of ordinary calculus. At the same time, the examples suggest that, in some cases at least, the complex integral result may be independent of the path connecting the end points of the curve along which the integral is evaluated. The complex integral version of the fundamental theorem of calculus may be easily established under the assumption that the integrand is the (complex) derivative of an analytic function. Assuming that $f(z) = F'(z)$ where F is a function regular in a region R of the complex plane containing the path C, and that f is continuous there, the indicated integral may be explicitly evaluated. For a function F given as

$$F(x + i\,y) = U(x, y) + i\,V(x, y)$$

the derivative takes the form (computing with real increments)

$$f(x + iy) = F'(x + iy) = \frac{\partial U}{\partial x}(x, y) + i\frac{\partial V}{\partial x}(x, y).$$

Writing the complex integral as a line integral gives

$$\int_C f(x + iy)\, d(x + iy) = \int_C \frac{\partial U}{\partial x} + i\frac{\partial V}{\partial x}(dx + i\, dy)$$

$$= \int_C \frac{\partial U}{\partial x}\, dx - \frac{\partial V}{\partial x}\, dy + i \int_C \frac{\partial V}{\partial x}\, dx + \frac{\partial U}{\partial x}\, dy.$$

Since F is analytic, the Cauchy–Riemann equations provide the equalities

$$\frac{\partial U}{\partial x} = \frac{\partial V}{\partial y}$$

$$\frac{\partial V}{\partial x} = -\frac{\partial U}{\partial y}$$

so that the original complex integral finally takes the form

$$\int_C f(x + iy)\, d(x + iy) = \int_C \frac{\partial U}{\partial x}\, dx + \frac{\partial U}{\partial y}\, dy + i \int_C \frac{\partial V}{\partial x}\, dx + \frac{\partial V}{\partial y}\, dy.$$

The integrands above are total differentials, and so may be integrated exactly. The conclusion is that (as one might expect)

$$\int_C f(z)\, dz = F(z_1) - F(z_0)$$

where C is any piecewise smooth path in R connecting z_0 to z_1. This result is essentially the anti-derivative rule in the context of contour integration.

The question of path independence of line integrals arises in vector calculus discussions, where Green's Theorem is utilized. This theorem may be used for the same purpose in the case of contour integrals. A version of Green's Theorem suitable for the present purposes may be recalled as follows:

Theorem. *Let R be a closed bounded domain of the plane whose boundary B consists of a finite number of simple (i.e., non-self intersecting) closed rectifiable curves. Assume that L and M have continuous first partial derivatives in R. Then*

$$\int_B L\, dx + M\, dy = \int \int_R \left(\frac{\partial M}{\partial x} - \frac{\partial L}{\partial y} \right) dA.$$

The question of path independence of line integrals of the form

$$\int_B L\, dx + M\, dy$$

can be considered through careful use of Green's Theorem above. The successful application of Green's Theorem to the problem requires an assumption that the domain[4] in question be free of holes[5] in order to obtain the result. Simple examples show that this hypothesis is in fact essential, and not an artifact of the proof procedure. The basic result is the following.

Theorem. *Suppose that D is a simply connected open domain in the plane, and that L and N have continuous first partial derivatives in D. Then*

$$\int_B L\,dx + M\,dy$$

is independent of the path in D if and only if in D,

$$\frac{\partial L}{\partial y} = \frac{\partial M}{\partial x}.$$

This result may be directly applied to the question of the path independence of the contour integral

$$\int_C f(z)\,dz. \tag{5.4}$$

We make the assumptions that f is an analytic function in a simply connected open region D, and consider piecewise smooth contours $C1$, $C2$ in D connecting z_0 and z_1. (See Figure 5.11).

Writing the contour integral as the sum of two line integrals and invoking the Green's Theorem quoted above, we conclude that (5.4) is independent of the path (in D) connecting z_0 to z_1 if and only if

$$\frac{\partial u}{\partial y} = -\frac{\partial v}{\partial x}$$

and

$$\frac{\partial v}{\partial y} = \frac{\partial u}{\partial x}$$

throughout D. However, these equalities are satisfied by virtue of the Cauchy–Riemann equations for the analytic function f. The contour integral is therefore independent of the path in D provided that f is analytic in the simply connected domain D.

The usual statement of the Cauchy Integral Theorem may be deduced from the path independence result above.

---

[4]Recall that a domain is defined as an open set, in which any two points may be joined by a polygonal line lying in the set. A closed domain is a domain together with its boundary.

[5]The requirement is that the domain be simply connected: every simple closed curve in the domain must have its interior in the region.

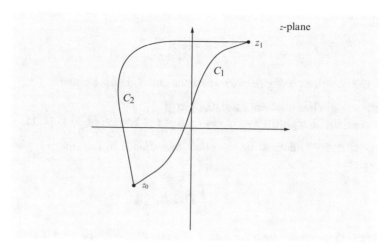

Fig. 5.11 Paths

Theorem. *Suppose that f is an analytic function in the simply connected open domain D, and let C be a simple closed piecewise smooth contour in D. Then*

$$\int_C f(z)\, dz = 0.$$

This result may be derived by reference to Figure 5.11. Let the closed contour C consist of C_1 followed by C_2 (traversed "backwards" from z_1 to z_0). Then (since path reversal introduces a minus sign)

$$\int_C f(z)\, dz = \int_{C_1} f(z)\, dz - \int_{C_2} f(z)\, dz.$$

By the path independence, this difference vanishes, and so

$$\int_C f(z)\, dz = 0$$

under the hypotheses.

Problems 5.3.

1. Evaluate (for integer $n > 0$)

$$\int_C z^n\, dz$$

where C is

a. the parabolic segment $x = t$, $y = t^2$, $0 \le t \le 1$,

b. the quarter-arc of the unit circle lying in the first quadrant.

2. Evaluate

$$\int_C \bar{z}\, dz$$

($\bar{z}$ is the complex conjugate of z) where the path C in the z-plane is

a. the straight line segment from $(0,0)$ to $(1,1)$,
b. the parabolic segment $x = t$, $y = t^2$, $0 \le t \le 1$, from $(0,0)$ to $(1,1)$.

3. Suppose that f is defined on the circle of radius R in the complex plane, and consider

$$\int_C f(z)\, dz$$

where C is the closed path consisting of a counterclockwise circuit of the circle of radius R centered at the origin in the z-plane. Show that

$$\int_C f(z)\, dz = \int_0^{2\pi} f(Re^{i\theta})\, i\, R\, e^{i\theta}\, d\theta.$$

4. Suppose that f is a function defined for complex values, but such that f assumes real values for real arguments. Show that

$$\int_a^b f(x)\, dx = \int_{L_1} f(z)\, dz$$

where L_1 is the straight-line path in the complex plane from $z = a + i0$ to $z = b + i0$ (i.e., along the real axis from a to b). Show also that

$$\int_a^b f(x)\, dx = \int_{L_2} f\left(\frac{z}{i}\right) d\left(\frac{z}{i}\right)$$

where L_2 is the path from $z = ia$ to $z = ib$ (along the imaginary axis).

5. a. Evaluate

$$\int_C \frac{1}{z}\, dz$$

where C is the unit circle in the complex plane traversed counter-clockwise.

b. Express the complex integral

$$\int_C \frac{1}{z}\, dz$$

in the form of real line integrals. For the real and imaginary parts, compute the corresponding L, M, and $\frac{\partial L}{\partial x} - \frac{\partial M}{\partial y}$.

c. Explain why the result of 5a is nonzero, in spite of the computation in 5b.

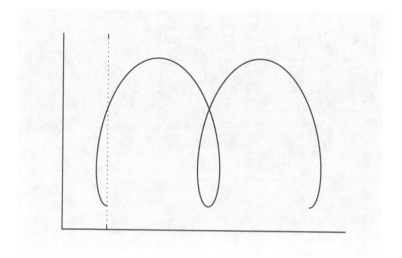

Fig. 5.12 Cycloid curve

6. Find the solution of the (complex) differential equation

$$\frac{dy}{dz} = f(z), \; y(0) = a,$$

where f is analytic in some domain D including the origin $z = 0$.

7. A prolate cycloid curve is given in parametric form by (Figure 5.12)

$$x = a\theta - b \sin \theta,$$
$$y = a - b \cos \theta, \; a < b.$$

Evaluate the line integral

$$\int_L \frac{(x+1)\,dx + y\,dy}{x^2 + 2x + 1 + y^2}$$

where L is the path of the cycloid from $\theta = 0$ to $\theta = 2\pi$.

5.4   Cauchy's Integral Formula, Taylor Series, and Residues

The usefulness of Cauchy's Integral Theorem lies not so much in the identification of certain vanishing integrals, but rather as a tool in the conversion of one integral to another more easily evaluated one. (Often, the latter is evaluated by inspection.)

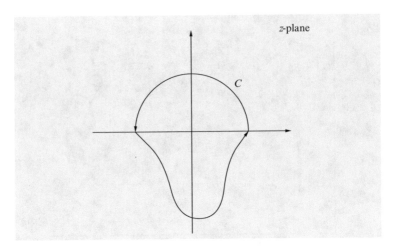

Fig. 5.13 Closed contour

   The basic technique employed may be described as "contour deformation." This can be illustrated by consideration of the integral involved in Cauchy's integral formula. This integral is

$$\int_C \frac{f(z)}{z - \zeta}\, dz.$$

The hypotheses are that f is analytic in a bounded simply connected region D and that C is a simple closed contour in D, traversed in the counterclockwise (with the interior to the left) sense. This hypothesis on the contour is indicated by writing the integral in the form

$$\oint_C \frac{f(z)}{z - \zeta}\, dz$$

indicating the direction (and the fact that the contour is closed) in the embellished integral sign. See Figure 5.13.
   The assumption is that ζ does not lie on C. If ζ lies outside of C, then $\frac{f(z)}{z-\zeta}$ is analytic on and inside C, and Cauchy's Integral Theorem implies that

$$\oint_C \frac{f(z)}{z - \zeta}\, dz = 0.$$

If ζ lies inside C, then the integrand has a singularity at ζ, and Cauchy's Theorem does not directly apply. However, the only singularity of the integrand is at $z = \zeta$, and so the path independence results (i.e., Cauchy's Integral Theorem) may be applied along a path which ensures that the singularity is outside of the closed

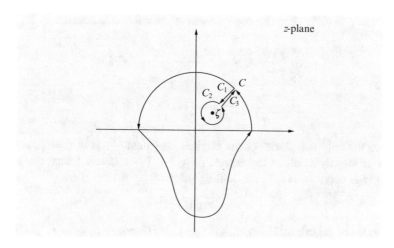

Fig. 5.14 Path with added components

contour. Such a path can be constructed by "excising" the singularity with an almost closed circle, and "access paths" from a point on the original contour C. The added path components are illustrated in Figure 5.14.

$$\oint_C \frac{f(z)}{z-\zeta}\, dz = \int_{C_1} \frac{f(z)}{z-\zeta}\, dz + \int_{C_2} \frac{f(z)}{z-\zeta}\, dz + \int_{C3} \frac{f(z)}{z-\zeta}\, dz,$$

(by path independence)

$$= \int_{C_2} \frac{f(z)}{z-\zeta}\, dz$$

if C_2 is deformed to a circle.

This last simplification occurs since f is analytic in the domain between the C_1, C_3 paths. The C_1, C_3 paths may be deformed until they coincide, and then cancel because they are traversed in opposite directions. The circular path C_2 may then be deformed to explicitly evaluate the contour integral.

If the path C_2 is chosen to form a circle C_ϵ of radius ϵ about the point $z = \zeta$ we have

$$\oint_C \frac{f(z)}{z-\zeta}\, dz = \oint_{C_\epsilon} \frac{f(z)}{z-\zeta}\, dz,$$

so that C has been deformed to a small circle containing at its center the singular point of the original integrand.

This last integral may be evaluated by writing it in terms of a polar coordinate parameterization of the boundary of C_ϵ. This is

$$z - \zeta = \epsilon\, e^{i\theta},\ 0 \le \theta < 2\pi,$$

so that $z = \zeta + \epsilon e^{i\theta}$. Then the explicit form of the integral in question is

$$\oint_C \frac{f(z)}{z - \zeta}\, dz = \int_0^{2\pi} \frac{f(\zeta + \epsilon e^{i\theta})}{\epsilon e^{i\theta}} \epsilon e^{i\theta}\, i\, d\theta$$

$$= i \int_0^{2\pi} f(\zeta + \epsilon e^{i\theta})\, d\theta.$$

Since the only singularity of the original integrand $\frac{f(z)}{z-\zeta}$ is at $z = \zeta$, ϵ may be chosen arbitrarily small in the above. As $\epsilon \to 0$ we obtain (since the analytic function f is continuous at ζ) the limiting value of

$$2\pi i f(\zeta)$$

as the value of the original integral. This provides *Cauchy's Integral Formula*:

$$\oint_C \frac{f(z)}{z - \zeta}\, dz = \begin{cases} 2\pi i f(\zeta) & \zeta \text{ inside } C, \\ 0 & \zeta \text{ outside } C. \end{cases}$$

In the above, C is a simple closed contour inside a simply connected domain D on which f is analytic.

Example. Suppose that $R(z) = q(z)/p(z)$ is a proper rational function, and that the zeroes of the denominator polynomial p are all distinct. Then (by the partial fraction algorithm)

$$R(z) = \sum_{i=1}^n \frac{r_i}{z - z_i}.$$

If C is a simple contour enclosing all of the points $\{z_j\}$ in the positive sense, then

$$\oint_C R(z)\, dz = 2\pi i \sum_{i=1}^n r_i.$$

In the application of Cauchy's integral formula to the problem of integral evaluation, it is often the case that the given integral appears in "real form." In order to apply Cauchy's result, it is necessary to interpret the given integral (or one closely related) as the parametric form of a contour integral.

Example. Evaluate

$$\int_0^{2\pi} \frac{1}{a\cos\theta + b}\, d\theta, \quad (a,b \text{ real}).$$

Note first that the integral is singular unless $\left|\frac{b}{a}\right| > 1$.

The integral may be interpreted as a complex integral with contour C taken as the unit circle in the complex plane. The parameterization is then

$$z = e^{i\theta}, \ 0 \le \theta \le 2\pi$$

with $dz = i\, e^{i\theta}\, d\theta$ so that $d\theta = \frac{dz}{iz}$, $|z| = 1$. At the same time

$$a\cos\theta + b = a\frac{(e^{i\theta} + e^{-i\theta})}{2} + b$$

$$= \frac{a}{2}\left(z + \frac{1}{z}\right) + b, \ |z| = 1.$$

Thus

$$\int_0^{2\pi} \frac{1}{a\cos\theta + b}\, d\theta = \oint_C \frac{1}{\frac{a}{2}\left(z + \frac{1}{z}\right) + b}\frac{dz}{iz}$$

$$= \frac{1}{i}\oint_C \frac{1}{\frac{a}{2}z^2 + bz + \frac{a}{2}}\, dz$$

$$= \frac{2}{ia}\oint_C \frac{1}{z^2 + \frac{2b}{a}z + 1}\, dz$$

$$= \frac{2}{ia}\oint_C \frac{1}{(z - z_1)(z - z_2)}\, dz.$$

Here, z_1, z_2 denote the zeroes of the polynomial $z^2 + \frac{2b}{a}z + 1$. If $\left|\frac{b}{a}\right| < 1$, the zeroes are complex, and lie on the unit circle (in this case the integral is singular, and direct evaluation is not expected). If $\left|\frac{b}{a}\right| > 1$ the zeroes are real, and since (from the form of the polynomial) $z_1 \cdot z_2 = 1$, exactly one of them lies inside of the unit circle. If z_1 is this smaller zero, then

$$f(z) = \frac{1}{z - z_2}$$

is analytic inside the unit circle. Therefore (by Cauchy's formula)

$$\frac{2}{ia}\oint_C \frac{1}{(z - z_1)(z - z_2)}\, dz = \frac{2}{ia}\, 2\pi i \frac{1}{z_1 - z_2}.$$

The roots of the quadratic are given by

$$-b/a \pm \sqrt{(b/a)^2 - 1}.$$

If $b/a > 0$, then $z_1 = -b/a + \sqrt{(b/a)^2 - 1}$ and

$$z_1 - z_2 = 2\sqrt{(b/a)^2 - 1}.$$

If $b/a < 0$, then $z_1 = -b/a - \sqrt{(b/a)^2 - 1}$ and $z_1 - z_2 = -2\sqrt{(b/a)^2 - 1}$. Finally, combining these results and restrictions,

$$\int_0^{2\pi} \frac{1}{a\cos\theta + b}\, d\theta = \begin{cases} \dfrac{4\pi}{\sqrt{b^2-a^2}} & \dfrac{b}{a} > 1, \\[2mm] \dfrac{-4\pi}{\sqrt{b^2-a^2}} & \dfrac{b}{a} < -1. \end{cases}$$

Many integrals occurring in the use of transform methods may be evaluated by use of contour integrals. This topic is discussed in Section 6.5.

Integrals involving singularities arising from simple zeroes of rational functions may generally be handled by techniques similar to those of the above examples. More complicated problems require the use of power series representations of analytic functions, a topic we now consider.

Mention has been made above of the fact that functions regular in a region are infinitely differentiable there. More than this is true; such functions may be represented by convergent Taylor series expansions. It is even the case that analytic functions have a series representation (a Laurent series) in the neighborhood of an isolated singularity. These series results may be derived on the basis of the Cauchy integral formula.

A Taylor series representation in powers of $(z - a)$ may be derived for an analytic function f which is regular in a disc $|z - a| < R$. Given such a function, choose $R_0 < R$, and note that for $|z - a| < R_0$ we have

$$f(z) = \frac{1}{2\pi i}\oint_C \frac{f(\zeta)}{\zeta - z}\, d\zeta, \tag{5.5}$$

where C is the circle $|z - a| = R_0 < R$. Now the singular term in the above integral may be written in the form

$$\frac{1}{\zeta - z} = \frac{1}{(\zeta - a) - (z - a)}$$

$$= \frac{1}{\zeta - a}\frac{1}{1 - \frac{(z-a)}{(\zeta-a)}}.$$

Along the contour C we have $|\zeta - a| = R_0$, while $|z - a| < R_0$ holds for z inside C. Hence along C, $\left|\frac{(z-a)}{(\zeta-a)}\right| < 1$, and so the expression may be regarded as the sum of a convergent geometric series

$$\frac{1}{1 - \frac{(z-a)}{(\zeta-a)}} = \sum_{n=0}^{\infty} \frac{(z-a)^n}{(\zeta-a)^n},$$

giving

$$\frac{1}{\zeta - z} = \sum_{n=0}^{\infty} \frac{(z-a)^n}{(\zeta-a)^{n+1}}.$$

For fixed z inside C this series is uniformly convergent (as a function of z) for z on the circle C. This means that the expansion may be used in (5.5), and that the integral in turn may be evaluated by termwise integration. The result is

$$f(z) = \frac{1}{2\pi i} \oint_C \sum_{n=0}^{\infty} \frac{(z-a)^n}{(\zeta-a)^{n+1}} f(\zeta) \, d\zeta$$

$$= \sum_{n=0}^{\infty} a_n (z-a)^n,$$

where

$$a_n = \frac{1}{2\pi i} \oint_C \frac{f(\zeta)}{(\zeta-a)^{n+1}} \, d\zeta.$$

Since a convergent power series representation of a function is necessarily the (in this case, complex) Taylor series of the function, we conclude that the Taylor coefficients are [6]

$$a_n = \frac{f^{(n)}(a)}{n!} = \frac{1}{2\pi i} \oint_C \frac{f(\zeta)}{(\zeta-a)^{n+1}}. \tag{5.6}$$

The power series representation above is a powerful tool in analytic function theory, which at the same time allows computations with analytic functions to be done on the basis of simple power series algebraic manipulations.

A representation for an analytic function in the neighborhood of an isolated singularity may be derived by an argument similar to the above.

Suppose that f is regular inside of the region $|z - a| < R$ at each point with the exception of the point $z = a$. (If f is analytic at a, the previous expansion result holds.). Given a point $z \neq a$, $|z| < R$ consider the contour C of Figure 5.15. Since the only singularity of f inside $|z - a| < R$ is at $z = a$, f is analytic on the region enclosed by C. Hence

$$f(z) = \frac{1}{2\pi i} \oint_C \frac{f(\zeta)}{\zeta - z} \, d\zeta.$$

Let C and c_0 denote the (positively oriented) contours $|\zeta - a| = R_0$, $|\zeta - a| = r_0$ where $0 < r_0 < |z - a| < R_0 < R$.

Adjusting the contour C so that the segments of C connecting the inner and outer circles coincide, we obtain

$$f(z) = \frac{1}{2\pi i} \oint_{C_0} \frac{f(\zeta)}{\zeta - z} \, d\zeta - \frac{1}{2\pi i} \oint_{c_0} \frac{f(\zeta)}{\zeta - z} \, d\zeta.$$

---

[6]To verify, simply differentiate termwise, and evaluate at the point $z = a$.

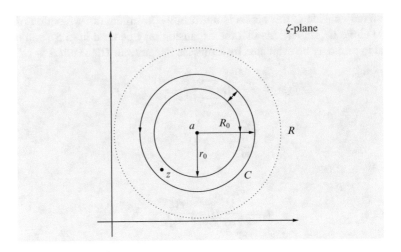

Fig. 5.15 Laurent expansion contours

Arguing as above, we find that the first term takes the form

$$\frac{1}{2\pi i} \oint_{C_0} \frac{f(\zeta)}{\zeta - z} \, d\zeta = \sum_{n=0}^{\infty} a_n \, (z - a)^n,$$

where

$$a_n = \frac{1}{2\pi i} \oint_C \frac{f(\zeta)}{(\zeta - a)^{n+1}} \, d\zeta.$$

To evaluate the second term note that along the inner circle c_0, $|\zeta - a| < |z - a|$, so that the expansion

$$\frac{1}{\zeta - z} = \frac{1}{(\zeta - a) - (z - a)}$$

$$= \frac{-1}{z - a} \frac{1}{1 - \frac{(\zeta - a)}{(z - a)}}$$

$$= -\sum_{n=0}^{\infty} \frac{(\zeta - a)^n}{(z - a)^{n+1}}$$

$$= -\sum_{n=1}^{\infty} \frac{(\zeta - a)^{n-1}}{(z - a)^n}$$

is appropriate. Substitution of this uniformly convergent expansion, followed by termwise integration, gives

$$\frac{-1}{2\pi i} \oint_{c_0} \frac{f(\zeta)}{\zeta - z} d\zeta = \sum_{n=-\infty}^{-1} a_n (z - a)^n, \tag{5.7}$$

where

$$a_n = \frac{1}{2\pi i} \oint_{c_0} f(\zeta) (\zeta - a)^{-n-1} d\zeta, \ n < 0,$$

$$= \frac{1}{2\pi i} \oint_{c_0} \frac{f(\zeta)}{(\zeta - a)^{n+1}} d\zeta. \tag{5.8}$$

Note that the two coefficients integrals in (5.6) and (5.7) for a_n have the same form. They differ only in the position of the contour. However, for each integral, the integrand is regular in the annulus $r_0 < |\zeta - a| < R_0$.

The respective contours of integration therefore may be deformed to any common intermediate contour in the annulus. For such a contour C,

$$f(z) = \sum_{n=-\infty}^{\infty} a_n (z - a)^n \tag{5.9}$$

where

$$a_n = \frac{1}{2\pi i} \oint_C \frac{f(\zeta)}{(\zeta - a)^{n+1}} d\zeta, \ -\infty < n < \infty.$$

This expansion representing an analytic function f in the neighborhood of an isolated singularity is called the *Laurent series* of the function.

The nature of the singularity of the function f at the point a may be classified on the basis of the form of its Laurent series. If the function f is such that $a_n = 0$ for $n < -k$ but $a_{-k} \neq 0$, then the Laurent series takes the form

$$f(z) = \frac{a_{-k}}{(z - a)^k} + \cdots + \frac{a_{-1}}{z - a} + \sum_{n=0}^{\infty} a_n (z - a)^n,$$

and f is said to have a *pole of order k* at the point $z = a$. The terms

$$\frac{a_{-k}}{(z - a)^k} + \cdots + \frac{a_{-1}}{z - a}$$

are referred to as the *principal part* of f at the singular point $z = a$. The difference between f and its principal part is expressible as a convergent power series, and so defines an analytic function regular in a neighborhood of $z = a$.

If no such k exists, f is said to have an essential singularity at the point $z = a$.

Example. Let $R(z)$ be a rational function, with no common factors. Then $R(z)$ has a pole at each of the zeroes of the denominator polynomial. The order of each of those poles is the multiplicity of the zero in the factorization of the denominator polynomial.

Example. Let $f(z) = e^{-\frac{1}{z^2}}$. The origin is an essential singularity for this function.

Example. Let $f(z) = \frac{1}{\sin(z)}$ Then f has a pole of first order at each of the points $z = n\pi$, $n = 0, \pm 1, \pm 2, \ldots$. To see this, expand $\sin(z)$ in a Taylor series about the point $z = n\pi$. Then

$$\sin z = (z - n\pi) - \frac{(z - n\pi)^3}{3!} + \frac{(z - n\pi)^5}{5!} + \cdots +$$

$$= (z - n\pi)\left(1 - \frac{(z - n\pi)^2}{3!} + \frac{(z - n\pi)^4}{5!} + \cdots +\right).$$

For $z \neq n\pi$,

$$\frac{1}{\sin z} = \frac{1}{z - n\pi}\left(1 - \frac{(z - n\pi)^2}{3!} + \frac{(z - n\pi)^4}{5!} + \cdots +\right)^{-1}$$

and for small values of $z - n\pi$, the above takes the form (using the geometric series)

$$\frac{1}{z - n\pi}\sum_{n=0}^{\infty}\left(\frac{(z - n\pi)^2}{3!} - \frac{(z - n\pi)^4}{5!} + \cdots\right)^n$$

$$= \frac{1}{z - n\pi} + \sum_{n=1}^{\infty} b_k (z - n\pi)^k.$$

This convergent series represents the Laurent series for $\frac{1}{\sin(z)}$ in the neighborhood of $n\pi$. The principal part is $\frac{1}{(z-n\pi)}$ and so the order of the pole is 1.

This last example illustrates the manner in which series expansion coefficients (at least in cases in which the singularities involved are poles) can be calculated by "algebraic" means.

Special interest attaches to the evaluation of the Laurent series coefficient a_{-1}, known as the *residue* of the function f at the singular point in question. For the case $n = -1$, the coefficient formula for a_n becomes

$$a_{-1} = \frac{1}{2\pi}\oint_C f(\zeta)\, d\zeta. \tag{5.10}$$

If the residue a_{-1} can be explicitly calculated by independent means, then this expression can be regarded as a means of evaluation of the indicated contour integral. The extension of the formula to the case of a contour containing a finite number of singularities of the function f is known as the Residue Theorem.

Theorem (Residue Theorem). *Let C be a simple piecewise smooth closed contour, and let f be an analytic function, regular in a neighborhood of the contour C. Suppose further that the domain enclosed by C contains a finite number of isolated singularities of the function f at the points $\{z_n\}$. Denote the residue of f at z_j by*

$$d_{-1} = Res\left(f(z)\Big|z = z_j\right);$$

then

$$\oint_c f(z)\, dz = 2\pi i \sum_{j=1}^{n} d_{-1} = 2\pi i \sum_{j=1}^{n} Res\left(f(z)\Big|z = z_j\right). \tag{5.11}$$

The derivation of this result is accomplished by contour deformation and is left as an exercise.

Example. Suppose that $R(z)$ is a rational function, and that f is regular in a region containing the simple closed contour C which in turn encloses all of the poles of R. The problem is to evaluate

$$\oint_C R(z) f(z)\, dz.$$

Since R is rational,

$$R(z) = q(z) + \frac{r(z)}{p(z)}$$

where q is the quotient, and $\frac{r}{p}$ is proper. Now

$$\oint_C R(z) f(z)\, dz = \oint \frac{r(z)}{p(z)} f(z)\, dz$$

since q is regular in the enclosing C. By partial fractions

$$\frac{r(z)}{p(z)} = \sum_{i=1}^{m} \sum_{j=1}^{n_i} \frac{\alpha_{ij}}{(z - z_i)^j},$$

and since f is regular, the possible poles of $\frac{r}{p} \cdot f$ are at the points $\{z_i\}$.[7]
By the Residue Theorem,

$$\oint \frac{r(z)}{p(z)} f(z)\, dz = 2\pi i \sum_{j=1}^{m} Res\left(\frac{r(z)}{p(z)} * f(z)\Big|z = z_j\right).$$

---

[7]The poles may not arise due to cancellation from a zero of appropriate order in f.

Example. For the example considered above (and others), a formula for the indicated residue may be given, provided that the order of the pole is known (this can usually be determined by algebraic means from the "leading" expansion terms). If a Laurent series is given

$$g(z) = \frac{a_{-k}}{(z-\alpha)^k} + \frac{a_{-k+1}}{(z-\alpha)^{k-1}} + \cdots + \frac{a_{-1}}{(z-\alpha)} + \sum_{n=0}^{\infty} a_n (z-\alpha)^n,$$

then

$$(z-\alpha)^k g(z) = a_{-k} + a_{-k+1}(z-\alpha) + \cdots + a_{-1}(z-\alpha)^{k-1} + \sum_{n=0}^{\infty} a_n (z-\alpha)^{n+k},$$

and

$$(k-1)!\, a_{-1} = \left(\frac{d}{dz}\right)^{k-1} \{(z-\alpha)^k g(z)\}\bigg|_{z=\alpha},$$

giving the residue formula

$$a_{-1} = \frac{1}{(k-1)!} \left(\frac{d}{dz}\right)^{k-1} \{(z-\alpha)^k g(z)\}\bigg|_{z=\alpha}$$

for the residue at a pole of order k.

Problems 5.4.

1. Evaluate

$$\oint_C \frac{1}{z^2+1}\, dz$$

   where C is a circle of radius $1/2$ centered at the origin.
2. Prove the Residue Theorem for the case in which a finite number of isolated singularities are enclosed by the simple closed rectifiable contour C.
   Hint: Use induction on the number of singularities, and contour deformation. See Figure 5.16.
3. Let C be a circle of radius 3 centered at the origin. Evaluate

$$\oint_C \frac{1}{(z+1)(z+2)}\, dz.$$

4. Evaluate

$$\oint \frac{\sin z\, e^z}{(z-1)(z+1)}\, dz$$

   where C is illustrated in Figure 5.17.

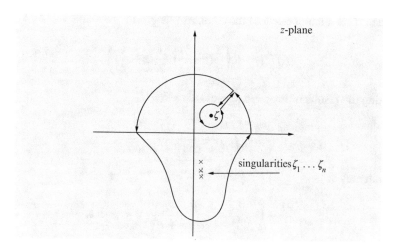

Fig. 5.16 Multiple singularities in C

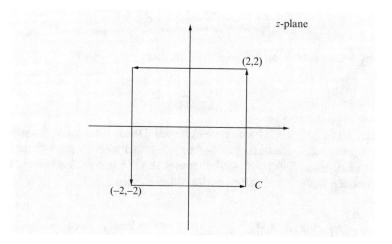

Fig. 5.17 Square path

5. Suppose that the integral

$$\frac{1}{2\pi} \int_0^{2\pi} F(\cos\theta, \sin\theta) \, d\theta$$

is to be evaluated. Show that

$$\frac{1}{2\pi} \int_0^{2\pi} F(\cos\theta, \sin\theta) \, d\theta = \frac{1}{2\pi i} \oint_C f(z) \frac{dz}{z},$$

where C is the unit circle of the complex plane, and

$$f(z) = F\left(\frac{1}{2}(z + \frac{1}{z}), \frac{1}{2i}(z - \frac{1}{z})\right).$$

6. Using the results of Problem 5, evaluate

$$\frac{1}{2\pi} \int_0^{2\pi} \frac{d\theta}{a^2 \cos \theta + b^2 \sin \theta}.$$

7. Evaluate

$$\frac{1}{2\pi} \int_0^{2\pi} \frac{d\theta}{1 - 2a \sin \theta + a^2}, \quad 0 < a < 1.$$

8. Evaluate

$$\frac{1}{2\pi} \int_0^{2\pi} \frac{\cos 3\theta}{1 - 2a \sin \theta + a^2} d\theta, \quad 0 < a < 1.$$

9. Consider the evaluation of integrals of the form

$$\int_{-\infty}^{\infty} \frac{1}{p(x)} dx$$

where p is a polynomial with no real zeroes. Define a contour C consisting of the segment of the real axis from $-b$ to b, together with the semi-circle $|z| = b$ in the half-plane $Im(z) > 0$. (See Figure 5.18.) If b is sufficiently large that all zeroes of p in the upper half-plane lie within C, then

$$\int_{-b}^{b} f(x) dx + \int_0^{\pi} \frac{1}{p(b e^{i\theta})} b i e^{i\theta} d\theta = 2\pi i \sum_{i=1}^{N} Res(\frac{1}{p(z)}),$$

where the sum is over the residues at poles of p in the upper half-plane. Show that

$$\int_{-\infty}^{\infty} f(x) dx = 2\pi i \sum_{i=1}^{n} Res(\frac{1}{p(z}\left| Re(z_i) > 0 \right),$$

provided

$$\lim_{b \to \infty} \int_0^{\pi} \frac{1}{p(b e^{i\theta})} b e^{i\theta} d\theta = 0.$$

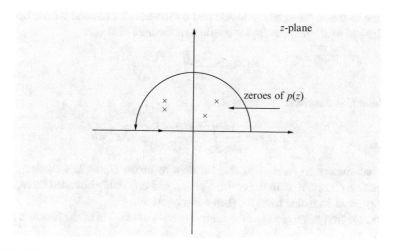

Fig. 5.18 Upper half-plane singularities

10. Use the Residue Theorem and Problem 9 to evaluate

$$\int_{-\infty}^{\infty} \frac{1}{x^2 + a^2} \, dx.$$

11. Use Problem 9 to evaluate

$$\int_{-\infty}^{\infty} \frac{1}{(x^2 + a^2)(x^2 + b^2)} \, dx.$$

12. Extend the argument of Problem 9 to evaluate

$$\int_{-\infty}^{\infty} \frac{e^{ix}}{x^2 + a^2} \, dx$$

and hence

$$\int_{-\infty}^{\infty} \frac{\cos x}{x^2 + a^2} \, dx.$$

13. If f and g are analytic in $|z| < R$, and have power series representations

$$f(z) = \sum_{n=0}^{\infty} a_n z^n,$$

$$g(z) = \sum_{n=0}^{\infty} b_n z^n,$$

find the coefficients of the power series for the product $h(z) = f(z) \cdot g(z)$.

14. If the function f is analytic inside and on the circular domain whose boundary is $|z - a| = R$, then the Taylor coefficient formula provides

$$f'(a) = \frac{1}{2\pi i} \oint \frac{f(z)}{(z - a)^2} \, dz$$

where C is the circle $|z - a| = R$, or

$$f'(a) = \frac{1}{2\pi} \int_0^{2\pi} \frac{f(a + Re^{i\theta})}{(Re^{i\theta})^2} Re^{i\theta} \, d\theta$$

in parameterized form. Use this formula to prove *Liouville's theorem*: If the function f is analytic in the entire z-plane, and uniformly bounded there, so that $|f(z)| < M$ for some fixed M, then f is a constant.

15. Find the first three terms of the Laurent series expansion of the function

$$\frac{z^2 + 1}{\sin z}$$

in the neighborhood of the point $z = \pi$.

16. Let f be a function analytic inside and on the boundary of the circle $|z| = R$. Define a function on the unit circle by

$$g(\theta) = f(re^{i\theta}), \quad \text{where } r < R.$$

Use Taylor series to show that the function g has no nonzero Fourier coefficients c_n for $n < 0$.

17. Define the function g as in problem (16). Find a Fourier series expansion for

$$\text{Re}\left(f(re^{i\theta}) \right)$$

and show that f may be reconstructed, given only the data

$$\text{Re}\left(f(re^{i\theta}) \right), \ 0 \le \theta \le 2\pi.$$

18. Evaluate

$$\oint_C \frac{z^2 + 1}{\sin^2(z)} \, dz$$

where C is the circle $|z| = 4$.

5.5 Complex Variables and Fluid Flows

Complex integration and the Residue Theorem find much use in problems of transform inversion (see Chapters 7, 8). Even more basic notions from complex function theory may be used to advantage in problems connected with potential

Fig. 5.19 Flow streamlines

theory, especially in the theory of two-dimensional ideal fluids. This theory has a compelling mathematical elegance, and provides both useful insights and a base for more realistic models of fluid behavior.

The basic physical variables of this theory are the components of a velocity field associated with the fluid flow. The variables u, v represent the x and y components of the velocity of the fluid particles located at coordinate (x, y) (and time t in time dependent problems).[8] See Figure 5.19.

In order to derive a set of governing equations, it is necessary to make assumptions about the fluid under consideration. We assume that the flow is time-invariant (steady) and incompressible. Under these assumptions, an equation of conservation of mass may be easily derived. The net flux of mass through the boundary of a control volume must vanish in a region without sources or sinks of fluid. Hence (for such a volume)

$$\int_{\partial V} (u, v) \cdot n \, dS = \int \int_V \text{div}(u, v) \, dV = 0.$$

---

[8] A consequence of the definition of (u, v) as the velocity field is that the x-acceleration of a particle at (x, y) is $\frac{\partial u}{\partial t} + u \frac{\partial u}{\partial x} + v \frac{\partial u}{\partial y}$ rather than $\frac{\partial u}{\partial t}$.

Since V is arbitrary, the relation

$$\frac{\partial u}{\partial x} + \frac{\partial v}{\partial y} = 0 \tag{5.12}$$

is obtained. If the flow contains distributed sources, the source terms appear on the right side of equation (5.12).

The relation (5.12) is a conservation law which is independent of the mechanical properties (save compressibility) of the fluid. Further mechanical assumptions are invoked in writing a conservation of momentum (Newton's law) relation for the fluid. For the case of an *ideal fluid*, the assumption is made that the velocity field is curl-free (irrotational)

$$\nabla \times (u, v) = \frac{\partial u}{\partial y} - \frac{\partial v}{\partial x} = 0,$$

and that viscous effects are absent.

These assumptions have the effect of making the conservation of mass relation (5.12) the primary governing equation. Under those assumptions, the momentum equation (i.e., Newton's second law) provides Bernoulli's law, relating the fluid velocity and the fluid pressure.

In most physical fluid flows, viscous effects play a role, at least in some regions of the flow field. In such regions the ideal fluid model is inappropriate, and a form of the momentum equation incorporating viscous force effects must also be considered. On the other hand, there are many flows in which viscous effects are confined to restricted regions of the flow field, and the ideal fluid model is relevant to the remainder.

The connection between ideal fluids and complex function theory arises from consideration of two line integrals derived from the velocity field. The first of these is simply the line integral of the velocity, given by

$$\int_L u \, dx + v \, dy.$$

For an ideal fluid, it is assumed that (the flow is irrotational)

$$\frac{\partial u}{\partial y} - \frac{\partial v}{\partial x} = 0,$$

so that (at least on a simply connected subregion of the flow) the line integral of the velocity field is independent of the path, and so may be used to define a potential function (called the velocity potential)

$$\phi(x, y) = \int_{(x_0, y_0)}^{(x, y)} u \, dx + v \, dy. \tag{5.13}$$

The velocity is then the gradient of this potential

$$\frac{\partial \phi}{\partial x} = u, \quad \frac{\partial \phi}{\partial y} = v,$$

and the requirement of irrotationality becomes essentially automatic, since

$$\frac{\partial u}{\partial y} - \frac{\partial v}{\partial x} = \frac{\partial^2 \phi}{\partial y \partial x} - \frac{\partial^2 \phi}{\partial x \partial y} = 0.$$

The governing equation then is simply the continuity equation (5.12). Writing this in terms of the velocity potential,

$$\frac{\partial u}{\partial x} + \frac{\partial v}{\partial y} = \frac{\partial^2 \phi}{\partial x^2} + \frac{\partial^2 \phi}{\partial y^2} = 0,$$

so that the velocity potential satisfies the potential equation (Laplace's equation).

A second line integral of interest in fluid flow arises from the computation of the flux of (fluid) mass across a path in the flow field. This is illustrated in Figure 5.20.

The flux is given by the integral of the component of fluid velocity normal to the path along the length of the curve.

To express this as a line integral, introduce the arc length s as a parameterization of the path. Then the path is

$$L = \left\{ (x(s), y(s)) \,\middle|\, 0 \le s \le l \right\},$$

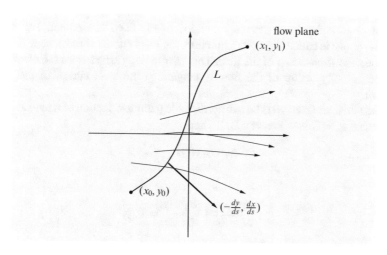

Fig. 5.20 Flow across a path

and $(\frac{dx}{ds}, \frac{dy}{ds})$ is the unit tangent vector along the path. The unit outward normal vector $\mathbf{n}$ is then $-\frac{dy}{ds}, \frac{dx}{ds}$, and the flux is given by

$$\int_0^l (-u\frac{dy}{ds} + v\frac{dx}{ds})\, ds.$$

Expressed as a line integral this is

$$\int_L v\, dx - u\, dy.$$

In order that this line integral be independent of path in some simply connected subregion of the fluid flow region, it is necessary and sufficient[9] (according to Green's Theorem) that

$$\frac{\partial}{\partial y}(v) - \frac{\partial}{\partial x}(-u) = 0,$$

or

$$\frac{\partial u}{\partial x} + \frac{\partial v}{\partial y} = 0.$$

However, this is simply the equation of continuity (5.12), which holds for any incompressible flow. We conclude that the line integral for the fluid flux defines another potential function

$$\psi(x, y) = \int_{(x_0, y_0)}^{(x, y)} v\, dx - u\, dy \tag{5.14}$$

at least on any simply connected subregion of the flow. The potential ψ defined in (5.14) above is called the stream function; the level curves of the stream functions are called the *streamlines* of the flow. This terminology arises from the observation that the velocity vector of the flow is tangent to the level curves of the stream function.

To see this, let $(x(t), y(t))$ be a differentiable parameterization of a path along the level curve $\psi = C$ (see Figure 5.21). Then

$$\psi(x(t), y(t)) = C,$$

and

$$\frac{\partial \psi}{\partial x}\frac{dx}{dt} + \frac{\partial \psi}{\partial y}\frac{dy}{dt} = 0.$$

---

[9]Here the flow field is assumed to have continuous partial derivatives.

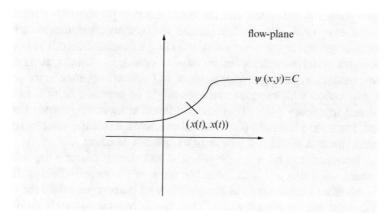

Fig. 5.21 Parameterized path

From the definition of ψ as a line integral in (5.14) we have

$$\frac{\partial \psi}{\partial x} = v,$$

and

$$\frac{\partial \psi}{\partial y} = -u.$$

Therefore the derivative relation calculated by the chain rule above becomes

$$v \frac{dx}{dt} = u \frac{dy}{dt},$$

which shows that the tangent vector to the streamline is parallel to the velocity field.

The stream function plays a central role in ideal fluid theory, based on the calculation made above. Since fluid flows parallel to the streamlines, a level curve of the stream function models a solid body (free of viscous drag in the case of an ideal fluid), since it marks a boundary not crossed by the fluid.

The governing equation for the stream function in the case of an ideal fluid (an irrotational flow) is obtained by substituting the stream function derived velocities in the condition

$$\frac{\partial v}{\partial x} - \frac{\partial u}{\partial y} = 0.$$

The result is that the stream function is also governed by Laplace's equation, so that

$$\frac{\partial^2 \psi}{\partial x^2} + \frac{\partial^2 \psi}{\partial y^2} = 0.$$

At this point it might appear that the ideal fluid flow problem has been reduced to the solution of Laplace's equation (subject to boundary conditions ensuring that the flow is tangential to any solid boundaries in the flow). While this is in a sense true, complex variable methods can be used to produce a virtual catalog of ideal fluid flow models, bypassing explicit solution of Laplace's equation in the process.

The connection with complex variables might be expected from the fact that the real and imaginary parts of an analytic function separately satisfy Laplace's equation. From the ideal fluid discussion above, we see that the velocity potential and stream function of an ideal flow satisfy Laplace's equation.

Note, however, that the velocity potential and stream function are not really independent, since they are both defined in terms of the velocity field of the flow. There is therefore a requirement of compatibility on the two potentials: the velocity field as derived from each must match. This compatibility condition is simply

$$\frac{\partial \phi}{\partial x} = u = -\frac{\partial \psi}{\partial y}, \tag{5.15}$$

$$\frac{\partial \phi}{\partial y} = v = \frac{\partial \psi}{\partial x}.$$

Associated with an ideal fluid flow problem therefore is a *pair* ϕ, ψ of solutions of Laplace's equation, satisfying the conditions in (5.15). Conversely, if a pair of solutions ϕ, ψ of Laplace's equation can be found which happen to satisfy (5.15), they may be interpreted as the velocity potential and stream function of an ideal fluid flow problem.

It might be thought far fetched to contemplate the solution of such a set of "cross coupled" partial differential equations. However, it happens that the conditions (5.15) are exactly the same form as the Cauchy–Riemann relations between the real and imaginary parts of an analytic function. Pairs of functions satisfying Laplace's equation and (5.15) therefore abound in the guise of the real and imaginary parts of analytic functions.

The problem of finding such pairs may be phrased in terms of the search for a complex potential function,

$$f = \phi + i\psi.$$

The desired relations hold for each analytic (potential) f: the velocity potential is $\mathrm{Re}(f)$, the stream function is $\mathrm{Im}(f)$.

Example. The simplest complex potential is

$$f(z) = Uz = U(x + iy).$$

Assuming U real,

$$\phi(x, y) = Ux,$$
$$\psi(x, y) = Uy,$$

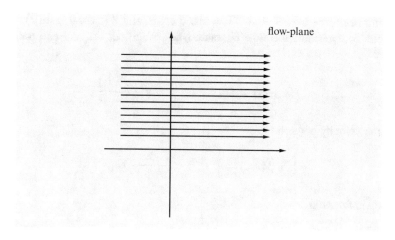

Fig. 5.22 Horizontal flow

so that $u = U$, $v = 0$, and the streamlines are lines of constant y. The flow is parallel to the x-axis and uniform. The flow is illustrated in Figure 5.22.

Example. For

$$f(z) = U e^{ia} z = (x\cos(a) + y\sin(a)) + i(y\cos(a) - x\sin(a))$$

the streamlines are straight lines at angle a, and the flow is again uniform.

In order to model the flow of a uniform stream past a solid body, it is necessary to construct a complex potential with constant imaginary part on the boundary of the body (so that the boundary is a streamline) and which tends to Uz for large $|z|$ (to obtain a uniform horizontal stream at infinity.)

Example. Construct a complex potential for uniform flow past a circle of radius a.

The potential must contain a term Uz to obtain the uniform stream at infinity, so we seek

$$f = Uz + correction.$$

The "correction" is to be chosen so that f has constant imaginary part for $|z| = a$, i.e., for $z = ae^{i\theta}$, $0 < \theta < 2\pi$. The constant imaginary part may be chosen as zero, since constants may be added to the potential functions without affecting the velocity. To cancel the imaginary part, we require that the correction value for $|z| = a$ be the complex conjugate of the free-stream contribution. Hence, on this basis an appropriate potential is

$$f(z) = Uz + U\frac{a^2}{z}.$$

Then $f(a\,e^{i\theta}) = Ua\,(e^{i\theta} + e^{-i\theta}) = 2\,U\,a\,\cos\theta$, and $\psi(a\cos\theta, a\sin\theta) = 0$ as required. To determine the flow characteristics, write out the real and imaginary parts of

$$f(x + iy) = U\,(x + iy) + U\,a^2\,\frac{(x - iy)}{x^2 + y^2}$$

to get the velocity potential

$$\phi(x, y) = U\,(x + a^2\,\frac{x}{x^2 + y^2}),$$

and stream function

$$\psi(x, y) = U\,(y - a^2\,\frac{y}{x^2 + y^2}).$$

The streamlines are given by the family of curves

$$y\,(x^2 + y^2 - a^2) = C\,(x^2 + y^2),$$

and tend to horizontal lines at large distances. The flow is illustrated in Figure 5.23.

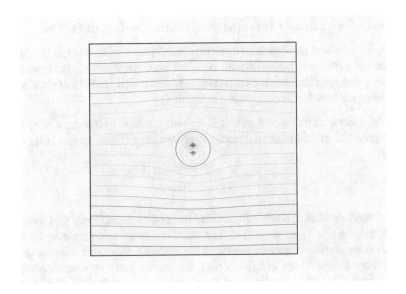

Fig. 5.23 Ideal flow past a circle

Software

Since Octave handles complex-valued variables, and at the same time can construct contour plots, it is very easy to visualize complex variable ideal fluid solutions. The circle flow problem can be plotted by defining a potential function and a region of the complex plane.

OCTAVE

```
[X Y]  =  meshgrid ( −3:.01:3 ,  −3:.01:3);
Z = X + i*Y;

W = circleflow (Z,  .5);
contour (X,  Y,  clip (imag (W),  [−4  4]),  20);

hold  on
t  =  0:.01:  2*pi;
plot (.5*cos (t),  .5*sin (t));
axis  equal
```

The circle has been put in for visibility, as the flow (by construction) avoids the circle. Notice that the data is clipped before the contours are plotted. This is virtually always required, since there will be singularities outside of the flow domain in almost all problems. A complex potential that is everywhere analytic is constant, and so does not correspond to any flow of interest.

It might be noted that the potential constructed above is singular at the origin. This causes no problem in interpretation, since the flow considered is exterior to the circle $z = a$. The singularity at the origin is outside of the flow field. In view of Liouville's theorem, constructed potentials of physical interest will inevitably have singularities in the finite part of the flow field[10].

Example. The logarithm function provides a potential with a "point source" interpretation. For the potential

$$f = \frac{\kappa}{2\pi} \log(z), (\kappa \text{ real}),$$

the streamlines are straight lines through the origin. The velocity field is

$$u = \frac{\kappa}{2\pi} \frac{x}{x^2 + y^2},$$

$$v = \frac{\kappa}{2\pi} \frac{y}{x^2 + y^2},$$

---

[10]The introduction of "solid bodies" defined by streamlines into the flow can also result in flow regions which are not simply connected. In this case, it may occur that the line integrals defining (locally) the velocity potential and stream function may fail to give a globally defined single-valued potential. This, however, causes no problem since the potentials are ambiguous only to the extent of a line integral around the obstacle boundary, and the physically relevant quantities are given by the gradient of the potentials.

from which one may calculate that the net fluid flux

$$\int_C v\, dx - u\, dy$$

through any closed contour C encircling the origin is exactly κ. This example is therefore a point source of fluid; a negative value of κ corresponds to a "sink." One may also consider imaginary values of κ. Such a potential corresponds to a point source of *vorticity* (curl of the velocity field.)

Example. As an example of the combination of potentials, we consider the uniform flow past a circular cylinder with an added point source of vorticity. In order to preserve the identity of the boundary of the cylinder as a streamline, it is necessary to add a vorticity source in a form which produces no change in the imaginary component of the potential on the circle boundary $|z| = a$. This suggests[11]

$$f = U\left(z + \frac{a^2}{z}\right) + iK\log\left(\frac{z}{a}\right).$$

Since $\left|\frac{z}{a}\right| = 1$ on the boundary, $i\log(\frac{z}{a})$ is real on $|z| = a$ and the circle will be a streamline for this potential.

For this choice we may compute the net circulation

$$\int_C u\, dx + v\, dy = K$$

for a simple closed contour encircling the origin. Typical streamlines associated with this potential are shown in Figure 5.24.

Various techniques are employed for the construction of fluid potentials. The examples above use the "cut and try" approach. A more systematic technique can be based on a study of conformal mappings. This topic is introduced in the following section.

Problems 5.5.

1. Let u and v be the x and y components of a two-dimensional fluid flow velocity field. Show that the x component of the acceleration of a fluid particle moving with the flow is

$$\frac{\partial u}{\partial t} + u\frac{\partial u}{\partial x} + v\frac{\partial u}{\partial y}.$$

---

[11] This potential is singular at the origin. Again, the flow region does not include the singular point, so no problem arises.

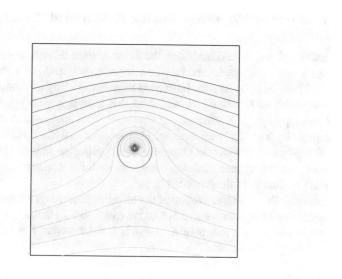

Fig. 5.24 Circle flow with circulation

2. Let $\rho(x, y, z, t)$ denote the density function of a three-dimensional fluid flow, with velocity vector field $\mathbf{u}(x, y, z, t)$. Show that ρ satisfies (the equation of continuity)

$$\frac{\partial \rho}{\partial t} + \nabla \cdot (\rho u) = 0$$

   in any source-/sink-free region.

3. Define a velocity potential for a two-dimensional flow by

$$f(z) = z + \frac{1}{z}.$$

   Find the equations of the streamlines of this flow.

4. Let $F = \phi + i\psi$ be a complex potential. Show that the streamlines are orthogonal to the lines of constant velocity potential.

5. Show that for a flow with complex potential f the fluid speed is given by $|f'(z)|$.

6. The stagnation points of a flow are the points at which the velocity of the fluid vanishes. Find the stagnation points for the case of uniform flow past a circular cylinder.

7. For uniform flow past a circular cylinder, find the fluid velocity along the surface of the cylinder. At what point is this a maximum?

8. Suppose that the stream function of an ideal flow is given. Find an expression for the associated velocity potential in terms of the stream function.

9. For the case of uniform flow past a circular cylinder of radius 1, compute the fluid flux through a straight line from $(1, 2)$ to $(4, 4)$.

5.6 Conformal Mappings and the Principle of the Argument

In the context of real valued functions, the flavor of most discussions of functions as mappings is local in character and based on linear approximations to the functions in question. These techniques may be carried over to the study of complex mappings defined by analytic functions. However, in this case it is also possible to obtain results on mappings of a more global character. That such results are possible is testimony to the "rigidity" of the allowed behavior of analytic functions.

Results on analytic function mappings find application in areas as (apparently) diverse as control system stability analysis, and the construction of complex potentials for ideal fluid flow problems.

We consider first the linearization of the mapping defined by a function f analytic in some domain D. It was pointed out above (Section 5.2) that a complex function may be considered as a mapping $R^2 \mapsto R^2$. If the function f is analytic, then it defines a "complex differential" according to

$$df = f'(z)\, dz$$

$$= f'(z)\, (dx + i\, dy).$$

The interpretation of this differential is the same as in the real variable case, in that it represents a (complex) linear approximation to the function f. Taking

$$f = u + i\, v$$

we have (for analytic f)

$$df = \left(\frac{\partial u}{\partial x} + i\, \frac{\partial v}{\partial x}\right)(dx + i\, dy)$$

$$= \left(\frac{\partial u}{\partial x}\, dx - \frac{\partial v}{\partial x}\, dy\right) + i\left(\frac{\partial v}{\partial x}\, dx + \frac{\partial u}{\partial x}\, dy\right),$$

so that the matrix representation of the differential mapping, regarded as a map $R^2 \mapsto R^2$ is (splitting real and imaginary parts)

$$\begin{bmatrix} du \\ dv \end{bmatrix} = \begin{bmatrix} \frac{\partial u}{\partial x} & -\frac{\partial v}{\partial x} \\ \frac{\partial v}{\partial x} & \frac{\partial u}{\partial x} \end{bmatrix} \begin{bmatrix} dx \\ dy \end{bmatrix}.$$

Note that the determinant of the coefficient matrix (the Jacobian of the mapping f) is

$$\left(\frac{\partial u}{\partial x}\right)^2 + \left(\frac{\partial v}{\partial x}\right)^2 = |f'(z)|^2,$$

so that the matrix is invertible if and only if $f'(z) \neq 0$. A point at which $f'(z) = 0$ is called a *critical point* of the analytic function f, in analogy with the terminology used in the minimization problems of elementary calculus.

At a noncritical point of an analytic function f the columns of the coefficient matrix are orthogonal, since

$$\left(\begin{bmatrix} \frac{\partial u}{\partial x} \\ \frac{\partial v}{\partial x} \end{bmatrix}, \begin{bmatrix} -\frac{\partial v}{\partial x} \\ \frac{\partial u}{\partial x} \end{bmatrix} \right) = \frac{\partial u}{\partial x} \frac{\partial v}{\partial x} - \frac{\partial v}{\partial x} \frac{\partial u}{\partial x} = 0.$$

At such a point, the columns may be scaled by their magnitude $|f'|(z)$ to produce an orthogonal matrix. The differential mapping then takes the form

$$\begin{bmatrix} du \\ dv \end{bmatrix} = |f'(z)| \left\{ \frac{1}{|f'(z)|} \begin{bmatrix} \frac{\partial u}{\partial x} & -\frac{\partial v}{\partial x} \\ \frac{\partial v}{\partial x} & \frac{\partial u}{\partial x} \end{bmatrix} \right\} \begin{bmatrix} dx \\ dy \end{bmatrix}, \qquad (5.16)$$

which represents multiplication by an orthogonal transformation matrix, followed by a scaling by $|f'|(z)$.

Because of the orthogonality, the angle between two different differential vectors is preserved in the image. Another description of this situation is to say that at a noncritical point of the analytic function f, two intersecting smooth curves are mapped to curves intersecting at the same angle. This is illustrated in Figure 5.25.

A complex mapping which preserves angles in this fashion is called *conformal*. The calculation above shows that an analytic function in a domain D is conformal at all points at which $f'(z) \neq 0$ (i.e., all noncritical points). Note that this result rests on the Cauchy–Riemann relations, which underlie the computation that the columns of the coefficient matrix are orthogonal. It is also possible to prove a converse of this result. Assuming that a smooth mapping g is conformal will force the orthogonality of matrix columns corresponding to the linearized g, in turn leading to the Cauchy–Riemann relations for the partials of g. This means that any smooth conformal mapping is therefore a mapping given by an analytic function.

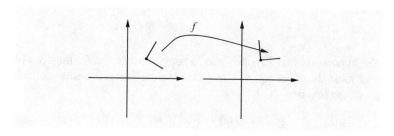

Fig. 5.25 Angle preservation

A consequence of this observation is that it is possible to prove the local existence of analytic inverse functions. The problem is, in essence, to solve the equation (with analytic f given)

$$w = f(z) \tag{5.17}$$

for z in terms of w. The assumption is that $f'(z_0) \neq 0$, so that the mapping f is conformal at z_0. In a neighborhood of z_0, the relation between w and z can be written in terms of a Taylor series as ($w_0 = f(z_0)$)

$$w = w_0 + f'(z_0)(z - z_0) + \frac{f''(z_0)}{2!}(z - z_0)^2 + \cdots. \tag{5.18}$$

We wish to invert this relation and find z expressed as a power series (analytic function) in w

$$(z - z_0) = g'(w_0)(w - w_0) + \frac{g''(w_0)}{2!}(w - w_0)^2 + \cdots.$$

But given that $f'(z_0) \neq 0$, it follows from the two-dimensional (real variable) implicit function theorem that (5.17) has a local inverse, and, moreover, that the linearization of the inverse function has as coefficient matrix the inverse of the coefficient matrix associated with the linearized f. That is, $g'(w_0)$ corresponds to

$$\frac{1}{|f'(z)|}\left\{\frac{1}{|f'(z)|}\begin{bmatrix} \frac{\partial u}{\partial x} & -\frac{\partial v}{\partial x} \\ \frac{\partial v}{\partial x} & \frac{\partial u}{\partial x} \end{bmatrix}\right\}^{-1}.$$

But this means that the inverse function is conformal as well, and from the result quoted above it must be analytic. The expansion therefore holds. (See the problems below for calculation methods.)

This inverse function result is still of a local character and differs from the real variable analogue only in the conclusion that the analyticity of the original function f is inherited by the inverse function g. More global information may be found about the equation

$$w = F(z)$$

under the hypothesis that F is analytic in some domain D. We inquire about the number of solutions of the equation in such a region. This is the same as the number of zeroes of the function

$$f(z) = F(z) - w$$

in the domain D. A formula for this may be derived from the Residue Theorem. The basis of the calculation lies in the following observation.

If the function f analytic in D has a zero at the point z_0, then the Taylor series of f about z_0 takes the form

$$f(z) = a_k (z - z_0)^k + a_{k+1} (z - z_0)^{k+1} + \cdots$$

where $k \geq 1$. We say f has a zero of order k in this case. The derivative of f takes the form

$$f'(z) = a_k \cdot k(z - z_0)^{k-1} + a_{k+1}(k + 1)(z - z_0)^k + \cdots$$

and has a zero of order $k - 1$. This means that the function $\frac{f'(z)}{f(z)}$ has a pole of order 1 at $z = z_0$. In a neighborhood of z_0, the Laurent expansion is

$$\frac{f'(z)}{f(z)} = \frac{a_k \cdot k(z - z_0)^{k-1} + a_{k+1}(k + 1)(z - z_0)^k + \cdots}{a_k(z - z_0)^k + a_{k+1}(z - z_0)^{k+1} + \cdots}$$

$$= \frac{k}{z - z_0} + \sum_{k=0}^{\infty} b_k (z - z_0)^k,$$

where $\sum_{k=0}^{\infty} b_k(z-z_0)^k$ is analytic in a neighborhood of z_0. The residue of $\frac{f'}{f}$ at z_0 is equal to k, the multiplicity (order) of the zero in question. Finally, the function $\frac{f'(z)}{f(z)}$ is analytic at all points in D different from the zeroes of f.

Let C be a simple, closed, positively oriented contour in the domain D, where f is analytic. Let the zeroes of f inside C be $z_1, z_2, \ldots z_N$, with respective orders k_j and suppose that $f \neq 0$ on C. [12]

Then the total number of zeroes inside C is given by

$$\sum_{j=1}^{N} k_j = \frac{1}{2\pi i} \oint_C \frac{f'(z)}{f(z)} \, dz.$$

At this point it might appear that this formula is of little use, since evaluation of the integral by residues, for example, would involve explicitly finding all the zeroes of f.

It turns out, however, that evaluating the integral as a line integral leads to an expression that has a simple geometric interpretation and depends only on the values of f on the boundary C.

To obtain this alternative form, write

$$f = u + i v$$

---

[12] A property of analytic functions is that it is impossible to have an infinite number of zeroes within a bounded domain, without having the function vanish identically. It is therefore not necessary to explicitly assume a finite number of zeroes. This is a consequence of the other assumptions imposed.

so that (using the Cauchy–Riemann equations)

$$
\frac{f'(z)}{f(z)} = \frac{\frac{\partial u}{\partial x} + i\,\frac{\partial v}{\partial x}}{u + iv}\,(dx + i\,dy)
$$

$$
= \frac{\left(\frac{\partial u}{\partial x}\,dx - \frac{\partial v}{\partial x}\,dy\right) + i\left(\frac{\partial u}{\partial x}\,dy + \frac{\partial v}{\partial x}\,dx\right)}{u + iv}
$$

(again using the Cauchy–Riemann relations)

$$
= \frac{\left(\frac{\partial u}{\partial x}\,dx + \frac{\partial u}{\partial y}\,dy\right) + i\left(\frac{\partial v}{\partial y}\,dy + \frac{\partial v}{\partial x}\,dx\right)}{u + iv}
$$

$$
= \frac{(u - iv)\left(\frac{\partial u}{\partial x}\,dx + \frac{\partial u}{\partial y}\,dy\right) + i(u - iv)\left(\frac{\partial v}{\partial y}\,dy + \frac{\partial v}{\partial x}\,dx\right)}{u^2 + v^2}
$$

so that

$$
\frac{f'(z)}{f(z)}\,dz = \frac{u\,du + v\,dv}{u^2 + v^2} + i\,\frac{u\,dv - v\,du}{u^2 + v^2}.
$$

Assuming that the closed curve C is parameterized by t, $0 \le t \le 1$, the integration of the first term in the path integral form gives

$$
\frac{1}{2\pi i}\int_0^1 \frac{u\frac{du}{dt} + v\frac{dv}{dt}}{u^2 + v^2}\,dt = \left.\frac{\ln\sqrt{u^2 + v^2}}{2\pi i}\right|_0^1 = 0,
$$

since the curve is closed, and $u^2 + v^2 \ne 0$ on C. For the second term, note that wherever $u \ne 0$,

$$
\frac{u\frac{dv}{dt} - v\frac{du}{dt}}{u^2 + v^2} = \frac{d}{dt}\arctan\left(\frac{v}{u}\right),
$$

while for $v \ne 0$,

$$
\frac{u\frac{dv}{dt} - v\frac{du}{dt}}{u^2 + v^2} = -\frac{d}{dt}\operatorname{arc\,cotan}\left(\frac{v}{u}\right),
$$

making possible the identification

$$
\frac{u\frac{dv}{dt} - v\frac{du}{dt}}{u^2 + v^2} = \frac{d}{dt}\arg\left(u + iv\right)
$$

(see Figure 5.26).

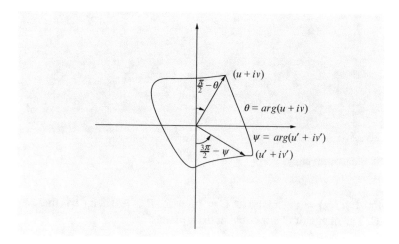

Fig. 5.26 Principle of the argument angles

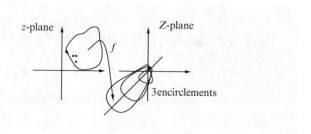

Fig. 5.27 Encirclements

The result is that

$$\frac{1}{2\pi i} \oint_C \frac{f'(z)}{f(z)} dz = \frac{1}{2\pi} \int_0^1 \frac{d}{dt} \arg (u + iv) \, dt$$

and so we can conclude that the number of zeroes inside C is given by

$$\sum_{j=1}^{N} k_j = \frac{1}{2\pi} \int_0^1 \frac{d}{dt} \arg (u + iv) \, dt. \tag{5.19}$$

This result is known as the *Principle of the Argument*. In words, (5.19) states that the number of zeroes of the function f (analytic on and inside C) inside the simple closed contour C is given by the number of counterclockwise encirclements of the origin by the locus of $f(z)$, as z traverses C in the positive sense. See Figure 5.27.

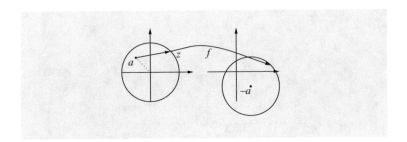

Fig. 5.28 Simple zero image

Example. Let $f(z) = z - a$, and let $C = |z| = R$, $R > a$. Then f has one zero (at a) inside C. The image of C is a circle in the z-plane

$$z = R\,e^{it} - a,$$

and the loci are as illustrated in Figure 5.28.

Example. Prove the fundamental theorem of algebra, that is, a complex polynomial $p(z)$ has exactly $\deg(p)$ zeroes.
    Choose a contour C of the form $|z| = R$. For R sufficiently large,

$$p(z) = z^n + a_{n-1}\,z^{n-1} + \cdots + a_0$$

cannot vanish on C since R may be chosen so that

$$|R|^n > |a_{n-1}\,R^{n-1} + \cdots + a_0|.$$

Now apply the principle of the argument to $p(z)$ Along C,

$$p(z) = z^n \left(1 + \frac{a_{n-1}}{z} + \cdots + \frac{a_0}{z^n}\right),$$

and

$$\arg p(z) = \arg\left(z^n\left(1 + \frac{a_{n-1}}{z} + \cdots + \frac{a_0}{z^n}\right)\right)$$

$$= \arg z^n + \arg\left(1 + \frac{a_{n-1}}{z} + \cdots + \frac{a_0}{z^n}\right).$$

For R sufficiently large,

$$\arg\left(1 + \frac{a_{n-1}}{z} + \cdots + \frac{a_0}{z^n}\right) \approx 0,$$

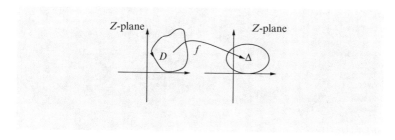

Fig. 5.29 Domain mapping

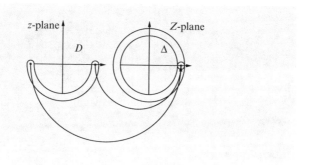

Fig. 5.30 A two-to-one mapping

and the arg of p as C is traversed once is indistinguishable from that of z^n. Hence

$$\frac{1}{2\pi} \int_0^{2\pi} \frac{d}{dt} (\arg(p)) \, dt = n,$$

and p has n zeroes inside C (counting multiplicities).

The principal results in the theory of conformal mapping concern the possibility of constructing an analytic function f which conformally maps a given domain D in the z-plane to a second given domain Δ in the Z-plane, see Figure 5.29.

It is desired to construct the function f in such a way that the mapping of D onto Δ is one-to-one. This problem concerns the behavior of f "in the large." The hypothesis that $f'(z) \neq 0$ in D serves only to guarantee that f is locally one-to-one, and does not prevent widely separated points from being mapped to the same point in the plane under f. An example of the problem is provided by the function $f(z) = z^2$ and a domain D which excludes the origin, but includes both $z = 1$ and $z = -1$. See Figure 5.30.

A function f that maps a domain D conformally onto a domain Δ in a one-to-one fashion is said to be *univalent* or *simple* in D. For applications univalent conformal mappings are desired, and it is therefore useful to be able to determine that a given analytic function defines a univalent conformal mapping on a given

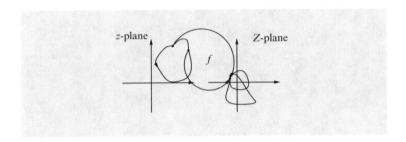

Fig. 5.31 Bivalent mapping

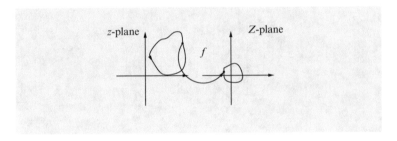

Fig. 5.32 Univalent mapping

domain. In view of the principle of the argument derivation above, it should be possible to determine the univalence (or lack thereof) of an analytic function on a domain bounded by a simple closed contour C from the behavior of the function on the boundary contour C.

To derive such a criterion consider a function f, analytic on and inside a simple closed contour C. Define a mapping to the Z-plane by

$$Z = f(z),$$

defined for z on and inside C.

Under the mapping f the image of the closed contour C is a smooth path Γ in the Z-plane. If the path Γ contains self intersections (Figure 5.31), then f is at least two-to-one at the intersection points and is not univalent on the boundary C. Suppose then that Γ is simple (no self intersections.) It is now possible to show that f maps the domain enclosed by C in the z-plane onto the domain enclosed by Γ in the Z-plane, and that f is univalent on this domain (Figure 5.32).

(In short if f in analytic on C and the enclosed domain, univalence on the boundary C guarantees univalence throughout the domain.)

To see this, notice that if Z_0 is any point of the Z-plane not on Γ, then (from our previous discussion above)

$$m = \frac{1}{2\pi i} \oint_C \frac{f'(z)}{f(z) - Z_0} \, dz$$

is the number of times within C that f assumes the value Z_0. Using the mapping relation

$$Z = f(z),$$

the variable in the line integral(s) may be changed to give

$$m = \frac{1}{2\pi i} \oint_C \frac{f'(z)}{f(z) - Z_0} \, dz = \frac{\pm 1}{2\pi i} \oint_\Gamma \frac{dZ}{Z - Z_0},$$

where the sign is adjusted to account for the positive contour traverse along C. However, by the Residue Theorem,

$$\frac{1}{2\pi i} \oint_\Gamma \frac{dZ}{Z - Z_0} = \begin{cases} 1 & Z_0 \text{ inside } \Gamma, \\ 0 & Z_0 \text{ outside } \Gamma. \end{cases}$$

This shows that within C, f assumes each value inside Γ exactly once, and values outside of Γ not at all.

To complete the argument, it must be shown impossible for a point on Γ to be the image of a point from the interior of C. If there were such a z_1, with

$$f(z_1) = Z_1 \in \Gamma,$$

then since f is analytic at z_1, points in a neighborhood of z_1 would be mapped to a neighborhood of Z_1 (see Figure 5.33), and in particular, points *outside* of Γ would be in the image of the interior of C, contrary to what was established above. Therefore no such z_1 exist.

The importance of the above result lies in the typical mode of application of conformal mapping in problems. In the typical situation (e.g., fluid flow problems, see below), the contours C and Γ are known from the problem specification, and f is essentially constructed to map C onto Γ in a one-to-one fashion. The univalence result then applies automatically to an analytic f so constructed.

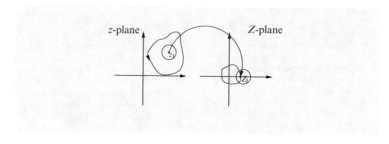

Fig. 5.33 Interior point mapped to boundary

A natural question to consider at this point is what pairs of domains D and Δ as above may be related by a univalent conformal mapping f. The central result on this question is the *Riemann Mapping Theorem*.

Proofs of this are beyond the scope of this text although it is useful to know such a result exists. A version of the theorem relevant to our discussion may be stated as follows.

Theorem. *Let D be the domain bounded by a simple, closed, Jordan curve C.*[13] *Then there exists a function f regular in D, such that $Z = f(z)$ maps D conformally onto $|z| < 1$.*

Proofs of this result (as well as being difficult) are nonconstructive, so that the theorem may be taken more as inspiration to continue the search for conformal mappings than as a computational aid in any particular situation. There exist "catalogs" of conformal mappings for many geometries of interest, and these may be consulted if the need arises.

We end this overview of conformal mapping with a discussion of its application in the theory of ideal fluids.[14]

Recall from Section 5.5 that an ideal fluid flow problem is "solved" through the construction of a complex velocity potential. The real part of this complex potential provides the velocity potential of the flow, with the imaginary part providing the corresponding stream function. The level curves of the stream function provide models of solid flow boundaries.

Conformal mapping may be used to derive velocity potentials for "new flow boundaries" from already known cases (Figure 5.34). The conventional "known case" is that of uniform flow past a circular cylinder (corresponding no doubt to the fact that one of the domains in the standard Riemann mapping theorem is the unit disk.)

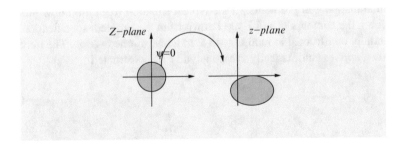

Fig. 5.34 Mapping a complex flow

---

[13] A Jordan curve divides the plane into an inside and an outside.

[14] Completely parallel applications can be made in electrostatics and steady state heat transfer. What matters is that the governing partial differential equation is the two-dimensional Laplace equation.

Suppose that a complex potential is known for a flow in terms of (X, Y) coordinates, that is, using the complex $Z = X + iY$-plane as a model of the flow region. We are given the potential for this flow

$$\Phi(Z) = \phi + i\psi,$$

and a streamline curve Γ, on which ψ vanishes: for $\gamma \in \Gamma$

$$\Phi(\gamma) = \phi(\gamma) + i0.$$

Suppose that it is possible to find a mapping F, conformal and univalent in the flow region, and under which the image of Γ is the (desired) curve C in the z-plane:

$$z = F(Z) \tag{5.20}$$

with

$$C = F(\Gamma).$$

The problem is to define a potential in the z-plane, in such a way that C is a stream line. This is done by defining the z-potential values to be those corresponding to the Z-plane values under the conformal transformation F. If the transformation inverse to (5.20) is denoted by f, so that

$$z = F(Z) \Leftrightarrow Z = f(z),$$

the appropriate z-potential is simply

$$\Psi(z) = \Phi(f(z)).$$

Since f is conformal and maps the exterior of C onto the exterior of Γ, where (since Φ is a Z-potential) Ψ is analytic, this expression is analytic in the appropriate region. Along the curve C (since f maps C onto Γ),

$$\Psi(c) = \Phi(f(c)) = \Phi(\gamma) = \phi(\gamma) + i0,$$

so that C is indeed a streamline for the z-plane flow with complex potential $\Psi(z)$, as desired.

Example. The mapping F given by

$$z = Z + \frac{c^2}{4Z} \tag{5.21}$$

is known as the Joukowski transformation. Note that for the mapping (5.21) the variables z and Z are nearly equal for large $|Z|$. This has the effect that "uniform stream at infinity" potentials in the Z-plane are transformed to flows of the same character in the z-plane. In short, (5.21) transforms a uniform flow past a given shape (Γ) in the Z-plane to a uniform flow past the image shape (C) in the z-plane.

Interest then centers on whether the images of known flows in the Z-plane correspond to flows of interest in the z-plane. Using the expression (5.21) it can be shown that circles centered at the origin in the Z-plane are transformed into ellipses in the z-plane.

In particular, a circle of radius $\frac{1}{2}(a+b)$ at the origin in the Z-plane maps into an ellipse of semi-axes (a, b) centered at the origin, provided the parameter c in (5.21) satisfies $c^2 = a^2 - b^2$. In the extreme case $b = 0$, and the ellipse degenerates to a line along the real axis.

With these computations in hand, the method outlined above can be used to calculate the flow (complex potential) for a "uniform flow past an elliptic cylinder" or "uniform flow past a flat plate." All that is required is to invert the mapping (5.21) and substitute the result into the potential for a uniform stream past a circular cylinder of radius $(\frac{1}{2}(a+b))$ in the Z-plane.

Solving (5.21) for Z gives

$$Z = \frac{1}{2}\left(z \pm \sqrt{z^2 - c^2}\right).$$

The desired branch ($Z \sim z$ for large Z) corresponds to the $+$ sign above so that[15]

$$Z = f(z) = \frac{1}{2}\left(z + \sqrt{z^2 - c^2}\right). \tag{5.22}$$

A complex potential which corresponds to a uniform flow past a cylinder of radius $\frac{1}{2}(a+b)$ at an angle α is

$$\Phi = U\left(Ze^{-i\alpha} + \frac{\left[\frac{(a+b)}{2}\right]^2 e^{i\alpha}}{Z}\right). \tag{5.23}$$

The corresponding z-plane potential is obtained by substituting (5.22) into (5.23) to give ($c^2 = a^2 - b^2$)

$$\frac{U}{2}\left[e^{-i\alpha}\left(z + \sqrt{z^2 - c^2}\right) + \frac{(a+b)^2 e^{i\alpha}}{\left(z + \sqrt{z^2 - c^2}\right)}\right].$$

The streamlines resulting from this example are shown in Figure 5.35.

Further pursuit of related transformations leads to mappings which produce images in the z-plane with the shape of airfoils. The interested reader may consult, for example, [1].

---

[15]The function $\sqrt{z^2 - c^2}$ can be defined with a branch cut from $z = -c$ to $z = +c$, or using a two-sheeted Riemann surface by pasting two such cut planes together. The $+$ sign corresponds to the "upper" sheet.

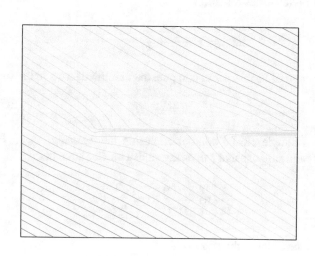

Fig. 5.35 Ideal flow past an elliptical obstacle

Note that the airfoil mapping cannot be conformal at the point corresponding to the trailing edge. It corresponds to a critical point of the mapping and is analyzed by considering the behavior in the neighborhood of this point separately.

Problems 5.6.

1. Suppose that f is a function analytic in some neighborhood of the origin, with Taylor series

$$f(z) = \sum_{n=0}^{\infty} a_n z^n.$$

Assuming that $f'(0) = a_1 \neq 0$, show that the equation

$$w = f(z)$$

has a formal Taylor series solution for z in terms of w,

$$z = \sum_{n=0}^{\infty} b_n w^n$$

in the sense that the unknown coefficients $\{b_n\}$ may be calculated by substituting $z = \sum_{n=0}^{\infty} b_n w^n$ and $f(z) = \sum_{n=0}^{\infty} a_n z^n$ into $w = f(z)$ and matching coefficients.

2. Given the Taylor series for $\sin(z)$

$$\sin(z) = z - \frac{z^3}{3!} + \frac{z^5}{5!} - \frac{z^7}{7!} + \cdots,$$

use the method of Problem 1 to compute the first three terms in the solution of

$$w = \sin(z).$$

3. Let C be a simple closed contour enclosing the zeroes $z_1, z_2, \ldots, z_n$ of the function f, and suppose that f does not vanish on C. Show that

$$\frac{1}{2\pi i} \oint_C \frac{z f'(z)}{f(z)} \, dz = \sum_{i=1}^{n} k_i z_i,$$

where k_i is the order of the ith zero of f.

4. Suppose that the equation

$$f(z) = w_0$$

with analytic function f has a solution z_0 for which $f'(z_0) = 0$. If C is a simple closed contour on which $f(z) \neq w_0$, show that (the integer)

$$\frac{1}{2\pi i} \oint_C \frac{f'(z)}{f(z) - w_0} \, dz \geq 2,$$

so that $f(z) = w_0$ has at least two solutions, given that $f'(z_0) = 0$.

5. Suppose that $f(0) = 0$, and $f'(0) \neq 0$, so that $f(z) - w = 0$ may be (locally) solved for z in terms of w in a unique fashion. Use the result of Problem 5.16 to show that the inverse function $z = F(w)$ may be represented as

$$F(w) = \frac{1}{2\pi i} \oint_C \frac{z f'(z)}{f(z) - w} \, dz$$

where C is a simple closed contour enclosing the unique root of $f(z) - w = 0$.

6. Suppose that f has an isolated pole of order s at the point z_0. Show that

$$\frac{1}{2\pi i} \oint_C \frac{f'(z)}{f(z)} \, dz = -s,$$

where C is the circle $|z - z_0| = \epsilon$, and ϵ is chosen sufficiently small.

7. Using the result of Problem 6, extend the principle of the argument to show that given a function f analytic on and inside of the simple closed contour C, with

the exception of the points $\zeta_1, \zeta_2, \ldots, \zeta_p$ at which f has a pole, and under the assumption that $f(z) \neq 0$ on C, we have

$$\frac{1}{2\pi} \int_C d\,(\arg(f(z))) = \sum_{i=1}^{N} k_i - \sum_{j=1}^{p} m_j,$$

where m_j is the order of the pole at ζ_j and k_i denotes the order of the zero of f at the interior point z_i.

8. Given a proper rational function

$$R(z) = \frac{q(z)}{p(z)},$$

$\deg(q) < \deg(p)$, the stability of a certain feedback control scheme depends on the absence of any solution of the equation

$$\frac{q(z)}{p(z)} + \frac{1}{k} = 0,$$

in the region $\mathrm{Re}(z) \geq 0$, where k is a (fixed) constant.

The Nyquist locus of the rational function R is the curve generated as the image of the imaginary axis under R:

$$\left\{ \frac{q(i\omega)}{p(i\omega)} \,\middle|\, -\infty < \omega < \infty \right\}.$$

Use the principle of the argument (Problem 7 above) to derive a condition on the Nyquist locus ensuring that all zeroes of

$$\frac{q(z)}{p(z)} + \frac{1}{k}$$

lie in the half-plane $\mathrm{Re}(z) < 0$.

Hint: Consider a finite segment of the Nyquist locus and close the contour with a "large" semi-circle in the half-plane $\mathrm{Re}(z) > 0$.

9. Use the results of Problem 8 to determine whether or not

$$f(z) = \frac{1}{z-1} + \frac{1}{2}$$

has a zero with $\mathrm{Re}(z) > 0$

10. The transformation

$$Z = \frac{az + b}{cz + d}, \quad ad - bc \neq 0,$$

is called a bilinear transformation. Show that this transformation is conformal for $z \neq \frac{-d}{c}$ and has the property that circles in the z-plane are transformed into circles in the Z-plane (allowing a straight line to be regarded as a circle through the point at infinity).

11. Prove that the inverse of the bilinear transformation of Problem 10 above is itself a bilinear transformation.

12. Show that the Joukowski transformation maps a circle at the origin in the Z-plane of radius $\frac{1}{2}(a+b)$ into an ellipse of semi-axes a, b in the z-plane, given the parameter relation $c^2 = a^2 - b^2$.

13. Find the critical points of the Joukowski transformation, and (using local polar coordinates) investigate the character of the transformation in the neighborhood of the singular point.

14. By writing out the complex integrals as line integrals, justify the apparent "change of variable" using

$$Z = f(z)$$

to obtain

$$\frac{1}{2\pi i} \oint_C \frac{f'(z)}{f(z) - Z_0} \, dz = \frac{\pm 1}{2\pi i} \oint_\Gamma \frac{dZ}{Z - Z_0}.$$

Reference

1. L.M. Milne-Thompson, *Theoretical Hydrodynamics*, 4th edn. (MacMillan, London, 1962)

Further Reading

L.V. Alfohrs, *Complex Analysis*, 2nd edn. (McGraw-Hill, New York, 1966)

R.V. Churchill, *Complex Variables and Applications*, 2nd edn. (McGraw-Hill, New York, 1960)

E.T. Copson, *Theory of Functions of a Complex Variable* (Oxford University Press, Oxford, 1962)

W. Fulks, *Advanced Calculus* (Wiley, New York, 1962)

J.F. Marsden, *Basic Complex Analysis* (W.H. Freeman and Company, San Francisco, 1975)

W. Miller, *Lie Theory and Special Functions* (Academic, New York, 1968)

Z. Nehari, *Introduction to Complex Analysis* (Allyn and Bacon, Boston, 1961)

E.C. Titchmarsh, *Eigenfunction Expansions Associated With Second Order Differential Equations* (Oxford University Press, London, 1946)

C.N. Watson, *A Treatise on the Theory of Bessel Functions*, 2nd edn. (Cambridge University Press, Cambridge, 1944)

E.T. Whittaker, C.N. Watson, *A Course of Modern Analysis*, 4th edn. (Cambridge University Press, Cambridge, 1927)

Chapter 6
Laplace Transforms

Laplace transforms associate a function of a complex variable with a function defined on a half-axis in "time." The formal properties of the Laplace transform lead to its use in solution of initial value problems for differential equations. The definition of convolution introduces input-output models and linear system responses. Impedance analysis for linear electrical circuits actually arises from the Laplace transform analysis of the circuit element governing equations.

Calculations with Laplace transforms are supported by various Maxima packages, and this approach is introduced.

6.1 Introduction

The Laplace transform is an operation useful in the solution and analysis of differential equations. The utility of the transform is based on the idea that its use replaces a differential equation problem by an algebraic exercise which is more easily solved. This aspect is particularly clear in the case of ordinary differential equations, although the same basic method carries over (with suitable complications) to the case of certain partial differential equations.

In this chapter we treat the basic properties of Laplace transforms together with some of the elementary applications.

Transform inversion in this section is treated at the "table look up" level. The so-called inversion integral for Laplace transforms is treated with the (logically precedent) Fourier inversion integral in the following chapter.

The sections below give first the definition and elementary mechanical properties of Laplace transforms, followed by applications to initial value problems and Fourier coefficient evaluation.

© Springer International Publishing Switzerland 2016
J.H. Davis, *Methods of Applied Mathematics with a Software Overview*,
Applied and Numerical Harmonic Analysis, DOI 10.1007/978-3-319-43370-7_6

These are followed by a discussion of the convolution theorem, with an introduction to linear system analysis. Finally, these notions are related to classical impedance methods for linear circuit analysts.

6.2   Definitions of the Laplace Transform

In view of the fact that the Laplace transform is most widely used in problems in which the independent variable is "time," we adopt this notation for the exposition below.

The Laplace transform associates with a given function (of time) f a second function F (of a complex variable s sometimes referred to as a complex frequency). The formal relationship is given by the integral formula

$$F(s) = \int_0^\infty f(t) \, e^{-st} \, dt. \tag{6.1}$$

The form is sometimes referred to as the one-sided Laplace transform to emphasize that only values of the argument "t" for $t > 0$ are involved in the construction of F. Indeed, it is useful to regard (6.1) as an operation appropriate only for functions f defined for positive values of their arguments. (The transform appropriate to functions defined on the whole real axis is the Fourier transform, considered in the following chapter.)

To actually define the Laplace transform, it is necessary to add to (6.1) a restriction on the functions f to which the integration operation is to be applied.

Definition. Let f be a (complex valued) function defined on $[0, \infty)$, with the property that for some real α, the integral

$$\int_0^\infty |f(t)| \, e^{-\alpha t} \, dt$$

converges. [1] For such f the Laplace transform is defined by

$$F(s) = \int_0^\infty f(t) \, e^{-st} \, dt$$

for all s such that $Re(s) \geq \alpha$.

---

[1] If the integral is considered as a Lebesgue integral, the restriction is that $g : g(t) = f(t)e^{-\alpha t}$ is an element of $L^1(0, \infty)$. In the case of a Riemann integral, the essential restriction is to exponential growth of f (at worst).

Note that the Laplace transform is (initially at least) only defined for values of s with sufficiently large real part. The infimum of $Re(s)$ such that the defining integral converges is called the *abscissa of convergence* of the transform, and is denoted here by σ_0.

If

$$s = \sigma + i\omega,$$

then

$$\mathcal{L}\{f(t)\} = F(s) = \int_0^\infty f(t)\, e^{-st}\, dt$$

exists for all s such that $\sigma = Re(s) > \sigma_0$.

Example. If $f(t) = e^{-at}$, $t > 0$, then

$$\mathcal{L}\{f(t)\} = \int_0^\infty e^{-(s+a)t}\, dt = \frac{1}{s+a},$$

for $Re(s) > -a$. Thus $\sigma_0 = -a$ is the abscissa of convergence.

Example. If $f(t) = 1$, $t > 0$, then

$$\mathcal{L}\{f(t)\} = \frac{1}{s}, \quad Re(s) > 0.$$

Example. If $f(t) = \sin\Omega t$, then

$$\mathcal{L}\{f(t)\} = \int_0^\infty e^{-st}\, \frac{e^{i\Omega t} - e^{-i\Omega t}}{2i}\, dt$$

$$= \frac{1}{2i}\lim_{T\to\infty}\left[\frac{e^{-(s-i\Omega)t}}{-(s-i\Omega)} - \frac{e^{-(s+i\Omega)t}}{-(s+i\Omega)} \right]_0^T$$

$$= \frac{1}{2i}\left[\frac{1}{s-i\Omega} - \frac{1}{s+i\Omega} \right]$$

$$= \frac{\Omega}{s^2 + \Omega^2}, \quad Re(s) > 0.$$

Example. If $f(t) = e^{t^2}$, then

$$\mathcal{L}\{f(t)\} = \int_0^\infty e^{t^2}\, e^{-st}\, dt = ?$$

diverges for all values of s. This f hence has no Laplace transform.

The Laplace transform associates with each function f (suitably restricted as above) defined on $[0, \infty)$, a function of the complex variable s,

$$\mathcal{L}\{f(t)\} = F(s) = \int_0^\infty f(t)\, e^{-st}\, dt, \ \text{Re}(s) > \sigma_0.$$

It is easy to see that for $\text{Re}(s) > \sigma_0$, the above F may be differentiated with respect to the variable s to obtain

$$F'(s) = \int_0^\infty (-t) f(t)\, e^{-st}\, dt, \ \text{Re}(s) > \sigma_0.$$

This calculation shows that F is an analytic function of the complex variable s in the region of convergence $\text{Re}(s) > \sigma_0$. This is illustrated in Figure 6.1.

The above examples show that the Laplace transform formula for any given function f may make sense as an analytic function on a region of the s-plane considerably larger than the original half-plane of definition of the transform. These formulas may be regarded as the analytic continuations of the original transforms. Such considerations are crucial in the use of the inversion integral for the Laplace transform (see the following chapter).

The usefulness of the Laplace transform as an operational tool for the solution of differential (or other) equations rests on the fact that functions (in the class entitled to transforms) are essentially uniquely identifiable by their transforms.

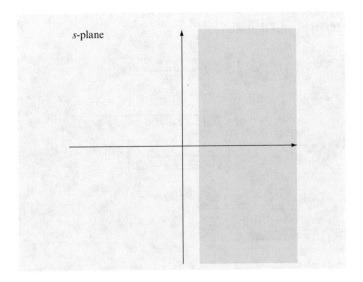

Fig. 6.1 Region of convergence

It is possible to prove a version of this result based on the complex variable ideas introduced above. An alternative approach (pursued here) is to throw the burden on the corresponding Fourier transform uniqueness theorem of the following chapter.

Assuming this result allows us to define the inverse Laplace transform.

Definition. Let F denote the Laplace transform of a function f, restricted as above. We then define the inverse Laplace transform by

$$\mathcal{L}^{-1}\{F(s)\} = f(t),$$

where f is such that[2]

$$\mathcal{L}\{f(t)\} = F(s).$$

The above definition is unsatisfactory in several respects. In the first place, it presupposes the ability to recognize a function F as a Laplace transform of an allowable function. In the second place, the definition is ambiguous in the pointwise sense, since functions may differ pointwise on even an infinite number of points, and yet have the same transform. A third objection is that the definition provides no clue as to how $\mathcal{L}^{-1}$ is to be computed.

In fact, all of these problems may be overcome by consideration of the inversion integral for Laplace transforms. This provides the required formula, and clears up the other ambiguities as well. The problems also may be sidestepped by the use of table lookup tactics in differential equation problems.

Obviously, one must recognize a transform in order to use a table, and use of the table amounts to implementing $\mathcal{L}^{-1}$ (by recognition). The pointwise ambiguity is in these cases removed by the observation that solutions of differential equations are (by definition) continuous functions.

Example. If $F(s) = \frac{1}{s+a}$, $\text{Re}(s) > a$, then

$$\mathcal{L}^{-1}\left\{\frac{1}{s+a}\right\} = e^{-at}, \ t \geq 0.$$

Example. If $F(s) = \frac{1}{s}$, then

$$\mathcal{L}^{-1}\left\{\frac{1}{s}\right\} = 1, \ t \geq 0.$$

Example. If $F(s) = \frac{\Omega}{s^2+\Omega^2}$, $\text{Re}(s) > 0$, then

---

[2] f here is evidently ambiguous to the extent that F remains invariant. In short, $\mathcal{L}^{-1}\{F\}$ should be regarded as a class of such functions.

$$\mathcal{L}^{-1}\left\{\frac{\Omega}{s^2 + \Omega^2}\right\} = \sin \Omega t, \ t \geq 0.$$

A table of transforms (and corresponding inverse transforms) may be easily constructed. One such table is included in Appendix B.

Problems 6.2.

1. Using the basic definition, compute $\mathcal{L}\{t\}$.
2. Define f by

$$f(t) = \begin{cases} 1, \ 0 \leq t < a, \\ 0, \quad t \geq a. \end{cases}$$

Compute $\mathcal{L}\{f(t)\}$.
3. Define g by

$$g(t) = \begin{cases} 1 - t, \ 0 \leq t < 1, \\ 0, \quad t \geq 1. \end{cases}$$

Compute $\mathcal{L}\{g(t)\}$.
4. Compute $\mathcal{L}\{h(t)\}$, where

$$h(t) = \begin{cases} e^{at}, \ 0 \leq t < 1, \\ 0, \quad t \geq 1. \end{cases}$$

Note: One of the objects of Sections 6.3 and 6.4 is to avoid the necessity for the explicit calculations involved in the above.
5. Find $\mathcal{L}\{\cos 97t\}$.
6. Suppose that the continuous function f vanishes identically outside of the interval $[0, T]$, for some fixed T. Show that the corresponding Laplace transform

$$F(s) = \int_0^T f(t)\, e^{-st}\, dt$$

is analytic for all finite values of s, and hence is an entire function of the complex variable s.

6.3 Mechanical Properties of Laplace Transforms

The Laplace transform enjoys a large number of formal properties which are useful in the solution of various problems by means of Laplace transform techniques.

Judicious use of these properties makes it possible to compute a large number of Laplace transforms without the explicit evaluation of integrals. Inverse transforms for a corresponding large number of problems may be computed by use of these tricks in conjunction with a suitable set of tables.

The basic property of the transform is that it is a linear operation.

Linearity

- If $\mathcal{L}\{f(t)\} = F(s)^3$ and $\mathcal{L}\{g(t)\} = G(s)$, then

$$\mathcal{L}\{\alpha f(t) + \beta g(t)\} = \alpha F(s) + \beta G((s).$$

- If $\mathcal{L}\{f(t)\} = F(s)$, $\mathcal{L}\{g(t)\} = G(s)$, then

$$\mathcal{L}^{-1}\{\alpha F(s) + \beta G(s)\} = \alpha f(t) + \beta g(t).$$

A change of time scale has a simple effect on Laplace transform computations.

Scaling Time

If $\mathcal{L}\{f(t)\} = F(s)$, then (for $c > 0$)

$$\mathcal{L}\{f(ct)\} = \int_0^\infty f(ct)\, e^{-st}\, dt$$

$$= \int_0^\infty f(\tau) e^{-s\frac{\tau}{c}}\, \frac{d\tau}{c}$$

$$= \frac{1}{c} F(\frac{s}{c}).$$

Further,

$$\mathcal{L}^{-1}\left\{\frac{1}{c} F(\frac{s}{c})\right\} = f(ct).$$

---

[3] Strictly speaking, $\mathcal{L}$ acts on functions to produce a function; however, the slight abuse of notation makes the formal properties clearer.

Example. Given $\mathcal{L}\{\sin t\} = \frac{1}{s^2+1}$, we compute

$$\mathcal{L}\{\sin \Omega t\} = \frac{1}{\Omega}\, \frac{1}{\frac{s}{\Omega}^2 + 1}$$

$$= \frac{\Omega}{s^2 + \Omega^2}.$$

One of the most useful properties of the transform is the exponential shift result.

Exponential Shift

If $\mathcal{L}\{f(t)\} = F(s)$, then

$$\mathcal{L}\{e^{at} f(t)\} = \int_0^\infty e^{-st} e^{at} f(t) = F(s - a),$$

and

$$\mathcal{L}^{-1}\{F(s - a)\} = e^{at} f(t) = e^{at} \mathcal{L}^{-1}\{F(s)\}.$$

Example. Exponential shift examples:
1.

$$\mathcal{L}\{e^{-at}\} = \mathcal{L}\{1\}(s + a)$$

$$= \frac{1}{s + a}.$$

2.

$$\mathcal{L}^{-1}\left\{\frac{1}{s^2 + 2s + 2}\right\} = \mathcal{L}^{-1}\left\{\frac{1}{(s + 1)^2 + 1}\right\}$$

$$= e^{-t} \sin t.$$

3.

$$\mathcal{L}\{\cos \Omega t\} = \mathcal{L}\left\{\frac{e^{i\Omega t} + e^{-i\Omega t}}{2}\right\}$$

$$= \frac{1}{2}\left(\frac{1}{s - i\Omega} + \frac{1}{s + i\Omega}\right)$$

$$= \frac{s}{s^2 + \Omega^2}.$$

Time Shift

Another useful property is that associated with changes of the time origin. Given a function f defined on $[0, \infty)$, we define a second function (called "f delayed by T") by

$$
g(t) = \begin{cases} 0, & 0 \leq t < T, \\ f(t - T), & t \geq T. \end{cases}
$$

This is illustrated in Figure 6.2.

The shifted function g may be concisely represented by the introduction of the Heaviside unit step function, defined (for all real t) by

$$
U(t) = \begin{cases} 1 & t \geq 0, \\ 0 & t < 0. \end{cases}
$$

With this notation, we have

$$
g(t) = U(t - T)f(t - T), \, t > 0.
$$

It should be noted that g vanishes identically on the interval $[0, T)$. This is connected with the fact that, for Laplace transform purposes, f is defined only for positive values of its argument. The fact that a given f may be represented by a *formula* that makes sense for negative values of its argument is largely irrelevant.

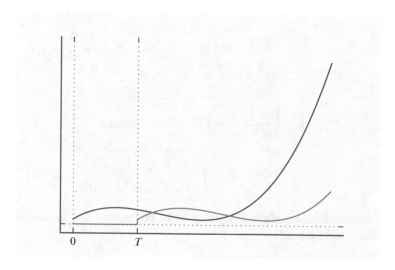

Fig. 6.2 Delayed function

The unit step factor in the definition of g serves to annihilate possible nonzero values associated with such f on the interval $[0, T)$.

The Laplace transform formulas associated with delayed functions are particularly simple.

$$\mathcal{L}\left\{U(t-T)f(t-T)\right\} = \int_0^\infty U(t-T)f(t-T)e^{-st}\,dt$$

$$= \int_T^\infty e^{-st}f(t-T)\,dt$$

$$= e^{-sT}\int_0^\infty e^{-s\tau}f(\tau)\,d\tau$$

$$= e^{-sT}\,F(s).$$

Conversely,

$$\mathcal{L}^{-1}\left\{e^{-sT}F(s)\right\} = U(t-T)f(t-T)$$

$$= \begin{cases} 0, & 0 \le t < T, \\ f(t-T), & t \ge T, \end{cases}$$

Example. Time delay instances:

1. Compute

$$\mathcal{L}\left\{U(t-10)\,e^{a(t-10)}\sin\Omega(t-10)\right\} = e^{-10s}\mathcal{L}\left\{e^{at}\sin\Omega t\right\}$$

$$= e^{-10s}\,\frac{\Omega}{(s-a)^2 + \Omega^2}.$$

2.

$$\mathcal{L}^{-1}\left\{\frac{e^{-s}s}{s^2+4}\right\} = U(t-1)\cos 2(t-1)$$

$$= \begin{cases} 0 & 0 \le t < 1, \\ \cos 2(t-1), & t \ge 1. \end{cases}$$

The discussion of transform analyticity in the previous section includes a result useful in various examples.

Multiplication by "t"

Various integrals can be evaluated by differentiating a "simpler" problem answer with respect to a parameter. Applying this idea to Laplace transforms gives

$$\mathcal{L}\{t \cdot f(t)\} = \int_0^\infty t \cdot f(t)\, e^{-st}\, dt$$

$$= -\frac{d}{ds} \int_0^\infty f(t)\, e^{-st}\, dt$$

$$= -\frac{d}{ds} \mathcal{L}\{f(t)\}.$$

Example. Time multiplication:

1.

$$\mathcal{L}\{t\, e^{-at}\} = -\frac{d}{ds}\frac{1}{s+a}$$

$$= \frac{1}{(s+a)^2}.$$

2.

$$\mathcal{L}\{t^n\, e^{-at}\} = \frac{(n)!}{(s+a)^{n+1}}.$$

3.

$$\mathcal{L}\left\{\frac{t^n}{n!}\right\} = \frac{1}{s^{n+1}}.$$

The connection between Laplace transforms and differentiation is particularly simple and forms the basis for applications to ordinary (or partial) differential equation problems. We present this result in a form which parallels the discussion of Fourier series differentiation problems, since essentially similar points arise.

Differentiation

We suppose that f is a function representable for $t > 0$ in the form

$$f(t) = f(0^+) + \int_0^t f'(t)\, dt, \tag{6.2}$$

where f' is a function such that

$$\int_0^\infty |f'(t)| e^{-\alpha t}\, dt < \infty$$

for some real $\alpha > 0$. This means that $\mathcal{L}\left\{f'(t)\right\}$ exists, and it follows from (6.2) that f also has a Laplace transform. From (6.2), f is continuous and has a limit as $t \to 0^+$, denoted in (6.2) as $f(0^+)$.

We compute $\mathcal{L}\{f(t)\}$ as

$$\lim_{T\to\infty} \int_0^T f(t)\, e^{-st}\, dt = \lim_{T\to\infty} \int_0^T e^{-st} \left[f(0^+) + \int_0^t f'(\tau)\, d\tau\right] dt.$$

Using integration by parts, this is

$$\lim_{T\to\infty} \left[f(0^+)\frac{1 - e^{-sT}}{s} - \frac{e^{-st}}{s} \int_0^t f'(\tau)\, d\tau \Big|_0^T + \frac{1}{s}\int_0^T e^{-st} f'(t)\, dt\right]$$

$$\lim_{T\to\infty} \left[f(0^+)\frac{1 - e^{-sT}}{s} + \frac{1}{s}\int_0^T e^{-st} f'(t)\, dt - \frac{e^{-sT}}{s}\int_0^T f'(\tau)\, d\tau\right].$$

The limits of the first two terms above are readily identifiable, and the desired result follows as soon as it can be shown that the limit of the third term vanishes. But we have

$$\int_0^T f'(\tau)\, d\tau = \int_0^T e^{\alpha\tau} e^{-\alpha\tau} f'(\tau)\, d\tau,$$

and so (assuming $\alpha > 0$)

$$\left|\int_0^T f'(\tau)\, d\tau\right| \le e^{\alpha T} \int_0^T e^{-\alpha\tau} |f'(\tau)|\, d\tau$$

$$\le e^{\alpha T} \int_0^\infty e^{-\alpha\tau} |f'(\tau)|\, d\tau.$$

Hence, for $\mathrm{Re}(s) > \alpha$, we have

$$\left|\frac{e^{-sT}}{s}\int_0^T f'(\tau)\, d\tau\right| \le \frac{1}{|s|}e^{T(\alpha - \mathrm{Re}(s))} \int_0^\infty e^{-\alpha\tau} |f'(\tau)|\, d\tau,$$

and the term in question hence tends to zero as $T \to \infty$. We therefore conclude that (for $\mathrm{Re}(s) > \alpha$, at least)

$$\mathcal{L}\{f(t)\} = \frac{1}{s}f(0^+) + \frac{1}{s}\mathcal{L}\left\{f'(t)\right\}.$$

This is more usually written in the form

$$\mathcal{L}\left\{f'(t)\right\} = s\mathcal{L}\{f(t)\} - f(0^+).\tag{6.3}$$

This result says essentially that differentiation in the time domain corresponds to multiplication by "s," diminished by the initial value, in terms of the Laplace transform domain. This should be compared with the corresponding Fourier series result.

Major uses of (6.3) appear in connection with ordinary differential equations and the transient analysis of circuits by impedance methods. These applications are discussed at length below. The result also can be used to compute various Laplace transforms.

Example. Since

$$\sin \Omega t = 0 + \int_0^t \Omega \cos \Omega \tau \, d\tau,$$

$$\mathcal{L}\{\sin \Omega t\} = 0 + \frac{1}{s}\mathcal{L}\{\Omega \cos \Omega t\}.$$

This relieves one of the burden of computing the transform of more than one of the above by hand.

Example. Since

$$t = 0 + \int_0^t 1 \, d\tau,$$

we have

$$\mathcal{L}\{1\} = \frac{1}{s} = s\mathcal{L}\{t\} - 0,$$

so that

$$\mathcal{L}\{t\} = \frac{1}{s^2}.$$

(This result was obtained above using "t multiplication." Neither method requires the integration by parts of a direct calculation).

Problems 6.3.

1. Recall that

$$\sinh at = \frac{e^{at} - e^{-at}}{2},$$

and

$$\cosh at = \frac{e^{at} + e^{-at}}{2}.$$

Find $\mathcal{L}\{\cosh at\}$ and $\mathcal{L}\{\sinh at\}$.

2. Find

$$\mathcal{L}^{-1}\left\{\frac{12\,s + 37}{9\,s^2 + 90\,s + 229}\right\}.$$

3. Compute

$$\mathcal{L}\{t\,\sin\Omega t\},$$

and

$$\mathcal{L}\{t\,\cos\Omega t\}.$$

4. Let g be defined by

$$g(t) = \begin{cases} t & 0 \le t < 1, \\ 0 & t \ge 1. \end{cases}$$

Compute $\mathcal{L}\{g(t)\}$ by noting that

$$g(t) = t\,(1 - U(t - 1)).$$

5. Compute $\mathcal{L}\{t^2\,e^{-t}\}$ in three distinct ways.

- By direct evaluation of the defining integral.
- By use of t multiplication on the result $\mathcal{L}\{e^{-t}\}$.
- By use of the exponential shift on $\mathcal{L}\{t^2\}$.

6. Compute $\mathcal{L}\{t^2\,\sinh t\}$ and $\mathcal{L}\{t^2\,\sinh^2 t\}$ by any means that seem appropriate.

7. Given the "well-known" Laplace transform result

$$\mathcal{L}\left\{\frac{1}{\sqrt{\pi t}} - a\,e^{a^2 t}\,\mathrm{erfc}(a\sqrt{t})\right\} = \frac{1}{\sqrt{s} + a},$$

find

- $\mathcal{L}^{-1}\left\{\frac{1}{\sqrt{s+1}+1}\right\}$,
- $\mathcal{L}^{-1}\left\{\frac{e^{-12s}}{\sqrt{s+2}+3}\right\}$.

8. Compute and *plot*

$$\mathcal{L}^{-1}\left\{\frac{se^{-s}}{(s+1)^2+1}\right\}.$$

9. Recall the Taylor series expansion

$$\sin t = \sum_{n=0}^{\infty} \frac{(-1)^n\, t^{2n+1}}{(2n+1)!}.$$

By transforming this series term by term, find the result

$$\mathcal{L}\{\sin(t)\} = \frac{1}{s^2+1}.$$

10. The Bessel function J_0 has the Taylor series expansion

$$J_0(t) = \sum_{n=0}^{\infty} \frac{(-1)^n\, (\frac{t}{2})^{2n}}{(n!)^2}.$$

By transforming this series term by term, derive the transform

$$\mathcal{L}\{J_0(t)\} = \frac{1}{\sqrt{s^2+1}}.$$

11. Show that

$$\mathcal{L}\{J_0(2\sqrt{at})\} = \frac{1}{s}e^{-\frac{a}{s}}.$$

12. Compute $\mathcal{L}\{f(t)\}$, where f is the periodic function

$$f(t) = \begin{cases} 1 & 0 \le t < 1, \\ 0 & 1 \le t < 2, \end{cases}$$

and $f(t+2) = f(t)$.

13. Compute

$$\mathcal{L}\{|\sin \Omega t|\}.$$

14. Suppose that f is periodic,

$$f(t + T) = f(t),$$

Show that

$$\mathcal{L}\{f(t)\} = \frac{\int_0^T f(t)\, e^{-st}\, dt}{1 - e^{-sT}}.$$

15. Digital/analog interface devices produce as outputs signals which are piecewise constant functions of time. Such an f is described as

$$f(t) = \{f_k, \; k\Delta \leq t < (k+1)\Delta,$$

where Δ is the sample interval. Find an expression for $\mathcal{L}\{f(t)\}$.

6.4   Elementary Transforms and Fourier Series Calculations

As a consequence of the familiarity of elementary functions, it seems inevitable that functions occurring in model problems will be constructed from such functions. It is often the case that for such problems the functions are described in terms of piecewise specified polynomials, exponential or trigonometric functions. This occurs in the specification of forcing functions for differential equations, as well as in Fourier series examples.

The basic properties of the Laplace transform described in the previous section are sufficient to allow near instantaneous calculation of transforms for such functions. The fact that both Laplace transform and (complex) Fourier series calculations involve exponentially weighted integrals also allows the Laplace methods to be adopted for Fourier series calculations.

These calculations may be approached from either a graphical or analytical point of view. We freely mix the methods in the examples below.

Example. Compute $\mathcal{L}\{f(t)\}$, where

$$f(t) = \begin{cases} e^t & 0 \leq t < 1, \\ 0 & t \geq 1. \end{cases}$$

A graphical solution to this problem is obtained by the following procedure. We first plot f, the desired function as in Figure 6.3, and as a first approximation, construct a simple exponential (Figure 6.4).

The exponential differs from the desired f only for $t > 1$. We correct the first approximation by subtracting an appropriate function. A moment's reflection produces the correction indicated in Figure 6.4.

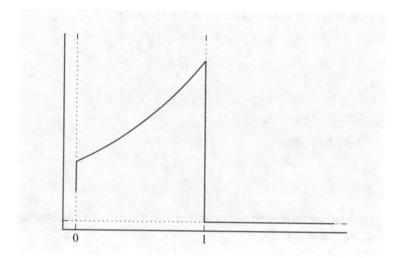

Fig. 6.3 The desired function

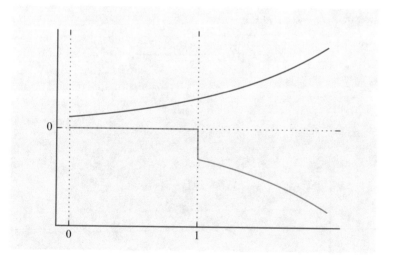

Fig. 6.4 Function decomposition

From this we compute (using the time shift formula)

$$\mathcal{L}\{f(t)\} = \frac{1}{s-1} - e\,\frac{e^{-s}}{s-1}. \tag{6.4}$$

This result may also be obtained analytically by writing

$$f(t) = e^t\,[1 - U(t - 1)]$$
$$= e^t - e^t\,U(t - 1)$$
$$= e^t - e\,e^{t-1}\,U(t - 1),$$

from which (6.4) follows immediately.

Example. Let f be defined by

$$f(t) = \begin{cases} A & 0 \le t < \Delta, \\ 0 & t \ge \Delta. \end{cases}$$

The graphical decomposition is as indicated in Figure 6.5. The corresponding analytical expression is that

$$f(t) = A\,[1 - U(t - \Delta)],$$
$$= A - A\,U(t - \Delta).$$

From either source, we obtain

$$\mathcal{L}\{f(t)\} = \frac{A}{s} - \frac{A\,e^{-s\Delta}}{s}.$$

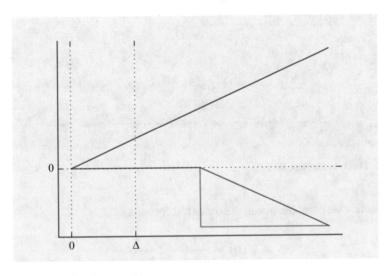

Fig. 6.5 Pulse function decomposition

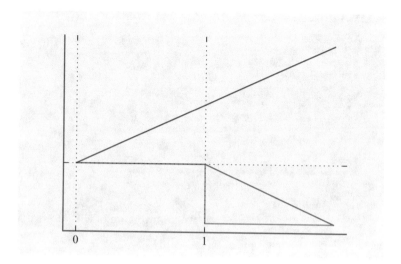

Fig. 6.6 Triangle decomposition

Example. Let f be defined by

$$f(t) = \begin{cases} t & 0 \leq t < 1, \\ 0 & t \geq 1. \end{cases}$$

Then $f(\cdot)$ may be decomposed as shown in Figure 6.6. The function "t," represents a first approximation. A first correction consisting of a "negative-going step" at $t = 1$ adjusts the value at $t = 1$ back to zero. The "negative ramp" correction holds the slope at zero for all $t > 1$. We therefore compute

$$\mathcal{L}\{f(t)\} = \frac{1}{s^2} - \frac{e^{-s}}{s} - \frac{e^{-s}}{s^2}.$$

The analytic derivation of this result is based on the computation

$$f(t) = t\,[1 - U(t-1)]$$
$$= t - [1 + (t-1)]\,U(t-1)$$
$$= t - U(t-1) - (t-1)\,U(t-1).$$

Example. Maxima handles Laplace transforms, including functions involving the "unit_step." The analytic derivation can be carried over to Maxima directly. The example also shows the "user control" of the simplification algorithms in Maxima.

WXMAXIMA

```
(%i7)       load(fourie)

            load("abs_integrate.mac")

            laplace(unit_step(t) - unit_step(t-1), t, s)

            define(f(t), t - (t-1)*unit_step(t-1) - unit_step(t-1))

            define(F(s), laplace(f(t), t, s))

            declare(n, integer)
(%o2)       /usr/share/maxima/5.38.0/share/calculus/fourie.mac
(%o3)       /usr/share/maxima/5.38.0/share/contrib/integration/abs_integrate.mac
```

$$(\%o4) \qquad \frac{1}{s} - \frac{\%e^{-s}}{s}$$

$$(\%o5) \qquad f(t) := t - \mathbf{unit\_step}\left(t-1\right)\left(t-1\right) - \mathbf{unit\_step}\left(t-1\right)$$

$$(\%o6) \qquad F(s) := \frac{1}{s^2} - \frac{\left(s+1\right)\%e^{-s}}{s^2}$$

```
(%o7)       done
```

```
(%i8)       F(2*n*%pi*%i)
```

$$(\%o8) \qquad \frac{\left(2\,\%i\,\pi\,n+1\right)\%e^{-2\,\%i\,\pi\,n}}{4\,\pi^2\,n^2} - \frac{1}{4\,\pi^2\,n^2}$$

```
(%i9)       demoivre(%o8)
```

$$(\%o9) \qquad \frac{2\,\%i\,\pi\,n+1}{4\,\pi^2\,n^2} - \frac{1}{4\,\pi^2\,n^2}$$

```
(%i10)      foursimp(%o9)
```

$$(\%o10) \qquad \frac{\%i}{2\,\pi\,n}$$

Example. Let $f(t)$ be defined by

$$f(t) = \begin{cases} t^2 & 0 \le t < 2, \\ 0 & t \ge 2. \end{cases}$$

The graphical decomposition for f is given in Figure 6.7. It is based on a first approximation, followed by adjustment of offset, slope, and finally curvature.

This results in the transform calculation

$$\mathcal{L}\{f(t)\} = \frac{2}{s^3} - \frac{4\,e^{-2s}}{s} - \frac{4\,e^{-2s}}{s^2} - \frac{2\,e^{-2s}}{s^3}.$$

Example. The (omitted) analytical treatment of the above can be regarded as a special case of the following problem. Let p be a polynomial of degree n (written in the useful form)

$$p(t) = p_0 + \frac{p_1\,t}{1!} + \frac{p_2\,t^2}{2!} + \cdots + \frac{p_n\,t^n}{n!}.$$

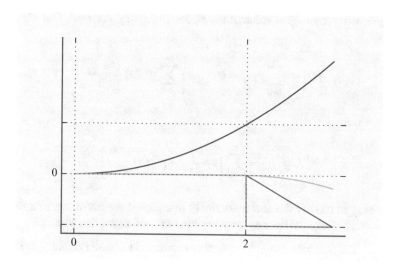

Fig. 6.7 Multiple corrections

Suppose that f is defined by

$$f(t) = \begin{cases} p(t) & 0 \le t < a, \\ 0 & t \ge a, \end{cases}$$

and that it is required to compute $\mathcal{L}\{f(t)\}$. Such problems can be handled by the graphical procedures suggested above, although adjustment of higher derivatives than the second may prove tedious. To treat this analytically, we write

$$f(t) = p(t)\,[1 - U(t - a)]$$
$$= p(t) - p(t)\,U(t - a).$$

The second term evidently represents the "corrections" associated with the graphical approach. In order to evaluate the transform of this term using the shift formula, it must be expressed in terms of "$t - a$." This is done using the Taylor series expansion of p about the point $t = a$.

This gives

$$p(t) = \sum_{j=0}^{n} p^{(j)}(a) \frac{(t - a)^j}{j!}.$$

Here, $p^{(j)}(a)$ represents the value of the jth derivative of p at $t = a$. The expression becomes

$$f(t) = \sum_{j=0}^{n} p_j \frac{t^j}{j!} - U(t-a) \sum_{j=0}^{n} p^{(j)}(a) \frac{(t-a)^j}{j!}.$$

This provides

$$\mathcal{L}\{f(t)\} = \sum_{j=0}^{n} \left[p_j \frac{1}{s^{j+1}} - p^{(j)}(a) \frac{e^{-as}}{s^{j+1}} \right].$$

It is easy to modify the above argument to compute the transform of a function defined as a polynomial over an arbitrary interval, and zero otherwise.

Example. The previous examples contain exponential and polynomial functions. To complete the standard repertoire we consider a trigonometric example.

The function f is defined by

$$f(t) = \begin{cases} \sin \omega_0 t & 0 \le t \le \frac{\pi}{\omega_0}, \\ 0 & t > \frac{\pi}{\omega_0}. \end{cases}$$

The graphical treatment of this example is illustrated in Figure 6.8. This gives

$$f(t) = \sin \omega_0 t + U\left(t - \frac{\pi}{\omega_0}\right) \sin \omega_0 \left(t - \frac{\pi}{\omega_0}\right)$$

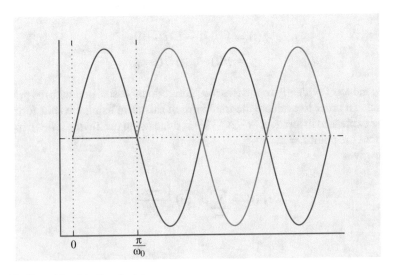

Fig. 6.8 Sinusoid and half-cycle delay

and

$$\mathcal{L}\{f(t)\} = \frac{\omega_0}{s^2 + \omega_0^2} + e^{-s\frac{\pi}{\omega_0}}\frac{\omega_0}{s^2 + \omega_0^2}.$$

The analytical treatment of this example follows from

$$f(t) = \sin(\omega_0 t)\left[1 - U(t - \frac{\pi}{\omega_0})\right]$$

$$= \sin(\omega_0 t) - \sin(\omega_0(t - \frac{\pi}{\omega_0} + \frac{\pi}{\omega_0}))\, U(t - \frac{\pi}{\omega_0})$$

$$= \sin(\omega_0 t) - \sin(\omega_0\,(t - \frac{\pi}{\omega_0}) + \pi)\, U(t - \frac{\pi}{\omega_0})$$

$$= \sin(\omega_0 t) + \sin(\omega_0(t - \frac{\pi}{\omega_0}))\, U(t - \frac{\pi}{\omega_0}).$$

By use of a similar method the transform of a segment of a sinusoid may be calculated. In fact, by combining the t-multiplication, time shift, and exponential shift methods, it is possible to compute (using no explicit integration) the Laplace transform of the function f defined by

$$f(t) = \begin{cases} p(t)\, e^{\alpha t}\, \cos(\beta t + \phi) & 0 \le a \le a \le b, \\ 0 & \text{otherwise.} \end{cases}$$

We refrain from producing the resulting explicit formula (see the problems below).

These considerations make it possible to easily calculate the transforms of functions given as piecewise sums of products of elementary (polynomial, exponential, or sinusoidal) functions. This skill suffices for the solution of most textbook examples.

These methods may also be used to calculate Fourier series expansions. The connection between the two problems may be described as follows. If f is a function which vanishes outside the interval $[0, T]$, then the associated Laplace transform is given by

$$\mathcal{L}\{f(t)\} = \int_0^T f(t)\, e^{-st}\, dt.$$

On the other hand, if a periodic function g is defined by

$$g(t) = \begin{cases} f(t) & 0 < t < T, \\ g(t - T) & \text{otherwise,} \end{cases}$$

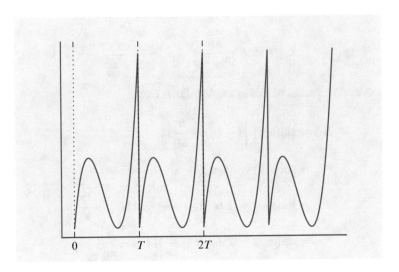

Fig. 6.9 Periodically extended function

(g is the T-periodic extension of f; see Figure 6.9), the complex Fourier series expansion coefficients for g may be calculated as

$$c_n = \frac{1}{T} \int_0^T g(t) e^{-in\omega_0 t}\, dt$$

$$= \frac{1}{T} \int_0^T f(t) e^{-in\omega_0 t}\, dt.$$

Comparing the transform and coefficient formulas, we deduce that

$$c_n = \frac{1}{T} F(in\omega_0) = \frac{1}{T} F(in\frac{2\pi}{T}). \tag{6.5}$$

Since for piecewise elementary f the Laplace transform may be calculated by inspection, the same is true of Fourier coefficients in the corresponding cases.

Example. Define g by

$$g(t) = \begin{cases} t & 0 \le t < 1, \\ g(t-1) & \text{otherwise.} \end{cases}$$

Using the previously calculated

$$F(s) = \frac{1}{s^2} - \frac{e^{-s}}{s} - \frac{e^{-s}}{s^2},$$

we compute

$$
\begin{aligned}
c_n &= \frac{1}{T}\left[\frac{1}{2n\pi i^2} - \frac{e^{-2n\pi i}}{2n\pi i} - \frac{e^{-2n\pi i}}{2n\pi i^2}\right] \\
&= \frac{(-1)}{2n\pi i},\ n \neq 0, \\
c_0 &= \frac{1}{2},\ \text{by inspection.}
\end{aligned}
$$

Hence

$$
g(t) = \frac{1}{2} + \sum_{\substack{n=-\infty \\ n\neq 0}}^{\infty} \frac{(-1)}{2n\pi i} e^{2n\pi\,it}.
$$

Example. Define g by

$$
g(t) = \begin{cases} \sin\Omega t & 0 \leq t \leq \frac{\pi}{\Omega}, \\ g(t - \frac{\pi}{\Omega}) & \text{otherwise.} \end{cases}
$$

(This is the "full wave rectified sinusoid"). From the computation above,

$$
F(s) = \frac{\Omega}{s^2 + \Omega^2} + e^{-\frac{s\pi}{\Omega}}\frac{\Omega}{s^2 + \Omega^2},
$$

and so (with $s = \frac{2n\pi i}{T} = 2ni\Omega$)

$$
\begin{aligned}
c_n &= \frac{\Omega}{\pi}\left[\frac{\Omega}{(2ni\Omega)^2 + \Omega^2} + e^{-\frac{(2ni\Omega)\pi}{\Omega}}\frac{\Omega}{(2ni\Omega)^2 + \Omega^2}\right] \\
&= \frac{2}{\pi}\frac{1}{1 - 4n^2}.
\end{aligned}
$$

Thus

$$
g(t) = \sum_{-\infty}^{\infty} \frac{2}{\pi}\frac{1}{1 - 4n^2} e^{(2ni\Omega\,t)}
$$

is the required expansion.

We make the obvious remark that these calculations are often considerably easier than direct evaluation of the Fourier coefficient integral. The reason for this is that standard anti-derivative methods are embedded in the Laplace transform manipulations.

Problems 6.4.

1. Let f be defined by

$$f(t) = \begin{cases} 1 - t & 0 \leq t < 1, \\ 0 & \text{otherwise.} \end{cases}$$

   Compute $\mathcal{L}\{f(t)\}$ using both a graphical and analytic decomposition of the given function.
2. Repeat Problem 1 for f defined by

$$f(t) = \begin{cases} t & 0 \leq t < 1, \\ 2 - t & 1 \leq t < 2, \\ 0 & t \geq 2. \end{cases}$$

3. Repeat Problem 1 for f given by

$$f(t) = \begin{cases} t(2 - t) & 0 \leq t < 2, \\ 0 & t \geq 2. \end{cases}$$

4. Let g be defined by

$$g(t) = \begin{cases} \cos t & 0 \leq t < \frac{\pi}{2}, \\ 0 & t \geq 2. \end{cases}$$

   Compute $\mathcal{L}\{g(t)\}$.
5. Suppose p is a polynomial of degree n, and f is defined by

$$f(t) = \begin{cases} p(t) & 0 \leq a \leq t < b, \\ 0 & \text{otherwise.} \end{cases}$$

   Find $\mathcal{L}\{f(t)\}$.
6. Suppose that

$$f(t) = \begin{cases} e^{\alpha t} p(t) & 0 \leq t < a, \\ 0 & t \geq a, \end{cases}$$

   where p is an nth-degree polynomial.
   Compute $\mathcal{L}\{f(t)\}$.

7. Let f consist of a sinusoidal segment:

$$f(t) = \begin{cases} \sin(\beta t + \phi) & 0 \le a \le t < b, \\ 0 & \text{otherwise.} \end{cases}$$

Find $\mathcal{L}\{f(t)\}$.

8. If f consists of a polynomial-multiplied sinusoidal segment, find its Laplace transform.

9. Suppose that f is close to a universal description of the typical final examination function:

$$f(t) = \begin{cases} p(t)\, e^{\alpha t}\, \sin(\beta t + \phi) & 0 \le a \le t < b, \\ 0 & \text{otherwise,} \end{cases}$$

where p is an nth-degree polynomial. Find a representation for $\mathcal{L}\{f(t)\}$. For that matter, find several, using different selections of "base function" and "rule application."

10. Define g by $g(t) = g(t - 1)$, with $g(t) = 1 - t$, $0 \le t < 1$. Find the real and complex forms of the Fourier series for g.

11. Let g be periodic of period 2, and coincide with f of Problem 2 above on the interval $[0, 2)$. Compute the Fourier series for g.

12. Construct the analytical heart of a computer-based Fourier series program, capable of constructing the Fourier series of a function specified by

$$g(t) = p_i(t)\, e^{\alpha_i t}\, \sin(\beta_i t + \phi_i),$$
$$a_i \le t < a_{i+1}, \quad i = 1, \dots, N,$$

where $a_1 = 0$, $a_{N+1} = T$. Take p_i as a polynomial of degree n_i.

6.5 Elementary Applications to Differential Equations

The systematic use of the derivative transform formula

$$\mathcal{L}\left\{f'(t)\right\} = s\mathcal{L}\{f(t)\} - f(0^+)$$

in connection with constant coefficient differential equations leads to rapid solutions of many problems.

A prototype problem is the inhomogeneous initial value problem

$$p\left(\frac{d}{dt}\right) x(t) = f(t), \quad t \ge 0,$$

$$x(0), \frac{dx}{dt}(0), \ldots \text{ all prescribed.}$$

The equation represents a constant coefficient ordinary differential equation of order n.

In order to apply the Laplace transform to the differential equation, it is necessary to derive the higher derivative version of the transform relation. Applying the first-order result to the function f' (in place of f) we obtain

$$\mathcal{L}\left\{f''(t)\right\} = s\mathcal{L}\left\{f'(t)\right\} - f'(0)$$

$$= s\left[s\mathcal{L}\{f(t)\} - f(0^+)\right] - f'(0)$$

$$= s^2\mathcal{L}\{f(t)\} - sf(0^+) - f'(0^+).$$

Arguing in a similar fashion (or constructing the obvious induction proof) we obtain

$$\mathcal{L}\left\{\frac{d^n f(t)}{dt^n}\right\} = s^n\mathcal{L}\{f(t)\} - s^{n-1}f(0^+) - s^{n-2}\frac{df}{dt}(0^+) - \cdots - \frac{d^{n-1}f}{dt^{n-1}}(0^+). \quad (6.6)$$

The result holds as long as $\frac{d^{n-1}f(t)}{dt^{n-1}}$ (see Section 6.2) is representable as the anti-derivative of a transformable function. In the case of application to the differential equation above, the form of the equation essentially guarantees the validity of the use of (6.6), as long as the forcing function is Laplace transformable. As a result, calculations for such problems proceed on a formal basis.

Application of the Laplace transform to both sides of the differential equation gives (with $\mathcal{L}\{x\} = X$)

$$\mathcal{L}\left\{p(\frac{d}{dt})x(t)\right\}$$

$$= \mathcal{L}\{f(t)\} = F(s)$$

$$= \left[s^n X(s) - s^{n-1}x(0^+) - s^{n-2}\frac{dx}{dt}(0^+) - \cdots - \frac{d^{n-1}x}{dt^{n-1}}(0^+)\right]$$

$$= p_{n-1}\left[s^{n-1}X(s) - s^{n-2}x(0^+) - s^{n-3}\frac{dx}{dt}(0^+) - \cdots - \frac{d^{n-2}x}{dt^{n-2}}(0^+)\right]$$

$$\vdots$$

$$+ p_0\left[X(s)\right].$$

This may be represented in the form

$$p(s)X(s) = \alpha(s) + F(s),$$

where

$$\alpha(s) = \left\{ s^{n-1} + p_{n-1} s^{n-2} + \cdots + p_1 \right\} x(0^+)$$

$$+ \left\{ s^{n-2} + p_{n-1} s^{n-3} + \cdots + p_2 \right\} \frac{dx}{dt}(0^+) + \cdots + \frac{d^{n-1}x}{dt^{n-1}}(0^+)$$

is a polynomial of degree $n - 1$ depending on the equation coefficients and the solution initial values.

The solution is obtained from

$$X(s) = \frac{\alpha(s)}{p(s)} + \frac{F(s)}{p(s)}, \tag{6.7}$$

$$x(t) = \mathcal{L}^{-1}\left\{ \frac{\alpha(s)}{p(s)} \right\} + \mathcal{L}^{-1}\left\{ \frac{F(s)}{p(s)} \right\}.$$

In the case of forcing functions of the sort discussed in the previous section, the inverse transforms of (6.7) may be evaluated by recourse to algebra (partial fractions), transform tables, and the shift formulas discussed previously. The case of a more general forcing function f is treated in the following section.

Example. Consider the circuit of Figure 6.10, and let the forcing function e (voltage source) be given by

$$e(t) = \begin{cases} A & 0 \le t < \Delta, \\ 0 & t \ge \Delta. \end{cases}$$

The governing equation is

$$RC \frac{dv}{dt} + v = e(t), \quad v(0) = v_0.$$

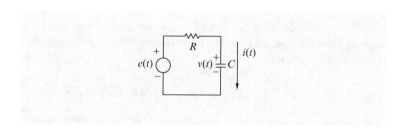

Fig. 6.10 RC circuit

We obtain

$$(RC\,s + 1)\,V(s) = RC\,v_0 + \frac{A\,(1 - e^{-s\Delta})}{s},$$

$$V(s) = \frac{1}{s + \frac{1}{RC}}\,v_0 + \frac{A\,(1 - e^{-s\Delta})}{(RC\,s + 1)\,s}.$$

By partial fractions (see below, for example)

$$\frac{1}{(RC\,s + 1)\,s} = \frac{\alpha}{RC\,s + 1} + \frac{\beta}{s}$$

$$= \frac{-RC}{RC\,s + 1} + \frac{1}{s},$$

and so

$$V(s) = \frac{v_0}{s + \frac{1}{RC}} + A\left\{\frac{-1}{s + \frac{1}{RC}} + \frac{1}{s}\right\} - A\,e^{-s\Delta}\left\{\frac{-1}{s + \frac{1}{RC}} + \frac{1}{s}\right\}.$$

Computing the inverse transform, we obtain

$$v(t) = v_0\,e^{-\frac{t}{RC}} - A\left(e^{-\frac{t}{RC}} - 1\right) - A\,U(t - \Delta)\left(e^{-\frac{t-\Delta}{RC}} - 1\right)$$

$$= \begin{cases} v_0 e^{-\frac{t}{RC}} - A\left(e^{-\frac{t}{RC}} - 1\right) & 0 \le t < \Delta, \\ \left(v_0 + A\,[e^{\frac{\Delta}{RC}} - 1]\right)e^{-\frac{t}{RC}} & t \ge \Delta. \end{cases}$$

The solution of the problems of this section typically requires the use of the partial fractions technique to reduce algebraic expressions to recognizable forms. This method produces standard forms for proper rational functions

$$R(s) = \frac{q(s)}{p(s)}.$$

The method proceeds by first factoring the denominator polynomial. It is convenient to treat the case of possibly complex coefficient polynomials, so that we may assume a factorization

$$p(s) = (s - z_1)^{n_1}(s - z_2)^{n_2} \ldots (s - z_j)^{n_j},$$

where $\{z_i\}$ denotes the zeroes of p (complex valued, in general), and $\{n_i\}$ the multiplicities of the zeroes.

The partial fractions method results from the fact that a proper rational function may be represented in the form[4]

$$R(s) = \sum_{i=1}^{j} \sum_{k=1}^{n_i} \frac{a_{ik}}{(s - z_i)^k} \tag{6.8}$$

$$= \frac{a_{1n_1}}{(s - z_1)^{n_1}} + \frac{a_{1n_1-1}}{(s - z_1)^{n_1-1}} + \cdots + \frac{a_{11}}{(s - z_1)}$$

$$+ \frac{a_{2n_2}}{(s - z_2)^{n_2}} + \frac{a_{2n_2-1}}{(s - z_2)^{n_2-1}} + \cdots + \frac{a_{21}}{(s - z_2)}$$

$$+ \cdots + \frac{a_{j1}}{(s - z_j)}.$$

This form can be established using a complex variable argument by successively subtracting the highest-order poles (represented by the terms in (6.8)) from $R(s)$. This is also the motivation behind the computation of the constants $\{a_{jk}\}$ of (6.8) through the so-called Heaviside coverup method.

This method may be described as follows. The constant a_{in_i} in the expansion associated with the highest power of $(s - z_i)$ may be determined by multiplying both sides of the equation by a factor of $(s - z_i)^{n_i}$. This gives

$$R(s)(s - z_i)^{n_i} = \frac{q(s)}{(s - z_1)^{n_1} \cdots (s - z_{i-1})^{n_{i-1}}(s - z_{i+1})^{n_{i+1}} \cdots (s - z_j)}$$

$$= a_{in_i} + \text{ terms containing a factor of } (s - z_i).$$

We conclude that a_{in_i} may be determined from

$$a_{in_i} = R(s)\,(s - z_i)^{n_i} \Big|_{s=z_i}.$$

This is equivalent to covering up the term $(s - z_i)^{n_i}$ in a factored form of $R(s)$, and is the source of the name of the method.

With the constant a_{in_i} associated with the highest degree $(s - z_i)$ term known, we move that term to the left side of (6.8). This gives a revised rational function

$$R_1(s) = R(s) - \frac{a_{in_i}}{(s - z_i)^{n_i}}$$

to consider. By construction $R_1(s)$ has a lower degree of singularity (a pole of lower degree) at the point $s = z_i$, and the procedure may be repeated until the entire expansion is obtained. We illustrate this procedure with the following problem.

---

[4]The expansion is a consequence of Laurent expansions and Liouville's theorem.

Example. Consider the solution of

$$\frac{dx}{dt} + x = f(t), \ x(0) = 0, \ (t > 0),$$

where

$$f(t) = \begin{cases} t & 0 \le t < 1, \\ 0 & t \ge 1. \end{cases}$$

Transforming produces

$$(s+1) X = \mathcal{L}\{f(t)\} = \frac{1}{s^2} - \frac{e^{-s}}{s} - \frac{e^{-s}}{s^2},$$

$$X(s) = \frac{1}{s^2(s+1)} - \frac{e^{-s}}{s(s+1)} - \frac{e^{-s}}{s^2(s+1)}.$$

In order to use tabular methods, we require partial fractions expansions of both

$$\frac{1}{s^2(s+1)}$$

and

$$\frac{1}{s(s+1)}.$$

Now

$$\frac{1}{s(s+1)} = \frac{a_{11}}{s} + \frac{a_{21}}{s+1}$$

from which $a_{11} = 1$, $a_{21} = -1$.

The expression of the other term is obtained most easily by multiplying the above result by $\frac{1}{s}$ to obtain

$$\frac{1}{s^2(s+1)} = \frac{1}{s^2} - \left(\frac{1}{s(s+1)}\right)$$

$$= \frac{1}{s^2} - \left(\frac{1}{s} - \frac{1}{s+1}\right)$$

$$= \frac{1}{s^2} - \frac{1}{s} + \frac{1}{s+1}.$$

To obtain this by the general method advocated above, write

$$\frac{1}{s^2(s+1)} = \frac{a_{12}}{s^2} + \frac{a_{11}}{s} + \frac{a_{21}}{s+1}.$$

Since determination of a_{21} first gives the final result directly, we find (by cover-up)

$$a_{12} = 1,$$

and compute

$$R_1(s) = \frac{1}{s^2(s+1)} - \frac{1}{s^2}$$

$$= \frac{-s}{s^2(s+1)} = \frac{-1}{s(s+1)}$$

$$= \frac{a_{11}}{s} + \frac{a_{21}}{s+1}.$$

Hence $R_1(s) = \frac{-1}{s} + \frac{1}{s+1}$ and (again)

$$R(s) = \frac{1}{s^2} - \frac{1}{s} + \frac{1}{s+1}.$$

Using these results in the differential equation solution,

$$X(s) = \left(\frac{1}{s^2} - \frac{1}{s} + \frac{1}{s+1}\right) - e^{-s}\left(\frac{1}{s} - \frac{1}{s+1}\right) - e^{-s}\left(\frac{1}{s^2} - \frac{1}{s} + \frac{1}{s+1}\right).$$

Finally,

$x(t)$

$$= t - 1 + e^{-t} - U(t-1)\left(1 - e^{-(t-1)}\right) - U(t-1)\left((t-1) - 1 + e^{-(t-1)}\right)$$

$$= \begin{cases} t - 1 + e^{-t} & 0 \le t < 1, \\ e^{-t} & t \ge 1. \end{cases}$$

It should be noted that the partial fractions method applies only to the rational functions arising in the above examples. The "delay factors" participate only in the final inversion process, and not in the partial fraction expansions.

Numerical Partial Fractions

Octave has a routine for numerically calculating partial fraction expansions. Quite properly, the documentation warns that the procedure is numerically unstable. The problem is that arbitrarily small perturbations in polynomial coefficients can make the difference between multiple and "merely close" roots in a denominator.

The algorithm uses a tolerance to decide when two roots are the same. The procedure produces the "correct" result when applied to our example

$$\frac{1}{s^3 + s^2 + 0 + 0}.$$

The Octave notation is that R consists of the numerators of the partial fraction expansion, while P contains the roots, repeated for the multiple roots case. K is the quotient, and only is nonzero in the case of an improper fraction.

OCTAVE

```
[R P K]  =  residue([1],  [1  1  0  0])

R  =

        1
       -1
        1


P  =

       -1
        0
        0


K  =

       []
```

Example. Maxima will make short work of the problem handled with Octave above, requiring only that the user type in the expression, rather than translate the expressions to an array format.

WXMAXIMA

(%i2)   partfrac(1/(s^3 +s^2), s)

(%o2) $\dfrac{1}{s+1} - \dfrac{1}{s} + \dfrac{1}{s^2}$

(%i5)   ilt((%o2), s, t)

(%o5) $\%e^{-t}+t-1$

Maxima can do partial fraction expansions on a symbolic basis, even including symbolic values for the pole locations, if that is desired. A related capacity is the "residue" procedure, which actually does calculations based on Laurent series expansions. The syntax for partial fractions can be found in the corresponding help screen, shown on page 385. The help for the "residue" procedure is on page 760.
WXMAXIMA

(%i1)   ?? partfrac

-- Function: partfrac (<expr>, <var>)
   Expands the expression <expr> in partial fractions with respect to
   the main variable <var>. 'partfrac' does a complete partial
   fraction decomposition. The algorithm employed is based on the
   fact that the denominators of the partial fraction expansion (the
   factors of the original denominator) are relatively prime. The
   numerators can be written as linear combinations of denominators,
   and the expansion falls out.

(%i1) 1/(1+x)^2 - 2/(1+x) + 2/(2+x);

(%o1) $\dfrac{2}{x+2} - \dfrac{2}{x+1} + \dfrac{1}{(x+1)^2}$

(%i2) ratsimp (%);

(%o2) $- \dfrac{x}{x^3 + 4x^2 + 5x + 2}$

(%i3) partfrac (%, x);

(%o3) $\dfrac{2}{x+2} - \dfrac{2}{x+1} + \dfrac{1}{(x+1)^2}$

Problems 6.5.

1. Find a partial fractions decomposition for the following rational functions.

   a.

$$R(s) = \frac{s}{(s+1)^2\,(s+2)},$$

   b.

$$R(s) = \frac{s^2 + 11\,s + 2}{s^3 + 2\,s^2 + 5\,s},$$

   c.

$$R(s) = \frac{1}{(s+1)^3\,s^2},$$

   d.

$$R(s) = \frac{(s+2)}{(s^2 + 2\,s + 5)(s+1)^2}.$$

2. Find the solutions of the following ordinary differential equations by use of Laplace transform methods.

   a.

$$\frac{dx}{dt} + x = e^{-t},\ x(0) = 1.$$

   b.

$$\frac{dx}{dt} + x = f(t),\ x(0) = 0,$$

$$f(t) = \begin{cases} t(2-t) & 0 \le t < 2, \\ 0 & t \ge 2. \end{cases}$$

   c.

$$\frac{dx}{dt} + x = f(t),\ x(0) = 0,$$

$$f(t) = \begin{cases} \sin \Omega t & 0 \le t < \frac{\pi}{\Omega}, \\ 0 & t \ge \frac{\pi}{\Omega}. \end{cases}$$

3. Find the solution of the problem

$$\frac{d^2x}{dt^2} + 9x = f(t), \ x(0) = \frac{dx}{dt}(0) = 0$$

where

$$f(t) = \begin{cases} \frac{1}{\Delta} & 0 \le t < \Delta, \\ 0 & t \ge \Delta. \end{cases}$$

What is the form of solution obtained as $\Delta \to 0$?
4. Show that in the case of a real coefficient rational function, the partial fractions decomposition may be written as a sum of terms of the form

$$\frac{(c_i s + d_i)}{(s^2 + e_i s + f_i)^{n_i}}$$

where $(s - z_i)(s - \bar{z}_i) = s^2 + e_i s + f_i$ is the quadratic factor associated with the complex conjugate pair of zeroes $\{z_i, \bar{z}_i\}$.
5. Suppose that in the proper rational function

$$R(s) = \frac{q(s)}{p(s)}$$

the zeroes of the polynomial p are all distinct.

$$p(s) = (s - z_1)(s - z_2) \dots (s - z_n).$$

Find an explicit representation of

$$\mathcal{L}^{-1} \left\{ \frac{q(s)}{p(s)} \right\}$$

for this special case.

6.6   Convolutions, Impulse Responses, and Weighting Patterns

The methods of the previous section provide explicit solutions to certain differential equations problems, provided that the forcing functions involved are specified in some sufficiently simple form. If the forcing function is sufficiently exotic, it is likely that our ability to compute transforms (or inverse transforms) will be exceeded. It would be useful to be able to obtain a solution representation for such

a case. We see below that this is possible independent of our ability to handle a
Laplace transform of the forcing function.

Another aspect of this issue is that it is often a *general* solution representation that
is most useful. Such a representation is required in order to estimate the effects of
uncertainty in the forcing function in a given model. The use of such representations
is also at the heart of various analyses of engineering systems (especially in the
control and communication areas) and other input-output models.

In the previous section, we obtained the form

$$x(t) = \mathcal{L}^{-1} \left\{ \frac{q(s)}{p(s)} \right\} + \mathcal{L}^{-1} \left\{ \frac{F(s)}{p(s)} \right\}$$

as a solution of the initial value problem

$$p\left(\frac{d}{dt}\right) x(t) = f(t), \ x(0^+), \dots, \frac{d^{n-1}x}{dt^{n-1}}(0^+) \text{ given.} \tag{6.9}$$

The first term of the solution may be evaluated by a partial fractions exercise, and
hence may be regarded as "known." If all of the initial conditions of (6.9) vanish,
then the solution reduces to the "forced solution"

$$x(t) = \mathcal{L}^{-1} \left\{ \frac{F(s)}{p(s)} \right\}. \tag{6.10}$$

This form represents the solution x in terms of the forcing function f. In fact,
this relationship is linear, as can be easily seen from the linearity of the Laplace
transform.

It is useful to give this a "system" interpretation. The equation (6.9) (or some
physical apparatus of which the equation is a model) is regarded as an object which
accepts the forcing function f as an input, and generates the solution x as an output.
The solution form shows that in the absence of initial conditions the input-output
relationship is a linear one.

The "inputs" and "outputs" referred to above are of course functions. If we
assume that the systems under discussion are physically realizable (or causal), then
a natural way to represent a causal linear input-output relationship is by means of
an integral of the form

$$x(t) = \int_0^t w(\tau) f(\tau) \, d\tau^5.$$

---

[5]With some care about the class of functions allowed as inputs, and continuity hypotheses,
representations of this form must hold.

This intuitively represents the value of the output at the (temporarily fixed) point t as a weighted average of the past values of the input function (a causal system does not respond before the kick is delivered). The function w provides the weighting of the past inputs.

This argument applies to a fixed time t. In general, the weighting function w must be allowed to vary with time. This gives the (revised form) input-output relationship

$$x(t) = \int_0^t g(t, \tau) f(\tau) \, d\tau \tag{6.11}$$

where g now represents the input weighting function associated with time t with explicit notice of its time dependence.

We expect that this relation identifies the form of the inverse transform in the forced response. However, (6.11) is still too general for the problem at hand, in that we have yet to exploit the fact that the model system has constant coefficients (is "time-invariant").

A consequence of this notion of time invariance is that a delay in the input function must produce a corresponding delay in the output (and no shape change).

A formal statement of this in terms of (6.11) is the requirement that (for $T > 0$)

$$u(t - T)x(t - T) = \int_0^t g(t, \tau) \, U(\tau - T) f(\tau - T) \, d\tau.$$

For $0 < t < T$, the formula reduces to

$$0 = \int_0^t g(t, \tau) \, 0 \, d\tau = 0,$$

since $U(\tau - T) = 0$ for $\tau < t < T$. For $t > T$, the formula expression becomes

$$x(t - T) = \int_T^t g(t, \tau) f(\tau - T) \, d\tau.$$

Changing variables of integration,

$$x(t - T) = \int_0^{t-T} g(t, \sigma + T) f(\sigma) \, d\sigma.$$

But since from (6.11) we can also represent this in the form

$$x(t - T) = \int_0^{t-T} g(t - T, \sigma) f(\sigma) \, d\sigma,$$

the hypothesis of time invariance evidently requires that the weighting function satisfy (for $t > T$)

$$g(t - T, \sigma) = g(t, \sigma + T)$$

or (putting $\sigma = 0$)

$$g(t - T, 0) = g(t, T).$$

The relation shows that g, ostensibly a function of two variables, in fact depends only on their difference. We write

$$g(t - T, 0) = g(t, T) = h(t - T),$$

to avoid carrying an extra variable in the input-output relation. We then obtain the forced response (6.11) in the (final) form

$$x(t) = \int_0^t h(t - \tau) f(\tau) \, d\tau. \tag{6.12}$$

The function g (or its specialization h to the time-invariant case) is referred to as the weighting pattern associated with the system (6.9).

The expression (6.12) is referred to as the convolution of the functions h and f. The discussion above should strongly suggest the inevitability of the occurrence of convolutions in connection with linear, time-invariant systems. This general discussion, however, does not yet identify the weighting pattern function h in (6.12) in terms of the original differential equation (6.9). To do this, we consider the effect of the Laplace transform on the convolution (6.12).

Strictly speaking, before attempting to transform (6.12) it should be verified that the function x defined by the convolution expression (6.12) is a function which *has* a Laplace transform (assuming h and f transformable). Since the arguments for this are closely related to those used below to make the computation, we leave this issue as one of the problems below.

Assuming x transformable, the transform of (6.12) is

$$\mathcal{L}\{x(t)\} = \int_0^\infty e^{-st} \left\{ \int_0^t h(t - \tau) f(\tau) \, d\tau \right\} dt.$$

The expression may be identified (see Figure 6.11) as an iterated-integral evaluation of the double integral

$$I = \int_0^\infty \int_0^\infty m(t, \tau) \, d\tau \, dt,$$

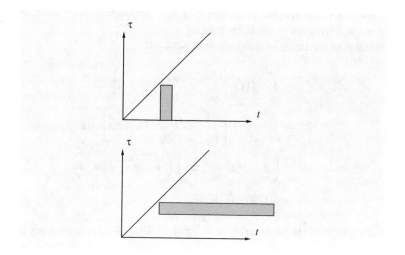

Fig. 6.11 Iterated integral interchange

where (for $t, \tau > 0$)

$$m(t, \tau) = U(t - \tau)h(t - \tau)f(\tau)e^{-st}.$$

The desired result follows by interchanging the order of integration in the integral. In order to justify such an interchange, we verify that the double integral I is in fact convergent.

Our assumption is that h and f are such that for some real α

$$|f(t)e^{-\alpha t}| = a(t),$$
$$|h(t)e^{-\alpha t}| = b(t),$$

with

$$\int_0^\infty a(t)\,dt < \infty,$$
$$\int_0^\infty b(t)\,dt < \infty.$$

Combining the bounds, we have

$$|m(t, \tau)| = |U(t - \tau)h(t - \tau)f(\tau)e^{-st}|$$
$$\leq U(t - \tau)b(t - \tau)e^{\alpha(t-\tau)}a(\tau)e^{\alpha \tau}e^{-\mathrm{Re}(s)t}$$
$$= U(t - \tau)b(t - \tau)e^{[\alpha - \mathrm{Re}(s)]t}a(\tau).$$

This estimate is sufficient to ensure that the double integral I converges for $\mathrm{Re}(s) > \alpha$, justifying the order interchange.

Interchanging the order of integration, we obtain

$$
\begin{aligned}
\mathcal{L}\{x(t)\} &= \int_0^\infty f(\tau) \left\{ \int_\tau^\infty e^{-st} h(t-\tau)\, dt \right\} d\tau \\
&= \int_0^\infty f(\tau) \left\{ \int_0^\infty e^{-s(\tau+\sigma)} h(\sigma)\, d\sigma \right\} d\tau \\
&= \left(\int_0^\infty f(\tau) e^{-s\tau}\, d\tau \right) \left(\int_0^\infty e^{-s\sigma} h(\sigma)\, d\sigma \right) \\
&= \mathcal{L}\{f(t)\}\, \mathcal{L}\{h(t)\}.
\end{aligned}
$$

This computation establishes (in the notation of Section 5.3) the property desired.

Theorem (Convolution). *If f and h are transformable, and x is defined by the* convolution

$$
x(t) = \int_0^t h(t-\tau) f(\tau)\, d\tau,
$$

then x has a Laplace transform, and

$$
\mathcal{L}\{x(t)\} = \mathcal{L}\{f(t)\}\, \mathcal{L}\{h(t)\},
$$

so that the transform of a convolution is the product of the transforms of the functions convolved.

   Conversely, if it is known that

$$
X(s) = H(s) F(s)
$$

where

$$
H(s) = \int_0^\infty h(t) e^{-st}\, dt,
$$

$$
F(s) = \int_0^\infty f(t) e^{-st}\, dt,
$$

for transformable functions f, h, (so that x factors as the product of two transforms), *then*

$$
\begin{aligned}
x(t) &= \int_0^t h(t-\tau) f(\tau)\, d\tau \\
&= \int_0^t f(t-\tau) h(\tau)\, d\tau,
\end{aligned}
$$

where

$$h(t) = \mathcal{L}^{-1}\{H(s)\},$$

and

$$f(t) = \mathcal{L}^{-1}\{F(s)\}.$$

(The possibility of interchanging the order of the functions above follows from the observation that the transform factors may be interchanged).

Example. Although the major utility of convolutions is in their connection with differential equations, the results above may also be used to compute certain inverse transforms.

If

$$X(s) = \frac{1}{s^2 + 3s + 2} = \frac{1}{(s+1)(s+2)},$$

then since

$$\mathcal{L}^{-1}\left\{\frac{1}{s+1}\right\} = e^{-t},$$

$$\mathcal{L}^{-1}\left\{\frac{1}{s+2}\right\} = e^{-2t},$$

we obtain

$$x(t) = \int_0^t e^{-(t-\tau)} e^{-2\tau}\, d\tau$$

$$= e^{-t}\int_0^t e^{-2\tau}\, d\tau$$

$$= e^{-t} - e^{-2t}.$$

This result may be verified by use of the partial fractions method.

The convolution result finally allows us to make the connection between (6.9) and (6.12) above. From

$$X(s) = \frac{1}{p(s)} F(s)$$

we immediately deduce that

$$x(t) = \int_0^t h(t - \tau) f(\tau) \, d\tau,$$

with

$$h(t) = \mathcal{L}^{-1} \left\{ \frac{1}{p(s)} \right\}.$$

This shows that the weighting pattern associated with (6.9) is determined by the characteristic polynomial for the differential equation. The function

$$H(s) = \frac{1}{p(s)}$$

is often referred to as the transfer function of the system (6.9). This is simply the Laplace transform of the weighting pattern function h.

Example. Consider the simple mechanical system of Figure 6.12. The governing equation is

$$\frac{d^2x}{dt^2} + 9x = f(t), \quad x(0) = \frac{dx}{dt}(0) = 0.$$

If the system starts at rest at the origin at the initial time $t = 0$, then we obtain the transform relation

$$X(s) = \frac{1}{s^2 + 9} F(s).$$

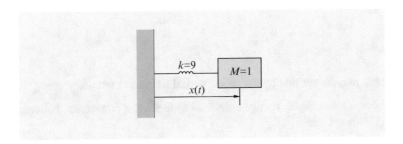

Fig. 6.12 Simple spring mass system

This identifies the transfer function

$$H(s) = \frac{1}{s^2 + 9}$$

and weighting pattern function

$$h(t) = \mathcal{L}^{-1}\left\{\frac{1}{s^2 + 9}\right\} = \frac{1}{3}\sin 3t.$$

The solution of the equation is therefore

$$x(t) = \int_0^t \frac{1}{3}\sin 3(t - \tau)f(\tau)\,d\tau,$$

representing the convolution between the weighting pattern and forcing function.

Although we have defined the notion of transfer function above in the context of the simple differential equation (6.9), the notion extends to more interesting situations. A natural extension is to the case of vector-matrix differential equations discussed in the following section. There are also (fairly obvious) connections with impedance analysis of circuits. See Section 6.8.

The weighting pattern defined above is also referred to as the impulse response of the system under discussion.

The intuitive idea of an impulse arises from consideration of various mechanical (or electrical) problems. The essential notion is that of a short, sharp forcing function (c.f. the proverbial "swift kick") which serves to instantaneously transfer energy to the system upon which it is impressed. These (and related) physical ideas have led to the development of a theory of generalized functions as a mathematical means of analyzing such phenomena. Rather than develop such a theory here, we concentrate on the intuitive physical basis. Generalized functions are discussed in Section 7.8.

The basic idea can be illustrated with the spring mass model constructed above. The input-output relation of that system was derived as

$$x(t) = \int_0^t \frac{1}{3}\sin 3(t - \tau)f(\tau)\,d\tau.$$

We assume that the forcing function f_Δ takes the form of Figure 6.13. The function f_Δ differs from zero only on the short interval $[0, \Delta]$, $f_\Delta(t) > 0$, and is scaled so that

$$\int_0^\Delta f_\Delta(\tau)\,d\tau = 1.$$

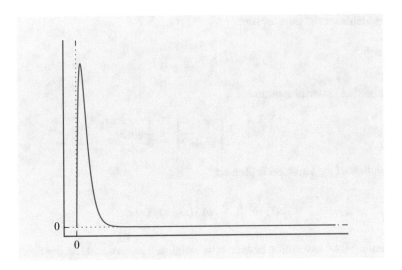

Fig. 6.13 Approximate delta function

The response to such an f_Δ is in principle given by the convolution integral. For times $t > \Delta$ (the duration of the forcing function), the response is (since $f_\Delta(\tau) = 0$ for $\tau > \Delta$)

$$x(t) = \int_0^t \frac{1}{3} \sin 3(t - \tau) f_\Delta(\tau) \, d\tau$$

$$\approx \frac{1}{3} \sin 3t \int_0^\Delta f_\Delta(\tau) \, d\tau$$

$$= \frac{1}{3} \sin 3t.$$

This approximation is based on the supposition that Δ is "short," so that

$$\frac{1}{3} \sin 3(t - \tau) \approx \frac{1}{3} \sin 3t,$$

for $0 \le \tau \le \Delta$. This emphasizes the idea that (for practical purposes) shortness of Δ is a term that makes sense only relative to the scale of variation in the weighting pattern term of the convolution. The approximation involved is illustrated in Figure 6.14. (The degree of approximation can also be estimated by a Taylor series exercise, if such is desired.) This result shows that (at least for times greater than the forcing function duration) the x response is approximated by the previously calculated weighting pattern function.

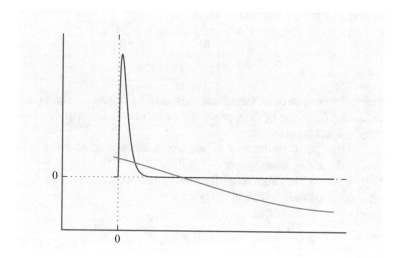

Fig. 6.14 Time scale comparison

It is a simple matter to argue that if the forcing function f_Δ is now delayed by a time T (that is, replaced by an input given by

$$f_1(t) = U(t - T)f_\Delta(t - T),$$

the corresponding response is

$$x(t) = 0, \ 0 \le t < T,$$

$$x(t) \approx \frac{1}{3} \sin 3 \, (t - T), \ t > T + \Delta.$$

(The response in the interval $[T, T + \Delta]$ is dependent on the unspecified detailed shape of f_Δ.)

The intuitive notion of an impulse is obtained from the above by letting $\Delta \to 0$, at the same time retaining the requirement that $f_\Delta > 0$, and the normalization

$$\int_0^\Delta f_\Delta(\tau) \, d\tau = 1.$$

The approximations made above become increasingly exact as $\Delta \to 0$. The limiting response is simply

$$x(t) = \frac{1}{3} \sin 3t,$$

or in the case of a delayed forcing function,

$$x(t) = \begin{cases} 0 & 0 \le t < T, \\ \frac{1}{3} \sin 3(t - T) & T \le t. \end{cases}$$

In this limit the effect of the ambiguous detailed nature of the approximate impulse forcing functions f_Δ disappears; the overall result is determined solely by the system weighting pattern.

Arguments in the style leading to this result are of course not limited to a discussion of the spring mass system of Figure 6.12.

The approximate impulse forcing function f_Δ may be impressed on the general differential equation (6.9) to obtain the system

$$p\left(\frac{d}{dt}\right) x(t) = f_\Delta, x(0) = \cdots = \frac{d^{n-1}}{dt^{N-1}} = 0,$$

with general solution

$$x(t) = \int_0^t h(t - \tau) f_\Delta(\tau) \, d\tau.$$

As calculated above, the weighting pattern is given by

$$h(t) = \mathcal{L}^{-1}\left\{\frac{1}{p(s)}\right\}.$$

As $\Delta \to 0$, we again obtain the limiting form of the response as

$$x(t) = h(t), t > 0.$$

In the case of a delayed input, the response is

$$x(t) = \int_0^t h(t - \tau) \, U(\tau - T) f_\Delta(\tau - T) \, d\tau,$$

with limiting form

$$x(t) = \begin{cases} 0 & 0 \le t < T, \\ h(t - T) & t \ge T \end{cases}$$
$$= U(t - T) h(t - T).$$

These results may be calculated on a formal basis by assuming the existence of a "function" (called a delta-function, or impulse function) with the properties that

$$\mathcal{L}\{\delta(t)\} = 1,$$

and

$$\mathcal{L}\{\delta(t - T)\} = e^{-sT}.$$

This object is to be regarded for the time being as the "limiting form" of f_Δ as $\Delta \to 0$. The "limiting form" of the differential equation in this notation is simply

$$p\left(\frac{d}{dt}\right) x(t) = \delta(t), \, x(0) = \cdots = \frac{d^{n-1}}{dt^{N-1}} = 0.$$

Using the above properties for a formal Laplace transform calculation, we obtain

$$p(s) X(s) = 1,$$

$$X(s) = \frac{1}{p(s)}.$$

Inverting,

$$x(t) = \mathcal{L}^{-1}\left\{\frac{1}{p(s)}\right\} = h(t),$$

which is just the desired result. In a similar fashion, the limiting form of the solution for the delayed input case is

$$p\left(\frac{d}{dt}\right) x(t) = \delta(t - T), x(0) = \cdots = \frac{d^{n-1}}{dt^{N-1}} = 0,$$

which gives first in Laplace transform form

$$X(s) = \frac{e^{-sT}}{p(s)}$$

and in turn the correct limiting result

$$x(t) = U(t - T) h(t - T).$$

These results are the source of the alternative terminology *impulse response* for the function h.

The properties of "delta" suffice for the purposes of making many formal calculations with delta-functions. A discussion of the properties of this object is more easily made in the context of Fourier transforms. Among the issues that must be considered is the fact that a "δ-function" is not a function at all. This means that such notions as integration, differentiation, and transform must be treated with some delicacy. An introduction to such a treatment is given in Section 7.8 below.

Problems 6.6.

1. Use the linearity of the Laplace transform to show that the response of the model

$$p \left(\frac{d}{dt} \right) x(t) = f(t), \ x(0) = \cdots = \frac{d^{n-1}}{dt^{N-1}} = 0$$

   is linear in the forcing function $f(\cdot)$.

2. a. Compute $\mathcal{L} \{\sin t\}$, $\mathcal{L} \{e^{-t}\}$.
   b. Compute the convolution of the functions $h(t) = e^{-t}, f(t) = \sin t$.
   c. Compute the Laplace transform of the result of (b), and verify that it is identical to the product $\mathcal{L} \{\sin t\} \mathcal{L} \{e^{-t}\}$.
   d. Compute the inverse Laplace transform of the product

$$\mathcal{L} \{\sin t\} \mathcal{L} \{e^{-t}\}$$

   by the partial fractions method, and verify that the result is the same as the explicit convolution calculation (b).

3. Consider the convolution of

$$h(t) = e^{-t}, \ f(t) = e^{-2t}$$

   given by $\int_0^t e^{-(t-\tau)} e^{-2\tau} \, d\tau$.

   a. Graph the function $e^{-2\tau}$ (versus τ) over the interval (of the τ-axis) $[0, t]$.
   b. On the same set of axes graph (versus τ) the function given by $e^{-(t-\tau)}$ over the interval $[0, t]$.
   c. On the same set of axes as the above, graph the function defined by the product $e^{-(t-\tau)} e^{-2\tau}$ over the interval $[0, t]$.
   d. Give an interpretation of the convolution as an area associated with (a), (b), (c) above. Comment on the interpretation as t varies.
   e. Give a similar interpretation for the general convolution

$$\int_0^t h(t-\tau) f(\tau) \, d\tau.$$

4. Make a simple change of variables to show that the function order in a convolution is immaterial:

$$\int_0^t h(t-\tau) f(\tau) \, d\tau = \int_0^t f(t-\tau) h(\tau) \, d\tau.$$

5. Show that the convolution operation results in a function which is Laplace transformable.

a. Show that this is the case for

$$x(t) = \int_0^t h(t - \tau) f(\tau) \, d\tau,$$

provided that

$$|f(t)| \le M_1 e^{\alpha_1 t},$$
$$|h(t)| \le M_2 e^{\alpha_2 t}.$$

b. Show that this is also the case assuming only that

$$|f(t) e^{-\alpha t}| \le a(t),$$
$$|h(t) e^{-\alpha t}| \le b(t),$$

with

$$\int_0^\infty a(t) \, dt < \infty, \quad \int_0^\infty b(t) \, dt < \infty.$$

6. The solution

$$x(t) = \int_0^t h(t - \tau) f(\tau) \, d\tau,$$

to the problem

$$p\left(\frac{d}{dt}\right) x(t) = f(t), x(0) = \cdots = \frac{d^{n-1}}{dt^{N-1}} = 0$$

was found above under the hypothesis that f has a Laplace transform. This apparently excludes, for example,

$$f(t) = e^{t^2}.$$

Show that consideration of forcing functions of the form

$$f(t) \left[1 - U(t - T)\right]$$

extends the validity of the solution formula to functions whose restriction to any finite interval is Laplace transformable.

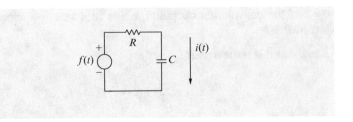

Fig. 6.15 Forced RC network

7. Consider the RC-network of Figure 6.15, and suppose that

$$f(t) = \begin{cases} \frac{1}{\Delta}, & 0 \le t < \Delta, \\ 0 & t \ge \Delta. \end{cases}$$

Find an explicit solution of this problem by transform methods, and find the limiting solution form as $\Delta \to 0$.

8. By use of Taylor series (or trigonometric identities) show that the solution form

$$\int_0^t \sin 3(t - \tau) f_\Delta(\tau) \, d\tau$$

approaches (for $t > 0$) the limiting form

$$\frac{1}{3} \sin 3t$$

as $\Delta \to 0$. We assume that the function f_Δ differs from zero only on the short interval $[0, \Delta]$, $f_\Delta(t) \ge 0$, and is scaled so that

$$\int_0^\Delta f_\Delta(\tau) \, d\tau = 1.$$

9. Let f_Δ be defined by

$$f_\Delta(t) = \begin{cases} \frac{1}{\Delta}, & 0 \le t < \Delta, \\ 0 & t \ge \Delta. \end{cases}$$

By use of the mean-value theorem of elementary calculus, show that for $t > 0$ we have the result that the solution of

$$p\left(\frac{d}{dt}\right) x(t) = f(t), x(0) = \cdots = \frac{d^{n-1}}{dt^{N-1}} = 0$$

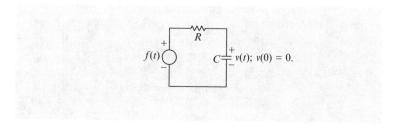

Fig. 6.16 RC circuit

tends to

$$\mathcal{L}^{-1}\left\{\frac{1}{p(s)}\right\}$$

as $\Delta \to 0$.

10. Find the transfer function relating the output voltage v to the input voltage f for the circuit of Figure 6.16
Find also the impulse response of this system.

6.7   Vector Differential Equations

The treatment of ordinary differential equations in the previous sections was based on the analysis of single equations (of high order).

A more useful analysis for many purposes can be based on systems of first-order differential equations. Such systems are often referred to as "state variable" models and were briefly referred to above in the treatment of periodic solutions of constant coefficient systems.

Such systems are written in the matrix-vector form

$$\frac{d\mathbf{x}}{dt} = \mathbf{A}\mathbf{x} + \mathbf{f}(t), \quad \mathbf{x}(0) = \mathbf{x}_0.$$

Here $\mathbf{x}$ is a vector-valued function of dimension n referred to as the *state vector*. The vector-valued function $\mathbf{f}$ is called the *forcing function*. The matrix $\mathbf{A}$ is of dimension $n \times n$ and is assumed independent of time.

The state vector $\mathbf{x}$ intuitively contains the "amount of information" required to specify the evolution of the system described by the equation. In physical terms, the state may often be formulated as the vector of values associated with the energy storage elements of a system. The notion is better illustrated by example than by an attempt to give an all inclusive definition of state.

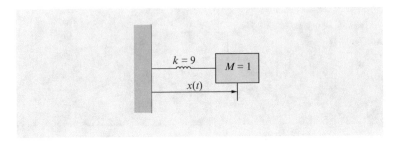

Fig. 6.17 Vibrating spring mass system

Example. Consider the spring mass system of Figure 6.17.

If we let v denote the velocity of the mass, x the displacement, then the governing equations may be written in the form

$$\frac{dx}{dt}(t) = v(t),$$

$$\frac{dv}{dt}(t) = -9\,x(t) + f(t).$$

If we define the state vector by

$$\begin{bmatrix} x(t) \\ v(t) \end{bmatrix},$$

then the coupled system may be rewritten in the vector matrix form

$$\frac{d}{dt}\begin{bmatrix} x(t) \\ v(t) \end{bmatrix} = \begin{bmatrix} 0 & 1 \\ -9 & 0 \end{bmatrix}\begin{bmatrix} x(t) \\ v(t) \end{bmatrix} + \begin{bmatrix} 0 \\ f(t) \end{bmatrix}.$$

(Recall that the total energy associated with this system is $\frac{9}{2}x^2(t) + \frac{1}{2}v^2(t)$. This may be taken to suggest the state vector used above.)

This system is in the required form.

Example. Consider the circuit of Figure 6.18.

The equations of motion associated with this circuit are given by

$$i(t) = C\frac{dv}{dt}, \tag{6.13}$$

$$e(t) = L\frac{di}{dt} + v(t).$$

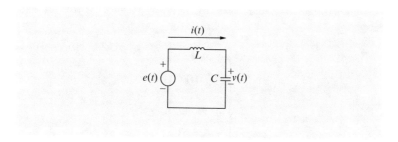

Fig. 6.18 LC circuit

Selecting

$$\begin{bmatrix} v(t) \\ i(t) \end{bmatrix}$$

as a state vector, we may write (6.13) in the standard vector matrix form

$$\frac{d}{dt}\begin{bmatrix} v(t) \\ i(t) \end{bmatrix} = \begin{bmatrix} 0 & \frac{1}{C} \\ -\frac{1}{L} & 0 \end{bmatrix}\begin{bmatrix} v(t) \\ i(t) \end{bmatrix} + \begin{bmatrix} 0 \\ \frac{e(t)}{L} \end{bmatrix}.$$

Example. The general second-order inhomogeneous differential equation

$$\frac{d^2}{dt^2}x(t) + \alpha\frac{d}{dt}x(t) + \beta x(t) = f(t)$$

may be put in a state variable form by a procedure analogous to that used for the spring mass problem above.

If we choose

$$\begin{bmatrix} x(t) \\ \frac{dx}{dt} \end{bmatrix}$$

as a state vector, then the system

$$\frac{d}{dt}\begin{bmatrix} x(t) \\ \frac{dx}{dt} \end{bmatrix} = \begin{bmatrix} 0 & 1 \\ -\beta & -\alpha \end{bmatrix}\begin{bmatrix} x(t) \\ \frac{dx}{dt} \end{bmatrix} + \begin{bmatrix} 0 \\ f(t) \end{bmatrix}$$

provides the required state variable representation.

The procedure of the last example extends readily to encompass the case of a single n^{th}-order equation (used as an example in the previous sections.) The nth-order equation

$$\frac{d^n}{dt^n}x(t) + p_{n-1}\frac{d^{n-1}}{dt^{n-1}}x(t) + \cdots + p_1\frac{d}{dt}x(t) + p_0x(t) = f(t)$$

is converted to state variable form by defining a state vector of successive derivatives:

$$\begin{bmatrix} x(t) \\ \frac{d}{dt}x(t) \\ \vdots \\ \frac{d^{n-1}}{dt^{n-1}}x(t) \end{bmatrix}.$$

The vector matrix equation takes the form

$$\frac{d}{dt}\begin{bmatrix} x(t) \\ \frac{d}{dt}x(t) \\ \vdots \\ \frac{d^{n-1}}{dt^{n-1}}x(t) \end{bmatrix}$$

$$= \begin{bmatrix} 0 & 1 & 0 & \cdots & 0 \\ 0 & 0 & 1 & 0 & \cdots \\ \vdots & \ddots & \ddots & \ddots & \vdots \\ -p_0 & -p_1 & -p_2 & \cdots & -p_{n-1} \end{bmatrix}\begin{bmatrix} x(t) \\ \frac{d}{dt}x(t) \\ \vdots \\ \frac{d^{n-1}}{dt^{n-1}}x(t) \end{bmatrix} + \begin{bmatrix} 0 \\ 0 \\ \vdots \\ f(t) \end{bmatrix}.$$

In the above, the last equation of the system is identifiable as the original differential equation. The remaining entries of the vector equation effectively define the components of the state vector as successive derivatives.

It should be noted that an initial condition statement for the single differential equation is translated immediately to a (vector) initial condition for the vector version.

With this background, it is useful to first discuss systems of the vector model in the case of a pure initial value problem. In this case the forcing function f vanishes, and the model reduces to

$$\frac{d\mathbf{x}}{dt} = \mathbf{A}\mathbf{x}, \ \mathbf{x}(0) = \mathbf{x}_0.$$

This problem may be readily solved by Laplace transforms, once it is realized that the differentiation formula

$$\mathcal{L}\left\{\frac{d}{dt}\mathbf{x}\right\} = s\mathcal{L}\{\mathbf{x}\} - \mathbf{x}(0^+)$$

is valid for vector valued (as well as scalar valued) functions of time. This is obvious, given that

$$X(s) = \mathcal{L} \left\{ \begin{bmatrix} x_1(t) \\ x_2(t) \\ \vdots \\ x_n(t) \end{bmatrix} \right\} = \begin{bmatrix} \mathcal{L}\{x_1(t)\} \\ \mathcal{L}\{x_2(t)\} \\ \vdots \\ \mathcal{L}\{x_n(t)\} \end{bmatrix}$$

is accepted as the only natural definition of the Laplace transform of a vector valued function of time.

Applying the Laplace transform to the vector equation produces

$$sX(s) - x(0^+) = \mathcal{L}\{Ax(t)\}$$

$$= A\mathcal{L}\{x(t)\}$$

$$= AX(s).$$

The fact that the transform and matrix multiplication are interchangeable follows easily from the integral definition of $\mathcal{L}\{\cdot\}$.

The relation provides a system of linear equations for the components of the solution transform.

This is

$$(sI - A)\,X(s) = x_0.$$

For all values of s different from the eigenvalues of A the coefficient matrix in the equation system is non-singular, and the solution is simply

$$X(s) = (sI - A)^{-1}x_0.$$

The classical representation of the inverse is given by

$$(sI - A)^{-1} = \frac{\text{Adj}\,(Is - A)}{\det\,(Is - A)}.$$

The matrix $\text{Adj}\,(Is - A)$ is the matrix of transposed signed cofactors associated with $(Is - A)$. Since this is a polynomial matrix in the variable s, the matrix inverse consists of a matrix of (proper) rational functions. The zeroes of the denominators of these rational functions (poles of the rational functions) occur only at zeroes of the polynomial $\det\,(sI - A)$. Since

$$p(s) = \det\,(sI - A)$$

is the characteristic polynomial of **A**, these coincide with the eigenvalues of the matrix **A**.

Example. For the spring-mass coefficient matrix,

$$(s\mathbf{I} - \mathbf{A}) = \begin{bmatrix} s & -1 \\ 9 & s \end{bmatrix}$$

and

$$(s\mathbf{I} - \mathbf{A})^{-1} = \frac{\begin{bmatrix} s & 1 \\ -9 & s \end{bmatrix}}{s^2 + 9}$$

$$= \begin{bmatrix} \frac{s}{s^2+9} & \frac{1}{s^2+9} \\ \frac{-9}{s^2+9} & \frac{s}{s^2+9} \end{bmatrix}.$$

Example. For the *LC* network above

$$(s\mathbf{I} - \mathbf{A}) = \begin{bmatrix} s & \frac{-1}{C} \\ \frac{1}{L} & s \end{bmatrix}$$

and

$$(s\mathbf{I} - \mathbf{A})^{-1} = \frac{\begin{bmatrix} s & \frac{1}{C} \\ \frac{-1}{L} & s \end{bmatrix}}{s^2 + \frac{1}{LC}}.$$

Example. For the general second-order equation case

$$(s\mathbf{I} - \mathbf{A}) = \begin{bmatrix} s & -1 \\ \beta & s + \alpha \end{bmatrix},$$

the inverse takes the form

$$(s\mathbf{I} - \mathbf{A})^{-1} = \frac{\begin{bmatrix} s + \alpha & 1 \\ -\beta & s \end{bmatrix}}{s^2 + \alpha s + \beta}.$$

For the matrix formulation of an nth-order differential equation, the matrix $\mathbf{A}$ constructed is what is known as the companion matrix associated with the polynomial $p(s)$. It is an elementary exercise to show that for such a companion matrix

$$\det(s\mathbf{I} - \mathbf{A}) = p(s).$$

The characteristic polynomial of the original differential equation hence appears in the denominator of the required inverse.

Since the inverse coefficient matrix is a matrix of proper rational functions, the method of partial fractions suffices to carry out the required inversion process. From

$$\mathbf{X}(s) = (s\mathbf{I} - \mathbf{A})^{-1}\,\mathbf{x}_0$$

we obtain the solution

$$\mathbf{x}(t) = \mathcal{L}^{-1}\left\{(s\mathbf{I} - \mathbf{A})^{-1}\,\mathbf{x}_0\right\}$$
$$= \mathcal{L}^{-1}\left\{(s\mathbf{I} - \mathbf{A})^{-1}\right\}\mathbf{x}_0.$$

It is conventional to denote the inverse transform in this expression as the *exponential of the matrix* $\mathbf{A}$,

$$e^{\mathbf{A}t} = \mathcal{L}^{-1}\left\{(s\mathbf{I} - \mathbf{A})^{-1}\right\}. \tag{6.14}$$

This has the effect of giving the solution to

$$\frac{d\mathbf{x}}{dt} = \mathbf{A}\,\mathbf{x}(t),\ \mathbf{x}(0) = \mathbf{x}_0$$

in the form (suggested from the scalar case)

$$\mathbf{x}(t) = e^{\mathbf{A}t}\,\mathbf{x}_0.$$

In fact, the term *is* the exponential of the matrix $\mathbf{A}$ in the sense that the matrix exponential may be represented as an exponential power series.

$$\mathcal{L}^{-1}\left\{(s\mathbf{I} - \mathbf{A})^{-1}\right\} = e^{\mathbf{A}t}$$

$$= \mathbf{I} + \mathbf{A}t + \mathbf{A}^2\frac{t^2}{2!} + \mathbf{A}^3\frac{t^3}{3!} + \cdots.$$

This infinite series may be shown to converge, to some matrix of functions. Further, the construction ensures that the differentiation law

$$\frac{d}{dt}e^{\mathbf{A}t} = \mathbf{A}e^{\mathbf{A}t}$$

holds in the matrix case, just as in the scalar one.

This means that

$$\mathbf{x}(t) = e^{\mathbf{A}t}\mathbf{x}_0$$

is a solution of the homogeneous vector differential equation, provided that the matrix exponential is defined by the series expression. Assuming uniqueness of solutions of the differential equation, the results calculated above identify (6.14) as a means of computation for the matrix exponential.

(For many problems, (6.14) is more suitable for hand calculation than an attempt to sum the corresponding infinite series.)

It is also easy to see that the matrix exponential satisfies the law of exponents

$$e^{\mathbf{A}t}e^{\mathbf{A}\tau} = e^{\mathbf{A}t+\tau}.$$

(See the problems below.)

Example. Since

$$\mathcal{L}^{-1}\left\{\begin{bmatrix} \frac{s}{s^2+9} & \frac{1}{s^2+9} \\ \frac{-9}{s^2+9} & \frac{s}{s^2+9} \end{bmatrix}\right\} = \begin{bmatrix} \cos 3t & \frac{1}{3}\sin 3t \\ -3\sin 3t & \cos 3t \end{bmatrix},$$

this is identified as

$$e^{\begin{bmatrix} 0 & 1 \\ -9 & 0 \end{bmatrix}t}.$$

Since we have

$$\frac{d}{dt}\begin{bmatrix} \cos 3t & \frac{1}{3}\sin 3t \\ -3\sin 3t & \cos 3t \end{bmatrix} = \begin{bmatrix} -3\sin 3t & \cos 3t \\ -9\sin 3t & -3\sin 3t \end{bmatrix}$$

it is easy to verify that the differentiation law holds by carrying out the required multiplication.

Properties and special cases of the matrix exponential may be pursued at length. Some of these aspects are included in the problems, where linear algebraic methods are emphasized. (Compare also Section 3.6.)

We next consider the effect of forcing functions in the vector equation model.

As the forced model is written, the forcing function $\mathbf{f}$ apparently consists of n ($= \dim \mathbf{x}$) independent functions. As is suggested by our examples above, it is often the case that the forcing function contains fewer than n independent functions. One way to emphasize this possibility is to introduce a vector function $\mathbf{u}$ (of dimension m, say) and to define $\mathbf{f}$ by

$$\mathbf{f} = \mathbf{B}\mathbf{u}$$

for some $n \times m$-dimensional matrix $\mathbf{B}$. In the examples constructed above, $\mathbf{B}$ may be taken as an $n \times 1$ matrix, and the forcing function is only one-dimensional. Higher-dimensional forcing functions of this sort are a common feature of multiple force mechanical systems, or multiple port electrical networks, to cite two obvious sources.

Accordingly, we consider the system

$$\frac{d\mathbf{x}}{dt} = \mathbf{A}\mathbf{x} + \mathbf{B}\mathbf{u}, \ \mathbf{x}(0) = \mathbf{x}_0.$$

Application of the Laplace transform gives

$$(s\mathbf{I} - \mathbf{A}) \, \mathbf{X}(s) = \mathbf{x}_0 + \mathbf{B}\,\mathbf{U}(s).$$

Here $\mathbf{X}$ denotes the transform of the solution, $\mathbf{U}$ that of the forcing function. For s different from an eigenvalue of $\mathbf{A}$ we obtain

$$\mathbf{X}(s) = (s\mathbf{I} - \mathbf{A})^{-1} \mathbf{x}_0 + (s\mathbf{I} - \mathbf{A})^{-1}\,\mathbf{B}\,\mathbf{U}(s).$$

The inversion of the first term has been treated above and gives

$$\mathcal{L}^{-1} \left\{ (s\mathbf{I} - \mathbf{A})^{-1} \mathbf{x}_0 \right\} = e^{\mathbf{A}t} \mathbf{x}_0.$$

The second term consists of products of $(s\mathbf{I} - \mathbf{A})^{-1}$ elements with transforms of the forcing term. This term may therefore be evaluated by use of the convolution theorem. This gives

$$\mathcal{L}^{-1} \left\{ (s\mathbf{I} - \mathbf{A})^{-1}\,\mathbf{B}\,\mathbf{U}(s) \right\} = \int_0^t e^{\mathbf{A}\,t-\tau}\,\mathbf{B}\,\mathbf{u}(\tau)\,d\tau.$$

(If this appears mysterious, consider in detail the matrix products of the transform expression. See the problems below.)

Combining those results, we find the solution

$$\mathbf{x}(t) = e^{\mathbf{A}t}\mathbf{x}_0 + \int_0^t e^{\mathbf{A}\,t-\tau}\,\mathbf{B}\,\mathbf{u}(\tau)\,d\tau.$$

The explicit form is referred to (for historical reasons) as the variation of constants formula. By analogy with the discussion of the previous section, the matrix

$$\mathbf{H}(s) = (s\mathbf{I} - \mathbf{A})^{-1}\mathbf{B}$$

(which appears as a factor between the forcing function transform and the solution transform) is called the *input-state transfer function matrix*. The associated inverse transform

$$e^{\mathbf{A}t}\mathbf{B} = \mathcal{L}^{-1}\left\{(s\mathbf{I} - \mathbf{A})^{-1}\mathbf{B}\right\}$$

appearing in the convolution term of the solution is called the *impulse response matrix* of the system.

Example. For the spring-block system considered above we have calculated

$$e^{\begin{bmatrix} 0 & 1 \\ -9 & 0 \end{bmatrix}t} = \begin{bmatrix} \cos 3t & \frac{1}{3}\sin 3t \\ -3\sin 3t & \cos 3t \end{bmatrix}.$$

From the original model equation we identify

$$\mathbf{B} = \begin{bmatrix} 0 \\ 1 \end{bmatrix}$$

so that

$$e^{\mathbf{A}t}\mathbf{B} = \begin{bmatrix} \frac{1}{3}\sin 3t \\ \cos 3t \end{bmatrix}.$$

The variation of constants solution is therefore

$$\begin{bmatrix} x(t) \\ v(t) \end{bmatrix} = \begin{bmatrix} \cos 3t & \frac{1}{3}\sin 3t \\ -3\sin 3t & \cos 3t \end{bmatrix} \begin{bmatrix} x(0) \\ v(0) \end{bmatrix} + \int_0^t \begin{bmatrix} \frac{1}{3}\sin 3(t-\tau) \\ \cos 3(t-\tau) \end{bmatrix} f(\tau)\,d\tau.$$

The corresponding transfer function matrix is given by

$$\begin{bmatrix} \frac{s}{s^2+9} & \frac{1}{s^2+9} \\ \frac{-9}{s^2+9} & \frac{s}{s^2+9} \end{bmatrix} \begin{bmatrix} 0 \\ 1 \end{bmatrix} = \begin{bmatrix} \frac{1}{s^2+9} \\ \frac{s}{s^2+9} \end{bmatrix}.$$

The reader should verify that these results contain those of the previous section, and in addition provide the system velocity.

Although we have emphasized the input-state transfer function above, there are instances in which it is useful to regard some linear combination of the state components as an output of a system. If such a linear combination is denoted by

$$\mathbf{y}(t) = \mathbf{C}\mathbf{x}(t),$$

then from the variation of constants solution we conclude that

$$\mathbf{y}(t) = \mathbf{C}\,e^{\mathbf{A}t}\,\mathbf{x}_0 + \int_0^t \mathbf{C}\,e^{\mathbf{A}t-\tau}\,\mathbf{B}\,\mathbf{u}(\tau)\,d\tau.$$

The relation in terms of Laplace transforms is

$$\mathbf{Y}(s) = \mathbf{C}\,(s\mathbf{I} - \mathbf{A})^{-1}\,\mathbf{x}_0 + \mathbf{C}\,(s\mathbf{I} - \mathbf{A})^{-1}\,\mathbf{B}\,\mathbf{U}(s).$$

The matrix of rational functions

$$\mathbf{G}(s) = \mathbf{C}\,(s\mathbf{I} - \mathbf{A})^{-1}\,\mathbf{B}$$

is referred to as the input-output *transfer function matrix*.

Transfer function matrices have their origin in impedance analysis of linear circuits. Some aspects of this connection are mentioned in the following section.

Problems 6.7.

1. Find a state variable model for a physical system consisting of the two coupled spring mass systems illustrated in Figure 6.19.
2. Find a state variable model for the circuit of Figure 6.20
3. Write the third-order ordinary differential equation problem

$$\frac{d^3x}{dt^3}(t) + 2\frac{dx}{dt}(t) + x(t) = e^{-t},\ t > 0,$$

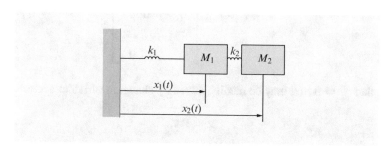

Fig. 6.19 Dual spring mass system

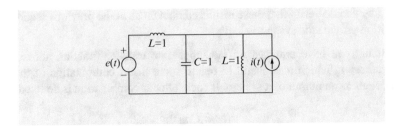

Fig. 6.20 "LCL" circuit

$$x(0) = 1, \ \frac{dx}{dt}(0) = 2, \ \frac{d^2x}{dt^2}(0) = 3$$

in state-variable form.

4. a. Show that in case the coefficient matrix of the system

$$\frac{d\mathbf{x}}{dt} = \mathbf{A}\,\mathbf{x}(t),$$

is diagonal

$$\mathbf{A} = \begin{bmatrix} \lambda_1 & 0 & \dots & 0 \\ 0 & \lambda_2 & 0 & \ddots \\ \vdots & \ddots & \ddots & \vdots \\ 0 & \dots & \dots & \lambda_n \end{bmatrix},$$

the associated matrix exponential is easily calculated.

b. Show that in case there exist a non-singular matrix **P** which diagonalizes **A** so that

$$\mathbf{P}\mathbf{A}\mathbf{P}^{-1} = \begin{bmatrix} \lambda_1 & 0 & \dots & 0 \\ 0 & \lambda_2 & 0 & \ddots \\ \vdots & \ddots & \ddots & \vdots \\ 0 & \dots & \dots & \lambda_n \end{bmatrix},$$

that $\frac{d\mathbf{x}}{dt} = \mathbf{A}\,\mathbf{x}(t)$ may be readily solved by changing variables according to

$$\mathbf{z} = \mathbf{P}\mathbf{x},$$

$$\frac{d\mathbf{z}}{dt} = \mathbf{P}\mathbf{A}\mathbf{P}^{-1}\mathbf{z}.$$

Conclude that for such a case

$$e^{\mathbf{A}t} = \mathbf{P}^{-1} e^{\mathbf{P}\mathbf{A}\mathbf{P}^{-1}t}\mathbf{P}$$

$$= \mathbf{P}^{-1} e^{\begin{bmatrix} \lambda_1 & 0 & \cdots & 0 \\ 0 & \lambda_2 & 0 & \ddots \\ \vdots & \ddots & \ddots & \vdots \\ 0 & \cdots & \cdots & \lambda_n \end{bmatrix} t} \mathbf{P}$$

$$= \mathbf{P}^{-1} e^{\begin{bmatrix} e^{\lambda_1 t} & 0 & \cdots & 0 \\ 0 & e^{\lambda_2 t} & 0 & \ddots \\ \vdots & \ddots & \ddots & \vdots \\ 0 & \cdots & \cdots & e^{\lambda_n t} \end{bmatrix} t} \mathbf{P}.$$

5. Show that in case $\mathbf{A}$ is in an elementary Jordan block form,

$$\mathbf{A} = \begin{bmatrix} \lambda & 1 & 0 & \cdots \\ 0 & \lambda & 1 & \cdots \\ \vdots & & \ddots & \ddots \\ 0 & 0 & \cdots & \lambda \end{bmatrix},$$

the matrix exponential $e^{\mathbf{A}t}$ may be readily calculated as

$$e^{\mathbf{A}t} = \begin{bmatrix} e^{\lambda t} & t e^{\lambda t} & \cdots & \frac{t^{n-1}}{(n-1)!} e^{\lambda t} \\ 0 & e^{\lambda t} & \cdots & \frac{t^{n-2}}{(n-2)!} e^{\lambda t} \\ \vdots & \ddots & \ddots & \vdots \\ 0 & 0 & 0 & e^{\lambda t} \end{bmatrix}$$

by either

a. Laplace transform methods, or
b. explicit summation of the infinite series.

6. Use the fact that any complex valued matrix may be put in Jordan canonical form by a similarity transformation (and the result of the previous problem) to show that the infinite series defining $e^{\mathbf{A}t}$ converges for each square matrix $\mathbf{A}$ and all values of t.

7. Show that the matrix exponential may be expressed in terms of polynomials and (possibly complex) exponential functions of the variable t.

8. Show that if A is the companion matrix associated with the polynomial p,

$$p(s) = s^n + p_{n-1}\, sn - 1 + p_{n-2}\, s^{n-2} + \cdots p_0,$$

$$\begin{bmatrix} 0 & 1 & 0 & \cdots \\ 0 & 0 & 1 & \cdots \\ \vdots & \vdots & \ddots & \vdots \\ -p_0 & -p_1 & \cdots & -p_{n-1} \end{bmatrix},$$

then

$$\det (I s - A) = p(s) = s^n + p_{n-1}\, sn - 1 + p_{n-2}\, s^{n-2} + \cdots p_0.$$

9. By use of Laplace transforms, find the matrix exponential for the 2×2 matrix

$$A = \begin{bmatrix} 0 & 1 \\ -\beta & -\alpha \end{bmatrix}$$

arising from the second-order ordinary differential equation

$$\frac{d^2 x}{st^2} + \alpha \frac{dx}{dt} + \beta x = 0.$$

(There are three distinct cases to consider.)

10. Show that the matrix exponential function satisfies

$$e^{A 0} = I,$$

and the composition law

$$e^{A t}\, e^{A \tau} = e^{A t + \tau}.$$

11. By expressing the matrix vector multiplications

$$(s I - A)^{-1} B U(s)$$

in terms of matrix element summations, verify in detail that

$$\mathcal{L}^{-1} \left\{ (s I - A)^{-1} B U(s) \right\} = \int_0^t e^{A (t-\tau)} B u(\tau)\, d\tau.$$

12. Calculate the input state transfer function matrix associated with the circuit of Figure 6.20.

13. Calculate the impulse response for the circuit of Figure 6.20

14. Find the variation of constants solution for the circuit of Figure 6.20.
15. Suppose that the state variable model

$$\frac{d\mathbf{x}}{dt} = \mathbf{A}\mathbf{x} + \mathbf{b}\,e(t)$$

is given, where e is a periodic function of period T.

Use the variation of constants formula to obtain an equation determining an initial condition such that the solution is periodic.

16. If the forcing function e has the form

$$e(t) = \sum_{n=-\infty}^{\infty} c_n \, e^{i n \omega_0 t},$$

show that

$$\mathbf{x}(0) = \sum_{n=-\infty}^{\infty} (i n \omega_0 \mathbf{I} - \mathbf{A})^{-1} \mathbf{b}\, c_n$$

is a solution of the initial condition equation of Problem 15.
Hint: Parseval's theorem is of use.

6.8 Impedance Methods

The impedance methods discussed in Section 2.8 find their use in the calculation of steady state periodic solutions for (stable) electric circuits.

That analysis is based on the differentiation law for Fourier series, in conjunction with the defining relations of the standard circuit elements.

The differentiation law for Laplace transforms leads to a form of analysis which is algebraically almost identical to that previously encountered.

The interpretation of Laplace transform impedance analysis is, however, quite distinct from the periodic forcing case (in spite of the algebraic connections). We will see below that such analysis is essentially equivalent to the solution of a system of the "state variable" form considered in the previous section. Since such models explicitly include both initial condition and forced responses of the system in question, Laplace transform analysis of linear circuits is often referred to as "transient analysis."

In order to keep the discussion at an elementary level, we consider only linear circuits consisting of interconnections of resistive, capacitive, and inductive elements with voltage and current sources (that is, conventional RLC circuits).

The defining relations for the conventional circuit elements of Figure 6.21 are

$$e(t) = R\,i(t),$$

$$e(t) = L\frac{di}{dt},$$

$$i(t) = C\frac{de}{dt}.$$

Assuming that the Laplace transform differentiation law is applicable to the current and voltage functions of a given circuit,[6] we may transform the equations to obtain

R:

$$E(s) = R\,I(s),$$

L:

$$E(s) = L\left(sI(s) - i(0^+)\right),$$

$$= s\,L\,I(s) - L\,i(0^+),$$

C:

$$I(s) = C\left(sE(s) - e(0^+)\right)$$

$$= s\,C\,I(s) - C\,e(0^+),$$

$$E(s) = \frac{1}{sC}\,I(s) + \frac{1}{s}\,e(0^+).$$

From experience with resistive networks (including current and voltage sources), it is possible to interpret the transformed circuit relations in a fashion analogous to Ohm's law. This interpretation is given in Figure 6.21

Since the Laplace transformed form of the Kirchoff current (or voltage) law is algebraically identical to the original, the conclusion is that a circuit may be analyzed using standard methods (e.g., loop and/or node analysis) provided that the circuit elements of the original (time domain) circuit are replaced by the (Laplace transform domain) equivalents of Figure 6.21.

This entails assigning an impedance (the transform equivalent of Ohm's law) of R to a resistor, sL to an inductor, and $\frac{1}{sC}$ to a capacitor, while inserting the appropriate initial condition voltage sources in series with the reactive devices.

---

[6]The fact that this is the case for a given circuit arrangement may be verified through a state-variable formulation of the model.

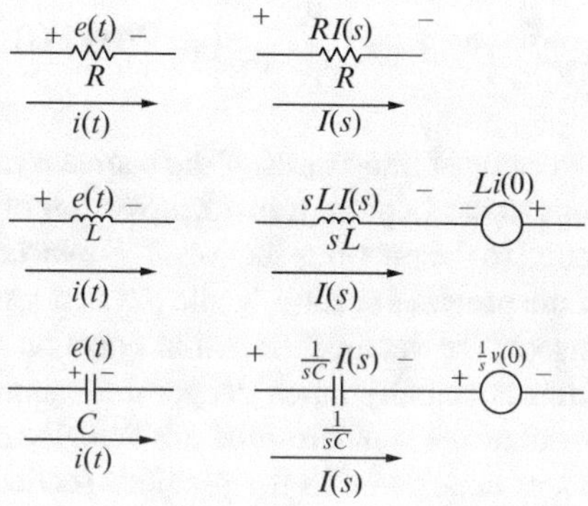

Fig. 6.21 Circuits and Laplace equivalents

Current or voltage sources in the original time domain circuit appear directly in transformed form in the transform domain equivalent.

It should be noted that this procedure is essentially the same as that encountered in the case of periodic solution analysis of linear circuits (Section 2.8). The differences are first that Laplace transforms (rather than Fourier coefficients) are the objects of the computation (in this sense the object of the calculation is different). In the second place, the Laplace domain equivalent circuit contains sources associated with the initial conditions of the circuit. A minor algebraic difference arises from the fact that "s" appears in the Laplace transform case in place of the "$jn\omega_0$" of the periodic analysis.

From these observations it is clear that the periodic case computations may be recovered from the algebraic effort of the Laplace analysis simply by replacing the initial conditions by zero, and "s" by "$jn\omega_0$" throughout.

We mention here that periodic solutions may also be approached by "purely" Laplace transform methods. Such a technique produces the explicit transient transition to a periodic solution (if such exists), and reveals the potential stability problems (mentioned in Section 2.8) in the normal course of events. The use of such a "pure" Laplace method falls beyond the techniques of partial fractions used above for inversion, and requires the use of the complex inversion integral for Laplace transforms (see Chapter 7). Application of standard loop and node analysis to the Laplace domain equivalent circuit (an exercise in linear algebra with coefficients dependent on the transform variable "s") gives an expression for the voltage and current transforms in terms of the initial conditions and transforms of the source functions.

Since the circuit loop (and node) equations are linear, it follows readily that the current and voltage transforms are expressible as linear combinations of the source terms with coefficients which are rational functions of the variable "s." If the network voltages, currents, initial capacitor voltages, and initial inductor currents are arrayed in vector form, this solution generally takes the form

$$\begin{bmatrix} I(s) \\ V(s) \end{bmatrix} = \mathbf{R}(s) \begin{bmatrix} \mathbf{i}(0^+) \\ \mathbf{v}(0^+) \end{bmatrix} + \mathbf{W}(s)\,\mathbf{F}(s),$$

where $\mathbf{F}(s)$ represents a vector of transforms of the source terms in the circuit.

The form is analogous to the Laplace transform version of the variation of constants formula encountered in the previous section. The matrix $\mathbf{W}(s)$ corresponds to the transfer function of the previous section, while the first term is analogous to the initial condition response of the vector differential equation case. The connection between the two formulas is actually much deeper than simple analogy. It can be shown that RLC networks of the standard type are actually representable by state variable models of the sort considered in the previous section. The demonstration of this requires the elimination of certain network degeneracies (capacitor loops, inductor cut sets) and involves considerations of network topology in an essential way. We refer the reader with a serious interest in circuit analysis to other sources for a complete discussion.

Example. We analyze the simple RLC network of Figure 6.22 as an example of the above considerations. The equivalent Laplace domain circuit is in Figure 6.23. The governing equations are

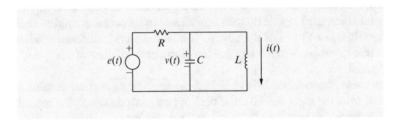

Fig. 6.22 RLC circuit

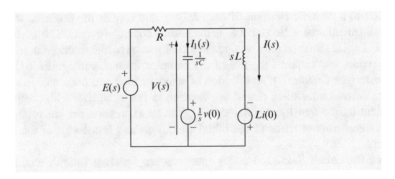

Fig. 6.23 RLC Laplace equivalent circuit

$$E(s) = V(s) + R(I(s) + I_1(s)),$$

$$V(s) = \frac{1}{sC} I_1(s) + \frac{1}{s} v(0^+),$$

$$V(s) = sL I(s) - L i(0^+).$$

Eliminating the currents from the first of the above,

$$E(s) = V(s) + R\left[sC V(s) - C v(0^+) + \frac{1}{sL} V(s) + \frac{1}{s} i(0^+) \right],$$

$$V(s)\left(1 + RC s + \frac{1}{s} \frac{R}{L} \right) = E(s) - \frac{R}{s} i(0^+) + RC v(0^+),$$

and finally

$$V(s) = \frac{E(s) - \frac{R}{s} i(0^+) + RC v(0^+)}{\left(1 + RC s + \frac{1}{s} \frac{R}{L} \right)}$$

$$= \frac{s}{s^2 RC + s + \frac{R}{L}} E(s) - \frac{RC i(0^+)}{s^2 RC + s + \frac{R}{L}} + \frac{RC s v(0^+)}{s^2 RC + s + \frac{R}{L}}.$$

The inductor current follows from

$$I(S) = \frac{i(0^+)}{s} + \frac{1}{sL} V(s).$$

These equations may be inverse transformed to express the capacitor voltage and inductor current as a combination of a convolution with the forcing voltage, together with an initial condition response.

Example. Not surprisingly, the symbolic calculation package Maxima has facilities for dealing with Laplace transforms. In addition to transform and inverse transforms, there are calculations for partial fraction expansions and evaluation of residues. That is, the "hand calculations" of Laplace transform problems are supported.

Somewhat more surprising is the fact that the differential equation solving routines will process systems of linear differential differential equations presented in "natural form" by means of Laplace methods, and return solutions in symbolic form.

An *LC* circuit solution is handled:
WXMAXIMA

(%i13)   del: C * 'diff(v(t), t) = i(t)
           de2: sin(omega * t) = L * 'diff(i(t), t) + v(t)

           desolve([del, de2], [v(t), i(t)])

(del) $C\left(\frac{d}{dt}\,v(t)\right)=i(t)$

(de2) $\sin(\omega\,t)=L\left(\frac{d}{dt}\,i(t)\right)+v(t)$

Is ω zero or nonzero? nonzero

Is C L positive, negative or zero? positive

(%o13) $[v(t)=$

$$\frac{\frac{C\,L\left(i(0)\,C\,L^2\,\omega^2+C\,L\,\omega-i(0)\,L\right)\sin\left(\frac{\sqrt{C\,L}\,t}{C\,L}\right)}{\sqrt{C\,L}\,(C\,L\,\omega^2-1)}+v(0)\,C\,L\cos\left(\frac{\sqrt{C\,L}\,t}{C\,L}\right)}{C\,L}-\frac{\sin(\omega\,t)}{C\,L\,\omega^2-1}\,,i(t)=$$

$$\frac{\frac{\left(i(0)\,C^2\,L^2\,\omega^2+C^2\,L\,\omega-i(0)\,C\,L\right)\cos\left(\frac{\sqrt{C\,L}\,t}{C\,L}\right)}{C\,L\,\omega^2-1}-\frac{v(0)\,C^2\,L\sin\left(\frac{\sqrt{C\,L}\,t}{C\,L}\right)}{\sqrt{C\,L}}}{C\,L}-\frac{C\,\omega\,\cos(\omega\,t)}{C\,L\,\omega^2-1}]$$

Problems 6.8.

1. The equivalent circuits of Figure 6.21   are constructed on the basis of an impedance and series voltage sources. Show that it is possible to construct the Laplace domain analogue of a conductance (called an admittance), and construct Laplace domain equivalents consisting of an admittance connected in parallel with an appropriate current source.
2. Find solutions of the following circuit problems using impedance (or admittance) methods.

   a. Figure 6.24.
   b. Figure 6.25.

3. Analyze the circuit of Figure 6.26. on the basis of impedance methods.
4. Show that the circuit Figure 6.27 may be analyzed on the basis of a state variable model with only two dimensions in spite of the fact that at first glance the circuit contains three energy storage elements.
   (This illustrates one small aspect of the conversion from impedance to state-variable models. This topic is treated in an article [1].)

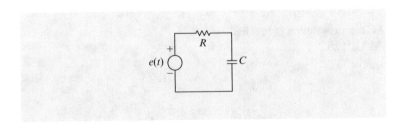

Fig. 6.24 RC voltage driven circuit

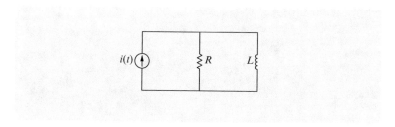

Fig. 6.25 RL circuit with current source

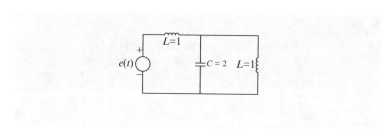

Fig. 6.26 LCL circuit with a voltage source

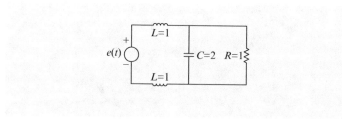

Fig. 6.27 Degenerate example

Reference

1. E.S. Kuh, R.A. Rohrer, The state variable approach to network analysis. Proc. IEEE **53**(7) (1965)

Further Reading

C.M. Close, *The Analysis of Linear Circuits* (Harcourt, Brace and World, New York, 1966)
P.M. DeRusso, R.J. Roy, C.M. Close, *State Variables for Engineers* (Wiley, New York, 1965)
C. Doetsch, *Guide to the Applications of Laplace Transforms* (D. Van Nostrand, Princeton, 1961)
N.V. Widder, *The Laplace Transform* (Princeton University Press, Princeton, 1946)

Chapter 7
Fourier Transforms

7.1 Introduction

Fourier transform methods find application in problems formally similar to those for which Laplace transform techniques are a suitable tool. Such applications include integral equations and partial and ordinary differential equations. The formal difference between the two classes of problems is that Laplace transforms are applied to functions defined on a half-line, while Fourier transforms apply to functions whose domain is the entire real axis. As a consequence, Laplace transforms are associated with initial value problems (transient responses), while Fourier transforms find more common application in input-output (forced response) models.

These distinctions are not hard and fast rules, as the two transforms are in fact closely related. This will become evident in the discussion of the formal manipulation rules (shift formulas, etc.) associated with the Fourier transform.

We make the connection between the two transforms explicit below in Section 7.6, with a discussion of the Laplace transform inversion integral.

In many ways the formal methods of Fourier transforms are virtually identical to the corresponding Laplace transform techniques. The main difference in the treatment of Fourier transforms is the introduction of an inversion integral for the calculation of inverse transforms. This is an alternative to "tables and tricks" methods for transform inversion in various applications, as well as a central feature of the theory and conceptual applications of Fourier transform methods. As mentioned above, the inversion integral for Fourier transforms provides an inversion integral for Laplace transforms (essentially with no extra effort). Use of these inversion integrals (in conjunction with various complex variable methods for integral evaluation) vastly increases the usefulness of transform methods.

The Fourier transform associates with a (suitably restricted) function defined on $(-\infty, \infty)$, a second function of a real variable.

© Springer International Publishing Switzerland 2016
J.H. Davis, *Methods of Applied Mathematics with a Software Overview*,
Applied and Numerical Harmonic Analysis, DOI 10.1007/978-3-319-43370-7_7

Definition. Let f be a function such that

$$\int_{-\infty}^{\infty} |f(t)|\, dt < \infty.\,^1$$

Then we define the Fourier transform of $f(\cdot)$ as

$$\mathcal{F}\{f(t)\} = \hat{f}(\omega) = \int_{-\infty}^{\infty} f(t)\, e^{-i\omega t}\, dt.$$

It turns out that $\hat{f}$ is essentially uniquely determined by f, so that (as in the case of Laplace transforms) we can define formally the inverse transform operation $\mathcal{F}^{-1}\{\cdot\}$ by

$$\mathcal{F}^{-1}\left\{\hat{f}(\omega)\right\} = f(t)$$

where f is such that

$$\mathcal{F}\{f(t)\} = \hat{f}(\omega).$$

This somewhat abstract definition of $\mathcal{F}^{-1}\{\cdot\}$ can justify "table lookup" inversion methods. More striking is the fact that the Fourier inversion process may be represented by an integral of the same sort as that defining the transform. This is the content of the Fourier inversion theorem.

Theorem (Fourier Inversion).
Suppose that

$$\hat{f}(\omega) = \mathcal{F}\{f(t)\} = \int_{-\infty}^{\infty} f(t)\, e^{-i\omega t}\, dt.$$

Then f may be recovered from

$$f(t) = \mathcal{F}^{-1}\left\{\hat{f}(\omega)\right\} = \frac{1}{2\pi} \int_{-\infty}^{\infty} \hat{f}(\omega)\, e^{i\omega t}\, d\omega.\,^2$$

This result asserts that the inversion process may be carried out simply by evaluating an integral, in principle of the same sort as that encountered in the

---

[1] The assumption that f is absolutely integrable is sufficient to ensure that $\hat{f}(\omega)$ is defined for all real ω. The issue of which functions have transforms is discussed further in Sections 7.2, 7.4.

[2] The class of transformable functions and sense of convergence of the inversion integral is discussed in the following Section 7.5.

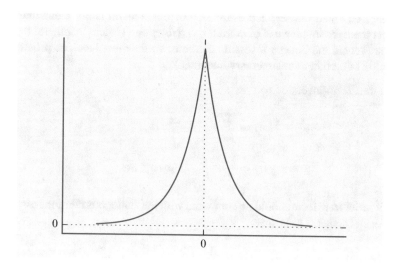

Fig. 7.1 Exponential tent

calculation of a transform. We illustrate the result with a simple example below. The similarity of the form of the transform and inversion integrals involved makes possible the construction of extensive tables of Fourier transform pairs.

Example. Define f by (assuming real $a > 0$)

$$f(t) = e^{-a|t|} = \begin{cases} e^{-at} & t \geq 0, \\ e^{at} & t < 0. \end{cases}$$

(See Figure 7.1.)
   Then

$$\hat{f}(\omega) = \int_{-\infty}^{\infty} f(t)\, e^{-i\omega t}\, dt$$

$$= \int_{-\infty}^{0} e^{at}\, e^{-i\omega t}\, dt + \int_{0}^{\infty} e^{-at}\, e^{-i\omega t}\, dt$$

$$= \frac{1}{-i\omega + a} + \frac{1}{i\omega + a}$$

$$= \frac{2a}{\omega^2 + a^2}.$$

The inversion integral then provides the result that

$$e^{-a|t|} = \frac{1}{2\pi} \int_{-\infty}^{\infty} \frac{2a}{\omega^2 + a^2}\, e^{i\omega t}\, d\omega.$$

(It should be noted that explicit evaluation of this integral is not elementary. The simplest methods involve use of complex variables and Cauchy's integral theorem. One may regard the Fourier inversion theorem, on the other hand, as providing an evaluation of certain nonelementary integrals.)

The transform/inversion pair

$$\hat{f}(\omega) = \int_{-\infty}^{\infty} f(t)\, e^{-i\omega t}\, dt, \tag{7.1}$$

$$f(t) = \frac{1}{2\pi} \int_{-\infty}^{\infty} \hat{f}(\omega)\, e^{i\omega t}\, d\omega$$

of the Fourier transform should be compared with the analogous relations associated with Fourier series. These are

$$c_n = c(n) = \frac{1}{2T} \int_{-T}^{T} f(t)\, e^{-in\omega_0 t}\, dt, \tag{7.2}$$

$$f(t) = \sum_{n=-\infty}^{\infty} c(n)\, e^{in\omega_0 t},$$

associated with the representation of a function f defined on the interval $[-T, T]$ as a weighted sum of complex exponential functions.

The relations (7.1) should be regarded as having the same intuitive content. The transform operation represents a decomposition into a continuum of harmonic components, while the inversion integral reassembles f from a continuous sum of harmonics.

The difference in the two cases arises from the fact that the functions for which the Fourier transform has been defined are not periodic (as are the Fourier series of (7.2), so that the description above should be regarded as more metaphorical than exact.

In spite of this, the Fourier inversion theorem can be given strong plausibility through an appeal to the Fourier series relations (7.2). While such a derivation lacks analytical rigor, it does illuminate the sense in which the Fourier transform/inversion cycle can be viewed as a harmonic decomposition/superposition process. Such a "pseudo derivation" may be carried out as follows.

The Fourier transform has been defined above for functions which are integrable, so that

$$\int_{-\infty}^{\infty} |f(t)|\, dt$$

is convergent. For such a function, it must be the case that the function essentially vanishes outside of some sufficiently large interval, say $[-T, T]$. For the purposes of this exposition, we assume that f vanishes identically outside of some interval $[-T, T]$. (See Figure 7.2.)

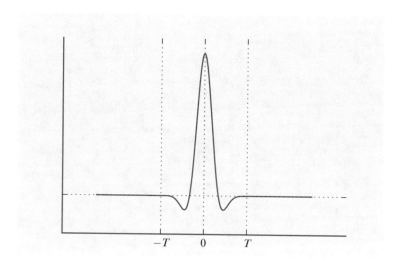

Fig. 7.2 Time limited signal

For such a function, the Fourier transform involves only an integral over the interval $[-T, T]$ (providing a connection with the Fourier series relations (7.2)). We have

$$\hat{f}(\omega) = \int_{-\infty}^{\infty} f(t)\, e^{-i\omega t}\, dt$$

$$= \int_{-T}^{T} f(t)\, e^{-i\omega t}\, dt.$$

The relations can connect the Fourier transform directly with Fourier series expansions for the interval $[-T, T]$.

However, such an expansion will be periodic of period $2T$, and hence is not easily related to the non-periodic f of Figure 7.2.

To make this transition, we consider not Fourier series expansions based on the interval $[-T, T]$, but expansions appropriate to the larger interval $[-KT, KT]$ (K denotes some large integer). Our tactic is to compute the expansions (as a function of K) and extract the desired result through a limiting process. Since the associated period will tend to infinity with K (and the fundamental frequency to zero), the results should at least resemble the equation (7.1). The process is illustrated in Figure 7.3.

On the interval $[-KT, KT]$, f may be represented as

$$f(t) = \sum_{n=-\infty}^{\infty} c(n)\, e^{i n \omega_0 t}$$

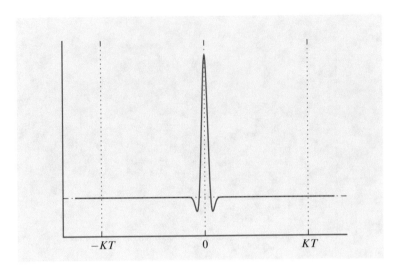

Fig. 7.3 The long view

where $\{c_n\}$ are defined by

$$c(n) = \frac{1}{2KT} \int_{-KT}^{KT} f(t) \, e^{-in\omega_0 t} \, dt$$

$$= \frac{1}{2KT} \int_{-T}^{T} f(t) \, e^{-in\omega_0 t} \, dt,$$

and the fundamental frequency is

$$\omega_0 = \frac{2\pi}{2KT} = \frac{\pi}{KT}.$$

We note that c_n is identifiable in terms of the Fourier transform of the original time-limited function as

$$c_n = \frac{1}{2KT} \hat{f}(n \frac{\pi}{KT}),$$

so that the Fourier series expansion can be rewritten as

$$f(t) = \sum_{n=-\infty}^{\infty} \frac{1}{2KT} \hat{f}(n \frac{\pi}{KT}) \, e^{in\frac{\pi}{KT}t}, \quad -KT \le t \le KT.$$

This expression is more readily identified by use of the suggestive substitution

$$\Delta\omega = \frac{\pi}{KT}.$$

With this change, it becomes

$$f(t) = \sum_{n=-\infty}^{\infty} \frac{1}{2\pi} \hat{f}(n\,\Delta\omega)\, e^{in\,\Delta\omega\, t}\, \Delta\omega.$$

As K tends to infinity, $\Delta\omega$ tends to zero, and this expansion is identified as a Riemann sum approximation to the inversion integral

$$f(t) = \frac{1}{2\pi} \int_{-\infty}^{\infty} \hat{f}(\omega)\, e^{i\omega t}\, d\omega$$

(see Figure 7.4).

This argument provides a heuristic derivation of the Inversion Theorem.

We complete this section with several examples of Fourier transform pairs. As in the case of Laplace transforms, the existence of various formulas relating to Fourier transform properties simplifies such calculations considerably. Further examples of transform calculation are reserved until such methods are available.

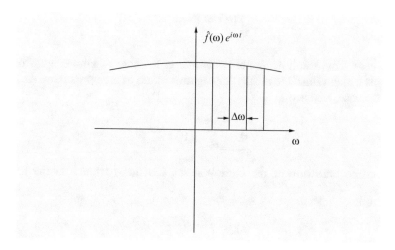

Fig. 7.4 Riemann sum approximation

7.2   Basic Fourier Transforms

Example. Let f be defined by

$$f(t) = \begin{cases} e^{-at} & t > 0, \\ 0 & t < 0 \end{cases}$$

assuming that $\text{Re}(a) > 0$. Then $\mathcal{F}\{f(t)\}$ exists, and

$$\hat{f}(\omega) = \int_{-\infty}^{\infty} f(t)\, e^{-i\omega t}\, dt \tag{7.3}$$

$$= \int_{0}^{\infty} e^{-(a+i\omega)t}\, dt$$

$$= \frac{1}{i\omega + a}.$$

The connection of this result with the Laplace transform calculation

$$F(s) = \int_{0}^{\infty} e^{-at}\, e^{-st}\, dt = \frac{1}{s + a}$$

should be noted. We see that

$$\hat{f}(\omega) = F(s) \Big|_{s=i\omega} \tag{7.4}$$

in this case. This result generally holds true in the case of Fourier transforms of functions which vanish identically for negative values of their argument. As long as this is the case, we have

$$\hat{f}(\omega) = \int_{0}^{\infty} f(t)\, e^{-i\omega t}\, dt.$$

The Laplace transform of the corresponding function (defined on the half-axis $t > 0$) is

$$F(s) = \int_{0}^{\infty} f(t)\, e^{-st}\, dt.$$

The assumption of existence of the Fourier transform is then sufficient to guarantee the existence of the Laplace transform for purely imaginary arguments, and to ensure validity of the identification.[3]

---

[3]This correspondence is one reason for defining the Fourier transform as we have above (alternatives displace the $\frac{1}{2\pi}$ inversion factor, or introduce minus signs).

Application of the inversion integral to the result calculated above gives

$$\frac{1}{2\pi} \int_{-\infty}^{\infty} \frac{1}{i\omega + a} e^{i\omega t} d\omega = \begin{cases} e^{-at} & t > 0, \\ 0 & t < 0, \end{cases}$$

for $\text{Re}(a) > 0$.

Example. Suppose that f is defined by

$$f(t) = \frac{1}{it + a}, \quad \text{Re}(a) > 0.$$

Then we have

$$\hat{f}(\omega) = \int_{-\infty}^{\infty} \frac{1}{it + a} e^{-i\omega t} dt.$$

This integral may be evaluated by reference to the inversion calculation made above. Rewriting this transform in the form

$$\hat{f}(\omega) = 2\pi \left(\frac{1}{2\pi} \int_{-\infty}^{\infty} \frac{1}{it + a} e^{it(-\omega)} dt \right)$$

we obtain

$$\hat{f}(\omega) = \begin{cases} 2\pi e^{-a(-\omega)} & -\omega > 0, \\ 0 & -\omega < 0. \end{cases}$$

This result may be made clearer by changing the dummy variable of integration in both of the integrals to a common neutral variable "λ." Then the inversion integral becomes (with t as parameter)

$$\frac{1}{2\pi} \int_{-\infty}^{\infty} \frac{1}{i\lambda + a} e^{i\lambda t} d\lambda = \begin{cases} e^{-at} & t > 0, \\ 0 & t < 0, \end{cases}$$

while the forward transform is

$$\hat{f}(\omega) = 2\pi \left(\frac{1}{2\pi} \int_{-\infty}^{\infty} \frac{1}{i\lambda + a} e^{i\lambda(-\omega)} d\lambda \right)$$

from which the result is apparent.

The example above illustrates the general result that Fourier transform formulas occur in pairs. In fact, functions such as the above (related through the Fourier transforms operation) are often referred to as Fourier transform pairs.

Example. The considerations employed above in connection with the $e^{-at} \Leftrightarrow \frac{1}{it+a}$ transform pair can be carried out in general to give a description of such pairs.

Let f be a transformable function, with Fourier transform

$$\hat{f}(\omega) = \int_{-\infty}^{\infty} f(t)\, e^{-i\omega t}\, dt.$$

Define a new function g by evaluating the transform function $\hat{f}$ as a function of "time."

$$g(t) = \hat{f}(t).$$

Then the Fourier transform of g is

$$\hat{g}(\omega) = \int_{-\infty}^{\infty} g(t)\, e^{-i\omega t}\, dt$$

$$= \int_{-\infty}^{\infty} \hat{f}(t)\, e^{-i\omega t}\, dt.$$

Application of the inversion formula to $\hat{f}$ gives

$$f(t) = \frac{1}{2\pi} \int_{-\infty}^{\infty} \hat{f}(\omega)\, e^{i\omega t}\, d\omega$$

$$= \frac{1}{2\pi} \int_{-\infty}^{\infty} \hat{f}(\lambda)\, e^{i\lambda t}\, d\lambda.$$

If the dummy variable λ is introduced into the forward transform, we have

$$\hat{g}(\omega) = \int_{-\infty}^{\infty} \hat{f}(\lambda) e^{-i\omega \lambda}\, d\lambda$$

$$= 2\pi \left(\frac{1}{2\pi} \int_{-\infty}^{\infty} \hat{f}(\lambda) e^{i\lambda(-\omega)}\, d\lambda \right).$$

Comparing these calculations, we obtain the general transform pair relation

$$\hat{g}(\omega) = \mathcal{F}\left\{ \hat{f}(t) \right\} = 2\pi f(-\omega).$$

Example. An example finding application in various communications problems is given by the rectangular pulse function, defined by

$$f(t) = \begin{cases} 1 & |t| \le a, \\ 0 & |t| > a. \end{cases}$$

(See Figure 7.5.)

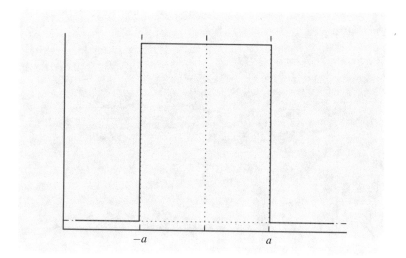

Fig. 7.5 Rectangular pulse

The Fourier transform is readily calculated as

$$\hat{f}(\omega) = \int_{-a}^{a} (1)\, e^{-i\omega t}\, dt$$

$$= \frac{e^{-i\omega t}}{-i\omega}\bigg|_{-a}^{a} = \frac{e^{-i\omega a} - e^{i\omega a}}{-i\omega}$$

$$= \frac{2\sin\omega a}{\omega}.$$

Application of the inversion formula gives

$$\frac{1}{2\pi}\int_{-\infty}^{\infty} \frac{2\sin\omega a}{\omega}\, e^{i\omega t}\, d\omega = \begin{cases} 1 & |t| \le a, \\ 0 & |t| > a \end{cases}.$$

The use of the transform pair result in this case provides

$$\mathcal{F}\left\{2\frac{\sin at}{t}\right\} = \begin{cases} 2\pi & |-\omega| \le a, \\ 0 & |-\omega| > a \end{cases}.$$

It is more conventional to express this last result in terms of the function

$$h(t) = \frac{\sin \Omega_0 t}{\Omega_0 t}$$

normalized to the value of unity at $t = 0$. This is referred to as a *sinc* function and illustrated in Figure 7.6.

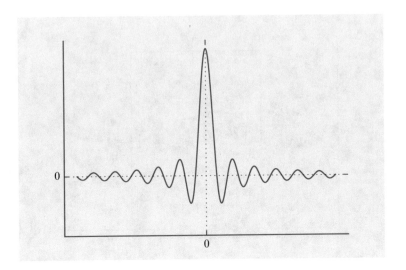

Fig. 7.6 Sinc function

From the pulse transform inversion we calculate

$$\mathcal{F}\left\{\frac{\sin \Omega_0 t}{\Omega_0 t}\right\} = \begin{cases} \frac{\pi}{\Omega_0} & |\omega| < \Omega_0, \\ 0 & |\omega| > \Omega_0. \end{cases} \tag{7.5}$$

Problems 7.2.

1. Compute the Fourier transform of the function f defined by

$$f(t) = \begin{cases} 1 & 0 \leq t < 2, \\ 0 & \text{otherwise.} \end{cases}$$

   Write out the form of the inversion integral for this case.
2. Repeat Problem 1 above for the function g defined by

$$g(t) = \begin{cases} t e^{-at} & t > 0 \ (a > 0), \\ 0 & t < 0. \end{cases}$$

   What is the relationship of these calculations to previously calculated Laplace transform examples?
3. Define a function f by

$$f(t) = \frac{1}{(it + a)^2} \ (a > 0).$$

   Compute $\mathcal{F}\{f(t)\}$ by use of the "pair relation" method.

4. Compute the Fourier transform and display the corresponding inversion formula for the function f defined by

$$f(t) = \begin{cases} t^2 - 1, & |t| < 1, \\ 0 & |t| > 1. \end{cases}$$

5. Let f be a Fourier transformable function which is even, in the sense that $f(t) = f(-t)$ for all t. Show that

$$\hat{f}(\omega) = \mathcal{F}\{f(t)\} = 2 \int_0^\infty f(t) \cos \omega t \, dt,$$

and hence that $\hat{f}$ is also an even function.

6. Let g be a transformable function which is odd, $g(-t) = -g(t)$. Show that

$$\hat{g}(\omega) = 2i \int_0^\infty g(t) \sin \omega t \, dt,$$

and hence conclude that $\hat{g}$ is odd in this case.

7. For a function f, defined on $[0, \infty)$, such that

$$\int_0^\infty |f(t)| \, dt < \infty,$$

the Fourier cosine transform is defined as

$$F_c(\omega) = \int_0^\infty f(t) \cos \omega t \, dt.$$

Use Problem 5 above and the Fourier inversion theorem to show formally that

$$f(t) = \text{constant} \left(\int_0^\infty F_c(\omega) \cos \omega t \, d\omega \right).$$

What is the appropriate value of the constant above?

8. For a function g defined on $[0, \infty)$ and such that

$$\int_0^\infty |g(t)| \, dt < \infty,$$

define the Fourier sine transform by

$$G_s(\omega) = \int_0^\infty g(t) \sin \omega t \, dt.$$

Use Problem 6 above and the Fourier inversion formula to show that (formally)

$$g(t) = \text{constant} \left(\int_0^\infty G_s(\omega) \sin \omega t \, d\omega \right)$$

determining the value of the constant involved in the process.

9. Alternative definitions of the Fourier transform are sometimes encountered. We have adopted the definition which leads to the simplest connections with the Laplace transforms. Alternative definitions have other advantages.

   A symmetric relation between transform and inversion is obtained by defining

$$\tilde{f}(\omega) = \frac{1}{\sqrt{2\pi}} \int_{-\infty}^\infty f(t) \, e^{-i\omega t} \, dt.$$

Show that

$$f(t) = \frac{1}{\sqrt{2\pi}} \int_{-\infty}^\infty \tilde{f}(\omega) \, e^{i\omega t} \, d\omega.$$

10. Another alternative Fourier transform definition arises from "hiding the 2π in the exponent." The usual interpretation is that "f" denotes the (real rather than radian) frequency variable, and

$$G(f) = \int_{-\infty}^\infty g(t) \, e^{-i2\pi f t} \, dt.$$

Show that the inversion formula takes the form

$$g(t) = \int_{-\infty}^\infty G(f) \, e^{i2\pi f t} \, df.$$

7.3   Formal Properties of Fourier Transforms

Since the Fourier transform is evidently closely related to the Laplace transform of the previous chapter, it is not surprising that it has a number of formal properties analogous to those encountered in the theory of Laplace transforms. Differences between the two cases arise from the facts that Fourier transforms are applied to functions defined on the whole real axis, and lack the exponential convergence factors associated with the defining integral in the half-axis Laplace transform case. The first property is that of linearity.

Linearity

If $\mathcal{F}\{f(t)\} = \hat{f}(\omega)$, $\mathcal{F}\{g(t)\} = \hat{g}(\omega)$, then

$$\mathcal{F}\{\alpha f(t) + \beta g(t)\} = \alpha \hat{f}(\omega) + \beta \hat{g}(\omega),$$

$$\mathcal{F}^{-1}\left\{\alpha \hat{f}(\omega) + \beta \hat{g}(\omega)\right\} = \alpha f(t) + \beta g(t).$$

These results follow directly from the integral definition of the transform operation. This also leads easily to the time scale property analogous to the Laplace transform case.

Time Scaling

If $\mathcal{F}\{f(t)\} = \hat{f}(\omega)$, then (for $c > 0$)

$$\mathcal{F}\{f(c\,t)\} = \frac{1}{c}\hat{f}(\frac{\omega}{c}).$$

In the case of Laplace transforms there exist both a time shift and an exponential multiplication formula. For the case of Fourier transforms, the multiplication result is usually given only for the case of purely imaginary exponents.[4] The two results then merge in the Fourier transform case.

Shift Formulas

The basic (and arguably only) result is the relationship of the Fourier transform and a delay operation. (The delay operation is illustrated in Figure 7.7).
    We compute the transform of the delayed function

$$\mathcal{F}\{f(t - t_0)\} = \int_{-\infty}^{\infty} f(t - t_0)\,e^{-i\omega t}\,dt$$

$$= \int_{-\infty}^{\infty} f(\tau)\,e^{-i\omega(\tau + t_0)}\,d\tau$$

$$= e^{-i\omega t_0}\,\mathcal{F}\{f(t)\}.$$

---

[4]Multiplication of a transformable function by real exponent exponential will often produce a function which has no transform in the sense described above.

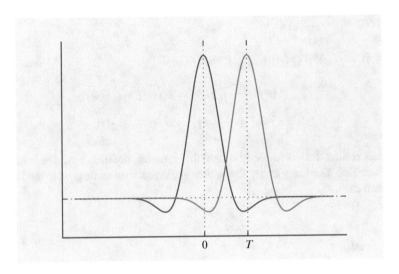

Fig. 7.7 Delayed waveform

The inverse transform relation corresponding to this is

$$\mathcal{F}^{-1}\left\{e^{-i\omega t_0}\hat{f}(\omega)\right\} = \mathcal{F}^{-1}\left\{\hat{f}(\omega)\right\}(t-t_0)$$

$$= f(t-t_0).$$

This should be compared with the Laplace transform result

$$\mathcal{L}\left\{U(t-T)f(t-T)\right\} = e^{-sT}F(s).$$

The results are directly comparable in the case of the Fourier transform of a function f which vanishes for negative values of its argument. As pointed out above, in this case the Fourier transform may be evaluated by substitution of "$i\omega$" for "s" in the Laplace transform. The two results then coincide.

In the case of a Fourier transformable function f which does not vanish for negative arguments, the unit step function intervenes in the Laplace transform case. The formulas are not directly comparable because of the essential difference in the domains of definition of the functions involved in the Laplace and Fourier transforms.

The Fourier transform version of the exponential shift formula may either be computed from the above using the transform pair notion of the previous chapter, or by direct evaluation. The result (obtained by either method) is

$$\mathcal{F}\left\{e^{i\omega_0 t}f(t)\right\} = \int_{-\infty}^{\infty} e^{i\omega_0 t}f(t)\,e^{-i\omega t}\,dt$$

$$= \hat{f}(\omega - \omega_0),$$

and conversely,

$$\mathcal{F}^{-1}\left\{\hat{f}(\omega - \omega_0)\right\} = e^{i\omega_0 t} f(t).$$

For reference, we accumulate these results as follows:

$$\mathcal{F}\left\{f(t - t_0)\right\} = e^{-i\omega t_0} \, \mathcal{F}\left\{f(t)\right\}, \tag{7.6}$$

$$\mathcal{F}^{-1}\left\{e^{-i\omega t_0} \hat{f}(\omega)\right\} = \mathcal{F}^{-1}\left\{\hat{f}(\omega)\right\}(t - t_0),$$

$$\mathcal{F}\left\{e^{i\omega_0 t} f(t)\right\} = \mathcal{F}\left\{f(t)\right\}(\omega - \omega_0),$$

$$\mathcal{F}^{-1}\left\{\hat{f}(\omega - \omega_0)\right\} = e^{i\omega_0 t} \, \mathcal{F}^{-1}\left\{\hat{f}(\omega)\right\}.$$

Example. Suppose that f is defined by

$$f(t) = \begin{cases} 1 & t_0 - a \le t \le t_0 + a, \\ 0 & \text{otherwise}, \end{cases}$$

with g defined by

$$g(t) = \begin{cases} 1 & |t| < a, \\ 0 & \text{otherwise}. \end{cases}$$

Then using the shift formula we have

$$\mathcal{F}\left\{f(t)\right\} = e^{-i\omega t_0} \, \mathcal{F}\left\{g(t)\right\}.$$

From our previous computations of the transform of g we find

$$\mathcal{F}\left\{f(t)\right\} = e^{-i\omega t_0} \frac{2 \sin \omega a}{\omega}.$$

Example. Define a transform through

$$\hat{f}(\omega) = \begin{cases} \pi \, e^{-i\omega 37} & |\omega| \le 26, \\ 0 & |\omega| > 26. \end{cases}$$

Then

$$f(t) = \mathcal{F}^{-1}\left\{\hat{f}(\omega)\right\}$$

$$= \mathcal{F}^{-1}\left\{\hat{g}(\omega)\right\}(t - 37)$$

where

$$\hat{g}(\omega) = \begin{cases} \pi \ |\omega| \le 26, \\ 0 \ |\omega| > 26. \end{cases}$$

Hence (inverting the transform g using our computations of the previous section),

$$f(t) = \frac{\sin 26(t - 37)}{(t - 37)}.$$

Differentiation Formulas

A version of the "transform of a derivative" result (required for applications to differential equations) may be derived as follows. We assume that f is an absolutely integrable function

$$\int_{-\infty}^{\infty} |f(t)| \, dt < \infty,$$

which is representable in the form of an anti-derivative of a function f' such that

$$\int_{-\infty}^{\infty} |f'(t)| \, dt < \infty.$$

That is,

$$f(t) = f(0) + \int_0^t f'(\tau) \, d\tau. \tag{7.7}$$

It then follows from (7.7) that f has a limit as $t \to \pm\infty$, and that (since f is integrable the limit has to be 0)

$$\lim_{t \to \pm\infty} f(t) = 0,$$

and that both f and f' are Fourier transformable. The problem is to relate the two transforms. We compute (integrating by parts)

$$\mathcal{F}\{f(t)\} = \lim_{M \to \infty} \int_{-M}^{M} f(t) e^{-i\omega t} \, dt$$

$$= \lim_{M \to \infty} \left[\frac{f(t) e^{-i\omega t}}{-i\omega} \Big|_{-M}^{M} - \int_{-M}^{M} f'(t) \frac{e^{-i\omega t}}{-i\omega} \, dt \right]$$

$$= 0 + \frac{1}{i\omega} \lim_{M\to\infty} \int_{-M}^{M} f'(t)\, e^{-i\omega t}\, dt$$

$$= \frac{1}{i\omega}\, \mathcal{F}\left\{f'(t)\right\},\ (\omega \neq 0).$$

Hence we obtain

$$\mathcal{F}\left\{f'(t)\right\} = i\omega\, \mathcal{F}\{f(t)\}$$

(initially for $\omega \neq 0$, then for all ω once one verifies that $\mathcal{F}\left\{f'(t)\right\}(0) = 0$ by direct integration).

This result differs from the corresponding Laplace transform case through the absence of an initial condition term. It may also be noted that the derivation above requires the explicit assumption that both f and f' are transformable, as well as the relation (7.7). In the analogous Laplace (and Fourier series) derivation only transformability of f together with the representation (7.7) is assumed.

A "t-multiplication" result may be derived in a manner analogous to the Laplace case. We assume that

$$\int_{-\infty}^{\infty} |f(t)|\, dt < \infty,$$

and also that

$$\int_{-\infty}^{\infty} |tf(t)|\, dt < \infty,$$

These assumptions are sufficiently strong to justify differentiation of

$$\hat{f}(\omega) = \int_{-\infty}^{\infty} f(t)\, e^{-i\omega t}\, dt$$

(under the integral sign) with respect to ω. This gives

$$\frac{d}{d\omega} \hat{f}(\omega) = \int_{-\infty}^{\infty} (-it) f(t)\, e^{-i\omega t}\, dt$$

or

$$\mathcal{F}\{tf(t)\} = i\, \frac{d}{d\omega} \hat{f}(\omega),$$

under the conditions assumed above.[5]

---

[5]By an appeal to the transform pair argument, the differentiation rule may be regarded as a version of the same result.

The combination of the results derived above with the identification of certain Fourier transforms in terms of Laplace transform calculations makes it possible to easily calculate transforms of a variety of elementary functions. In fact, it is possible to adapt the discussion of Section 6.4 in a wholesale fashion to compute transforms of functions given as piecewise products of elementary functions.

The considerations involved should be clear from the following example.

Example. Define f by

$$f(t) = \begin{cases} (t+1)^2 & -1 \le t < 0, \\ 1-t & 0 \le t < 1, \\ 0 & \text{otherwise.} \end{cases}$$

Then with

$$f_1(t) = \begin{cases} (t+1)^2 & -1 \le t < 0, \\ 0 & \text{otherwise,} \end{cases}$$

and

$$f_2(t) = \begin{cases} 1-t & 0 \le t < 1, \\ 0 & \text{otherwise,} \end{cases}$$

we have $f(t) = f_1(t) + f_2(t)$.

The function f_1 may be thought of as a shifted version of a parabolic segment as illustrated in Figure 7.8.

The Fourier transform of f_2 follows immediately from the corresponding Laplace calculation

$$F_2(s) = \frac{1}{s} - \frac{1}{s^2} + \frac{e^{-s}}{s^2}.$$

Hence (since f_2 is Fourier as well as Laplace transformable)

$$\mathcal{F}\{f_2(t)\} = \frac{1}{i\omega} - \frac{1}{(i\omega)^2} + \frac{e^{-i\omega}}{(i\omega)^2}.$$

The parabolic segment f_1 may be regarded as a left shifted (advanced) version of the function f_3 (Figure 7.9)

The Laplace transform of f_3 is readily seen to be (c.f. Figure 7.9)

$$F_3(s) = \frac{2}{s^3} - \frac{e^{-s}}{s} - \frac{2e^{-s}}{s^2} - \frac{2e^{-s}}{s^3}.$$

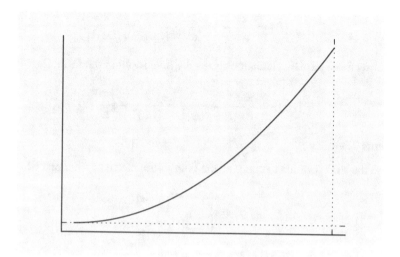

Fig. 7.8 Parabolic segment

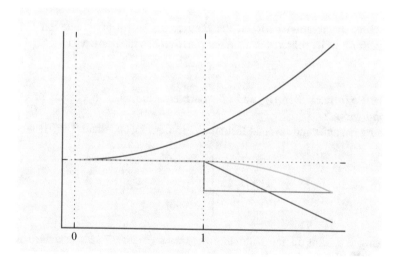

Fig. 7.9 Decomposition of parabolic segment

The corresponding Fourier transform is therefore

$$\mathcal{F}\{f_3(t)\} = \frac{2}{(i\omega)^3} - \frac{e^{-i\omega}}{i\omega} - \frac{2\,e^{-i\omega}}{(i\omega)^2} - \frac{2\,e^{-i\omega}}{(i\omega)^3}.$$

"Advancing" this result by means of the shift formula provides

$$\mathcal{F}\{f_1(t)\} = \frac{2\,e^{i\omega}}{(i\omega)^3} - \frac{1}{i\omega} - \frac{2}{(i\omega)^2} - \frac{2}{(i\omega)^3}.$$

Combining these partial calculations gives the final result in the form

$$\mathcal{F}\{f(t)\} = \frac{2\,e^{i\omega}}{(i\omega)^3} + \frac{3}{(\omega)^2} - \frac{2}{(i\omega)^3} - \frac{e^{-i\omega}}{\omega^2}.$$

Problems 7.3.

1. Use the shift formula to compute the Fourier transform of f defined by

$$f(t) = \begin{cases} e^{-(t-12)} & t \geq 12, \\ 0 & t < 12. \end{cases}$$

2. Define the "time reversal of the function f" by

$$g(t) = f(-t).$$

   Find the relation between the Fourier transform of a function f and the transform of its time reversal. What function has the transform

$$\overline{\hat{f}}$$

   where $\hat{f}(\omega) = \mathcal{F}\{f(t)\}$, and $^{-}$ represents the usual operation of complex conjugation?

3. Using the shift and reversal formulas (Problem 2), calculate $\mathcal{F}\{h(t)\}$ where

$$h(t) = \begin{cases} e^{\beta(t+a)} & t < -a, \\ 0 & |t| < a, \\ e^{-\beta(t-a)} & t > a. \end{cases}$$

4. Using the hints in the following problems (Figure 7.10), use the shift and "t-multiplication" results to compute the Fourier transform of the function f defined by

$$f(t) = \begin{cases} \frac{t+a}{a} & -a \leq t < 0, \\ \frac{a-t}{a} & 0 \leq t < a, \\ 0 & \text{otherwise.} \end{cases}$$

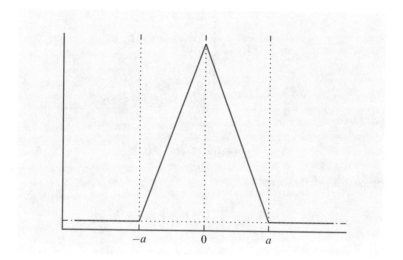

Fig. 7.10 Triangular waveform

5. Compute the Laplace transform of f, defined by

$$f(t) = \begin{cases} \frac{t}{a} & 0 \le t < a, \\ \frac{2a-t}{a} & a \le t < 2a, \\ 0 & t > 2a. \end{cases}$$

6. Compute the Fourier transform of the function defined by extending f above to be zero for $t < 0$. From this result, compute $\mathcal{F}\{g(t)\}$, where

$$g(t) = \begin{cases} \frac{t+a}{a} & -a \le t < 0, \\ \frac{a-t}{a} & 0 \le t < a, \\ 0 & \text{otherwise.} \end{cases}$$

(Construction of the graphs of the above functions greatly simplifies this problem.)

7. Compute the Fourier transforms of f and g defined by

$$f(t) = \begin{cases} 0 & t < -1, \\ 1+t & -1 < t < 0, \\ 1-t & 0 < t < 1, \\ 0 & t > 1. \end{cases}$$

$$g(t) = \begin{cases} 0 & t < -1, \\ 1 & -1 < t < 0, \\ -1 & 0 < t < 1, \\ 0 & t > 1. \end{cases}$$

Verify that

$$\mathcal{F}\{g(t)\} = i\omega\,\mathcal{F}\{f(t)\}.$$

8. Compute $\mathcal{F}\{h(t)\}$ where

$$h(t) = \begin{cases} 1-t & -1 < t < 0, \\ -1-t & 0 < t < 1, \\ 0 & \text{otherwise.} \end{cases}$$

Verify that h is differentiable (except at the isolated points $t = -1, 0, 1$), and (with the exception of those three points)

$$h'(t) = g(t),$$

with g as in Problem 7.
   Compute (or recall) $\mathcal{F}\{g(t)\}$, and show that

$$\mathcal{F}\{g(t)\} \neq i\omega\,\mathcal{F}\{h(t)\}.$$

Does this example violate the differentiation law for Fourier transforms? Why or why not?

9. This and the following several problems relate to the partial fractions inversion of rational Fourier transforms. The notation follows the discussion of the corresponding problem in the Laplace transform context, Section 6.5.
   Suppose that the polynomial p has distinct (non-repeated) zeroes, so that

$$p(s) = (s - z_1)(s - z_2)\ldots(s - z_n)$$

is a factorization displaying the (possibly complex) zeroes $\{z_i\}$.
   Then the usual partial fractions expansion

$$\frac{1}{p(s)} = \sum_{i=1}^{n} \frac{a_i}{s - z_i}$$

allows the ready evaluation of the inverse Fourier transform.

Suppose that $\text{Re}(z_1) > 0$, $i = 1, 2, \ldots, M$, and $\text{Re}(z_i) < 0$, $i = M + 1, \ldots, n$. Find

$$\mathcal{F}^{-1} \left\{ \frac{1}{p(i\omega)} \right\}$$

in terms of the partial fraction expansion coefficients $\{a_i\}$ and polynomial zeroes $\{z_i\}$.

10. The general partial fractions expansion of a proper rational function $\frac{q(s)}{p(s)}$ is

$$\frac{q(s)}{p(s)} = \sum_{i=1}^{j} \sum_{k=1}^{n_i} \frac{a_{ik}}{(s - z_i)^k},$$

where

$$p(s) = (s - z_1)^{n_1} (s - z_2)^{n_2} \ldots (s - z_j)^{n_j}$$

is a factorization of the denominator polynomial p.

Assuming that $\text{Re}(z_i) > 0$, $i = 1, \ldots, m$, $\text{Re}(z_i) < 0$, $i = m + 1, \ldots, j$, find an expression for

$$\mathcal{F}^{-1} \left\{ \frac{q(i\omega)}{p(i\omega)} \right\}$$

in terms of the partial fractions expansion coefficients and zeroes of p.

11. a. Compute

$$\mathcal{F} \left\{ U(-t) \, e^{at} \right\},$$

assuming $\text{Re}(a) > 0$.

b. Use (11a) to evaluate

$$\mathcal{F}^{-1} \left\{ \frac{1}{-\omega^2 - 2i\omega - 3} \right\}.$$

12. Suppose that f has a transform as in Problem 10.

a. Assuming $\text{Re}(z_i) < 0$, $i = 1, 2, \ldots, j$ evaluate

$$\mathcal{F}^{-1} \left\{ \hat{f}(\omega) \right\}.$$

b. Assuming $\text{Re}(z_i) > 0$, $i = 1, 2, \ldots, j$, evaluate

$$\mathcal{F}^{-1} \left\{ \hat{f}(\omega) \right\}.$$

c. Evaluate

$$\mathcal{F}^{-1}\left\{\hat{f}(\omega)\right\},$$

assuming $\text{Re}(z_i) < 0$, $i = 1,\dots,m$, while $\text{Re}(z_i) > 0$, $i = m+1,\dots,j$.

7.4   Convolutions and Parseval's Theorem

The concept of the convolution of two functions was introduced in the context of Laplace transforms in Section 6.6. The functions involved in that chapter were defined only on a half-line. A notion of convolution can also be defined for the Fourier transformable functions of the present chapter.

This convolution is defined by the formula

$$x(t) = \int_{-\infty}^{\infty} g(t-\tau)f(\tau)\,d\tau, \quad -\infty < t < \infty \tag{7.8}$$

often abbreviated as

$$x(t) = [g \star f]\,(t).$$

The form of the definition (7.8) is not an arbitrary construct. It is possible to argue that the formula inevitably arises from an attempt to construct input-output models of linear time-invariant physical systems.

If the function f represents the time history of the input to some (linear) system, then an output of the system which has the interpretation of a weighted average of the input values may be represented in the form of an integral

$$x(t) = \int_{-\infty}^{\infty} w(t,\tau)f(\tau)\,d\tau.$$

Here, the function w represents the input weighting function associated with the value of the output at time t.

If the underlying system is time-invariant,[6] then a delay (or advance) in the input history results in a corresponding delay (or advance) in the output, so that

$$x(t-T) = \int_{-\infty}^{\infty} w(t,\tau)f(\tau-T)\,d\tau.$$

(See Figure 7.11.)

---

[6]For purposes of exposition, we refer to the independent variable as "time." The discussion obviously applies as well to "spatially invariant" effects.

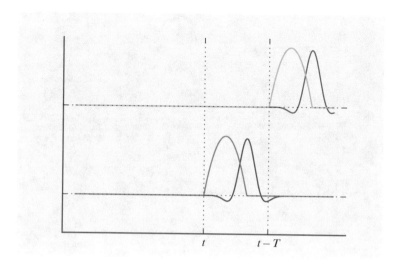

Fig. 7.11 Time invariant relation

Changing the variables of integration and argument according to

$$\tau' = \tau - T,$$
$$t' = t - T,$$

we obtain

$$x(t') = \int_{-\infty}^{\infty} w(t' + T, \tau' + T) f(\tau') \, d\tau',$$

$$x(t) = \int_{-\infty}^{\infty} w(t + T, \tau + T) f(\tau) \, d\tau.$$

Comparing the two expressions, we conclude that (for all t, τ, T)

$$w(t, \tau) = w(t + T, \tau + T)$$

(since f is arbitrary).

Setting $T = -\tau$, we conclude that the weighting pattern satisfies

$$w(t, \tau) = w(t - \tau, 0) = g(t - \tau),$$

so that the assumption of time invariance implies that the weighting function depends only on the difference of its arguments. The general input-output relationship then takes the form

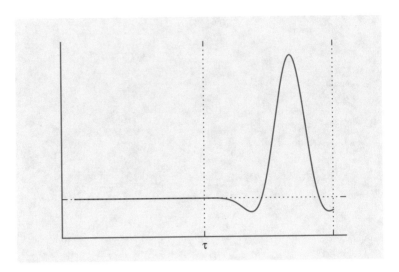

Fig. 7.12   Original function f

$$x(t) = \int_{-\infty}^{\infty} g(t - \tau) f(\tau)\, d\tau.$$

This is just the form of the convolution defined earlier.

The convolution formula has a ready graphical interpretation that is useful in the evaluation of certain convolutions (as well as essential to an understanding of convolution).

The convolution represents an integral of the product of two functions (of "τ"), while the argument variable "t" plays the role of a parameter in the integral.

One of the factors of the integrand is $f(\tau)$. This appears graphically as a plot of the original function f. This is illustrated in Figure 7.12.

The other factor may be visualized as resulting from a delay (assuming $t > 0$, graphically an advance if $t < 0$) of the function g by an amount t, followed by a reflection through the line $\tau = t$. This process is illustrated in Figure 7.13.

The value of the convolution function (at fixed t) is represented by the "area" under the product of the two functions of Figures 7.12 and 7.13. As t varies, the shift in Figure 7.13 gives rise to the convolution function variation. This process is illustrated in Figure 7.14.

Example. The graphical interpretation of the convolution process leads to simple convolution computations for piecewise polygonal functions. Such a computation rests on the ability to visualize the translation process outlined above and to calculate areas under products of polygonal segments by inspection.

We illustrate this process with the convolution of two rectangular pulses.

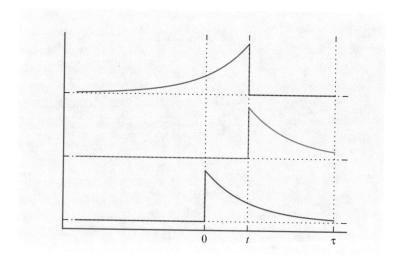

Fig. 7.13 Shifted and reflected g

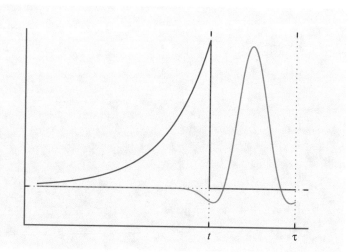

Fig. 7.14 The convolution product

Let f and g be defined by

$$f(t) = \begin{cases} 2 & |t| \le 2, \\ 0 & |t| > 2, \end{cases}$$

$$g(t) = \begin{cases} 1 & |t| \le 1, \\ 0 & |t| > 1. \end{cases}$$

The relevant diagram is given in Figure 7.15.

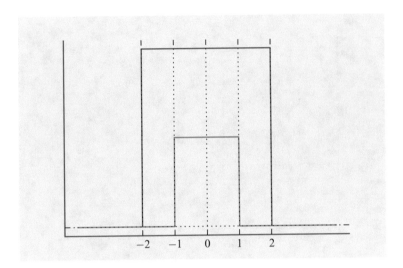

Fig. 7.15 Rectangular pulse functions

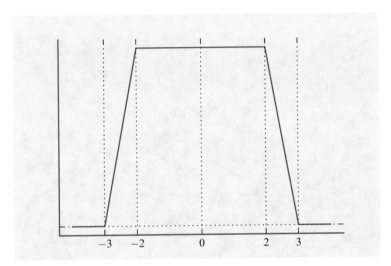

Fig. 7.16 Convolution of square pulses

From this diagram we deduce that the convolution function is constant (with value 4) as long as $|t| \leq 1$, and vanishes identically for $|t| > 3$ (since the rectangles of Figure 7.15 are then disjoint). It is also clear from 7.15 that the variation is linear with t in the intermediate regimes $-3 < t < -1$, $1 < t < 3$. We conclude that the graph of $[g \star f]$ is as illustrated in Figure 7.16.

An analytical expression of this result is read from 7.16:

$$[g \star f](t) = \begin{cases} 0 & |t| > 3, \\ 2(t+3) & -3 \le t \le 1, \\ 4 & |t| < 1, \\ 4 - 2(t-1) & 1 \le t \le 3. \end{cases}$$

Example. Let f and g be exponential functions, defined by (U denotes the usual unit-step function)

$$f(t) = \begin{cases} U(t)\, e^{-t} & t \ge 0, \\ 0 & t < 0, \end{cases}$$

$$g(t) = \begin{cases} U(t)\, e^{-2t} & t \ge 0, \\ 0 & t < 0. \end{cases}$$

We compute the convolution $[g \star f]$ analytically as follows:

$$[g \star f](t) = \int_{-\infty}^{\infty} U(t-\tau)\, e^{-2(t-\tau)}\, U(\tau)\, e^{-\tau}\, d\tau.$$

In the above integral, $U(\tau) = 0$ for $\tau < 0$, and hence

$$[g \star f](t) = \int_{0}^{\infty} U(t-\tau)\, e^{-2(t-\tau)}\, U(\tau)\, e^{-\tau}\, d\tau.$$

For $t < 0$, the argument of the unit step function in the integrand is negative over the entire range of integration. Hence the integrand vanishes, and we conclude that

$$[g \star f](t) = 0, \; t < 0.$$

If $t > 0$, the argument of the unit step is positive for $0 < \tau < t$, negative for $\tau > t$. The integral therefore reduces to

$$[g \star f](t) = \int_{0}^{t} U(t-\tau)\, e^{-2(t-\tau)}\, U(\tau)\, e^{-\tau}\, d\tau$$

$$= e^{-2t}\, e^{\tau} \Big|_{0}^{t}$$

$$= e^{-2t}\left(e^{t} - 1\right)$$

for $t > 0$.

The results are combined as

$$[g \star f](t) = U(t) \left[e^{-t} - e^{-2t} \right].$$

It should be noted that this calculation is essentially the same convolution encountered in the context of functions defined on a half-line in the previous chapter. In fact, it is always the case that this identification may be made when the functions f and g vanish for negative values of their arguments.

For such functions (since f vanishes for negative arguments)

$$f(t) = U(t)f(t),$$

$$g(t) = U(t)g(t),$$

and the reductions of the previous example may be carried out in general:

$$[g \star f](t) = \int_{-\infty}^{\infty} g(t - \tau)f(\tau)\,d\tau$$

$$= \int_{-\infty}^{\infty} U(t - \tau)\,g(t - \tau)\,U(\tau)f(\tau)\,d\tau$$

$$= \int_{0}^{\infty} U(t - \tau)\,g(t - \tau)f(\tau)\,d\tau$$

$$= 0,\ (t < 0)$$

$$= \int_{0}^{t} 1\,g(t - \tau)f(\tau)\,d\tau + \int_{t}^{\infty} 0\,d\tau$$

$$= \int_{0}^{t} g(t - \tau)f(\tau)\,d\tau\ (t > 0).$$

This observation plays a role in the discussion of Fourier transform solutions of ordinary differential equations.

The simple relationship between the Laplace transform and convolution has a complete analogue in the Fourier transform case. To derive this result, we should first verify that, in fact, the convolution of two transformable functions is transformable.

Given that

$$\int_{-\infty}^{\infty} |f(t)|\,dt < \infty,$$

$$\int_{-\infty}^{\infty} |g(t)|\,dt < \infty$$

we consider

$$x(t) = \int_{-\infty}^{\infty} g(t-\tau)f(\tau)\,d\tau.$$

Then

$$|x(t)| \le \int_{-\infty}^{\infty} |g(t-\tau)|\,|f(\tau)|\,d\tau$$

and hence

$$\int_{-\infty}^{\infty} |x(t)|\,dt \le \int_{-\infty}^{\infty} \left\{ \int_{-\infty}^{\infty} |g(t-\tau)|\,|f(\tau)|\,d\tau \right\}\,dt$$

$$= \int_{-\infty}^{\infty} \left\{ \int_{-\infty}^{\infty} |g(t-\tau)|\,|f(\tau)|\,dt \right\}\,d\tau$$

$$= \int_{-\infty}^{\infty} |g(t)|\,dt \int_{-\infty}^{\infty} |f(\tau)|\,d\tau$$

$$< \infty.$$

This verifies that the convolution x has a Fourier transform.

Theorem (Convolution Theorem). *If f and g are transformable functions restricted as above, we compute t the transform*

$$\int_{-\infty}^{\infty} x(t)\,e^{-i\omega t}\,dt = \int_{-\infty}^{\infty} e^{-i\omega t} \left\{ \int_{-\infty}^{\infty} g(t-\tau)f(\tau)\,d\tau \right\}\,dt.$$

We view this as an iterated integral and interchange the order of integration to give the conclusion that the transform of the convolution of two functions is the product of the transforms:

$$\mathcal{F}\{x(t)\} = \int_{-\infty}^{\infty} f(\tau) \left\{ \int_{-\infty}^{\infty} g(t-\tau)\,e^{-i\omega t}\,dt \right\}\,d\tau$$

$$= \int_{-\infty}^{\infty} f(\tau) \left\{ \int_{-\infty}^{\infty} g(\sigma)\,e^{-i\omega(\tau+\sigma)}\,d\sigma \right\}\,d\tau$$

$$= \left(\int_{-\infty}^{\infty} f(\tau)\,e^{-i\omega\tau}\,d\tau \right) \left(\int_{-\infty}^{\infty} g(\sigma)\,e^{-i\omega\sigma}\,d\sigma \right)$$

$$= \hat{f}(\omega)\,\hat{g}(\omega).$$

In the usual manner, this result may be converted to other (equivalent) forms by use of inversion and the Fourier pair relations. Inversion of the transform product gives directly the restatement of the above result.

$$\mathcal{F}^{-1}\left\{\hat{f}(\omega)\,\hat{g}(\omega)\right\} = [f \star g]\,(t)$$

$$= [g \star f]\,(t).$$

The interchangeability of the functions f and g on the right side follows readily from the observation that the transform product factors may be exchanged.

The simplest way to obtain the relevant pair relation is to construct the convolution of two transformed functions. We define

$$\hat{h}(\omega) = \int_{-\infty}^{\infty} \hat{g}(\omega - \lambda)\hat{f}(\lambda)\,d\lambda$$

$$= \left[\hat{g} \star \hat{f}\right](\omega),$$

and next apply the inversion theorem. The corresponding "time function" is given by

$$h(t) = \frac{1}{2\pi}\int_{-\infty}^{\infty}\hat{h}(\omega)\,e^{i\omega t}\,d\omega$$

$$= \frac{1}{2\pi}\int_{-\infty}^{\infty}e^{i\omega t}\left\{\int_{-\infty}^{\infty}\hat{g}(\omega - \lambda)\hat{f}(\lambda)\,d\lambda\right\}\,d\omega.$$

Changing the integration order as above, we obtain

$$\mathcal{F}^{-1}\left\{\hat{h}(\omega)\right\} = \frac{1}{2\pi}\int_{-\infty}^{\infty}\hat{f}(\lambda)\left\{\int_{-\infty}^{\infty}e^{i\omega t}\,\hat{g}(\omega - \lambda)\,d\omega\right\}\,d\lambda$$

$$= \frac{1}{2\pi}\int_{-\infty}^{\infty}\hat{f}(\lambda)\left\{\int_{-\infty}^{\infty}e^{i(\lambda+\mu)t}\,\hat{g}(\mu)\,d\mu\right\}\,d\lambda$$

$$= \left(\frac{1}{2\pi}\int_{-\infty}^{\infty}\hat{f}(\lambda)\,e^{i(\lambda)t}\,d\lambda\right)2\pi\left(\int_{-\infty}^{\infty}e^{i(\mu)t}\,\hat{g}(\mu)\,d\mu\right)$$

$$= 2\pi\,\mathcal{F}^{-1}\left\{\hat{f}(\omega)\right\}\,\mathcal{F}^{-1}\{\hat{g}(\omega)\}$$

$$= 2\pi\,f(t)\,g(t).$$

Transforming this last result gives (the complete list)

$$\mathcal{F}\{f(t)\,g(t)\} = \frac{1}{2\pi}\left[\hat{f} \star \hat{g}\right](\omega) \qquad\qquad (7.9)$$

$$f(t)\,g(t) = \mathcal{F}^{-1}\left\{\frac{1}{2\pi}\left[\hat{f} \star \hat{g}\right](\omega)\right\}$$

$$\mathcal{F}\{[f \star g]\,(t)\} = \hat{f}(\omega)\,\hat{g}(\omega)$$

$$[f \star g]\,(t) = \mathcal{F}^{-1}\left\{\hat{f}(\omega)\,\hat{g}(\omega)\right\}.$$

These results are usually stated as the transform of a convolution is the product of the transforms of the functions convolved, and the transform of a product is ($\frac{1}{2\pi}$ times) the convolution of the transforms of the factors.

Example. We previously calculated the convolution

$$[U(t)\, e^t \star U(t)\, e^{-2t}](t) = U(t)\, [e^{-t} - e^{-2t}].$$

Computing the Fourier transform of this result, we obtain readily

$$\mathcal{F}\{[U(t)\, e^{-t} \star U(t)\, e^{-2t}]\} = \frac{1}{i\omega + 1} - \frac{1}{i\omega + 2}$$

$$= \frac{1}{i\omega + 1}\, \frac{1}{i\omega + 2}$$

$$= \mathcal{F}\{U(t)\, e^{-t}\}\, \mathcal{F}\{U(t)\, e^{-2t}\}.$$

This verifies the convolution law for this particular case.

Example. Suppose that f is defined as the product

$$f(t) = e^{-|t|}\, U(t)\, e^{-2t}.$$

Then since

$$\mathcal{F}\{e^{-|t|}\} = \frac{2}{\omega^2 + 1},$$

$$\mathcal{F}\{U(t)\, e^{-2t}\} = \frac{1}{i\omega + 2},$$

we find that the transform of f has the representation

$$\hat{f}(\omega) = \frac{1}{2\pi} \int_{-\infty}^{\infty} \frac{2}{(\omega - \lambda)^2 + 1}\, \frac{1}{i\lambda + 2}\, d\lambda.$$

Since f actually has the simple representation

$$f(t) = U(t)\, e^{-3t},$$

we can compute directly that

$$\mathcal{F}\{f(t)\} = \frac{1}{i\omega + 3} = \frac{1}{2\pi} \int_{-\infty}^{\infty} \frac{2}{(\omega - \lambda)^2 + 1}\, \frac{1}{i\lambda + 2}\, d\lambda.$$

(The integral in this may be explicitly evaluated by use of complex variable methods. See Section 7.5 below).

The Convolution Theorem results (7.9) play a central role in the application of Fourier transforms to problems of differential and integral equations (Section 7.7). The result may also be applied to obtain the Fourier transform version of Parseval's Theorem.

We recall that Parseval's Theorem (in the case of Fourier series expansions) has the intuitive content of the decomposition of the power in a function into the powers associated with its harmonic components. In view of our pseudo-derivation of the Fourier transform-inversion relation from the Fourier series relations, it should be a natural conclusion that an analogous result (and interpretation) is available in the context of Fourier transforms.

The quantity of interest may be expressed as

$$E = \int_{-\infty}^{\infty} |f(\tau)|^2 \, d\tau.$$

If f represents the time history of a (real valued) voltage source, then E represents the total energy delivered (over all time) to a one-ohm resistor connected to this source. In a mechanical interpretation, f may be taken to represent a force record. If this force is applied to a (unit magnitude) ideal (dashpot) damper, then E similarly represents the total energy absorbed by the device. These interpretations are illustrated in Figure 7.17.

The energy expression may be related to the convolution operation by noting that formally

$$\int_{-\infty}^{\infty} |f(\tau)|^2 \, d\tau = \int_{-\infty}^{\infty} \bar{f}(\tau - t) f(\tau) \, d\tau \bigg|_{t=0}.$$

This takes the form of the evaluation at $t = 0$ of a convolution of the usual sort, provided that a function g is defined so that

$$g(t - \tau) = \bar{f}(\tau - t).$$

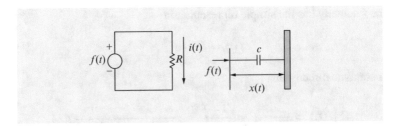

Fig. 7.17 Energy dissipation

That is

$$g(x) = \bar{f}(-x),$$

so that g is the complex conjugate reflection of the original f. With this identification, the convolution takes the form

$$h(t) = \int_{-\infty}^{\infty} g(t - \tau) f(\tau) \, d\tau,$$

and the usual convolution theorem is applicable. We require $h(0)$, and calculate it by Fourier inversion:

$$h(0) = \int_{-\infty}^{\infty} |f(\tau)|^2 \, d\tau$$

$$= \frac{1}{2\pi} \int_{-\infty}^{\infty} \hat{h}(\omega) e^{i\omega t} \, d\omega \bigg|_{t=0}$$

$$= \frac{1}{2\pi} \int_{-\infty}^{\infty} \hat{g}(\omega) \hat{f}(\omega) \, d\omega.$$

The transform $\hat{g}$ may be calculated from the definition in terms of f. We have

$$\hat{g}(\omega) = \int_{-\infty}^{\infty} g(x) e^{-i\omega x} \, dx$$

$$= \int_{-\infty}^{\infty} \bar{f}(-x) e^{-i\omega x} \, dx$$

$$= \int_{-\infty}^{\infty} \bar{f}(\tau) e^{i\omega \tau} \, d\tau$$

$$= \overline{\int_{-\infty}^{\infty} f(\tau) e^{-i\omega \tau} \, d\tau}$$

$$= \overline{\hat{f}(\omega)}.$$

Using this result we obtain the desired form of the theorem.

Theorem (Parseval's Theorem). *Suppose that f is a function such that*

$$\int_{-\infty}^{\infty} |f(\tau)|^2 \, d\tau < \infty.$$

Then

$$\int_{-\infty}^{\infty} |f(\tau)|^2 \, d\tau = \frac{1}{2\pi} \int_{-\infty}^{\infty} \hat{f}(\omega) \overline{\hat{f}(\omega)} \, d\omega \tag{7.10}$$

$$= \frac{1}{2\pi} \int_{-\infty}^{\infty} |\hat{f}(\omega)|^2 \, d\omega.$$

This result is usually paraphrased as the statement that the energy of the function is equal to ($\frac{1}{2\pi}$ times) the energy of the transform.

This is a "square form" of the Parseval relation. As in the case of Fourier series result, there exists an equivalent product form. This may be derived simply by treating the expression

$$\int_{-\infty}^{\infty} a(\tau) \overline{b}(\tau) \, d\tau = \int_{-\infty}^{\infty} a(\tau) \overline{b}(\tau - t) \, d\tau \bigg|_{t=0}$$

in a fashion parallel to that used above. The result is that the equality

$$\int_{-\infty}^{\infty} a(\tau) \overline{b}(\tau) \, d\tau = \frac{1}{2\pi} \int_{-\infty}^{\infty} \hat{a}(\omega) \overline{\hat{b}}(\omega) \, d\omega \tag{7.11}$$

holds for all a, b such that

$$\int_{-\infty}^{\infty} |a(\tau)|^2 \, d\tau < \infty, \quad \int_{-\infty}^{\infty} |b(\tau)|^2 \, d\tau < \infty.$$

If the left side of (7.11) is regarded as the inner product of the functions a, b, then the product form of Parseval's relation may be regarded as the statement that the inner product may be calculated "in the frequency domain" (i.e., in terms of the Fourier transforms).

Such inner products arise in various optimization problems, and (7.11) often provides a useful computational device for the solution of such problems.

Example. The form of the Parseval relation (7.10) may be illustrated by the example

$$f(t) = U(t) \, e^{-t}.$$

Then we have

$$\int_{-\infty}^{\infty} |f(\tau)|^2 \, d\tau = \int_{0}^{\infty} e^{-2t} \, dt = \frac{1}{2}.$$

Further,

$$\frac{1}{2\pi} \int_{-\infty}^{\infty} |\hat{f}(\omega)|^2 \, d\omega = \frac{1}{2\pi} \int_{-\infty}^{\infty} \frac{1}{1 + \omega^2} \, d\omega = \frac{1}{2\pi} \pi = \frac{1}{2},$$

which verifies (7.10) for this example.

As was mentioned in connection with the Fourier series version of Parseval's theorem, more serious application of the result occurs in the analysis of "signal energy" distribution. An example of such application is included in the problems below.

Problems 7.4.

1. Compute (by evaluation of the relevant integral) the convolution $[f \star g]$ of the functions f, g defined by

$$f(t) = e^{-2|t|},$$

$$g(t) = U(t) e^{-t} = \begin{cases} e^{-t} & t \geq 0, \\ 0 & t < 0. \end{cases}$$

   Compute the Fourier transform of your result, verifying the convolution theorem in this case.

2. Suppose that f is a rectangular pulse function

$$f(t) = \begin{cases} 1 & |t| \leq 1, \\ 0 & |t| > 1. \end{cases}$$

   Compute $[f \star f](t)$ by graphical means.

3. With f as defined above, compute

$$[f \star f \star f](t).$$

   What is the Fourier transform of this function?

4. Show that if f lives on the interval $|t| < T_f$, (that is, $f(t) = 0$ for $|t| > T_f$) and g lives on the interval $|t| < T_g$, that the convolution $[f \star g]$ lives on the interval $|t| \leq T_f + T_g$.

5. Show that the operation of convolution is commutative and associative. That is, that $[g \star f] = [f \star g]$, and

$$[[f \star g] \star h] = [f \star [g \star h]].$$

6. Compute

$$\mathcal{F}^{-1} \left\{ \frac{2 \sin \omega a}{\omega} \frac{1}{i\omega + 1} \right\}.$$

7. Suppose that it is known that (the Fourier transformable) function x satisfies the integral equation $(-\infty < t < \infty)$

$$x(t) + \int_{-\infty}^{\infty} e^{-(t-\tau)} x(\tau)\, d\tau = U(-t)\, e^{2t}.$$

   Find $x(\cdot)$.

8. Assuming that all formal manipulations are justified, find a solution of the integral equation

$$x(t) + \int_{-\infty}^{\infty} g(t-\tau)\, x(\tau)\, d\tau = f(t).^{7}$$

9. Use Parseval's Theorem to evaluate

$$\int_{0}^{\infty} \left(\frac{\sin x}{x}\right)^{2} dx.$$

10. If a voltage source $e(t) = U(t)\, e^{-t}$ is connected to a one-ohm resistor, the energy delivered to the resistor is

$$\int_{-\infty}^{\infty} [(U(t)e^{-t}]^{2}\, dt.$$

   If the same source is applied to the RC circuit of Figure 7.18, the governing equation is (with no initial charge on the capacitor)

$$e(t) = \frac{1}{1}\int_{-\infty}^{t} i(\tau)\, d\tau + 1\, i(t).$$

   The energy delivered to the one ohm resistor (the load) is now

$$\int_{-\infty}^{\infty} (i(t))^{2}\, dt.$$

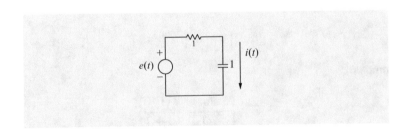

Fig. 7.18 Unit parameter RC circuit

---

[7] There exist examples of equations of this form for which the formal calculations are not justifiable.

Use Parseval's theorem to compute the fraction of the available energy delivered to the load with the circuit in Figure 7.18. Express this quantity in the case of an arbitrary source function e.

Hint:

$$\int \frac{x^2}{(x^2+1)^2} \, dx = \frac{-x}{2(x^2+1)} + \int \frac{1}{2(x^2+1)} \, dx.$$

7.5   Comments on the Inversion Theorem

We have initially defined the Fourier transform above for functions f which are integrable,

$$\int_{-\infty}^{\infty} |f(t)| \, dt < \infty.$$

This restriction ensures that the integral

$$\int_{-\infty}^{\infty} f(t) \, e^{-i\omega t} \, dt$$

converges for all real ω, so that the Fourier transform function is well defined. The inversion formula "guessed" in Section 7.1 takes the form

$$\frac{1}{2\pi} \int_{-\infty}^{\infty} \hat{f}(\omega) \, e^{i\omega t} \, d\omega.$$

Since the inversion integral takes the same form as the original transform integral, the integral in question will be convergent in the same sense as the original transform, provided that $\hat{f}$ is also an integrable function. That is, provided that the integral

$$\frac{1}{2\pi} \int_{-\infty}^{\infty} |\hat{f}(\omega)| \, d\omega$$

is convergent.

We have previously calculated as an example

$$\mathcal{F}\{U(t) \, e^{-at}\} = \frac{1}{i\omega + a}.$$

However, the absolute transform integral for this example is

$$\int_{-\infty}^{\infty} |\hat{f}(\omega)| \, d\omega = \int_{-\infty}^{\infty} \frac{1}{\sqrt{\omega^2 + a^2}} \, d\omega,$$

which is readily seen to be divergent in the usual sense. This simple example shows that the hypothesis that

$$\int_{-\infty}^{\infty} |f(t)| \, dt$$

is convergent, while convenient for the definition of the transform, fails in general to guarantee absolute convergence of the inversion integral. This example indicates that the convergence and evaluation of the inversion integral is likely to be a matter of some delicacy.

The discussion of the convolution and Parseval theorems in the previous section provides the keys for a consideration of such issues. The Parseval relation may be regarded as identifying a class of functions for which a symmetric transform-inverse transform exists, while the convolution result provides the basic analytical tool for handling this problem by a combination of approximation and limiting procedures.

Strictly speaking, the argument for the validity of the required limiting procedures requires the invocation of certain standard convergence theorems for Lebesgue-type integrals. Since we are not assuming familiarity with these results, we are content to outline the methods involved in this approach and emphasize the meaning of the relevant results.

The basic approximation procedure relies on the construction of functions which can be regarded as approximations to the delta function introduced in Section 6.6.

These "approximate delta functions" play the role of a smoothing device which serves to guarantee convergence of certain integrals.

A large variety of functions which may be used to fill this role exists. Out of these we select an example which later proves useful in the solution of certain boundary value problems.

The example in question arises from the well-known Gaussian probability density function. This is defined by the formula

$$W_\sigma(t) = \frac{1}{\sqrt{2\pi}\sigma} e^{-\frac{t^2}{2\sigma^2}}.$$

Since this function is a probability density function, we have the normalization

$$\int_{-\infty}^{\infty} \frac{1}{\sqrt{2\pi}\sigma} e^{-\frac{t^2}{2\sigma^2}} \, dt = 1,$$

independent of the (variance) parameter σ. The dependence of the function on σ is illustrated in Figure 7.19. As $\sigma \to 0$ the weight of the function becomes concentrated at 0.

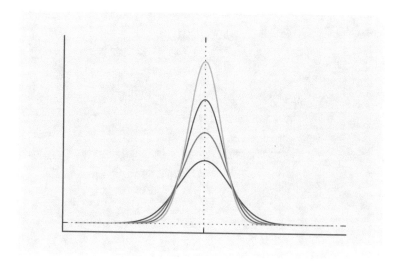

Fig. 7.19 Gaussian kernels

Using the properties of the Gaussian density, it is an easy exercise to obtain the following result. Suppose that f is a bounded, continuous function such that

$$\int_{-\infty}^{\infty} |f(t)|\, dt < \infty.$$

Then it follows that

$$\lim_{\sigma \to 0} \int_{-\infty}^{\infty} W_\sigma(t - \tau) f(\tau)\, d\tau = f(t).$$

This observation together with the fact that the "delta function" satisfies the formal relation

$$\int_{-\infty}^{\infty} \delta(t - \tau) f(\tau)\, d\tau = f(t)$$

leads us to refer to W_σ as a family of approximate delta functions.

We also require the Fourier transforms of this family. By direct calculation we have

$$\hat{W}_\sigma(\omega) = \lim_{A \to \infty} \int_A^A \frac{1}{\sqrt{2\pi}\sigma}\, e^{-\frac{t^2}{2\sigma^2}}\, e^{-i\omega t}\, dt$$

$$= \lim_{A \to \infty} \int_A^A \frac{1}{\sqrt{2\pi}\sigma}\, e^{-\frac{t^2}{2\sigma^2} + i\omega t}\, dt.$$

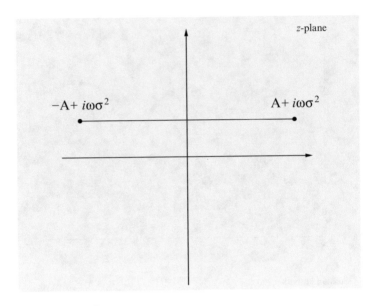

Fig. 7.20 Line integral path

We next complete the square in the exponent of the integrand. Since

$$\frac{t^2}{2\sigma^2} + i\omega t = \frac{1}{2\sigma^2}\left(t^2 + 2\sigma^2 i\omega t\right)$$

$$= \frac{1}{2\sigma^2}\left((t + i\omega\sigma^2)^2 + \omega^2\sigma^4\right),$$

the above integral reduces to the computation of

$$\lim_{A\to\infty}\left[e^{-\frac{\omega^2\sigma^2}{2}}\int_{-A}^{A}\frac{1}{\sqrt{2\pi}\sigma}e^{-\frac{1}{2\sigma^2}(t+i\omega\sigma^2)^2}\,dt\right].$$

This integral can be identified as a (line) integral in the complex plane with t as the parameter of integration. The relevant diagram is in Figure 7.20.

We identify in these terms

$$\int_{-A}^{A}\frac{1}{\sqrt{2\pi}\sigma}e^{-\frac{1}{2\sigma^2}(t+i\omega\sigma^2)^2}\,dt = \int_{-A+i\omega\sigma^2}^{A+i\omega\sigma^2}\frac{1}{\sqrt{2\pi}\sigma}e^{-\frac{1}{2\sigma^2}z^2}\,dz.$$

In view of the fact that the integrand is an analytic function [8] in the finite part of the z-plane, the integration contour of Figure 7.20 may be deformed to the contour of Figure 7.21. Hence we obtain

---

[8] See the following section and references therein.

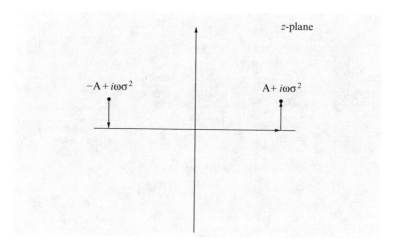

Fig. 7.21 Deformed inversion contour

$$\int_{-A}^{A} \frac{1}{\sqrt{2\pi}\sigma} e^{-\frac{1}{2\sigma^2}(t+i\omega\sigma^2)^2} \, dt$$

$$= \int_{C_1} \frac{1}{\sqrt{2\pi}\sigma} e^{-\frac{1}{2\sigma^2} z^2} \, dz + \int_{C_2} \frac{1}{\sqrt{2\pi}\sigma} e^{-\frac{1}{2\sigma^2} z^2} \, dz + \int_{C_3} \frac{1}{\sqrt{2\pi}\sigma} e^{-\frac{1}{2\sigma^2} z^2} \, dz$$

$$= \int_{-A}^{A} \frac{1}{\sqrt{2\pi}\sigma} e^{-\frac{1}{2\sigma^2} x^2} \, dx + \left[\int_{C_1} + \int_{C_3} \right] \frac{1}{\sqrt{2\pi}\sigma} e^{-\frac{1}{2\sigma^2} z^2} \, dz,$$

where we have used the observation that the contour C_2 is naturally parameterized by the $x (= \operatorname{Re} z)$-axis.

It is easy to see (see the problems below) that the integrals along C_1 and C_3 tend to zero as $A \to \infty$. In fact, an estimate of the form

$$\left| \int_{C_j} \frac{1}{\sqrt{2\pi}\sigma} e^{-\frac{1}{2\sigma^2} z^2} \, dz \right| \le M(\omega) \, e^{-\frac{A^2}{2b\sigma^2}}, \quad j = 1, 3$$

holds. This estimate shows that the estimate tends to zero as $A \to \infty$.

Using these results to evaluate the required limit in the transform expression, we find that the required Fourier transform is simply

$$\mathcal{F} \left\{ \frac{1}{\sqrt{2\pi}\sigma} e^{-\frac{t^2}{2\sigma^2}} \right\} = e^{-\frac{\omega^2\sigma^2}{2}} \int_{-\infty}^{\infty} \frac{1}{\sqrt{2\pi}\sigma} e^{-\frac{1}{2\sigma^2} x^2} \, dx \tag{7.12}$$

$$= e^{-\frac{\omega^2\sigma^2}{2}}.$$

Graphs of this transform (for various values of the parameter σ) are given in Figure 7.22.

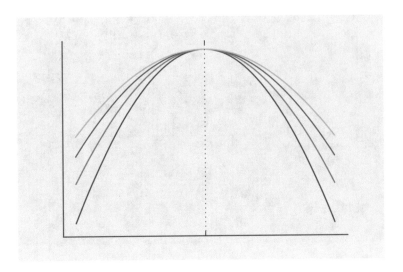

Fig. 7.22 Gaussian transforms

We notice in particular that

$$0 \leq \mathcal{F}\left\{\frac{1}{\sqrt{2\pi}\sigma}e^{-\frac{t^2}{2\sigma^2}}\right\} \leq 1,$$

and that the Gaussian transform tends to 1 as $\sigma \to 0$. This result together with the observation that the Gaussian kernel acts as an approximate delta function makes possible a discussion of the inversion theorem.

In effect, there exist two versions of the Fourier Inversion Theorem. These differ in the hypotheses placed upon the function being transformed.

The first version is referred to as the "L^1 theory" and deals with functions f such that

$$\int_{-\infty}^{\infty} |f(t)|\, dt < \infty.^9$$

It is relatively easy to see that the inversion theorem is valid for an L^1 function which has been smoothed by convolution with the "approximate delta function" W_σ.

---

[9]Such functions are said to be "of class L^1."

Theorem (Fourier Inversion Theorem (L^1)). *We have for f in L^1,*

$$[W_\sigma \star f](t) = \int_{-\infty}^{\infty} W_\sigma(t-\tau)f(\tau)\,d\tau$$

$$= \frac{1}{2\pi} \int_{-\infty}^{\infty} \mathcal{F}\{[W_\sigma \star f](t)\}\,e^{i\omega t}\,dt$$

$$= \frac{1}{2\pi} \int_{-\infty}^{\infty} \mathcal{F}\{W_\sigma(t)\}\,\hat{f}(\omega)\,e^{i\omega t}\,dt,$$

so that the inversion theorem holds for $W_\sigma \star f$.

Proof.

$$[W_\sigma \star f](t) = \int_{-\infty}^{\infty} f(t-\tau)\,W_\sigma(\tau)\,d\tau$$

$$= \int_{-\infty}^{\infty} f(t-\tau)\left[\frac{1}{2\pi}\int_{-\infty}^{\infty}\mathcal{F}\{W_\sigma(t)\}\,e^{i\omega\tau}\,d\tau\right]d\tau.$$

Interchanging the order of integration[10] we obtain

$$[W_\sigma \star f](t) = \frac{1}{2\pi}\int_{-\infty}^{\infty}\mathcal{F}\{W_\sigma\}\left[\int_{-\infty}^{\infty} f(t-\tau)\,e^{i\omega\tau}\,d\tau\right]d\omega$$

$$= \frac{1}{2\pi}\int_{-\infty}^{\infty}\mathcal{F}\{W_\sigma\}\,e^{i\omega t}\hat{f}(\omega)\,d\omega$$

(by the shift theorem)

$$= \frac{1}{2\pi}\int_{-\infty}^{\infty}\mathcal{F}\{[W_\sigma \star f]\}\,e^{i\omega t}\,d\omega$$

(by the convolution theorem).

The usual results concerning inversion of transforms of L^1 functions follow by letting $\sigma \to 0$ in the above result.

---

[10]Justified since the double integral exists.

Combining these results above with the approximate delta argument, we conclude that if f is in L^1, continuous, and bounded, then

$$f(t) = \lim_{\sigma \to 0} [W_\sigma \star f](t)$$

$$= \lim_{\sigma \to 0} \frac{1}{2\pi} \int_{-\infty}^{\infty} \mathcal{F}\{W_\sigma\} e^{i\omega t} \hat{f}(\omega)\, d\omega$$

$$= \lim_{\sigma \to 0} \frac{1}{2\pi} \int_{-\infty}^{\infty} e^{-\frac{\omega^2 \sigma^2}{2}} \hat{f}(\omega) e^{i\omega t}\, d\omega.$$

The factor $\mathcal{F}\{W_\sigma\} = e^{-\frac{\omega^2 \sigma^2}{2}}$ which appears may be regarded as a convergence factor in the integrand that compensates for the fact that (as noted above) $\hat{f}$ need not be integrable.

If in fact $\hat{f}$ is integrable, so that

$$\int_{-\infty}^{\infty} |\hat{f}(\omega)|\, d\omega < \infty,$$

then standard results[11] justify a passage to the limit under the integral sign in the integral. Since

$$\lim_{\sigma \to 0^+} \mathcal{F}\{W_\sigma\} = 1,$$

this gives a proof of the Fourier inversion theorem result

$$f(t) = \frac{1}{2\pi} \int_{-\infty}^{\infty} \hat{f}(\omega) e^{i\omega t}\, d\omega \qquad (7.13)$$

under the hypotheses (f integrable and continuous, $\hat{f}$ integrable) imposed. In the absence of the fortunate circumstance that $\hat{f}$ is integrable, the limit form of the inversion given above is all that may be expected. This weaker form of the inversion theorem provides a sense in which the inversion integral may be understood even in this case.[12]

A symmetric transform-inversion relation (referred to as the L^2 theory of Fourier transforms) arises from consideration of functions f which are square-integrable[13]

$$\int_{-\infty}^{\infty} |f(t)|^2\, dt < \infty.$$

---

[11] The result in question is the Dominated Convergence Theorem for Lebesgue integrals.

[12] As mentioned above, there are other choices of "approximate delta functions" which may be employed in such arguments. Many of these choices are related to the idea of a "summation method" for the (possibly divergent) inversion integral.

[13] The terminology is that f belongs to the class L^2.

The basic results for this case follow from a combination of the convolution smoothing method used above with the formal argument employed to obtain the Parseval relation in the previous section. In outline, we construct the function h according to

$$h(t) = \int_{-\infty}^{\infty} \bar{f}(\tau - t) f(\tau) \, d\tau.$$

Under the hypotheses that f is both integrable and square-integrable ($f \in L^1 \cap L^2$), it can be shown that h defined by the convolution expression is such that (i.e., bounded, continuous, and integrable) the inversion (7.13) is valid. Application of this in conjunction with the appropriate limiting argument provides a rigorous proof that the Parseval relation

$$\int_{-\infty}^{\infty} |f(t)|^2 \, dt = \frac{1}{2\pi} \int_{-\infty}^{\infty} |\hat{f}(\omega)|^2 \, d\omega$$

holds, provided that $f \in L^1 \cap L^2$.

If $f \in L^1 \cap L^2$, the Fourier transform is unambiguously defined by the usual integral formula

$$\int_{-\infty}^{\infty} f(t) e^{-i\omega t} \, dt$$

(convergent since $f \in L^1$).

If f belongs to L^2 (but not necessarily to L^1 as well), we define the Fourier transform by constructing an approximating sequence. Given $f \in L^2$, define a sequence of functions by replacing f by zero for arguments larger than N in magnitude. That is,

$$f_N(t) = \begin{cases} f(t) & |t| \leq N, \\ 0 & |t| > N. \end{cases}$$

Then f_N (see problems below) is in $L^1 \cap L^2$ (so that $\hat{f}_N$ is defined) and forms a Cauchy sequence converging to f in L^2. This means that

$$\lim_{N \to \infty} \int_{-\infty}^{\infty} |f_N(t) - f(t)|^2 \, dt = 0,$$

and

$$\int_{-\infty}^{\infty} |f_N(t) - f_M(t)|^2 \, dt \leq \epsilon^2$$

for all M, N sufficiently large, and any $\epsilon > 0$.

With $\hat{f}_N$ defined through the usual integration

$$\hat{f}_N(\omega) = \int_{-\infty}^{\infty} f_N(t)\, e^{-i\omega t}\, dt,$$

the Parseval relation shows that

$$\int_{-\infty}^{\infty} |f_N(t) - f_M(t)|^2\, dt = \frac{1}{2\pi} \int_{-\infty}^{\infty} |\hat{f}_N(\omega) - \hat{f}_M(\omega)|^2\, d\omega.$$

This equality shows that $\{\hat{f}_N\}$ forms a Cauchy sequence in L^2 (of the transform variable). Invoking the classical Riesz-Fischer result that L^2 is complete, we define the transform $\hat{f}$ as the L^2 limit of the $\{\hat{f}_N\}$.

We have

$$\hat{f}(\omega) = \lim_{N\to\infty} \int_{-N}^{N} f(t)\, e^{-i\omega t}\, dt, \tag{7.14}$$

where the limit is to be interpreted with respect to L^2 of the transform variable.

The major reward for this slightly indirect transform definition is in the symmetry of the result. Since the transform function is guaranteed to be square-integrable, the Fourier inversion integral has an interpretation in a sense identical to that encountered in (7.14). That is, as the limit (in the L^2 sense) of the sequence of functions defined by

$$f_M(t) = \frac{1}{2\pi} \int_{-M}^{M} \hat{f}(\omega)\, e^{i\omega t}\, d\omega.$$

Using the fact that the Fourier inversion integral is valid for sufficiently well-behaved functions, it can finally be shown that the above sequence has a limit actually equal to f (in the sense of equality in L^2). A statement of these results is given below.

Theorem (Fourier Inversion Theorem (L^2)). *Suppose that $f \in L^2$, so that*

$$\int_{-\infty}^{\infty} |f(t)|^2\, dt.$$

Then there exists a function $\hat{f}$ such that

$$\frac{1}{2\pi} \int_{-\infty}^{\infty} |\hat{f}(\omega)|^2\, d\omega = \int_{-\infty}^{\infty} |f(t)|^2\, dt$$

and

$$\hat{f}(\omega) = \lim_{N \to \infty} \int_N^N f(t) e^{-i\omega t} \, dt,$$

in the sense that

$$\int_{-\infty}^{\infty} \left| \hat{f}(\omega) - \int_{-N}^N f(t) e^{-i\omega t} \, dt \right|^2 \, d\omega \to 0$$

as $N \to \infty$. Further, the Fourier inversion holds in the sense that

$$\int_{-\infty}^{\infty} \left| f(t) - \frac{1}{2\pi} \int_{-M}^M \hat{f}(\omega) e^{i\omega t} \, d\omega \right|^2 \, dt$$

tends to zero as $N \to \infty$

It should be noted that each of the integrals whose limit is to be calculated above is convergent in the usual sense. This follows since the inequality

$$\int_{-N}^N |f(t)| \, dt \le \left(\int_{-N}^N |f(t)|^2 \, dt \right)^{\frac{1}{2}} \left(\int_{-N}^N 1 \, dt \right)^{\frac{1}{2}}$$

obtained readily from the Cauchy–Schwarz inequality. Since

$$\left| \int_{-N}^N f(t) e^{-i\omega t} \, dt \right| \le \int_{-N}^N |f(t)| \, dt,$$

the integral in question is convergent.

Even though this is the case, it is not often the case that the integral is "simple," in the sense that it may be evaluated explicitly in terms of an elementary anti-derivative. A practically more useful evaluation technique is based on complex variable methods and is discussed in the following section.

Problems 7.5.

1. The following problems outline the proof of the fact that W_σ supplies an "approximate delta function." Show first that

$$\lim_{\sigma \to 0} W_\sigma(t) = 0,$$

uniformly in t, as long as t is restricted away from the origin, $|t| > \delta > 0$.

2. Show that for any $\delta > 0$,

$$\lim_{\sigma \to 0} \int_{-\delta}^{\delta} W_\sigma(t) \, dt = 1.$$

3. Use the results of Problems 1, 2 above to show that for any f which is everywhere continuous and integrable, so that

$$\int_{-\infty}^{\infty} |f(t)|\, dt < \infty,$$

we have

$$\lim_{\sigma \to 0} [W_\sigma \star f] = f(t)$$

where

$$[W_\sigma \star f] = \int_{-\infty}^{\infty} W_\sigma(t - \tau) f(\tau)\, d\tau.$$

Hint: The graphical interpretation of convolution provides the key.

4. Another family of approximate delta functions can be constructed from the function $f(t) = e^{-|t|}$. Find a function g such that

$$\hat{g}(\omega) = e^{-\sigma^2 |\omega|}.$$

Show that the properties of Problems 1 and 2 hold with W_σ replaced by g_σ. Outline an inversion theorem for Fourier transforms of L^1 functions based on the use of g_σ in place of W_σ.

5. Compute the Fourier transform of the triangular function $h_\sigma(t)$ of Figure 7.23.

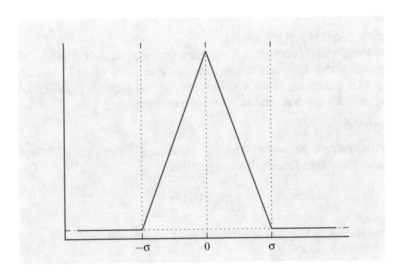

Fig. 7.23 Triangular pulse function

Outline a Fourier inversion theorem based on the use of h_σ as an approximate delta function.

6. Consider the function f defined by

$$f(t) = U(t)\, e^{-t}.$$

a. Define the L^2-approximant f_N, and explicitly compute

$$\int_{-\infty}^{\infty} |f_N(t) - f_M(t)|^2\, dt.$$

b. Use the known integral

$$\int_{-\infty}^{\infty} \frac{e^{i\omega t}}{1 + \omega^2}\, d\omega = \pi\, e^{-|t|}$$

to explicitly compute

$$\frac{1}{2\pi} \int_{-\infty}^{\infty} |\hat{f}_N - \hat{f}_M|^2\, d\omega,$$

and verify that $\{\hat{f}_N\}$ is a Cauchy sequence.

7.6 Fourier Inversion by Contour Integration

The use of complex variable methods as a tool for integral evaluation arose briefly in the course of the Gaussian density transform calculation of the previous section. As such methods are probably the single most powerful method for evaluation of Fourier inversion integrals, we treat the topic in some detail below.

The methods in question rely on the notions of contour integration and the Residue Theorem from complex analysis. These problems provide a major example for the discussion of the Residue Theorem in Chapter 5 above.

From the previous section we recall that the natural evaluation of the Fourier inversion integral requires consideration of limits of the form

$$\lim_{B \to \infty} \left[\frac{1}{2\pi} \int_{-B}^{B} e^{i\omega t} \hat{f}(\omega)\, d\omega \right].$$

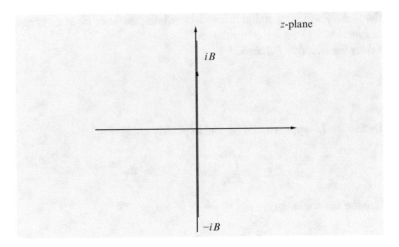

Fig. 7.24 Inversion integral path

One method for the evaluation of the above is to identify the integral as a (line) integral in the complex plane. The convention we adopt is to regard the integral as performed along the imaginary axis in the complex plane.[14]

The relevant diagram is given in Figure 7.24.

The appropriate identifications may be made from consideration of the example

$$\frac{1}{2\pi} \int_{-B}^{B} \frac{1}{i\omega + a} e^{i\omega t} \, d\omega$$

(whose limit should be $U(t) \, e^{-at}$, for $a > 0$).

The parameterization appropriate for the line of Figure 7.24 is given by

$$z = i\omega$$
$$dz = i \, d\omega.$$

The inversion integral can then be expressed in the form

$$\frac{1}{2\pi} \int_{-B}^{B} \frac{1}{i\omega + a} e^{i\omega t} \, d\omega = \frac{1}{2\pi i} \int_{-iB}^{iB} \frac{1}{z + a} e^{zt} \, dz,$$

with the adoption of the convention that the limits $\{-iB, iB\}$ on the complex line integral indicate use of the straight line path of Figure 7.24.

---

[14]This is in consonance with the usual Laplace transform inversion, and naturally introduces the $2\pi i$ factor associated with the Residue Theorem.

The same identifications may be made in the case of the general form

$$\frac{1}{2\pi} \int_{-B}^{B} \hat{f}(\omega)\, e^{i\omega t}\, d\omega.$$

Using the same parameterization and our notational convention regarding the path, we obtain

$$\frac{1}{2\pi} \int_{-B}^{B} \hat{f}(\omega)\, e^{i\omega t}\, d\omega = \frac{1}{2\pi i} \int_{iB}^{iB} \hat{f}(\frac{z}{i})\, e^{zt}\, dz.$$

To simplify notation, we introduce an auxiliary function g defined by

$$g(z) = \hat{f}(\frac{z}{i}),$$

and consider integrals of the form

$$\frac{1}{2\pi i} \int g(z)\, e^{zt}\, dz.$$

We omit mention of a path of integration associated with this integral at this time, since our next task (required in order to utilize the Residue Theorem) is to modify the path of Figure 7.24 in order to obtain a closed contour.

This extension of the original path of Figure 7.24 must be selected so that the integral so obtained is well behaved as $B \rightarrow \infty$ (since such a limit is required in order to calculate the inverse transform).

Considering the exponential factor of the integrand, we note that

$$|e^{zt}| = |e^{(x+iy)t}|$$

$$= |e^{xt}|$$

$$= e^{xt}.$$

Hence (for $t > 0$ for example) we have

$$|e^{zt}| \le 1,$$

for $x = \mathrm{Re}(z) < 0$. This suggests that for $t > 0$ it may be appropriate to close the contour of Figure 7.24 with an arc lying in the left-half plane. The resulting closed contour is denoted by Γ_B^+ and is illustrated in Figure 7.25.

If we denote by S_B the semi-circular arc of Γ_B^+ lying in the left-half plane, then we have

$$\frac{1}{2\pi i} \int_{\Gamma_B^+} g(z)\, e^{zt}\, dz = \frac{1}{2\pi i} \int_{-iB}^{iB} g(z)\, e^{zt}\, dz + \frac{1}{2\pi i} \int_{S_B} g(z)\, e^{zt}\, dz.$$

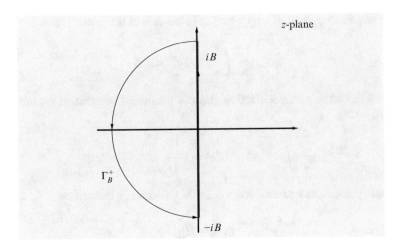

Fig. 7.25 Inversion contour

If it can be established that

$$\lim_{B\to\infty}\frac{1}{2\pi i}\int_{S_B} g(z)\, e^{zt}\, dz = 0,$$

then the closed contour integral may be used to compute the desired inverse transform as

$$\lim_{B\to\infty}\frac{1}{2\pi i}\int_{\Gamma_B^+} g(z)\, e^{zt}\, dz.$$

Since this integral is over a closed contour, the evident hope is that it may be simply evaluated by use of the Residue Theorem. (For simple cases, the value of the integral is in fact independent of B for sufficiently large B and the evaluation is straightforward.)

In order to carry this program out, it is necessary to verify that the condition that the added semi-circular path contribution vanishes actually holds. Since explicit evaluation of the additional integral

$$\frac{1}{2\pi i}\int_{S_B} g(z)\, e^{zt}\, dz$$

is usually out of the question, we seek verifiable estimates on the integrand g which will guarantee that this integral tends to zero as $B \to \infty$ as desired. Such estimates are obtained by simply parameterizing the integration in the natural fashion appropriate to the arc

$$z = B e^{i\theta}, \quad \frac{\pi}{2} \le \theta \le \frac{3\pi}{2},$$

and estimating the resulting integral

$$\frac{1}{2\pi i} \int_{\frac{\pi}{2}}^{\frac{3\pi}{2}} e^{B e^{i\theta} t} g(B e^{i\theta}) B e^{i\theta} \, i \, d\theta.$$

A Crude Estimate

Suppose that on the semi-circle S_B we have a bound on g dependent only on the radius B,

$$|g(B e^{i\theta})| \le M(B).$$

Then

$$\left| \frac{1}{2\pi i} \int_{\frac{\pi}{2}}^{\frac{3\pi}{2}} e^{B e^{i\theta} t} g(B e^{i\theta}) B e^{i\theta} \, i \, d\theta \right| \le \frac{1}{2\pi} \int_{\frac{\pi}{2}}^{\frac{3\pi}{2}} |g(B e^{i\theta})| \, B \, d\theta \le \frac{1}{2} M(B) B.$$

Hence, provided that

$$\lim_{B \to \infty} M(B) B = 0,$$

the limit of the integral along S_B vanishes as $B \to \infty$, and we will be able to evaluate the inversion integral by means of a contour integral calculation.
    For our initial example

$$g(z) = \frac{1}{z + a},$$

and so we obtain $M(B) = \frac{K}{B}$ as a natural estimate. For this example

$$\lim_{B \to \infty} M(B) B$$

fails to vanish, and so the crude estimate above is not sufficiently sharp to cover that example.

A Slightly More Clever Estimate

We still utilize a crude estimate of the form

$$|g(B\,e^{i\theta})| \le M(B),$$

but now take advantage of the fact that $\text{Re}(z) < 0$ along the contour S_B. In fact,

$$e^{zt} = e^{(x+iy)t} = e^{B(\cos\theta + i\sin\theta)t}$$

and

$$|e^{zt}| = e^{B(\cos\theta)t}.$$

Since $\cos\theta < 0$ throughout the range of integration, this substantially reduces the size of the integral. (Recall that we are initially considering only values of $t > 0$.)
From Figure 7.26 we see that

$$\cos\theta \le 1 - \frac{2\theta}{\pi}, \quad \frac{\pi}{2} \le \theta \le \pi,$$

consequently

$$|e^{zt}| = e^{B(\cos\theta)t} \le e^{Bt(1-\frac{2\theta}{\pi})t}$$

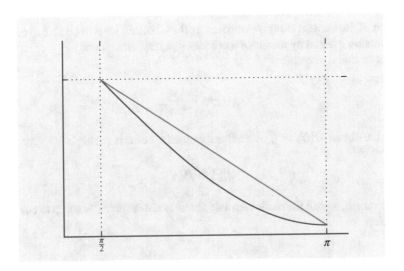

Fig. 7.26 Bounding the cosine function

for $\frac{\pi}{2} \le \theta \le \pi$, and

$$\left| \frac{1}{2\pi i} \int_{\frac{\pi}{2}}^{\frac{3\pi}{2}} e^{B e^{i\theta} t} g(B e^{i\theta}) B e^{i\theta} i\, d\theta \right| \le \frac{1}{2\pi} M(B) B \int_{\frac{\pi}{2}}^{\pi} \le e^{Bt(1-\frac{2\theta}{\pi})} d\theta$$

$$= \frac{1}{2\pi} M(B) B e^{Bt} \frac{\left[e^{-2Bt} - e^{-Bt} \right]}{-\frac{2Bt}{\pi}}$$

$$= M(B) \frac{\left[1 - e^{-Bt} \right]}{4t}.$$

Combining a similar estimate for the range $[\pi, \frac{3\pi}{2}]$ we find

$$\left| \frac{1}{2\pi i} \int_{S_B} g(z) e^{zt}\, dz \right| \le M(B) \frac{\left[1 - e^{-Bt} \right]}{2t}.$$

From this we conclude that (for fixed $t > 0$) the desired result follows from the hypothesis that $|g(z)| \to 0$ uniformly along the arc S_B as $B \to \infty$.

This latter condition is sufficient to handle a large number of cases.

The above discussion and treatment is based on the premise that $t > 0$. For the case that $t < 0$ the sign of the real part of the exponent in the e^{zt} factor is reversed. This is compensated by closing the original path with a semi-circular arc in the right-half plane (see Figure 7.27)

Arguing as above, we expect to calculate (for $t < 0$)

$$f(t) = \mathcal{F}^{-1} \left\{ \hat{f}(\omega) \right\} = \lim_{B \to \infty} \frac{-1}{2\pi i} \oint_{\Gamma_{B^-}} e^{zt} g(z)\, dz,$$

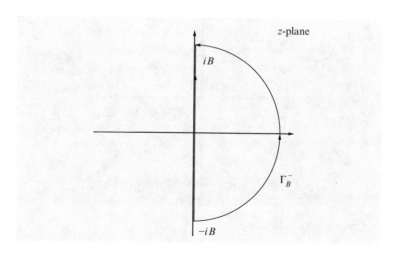

Fig. 7.27   The "$t < 0$" contour

as long as

$$\sup_{-\frac{\pi}{2} \leq \theta \leq \frac{\pi}{2}} |g(B\, e^{i\theta})| = M(B) \to 0, \text{ as } B \to \infty.$$

(The minus sign arises since the contour integral around Γ_{B-} is taken in the conventional counterclockwise sense, while the modified contour of Figure 7.22 is traversed in the opposite sense.)

The above general description of the contour integral method should be regarded as an illustration of the required technique. The details are subject to modification in some particular cases (an example is given below).

Example. Consider the example

$$f(t) = U(t)\, e^{-at}, \ (a > 0),$$

$$\hat{f}(\omega) = \frac{1}{i\omega + a}.$$

Following the procedure outlined above, we introduce the function g defined by

$$g(z) = \frac{1}{z + a}.$$

Note that g has a simple pole at the point $z = -a$ and that $g \to 0$ uniformly on a circle of radius B ($B > a$). The appropriate contours Γ_{B+}, Γ_{B-} are illustrated in Figure 7.28, together with the pole of the function g.

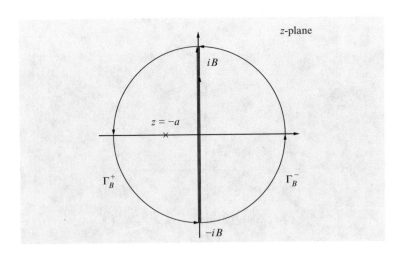

Fig. 7.28 Contours and pole location

Since $g(z) e^{zt}$ is analytic in the finite right-half plane (in fact, everywhere except at $z = -a$), we conclude that

$$\frac{-1}{2\pi i} \oint_{\Gamma_{B-}} g(z) e^{zt} \, dz = 0$$

(and for $B > a$)

$$\frac{1}{2\pi i} \oint_{\Gamma_{B+}} g(z) e^{zt} \, dz = \text{Residue} \left(e^{zt} g(z) \Big|_{} z = -a \right) = e^{-at}.$$

Combining these results with the general arguments above, we find that

$$\lim_{B \to \infty} \frac{1}{2\pi i} \int_{-B}^{B} \frac{1}{i\omega + a} e^{i\omega t} \, d\omega = \begin{cases} 0 & t < 0, \\ e^{-at} & t > 0. \end{cases}$$

These calculations are sufficient to identify (in the L^2 sense)

$$\lim_{B \to \infty} \frac{1}{2\pi i} \int_{-B}^{B} \frac{1}{i\omega + a} e^{i\omega t} \, d\omega = U(t) \, e^{-at}.$$

As a matter of interest, it is possible (see the problems below) to explicitly evaluate[15]

$$\lim_{B \to \infty} \frac{1}{2\pi i} \int_{-B}^{B} \frac{1}{i\omega + a} \, d\omega = \frac{1}{2}, \ (a > 0).$$

The fact that this differs from the declared value of the function at $t = 0$ may be noted.

The example above illustrates that the inversion integral

$$\frac{1}{2\pi} \int_{-B}^{B} \hat{f}(\omega) e^{i\omega t} \, d\omega$$

may converge to a function which differs (at isolated points, at least) from the original function defining the transform.

This phenomenon is closely related to the distinction between the notions of pointwise and mean-square convergence mentioned earlier in connection with the convergence of Fourier series.

---

[15]This amounts to the evaluation of the principal value at infinity of the indicated improper integral.

The situation is that the L^2 version of the Fourier inversion theorem (Section 7.4) asserts only that

$$f_B(t) = \frac{1}{2\pi} \int_{-B}^{B} \hat{f}(\omega) e^{i\omega t} d\omega$$

converges to f in the $L^2(-\infty, \infty)$ sense, so that

$$\lim_{B \to \infty} \int_{-\infty}^{\infty} \left[\left| f(t) - \frac{1}{2\pi} \int_{-B}^{B} \hat{f}(\omega) e^{i\omega t} d\omega \right|^2 \right] dt = 0.$$

On the other hand, the results obtained above (under suitable hypotheses) concerning contour integral calculations are assertions regarding the *pointwise convergence* of certain integrals.

Using the previously adopted notation, we have (for $t > 0$, with a similar representation for $t < 0$)

$$\frac{1}{2\pi} \int_{-B}^{B} \hat{f}(\omega) e^{i\omega t} d\omega + \frac{1}{2\pi i} \int_{S_B} g(z) e^{zt} dz = \frac{1}{2\pi i} \oint_{\Gamma_{B+}} g(z) e^{zt} dz. \qquad (7.15)$$

The first term of this integral is known to be convergent in L^2 to the original function $f \in L^2$. Under our motivating hypothesis, the second term approaches zero as $B \to \infty$ pointwise for each $t > 0$. In order to draw the desired conclusion from the equation, it is necessary to verify that two of the three terms involved are convergent in the same sense. Since the "object of the exercise" is to utilize an observation that the contour integral

$$\frac{1}{2\pi i} \int_{\Gamma_{B+}} g(z) e^{zt} dz$$

is readily evaluated, it is natural to place the additional burden on this term.

If

$$\frac{1}{2\pi i} \oint_{\Gamma_{B+}} g(z) e^{zt} dz$$

defines a family of functions convergent in L^2, then

$$\frac{1}{2\pi i} \int_{S_B} g(z) e^{zt} dz$$

(from (7.15)) is also a family convergent in L^2. Under our standing hypothesis that this last term approaches zero pointwise in t, it follows that this term tends to zero in L^2, and hence that

$$\lim_{B \to \infty} \frac{1}{2\pi} \int_{-B}^{B} \hat{f}(\omega) e^{i\omega t} \, d\omega = \lim_{B \to \infty} \frac{1}{2\pi i} \oint_{\Gamma_{B+}} g(z) e^{zt} \, dz$$

in the L^2 sense. (As written, this applies to the restriction of the original f to positive arguments. Duplicating the argument for the $t < 0$ case completes the argument.)

For the sake of completeness, we state the results of the above discussion as a theorem.

Theorem (Contour Inversion Theorem). *Suppose that $f \in L^2$, and let $\hat{f}$ be the Fourier transform of f. Suppose that*

1. there exists a function g of the complex variable z such that

$$\hat{f}(\omega) = g(i\omega)$$

(in L^2),

2.

$$\lim_{B \to \infty} \sup_{\theta} |g(B\, e^{i\theta})| = 0$$

and that

3. the limits

$$\lim_{B \to \infty} \frac{1}{2\pi i} \int_{\Gamma_{B+}} g(z) e^{zt} \, dz, \ t > 0,$$

$$\lim_{B \to \infty} \frac{1}{2\pi i} \oint_{\Gamma_{B-}} g(z) e^{zt} \, dz, \ t < 0$$

exist in $L^2(0, \infty)$ and $L^2(-\infty, 0)$ respectively.

Then

$$f(t) = \begin{cases} \lim_{B \to \infty} \frac{1}{2\pi i} \oint_{\Gamma_{B+}} g(z) e^{zt} \, dz & t > 0, \\ -\lim_{B \to \infty} \frac{1}{2\pi i} \oint_{\Gamma_{B-}} g(z) e^{zt} \, dz & t < 0. \end{cases}$$

It is perhaps worthwhile to remark on the content of the above result. In principle, the hypothesis that f belongs to L^2 is verifiable (through Parseval's Theorem) directly from $\hat{f}$. As should be clear from the preceding discussion, the choice of g is a straightforward matter. The verification of condition (2) is usually the only difficult part of the hypotheses, as it requires construction of suitable inequalities.

The condition 3 may appear difficult. However, in view of the fact that the indicated limits determine exactly the result of the inversion, this condition is in essence simply the requirement that the indicated answer makes sense. The result in effect is an invitation to make the formal contour integral calculations. If these formal calculations converge to a candidate for the answer, then the formal calculation is justified (after the fact).[16]

Example. We now consider the inversion of Fourier transforms which are rational functions of their argument ω. Such problems may be handled in principle by the partial fractions method. We express the result in terms of the contour integral results above.

We assume that we are given

$$\hat{f}(\omega) = \frac{q(i\omega)}{p(i\omega)}$$

where p, q are polynomial functions. (Writing the arguments in the rational function as "$i\omega$" facilitates identification of the function g above. It is also the natural form in which such transforms appear in various problems.)

In order that the transform should define a square integrable function of ω, it is necessary that $\deg(p) > \deg(q)$. This requirement for square integrability also places a restriction on the locations of the zeroes of the polynomial p. Assuming that the formula represents $\hat{f}$ "in lowest terms," it is easy to see that p may have no purely imaginary zero if $\hat{f}$ is to be square-integrable.

With these preliminary caveats, the previous considerations may be directly applied. The function g is identified as

$$g(z) = \frac{q(z)}{p(z)}.$$

Since $\deg(p) > \deg(q)$, it follows that

$$|g(B\,e^{i\theta})| \to 0$$

as $B \to \infty$, and so the inversion may be carried out by contour integration.

It is conventional to construct a z-plane diagram for this process, denoting by $\times$ the locations of the poles (singularities) of the function g. Since g has no poles on the imaginary axis, such a diagram takes the form of Figure 7.29.

Since the singularities of g consist of a finite number of poles, the required contour integrals may be evaluated by use of the Residue Theorem. If $\{z_j\}_{j=1}^{M}$ denotes the set of poles of g (zeroes of p) with $\mathrm{Re}(z_j) < 0$, then we obtain for $t > 0$, and $B > \max|z_j|$,

---

[16]From this view the result is analogous to some classical results concerning differentiation of series. Modulo a side condition, attempt the calculation; if a sensible result is obtained, it is the desired answer.

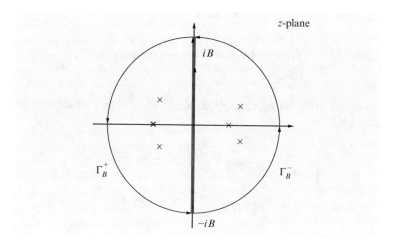

Fig. 7.29 Poles and contours

$$f(t) = \frac{1}{2\pi i} \oint_{\Gamma_{B+}} g(z)\, e^{zt}\, dz,$$

$$= \sum_{\text{Re}(z_j)<0} \text{Residue}\left(\frac{q(z)}{p(z)}\, e^{zt}\bigg|_{z = z_j}\right).$$

For the case that $t < 0$, again the contour integral required has constant value as long as $B > \max|z_k|$, where $\{z_k\}_{k=M+1}^{N}$ denotes the set of zeroes of p with $\text{Re}(z_k) > 0$. This gives

$$f(t) = \frac{-1}{2\pi i} \oint_{\Gamma_{B-}} g(z)\, e^{zt}\, dz,$$

$$= -\sum_{\text{Re}(z_k)>0} \text{Residue}\left(\frac{q(z)}{p(z)}\, e^{zt}\bigg|_{z = z_k}\right)$$

for $t < 0$. The above formulas are derived on the premise that g has poles in both of the half-planes $\text{Re}(z) < 0$, $\text{Re}(z) > 0$. If the poles are confined to one half-plane or the other, one of the contour integrals vanishes identically and the indicated sum should be interpreted as vanishing.

Example. The results of the previous example can be made more explicit by introducing the formulas for the indicated residues. Since the formula in question depends on the multiplicity of the pole in question, we define n_k as the multiplicity of the pole of g at $z = z_k$ This indicates that p has the factorization

$$p(z) = (z - z_1)^{n_1} (z - z_2)^{n_2} \ldots (z - z_M)^{n_M} (z - z_{M+1})^{n_{M+1}} \ldots (z - z_N)^{n_N}.$$

The required residue may be expressed as

$$
\text{Residue} \left(\frac{q(z)}{p(z)} e^{zt} \bigg| z = z_k \right) = \frac{1}{(n_k - 1)!} \frac{d}{dz}^{(n_k-1)} \left((z - z_k)^{n_k} \frac{q(z)}{p(z)} e^{zt} \right) \bigg|_{z=z_k}.
$$

Combining this with the previous calculation, we find that an explicit inversion formula is given by

$$
f(t) = \begin{cases} \displaystyle\sum_{\text{Re}(z_k < 0)} \frac{1}{(n_k-1)!} \frac{d}{dz}^{(n_k-1)} \left((z - z_k)^{n_k} \frac{q(z)}{p(z)} e^{zt} \right) \bigg|_{z=z_k} & t > 0, \\[2em] \displaystyle -\sum_{\text{Re}(z_k > 0)} \frac{1}{(n_k-1)!} \frac{d}{dz}^{(n_k-1)} \left((z - z_k)^{n_k} \frac{q(z)}{p(z)} e^{zt} \right) \bigg|_{z=z_k} & t < 0. \end{cases}
$$

Example. A numerical example of the above procedure is provided by the transform

$$
\hat{f}(\omega) = \frac{1}{(i\omega + 1)^2} \frac{1}{i\omega - 2}.
$$

We have

$$
g(z) = \frac{1}{(z + 1)^2} \frac{1}{z - 2}.
$$

The associated "pole pattern" is given in Figure 7.30.

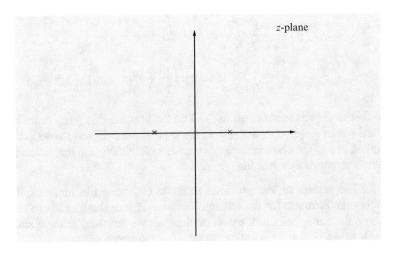

Fig. 7.30 Pole locations

From the general formula above, we find that for $t > 0$

$$f(t) = \frac{d}{dz}\left((z+1)^2 \frac{1}{(z+1)^2} \frac{1}{z-2} e^{zt}\right)\Bigg|_{z=-1}$$

$$= \left(-\frac{1}{(z-2)^2} e^{zt} + \frac{t\,e^{zt}}{(z-2)}\right)\Bigg|_{z=-1}$$

$$= -\frac{e^{-t}}{9} + \frac{t\,e^{-t}}{-3}.$$

For $t < 0$,

$$f(t) = -\left((z-2)\frac{1}{(z+1)^2}\frac{1}{z-2}e^{zt}\right)\Bigg|_{z=2}$$

$$= -\frac{1}{9} e^{2t}.$$

In various contexts (solutions of certain differential equations, for example) it is possible to encounter Fourier transforms of functions which are of limited duration in time. In such a case, the Fourier transform takes the form

$$\hat{f}(\omega) = \int_{-T_0}^{T_1} f(t)\, e^{-i\omega t}\, dt,$$

and if g is defined by the usual

$$g(z) = \int_{-T_0}^{T_1} f(t)\, e^{-zt}\, dt,$$

it is easy to see that g is an analytic (entire) function of the complex variable z for all finite z. From this observation, it is clear that our previous outline cannot be directly applied, since the indicated closed contour integrals all must vanish identically. It is the case, however, that the methods of our previous discussion can often be adapted to provide the desired results. We illustrate the typical considerations involved in an example.

Example. The Fourier transform of the "rectangular pulse function" defined by

$$f(t) = \begin{cases} 1 & |t| \le a, \\ 0 & |t| > a \end{cases}$$

is given by

$$\hat{f}(\omega) = \frac{2 \sin \omega a}{\omega}.$$

Since

$$\hat{f}(\omega) = \frac{e^{i\omega a} - e^{-i\omega a}}{i\omega},$$

the associated complex valued function is

$$g(z) = \frac{e^{az} - e^{-az}}{z}.$$

The inversion integral takes the form

$$f(t) = \lim_{B \to \infty} \frac{1}{2\pi i} \int_{-iB}^{iB} \frac{e^{az} - e^{-az}}{z} e^{zt} \, dz.$$

As noted above, $g(z)$ is an entire function of z. Further (with typical behavior), g is exponentially unbounded as $\text{Re}(z) \to \infty$. On the other hand, the exponential factor e^{zt} in the integrand may be used to "tame" the growth of the integrand. This observation is one of the keys to the evaluation of the inversion integral.

Specifically, if $t < -a$, then the overall integrand

$$g(z) e^{tz} = \frac{e^{az} - e^{-az}}{z} e^{zt}$$

vanishes as $\text{Re}(z) \to \infty$. If $t = -a - \delta$, $\delta > 0$, the integral becomes

$$f(-a - \delta) = \lim_{B \to \infty} \frac{1}{2\pi i} \int_{-iB}^{iB} \frac{e^{(a-a)z} - e^{-2az}}{z} e^{-\delta z} \, dz$$

$$= \lim_{B \to \infty} \frac{1}{2\pi i} \int_{-iB}^{iB} e^{-\delta z} \tilde{g}(z) \, dz.$$

Now the function $\tilde{g}$ satisfies the estimate

$$|\tilde{g}(B e^{i\theta})| \le \frac{2}{B},$$

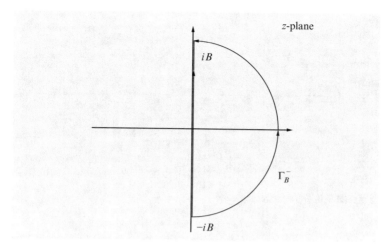

Fig. 7.31 Initial contour

and so tends to zero as $B \rightarrow \infty$. The integral has exactly the form encountered above in the evaluation of the usual inversion integral "for negative t." The integral in this case is also evaluated by closing the contour in the right half-plane (Figure 7.31)

Since $\tilde{g}$ is analytic on and inside Γ_{B-}, we conclude (from the previous arguments) that the integral vanishes identically. That is,

$$f(t) \equiv 0, \text{ for } t < -a.$$

In a similar fashion, for $t > a$, we write $t = a + \delta$, $\delta > 0$, and close the original path through the left half-plane. The resulting integral is readily seen to vanish, and

$$f(t) \equiv 0, \text{ for } t > a.$$

is concluded.

For the case that $|t| < a$, we consider again the integral

$$\frac{1}{2\pi i} \int_{-iB}^{iB} \frac{e^{az} - e^{-az}}{z} e^{zt} \, dz.$$

Since $|t| < a$, the terms $e^{(t+a)z}$ and $e^{(t-a)z}$ have exponents with real parts of opposite sign. This suggests breaking the integral into the sum of two integrals, and closing the contours in opposite directions to take advantage of the exponential decay factors.

To split the path integral as suggested is not quite feasible, since the resulting pair of integrands would contain a pole at the origin. We avoid this problem by modifying the path of the integral to include a detour around the origin. The original and modified paths are as illustrated in Figure 7.32.

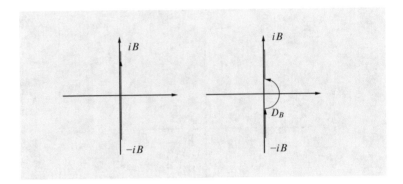

Fig. 7.32 Initial path deformation

Since the integrand is an entire function of z, the integral is *independent of the path* connecting the end points, and

$$\frac{1}{2\pi i} \int_{-iB}^{iB} \frac{e^{az} - e^{-az}}{z} e^{zt} \, dz = \frac{1}{2\pi i} \int_{D_B} \frac{e^{az} - e^{-az}}{z} e^{zt} \, dz.$$

Since the detour path D_B avoids the origin, the integrand may now be split as suggested to give

$$\frac{1}{2\pi i} \int_{D_B} \frac{e^{az} - e^{-az}}{z} e^{zt} \, dz = \frac{1}{2\pi i} \int_{D_B} \frac{e^{(t+a)z}}{z} \, dz - \frac{1}{2\pi i} \int_{D_B} \frac{e^{(t-a)z}}{z} \, dz.$$

We note that both integrands contain a pole at the origin, and that the assumption that $|t| < a$ implies that

$$t + a > 0,$$

$$t - a < 0.$$

In view of the signs of the exponents of the integrands, the limiting values of the integrals are evaluated by closing the path D_B with arcs lying in opposite half-planes, as illustrated in Figure 7.33.

For the second integral, the contour Γ_{B-} of Figure 7.33 is appropriate. Since the pole at the origin lies outside the contour Γ_{B-}, we have

$$\frac{1}{2\pi i} \oint_{\Gamma_{B-}} \frac{e^{z(t-a)}}{z} \, dz = 0$$

for the first integral. By evaluating the residue at the origin, we immediately find

$$\frac{1}{2\pi i} \oint_{\Gamma_{B+}} \frac{e^{z(t+a)}}{z} \, dz = 1.$$

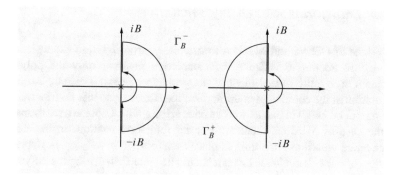

Fig. 7.33 Integration contours

Considering the inequalities on the exponents, together with the fact that $\left|\frac{1}{z}\right| \to 0$ as $|z| \to \infty$, we conclude that the arguments outlined at the beginning of the section are applicable to the evaluation of the limiting values of the integrals. In short, we conclude that (with S_B denoting the closing arc, as above)

$$\lim_{B \to \infty} \frac{1}{2\pi i} \int_{D_B} \frac{e^{az} - e^{-az}}{z} e^{zt} \, dz$$

$$= \lim_{B \to \infty} \frac{1}{2\pi i} \int_{\Gamma_{B+}} \frac{e^{(t+a)z}}{z} \, dz - \lim_{B \to \infty} \frac{1}{2\pi i} \int_{S_B} \frac{e^{(t-a)z}}{z} \, dz$$

$$= 1 - 0.$$

Combining this result with our previous calculations finally provides

$$\lim_{B \to \infty} \frac{1}{2\pi} \int_{-B}^{B} \frac{\sin \omega a}{\omega} e^{i\omega t} \, d\omega = \begin{cases} 0 & |t| > a, \\ 1 & |t| < a \end{cases}$$

as expected.

The examples treated above essentially involve only functions with a finite number of singularities, all of which are poles.

Certain other problems of a similar sort involve more complicated considerations (e.g., branch cuts, or an infinite number of singularities). Some examples of this are provided in the context of the closely related Laplace transform inversion integral in the following section.

Fourier Transforms with Octave/Maxima

The abilities of Octave and Maxima to calculate partial fraction expansions were mentioned in Section 6.5. Since the calculation involves only the polynomial coefficients, it is quite independent of the problem context. The only issue is the interpretation of the coefficient in the application situation, so that Fourier transform problems can be handled just as well as ones arising with Laplace transforms.

In the case of Maxima, there are Fourier transform routines in the `fourie` package along with those handling Laplace transforms.

In view of our calculations above, it should be clear that writing the polynomials in terms of $i\omega$ as the variable is advantageous. Then the partial fraction expansions are expressed with "$i\omega$ minus root" terms, and the inverse transform calculation is immediate.

Problems 7.6.

1. By explicit calculation of an anti-derivative, show that

$$\lim_{B \to \infty} \frac{1}{2\pi} \int_{-B}^{B} \frac{1}{i\omega + a}\, d\omega = \frac{1}{2}.$$

2. Define a sequence of functions $\{f_N\}$ by

$$f_N(t) = \begin{cases} N & 0 < t < \frac{1}{N}, \\ 0 & \text{otherwise.} \end{cases}$$

   Show that $\{f_N\}$ tends to zero pointwise, but that $\{f_N\}$ does not tend to zero in the sense of L^2.

3. Consider the rational function

$$\hat{f}(\omega) = \frac{q(i\omega)}{p(i\omega)}$$

   where p, q are polynomials. Show that $\hat{f}$ is the Fourier transform of some square-integrable function if and only if

   a. degree$(p) >$ degree(q), and (assuming $\hat{f}$ is in "lowest terms")
   b. no zero of p has zero real part.

4. Suppose that a function f is defined by

$$f(t) = \begin{cases} A\,e^{-at} & t \geq 0, \\ B\,e^{bt} & t < 0 \end{cases}$$

   with Re$(a) > 0$, Re$(b) > 0$. Show that

$$\int_{-\infty}^{\infty} |\hat{f}(\omega)|\, d\omega < \infty$$

if and only if the function f is everywhere continuous.

5. Let f be defined as in Problem 4. Show that

$$\frac{1}{2\pi} \int_{-\infty}^{\infty} \hat{f}(\omega)\, e^{i\omega t}\, d\omega = \begin{cases} A\, e^{-at} & t > 0, \\ \frac{A+B}{2} & t = 0^{17}, \\ B\, e^{bt} & t < 0. \end{cases}$$

6. Suppose that f is a (square-integrable) function expressible (separately for $t < 0$ and for $t > 0$) as a finite sum of polynomials times (complex) exponential functions, so that $\hat{f}$ is a rational function of ω. Show that results analogous to Problems 4, 5 above hold.

   That is,

$$\int_{-\infty}^{\infty} |\hat{f}(\omega)|\, d\omega < \infty$$

if and only if the function f is everywhere continuous, and

$$\frac{1}{2\pi} \int_{-\infty}^{\infty} \hat{f}(\omega)\, e^{i\omega t}\, d\omega = \begin{cases} f(t) & t > 0, \\ \frac{f(0^+)+f(0^-)}{2} & t = 0, \\ f(t) & t < 0. \end{cases}$$

   Hint: This is largely a bookkeeping exercise which can be made to follow from Problems 4 and 5 above in conjunction with either use of a partial fractions argument, or equivalently, invocation of the inversion form obtained above for rational transforms.

7. Find the inverse transforms of:

   a. $\hat{f}(\omega) = \frac{1}{(i\omega+1)(i\omega+2)}$,

   b. $\hat{g}(\omega) = \frac{1}{(-i\omega+1)(i\omega+2)}$,

   c. $\hat{h}(\omega) = \frac{1}{(-i\omega+1)(-i\omega+2)}$.

8. Find the inverse transforms of

   a. $\hat{f}(\omega) = \frac{i\omega+1}{\omega^2+4i\omega+4}$,

   b. $\hat{g}(\omega) = \frac{\omega^2}{(\omega^2+1)^2}$.

---

[17] In this case the integral should be interpreted as a "principal value at infinity," that is, $\displaystyle\lim_{B\to\infty} \int_{-B}^{B} \hat{f}(\omega)\, d\omega$.

9. Evaluate the integral

$$\int_{-\infty}^{\infty} \frac{\sin(ax+b)}{x^2+d^2}\,dx$$

given that a, b, d are all real parameters, $d \neq 0$.

10. Suppose that f is a function which vanishes identically outside of the (finite) interval $[-T_0, T_1]$, and is continuous on the interval $[-T_0, T_1]$.

Show that if a function of the complex variable z is defined by

$$g(z) = \hat{f}(\frac{z}{i})$$

where $\hat{f}$ is the Fourier transform of f, then g is everywhere analytic in the finite part of the complex plane (that is, is an entire) function of z.

Remark. The hypothesis of continuity is stronger than is required, but allows the result to be obtained from classical (Riemann) integration theorems.

11. Define f as

$$f(t) = \begin{cases} e^{-at} & 0 \le t \le 1, \\ 0 & \text{otherwise.} \end{cases}$$

a. Compute $\hat{f}$.
b. Explicitly evaluate the inversion integral

$$\frac{1}{2\pi} \int_{-\infty}^{\infty} \hat{f}(\omega)\, e^{i\omega t}\, d\omega$$

by complex variable methods. (For definiteness, assume $\mathrm{Re}(a) > 0$. Notice that the details of the inversion are dependent on the value of $\mathrm{Re}(a)$.)

12. Define f by

$$f(t) = \begin{cases} t & 0 \le t \le 1, \\ 0 & \text{otherwise.} \end{cases}$$

Compute $\hat{f}$, and recover f from the transform by explicitly evaluating the inversion integral by complex variable methods.

13. Repeat Problem 12 for the function g defined by

$$f(t) = \begin{cases} t^2 - 1 & |t| \le 1, \\ 0 & \text{otherwise.} \end{cases}$$

7.7 The Laplace Transform Inversion Integral

In this section we derive and illustrate the Laplace transform inversion integral. As previously mentioned, such a representation of the Laplace transform inversion process may be readily derived from the Fourier Inversion Theorem. The required result is closely connected with the complex-variable inversion methods of the previous section.

We recall that the formal definition of the Laplace transform is given by

$$F(s) = \mathcal{L}\{f(t)\} = \int_0^\infty f(t) \, e^{-st} \, dt.$$

Writing the complex variable s in terms of its real and imaginary parts

$$s = \sigma + i\omega,$$

the Laplace transform takes the readily identifiable form

$$\mathcal{L}\{f(t)\} = \int_0^\infty f(t) \, e^{-\sigma t} \, e^{-i\omega t} \, dt.$$

Defining the function h by

$$h(t) = \begin{cases} e^{-\sigma t} f(t) & t \geq 0, \\ 0 & t < 0 \end{cases}$$

identifies the Laplace transform of f in terms of the Fourier transform of h. In terms of our usual notation for Fourier and Laplace transforms

$$\hat{h}(\omega) = \int_{-\infty}^\infty h(t) \, e^{-i\omega t} \, dt = F(\sigma + i\omega). \tag{7.16}$$

In view of our previous discussion of the Fourier transform inversion theory (Section 7.4), we may immediately deduce classes of functions for which the Laplace transform (and inversion integral) are naturally defined.

One such class arises from the "L^1 theory" of the Fourier transform and consists of those functions f such that (for some σ) $h \in L^1$. That is,

$$\int_0^\infty |f(t) \, e^{-\sigma t}| \, dt < \infty.^{18}$$

---

[18]This is essentially the class for which we originally defined the Laplace transform in Chapter 6.

An inversion integral for the Laplace transform (in this class of functions) may be constructed by simply applying the L^1-theory results of the previous section to the function h.

A simpler form of the Laplace inversion theorem is obtained by applying the L^2-Fourier transform theory to the function h above. The natural class of functions to be considered consists of those f such that $h \in L^2$ (for some real σ)

$$\int_0^\infty |f(t)\, e^{-\sigma t}|^2 \, dt < \infty.$$

The Laplace transform of such a function f may be defined as the L^2-Fourier transform of h,

$$F(\sigma + i\omega) = \lim_{A\to\infty} \int_{-A}^A h(t)\, e^{-i\omega t} \, dt$$

$$= \lim_{A\to\infty} \int_0^A f(t)\, e^{-\sigma t}\, e^{-i\omega t} \, dt.$$

The above limit should in principle be interpreted as an L^2 limit. However, it is easily seen (using the Cauchy–Schwarz inequality) that for $\delta > 0$,

$$\int_0^A f(t)\, e^{-\sigma t}\, e^{-\delta t}\, e^{-i\omega t} \, dt$$

is actually an absolutely convergent integral. Since in the construction of the Laplace transform we may take $\mathrm{Re}(s)$ as large as desired, the convergence of the Laplace transform is not in practice as delicate a matter as the definition as a Fourier transform might appear to indicate.

Assuming that $\sigma(=\mathrm{Re}(s))$ has been chosen sufficiently large, we may obtain a Laplace transform inversion result by simply applying the usual Fourier Inversion Theorem to h. This gives the conclusion that

$$h(t) = f(t)\, e^{-\sigma t} = \lim_{B\to\infty} \frac{1}{2\pi} \int_{-B}^B \hat{h}(\omega)\, e^{i\omega t} \, d\omega.$$

To express the inversion formula in explicit terms involving the transform, we identify it with a path integral (cf. Section 7.5). To express the inversion integral in terms of a path integral, we require a function g of a complex variable z such that

$$g(i\omega) = \hat{h}(\omega).$$

Since we have

$$\hat{h}(\omega) = F(\sigma + i\omega)$$

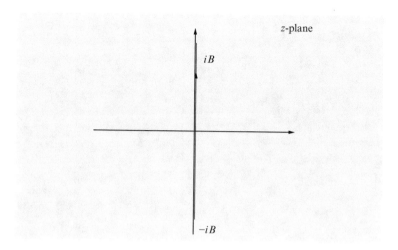

Fig. 7.34 Inversion path

we make the identification

$$g(z) = F(\sigma + z),$$

and write the Laplace inversion integral in the form

$$f(t)\, e^{-\sigma t} = \lim_{B\to\infty} \frac{1}{2\pi i} \int_{-iB}^{iB} F(\sigma + z)\, e^{zt}\, dz.$$

The path involved is illustrated in Figure 7.34.

The path integral may also be parameterized in such a way as to use the Laplace transform argument as the variable of integration. Since

$$\frac{1}{2\pi i} \int_{-iB}^{iB} F(\sigma + z)\, e^{zt}\, dz = \frac{1}{2\pi i} \int_{-iB}^{iB} F(\sigma + z)\, e^{(\sigma+z)t}\, e^{-\sigma t}\, dz,$$

the required integral may also be written in the form

$$f(t)\, e^{-\sigma t} = \lim_{B\to\infty} \frac{e^{-\sigma t}}{2\pi i} \int_{\sigma-iB}^{\sigma+iB} F(s)\, e^{st}\, ds. \tag{7.17}$$

The associated path is illustrated in Figure 7.35. (It should be emphasized that this is a simple matter of parameterization. There is no "contour deformation" involved in the change to the more usual form (7.17).)

As noted in the previous section, the limit in (7.17) should in principle be interpreted in the L^2-sense. If the indicated integral is actually convergent in the

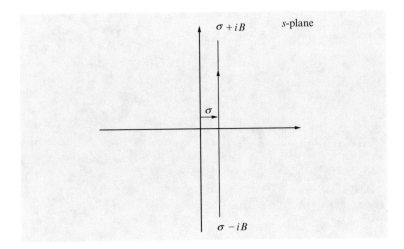

Fig. 7.35 Corresponding *s*-plane path

pointwise sense, then the "exponential convergence factors" in (7.17) may be removed. The inversion formula in this case takes the form

$$f(t) = \frac{1}{2\pi i} \int_{\sigma - i\infty}^{\sigma + i\infty} F(s)\, e^{st}\, ds.^{19} \tag{7.18}$$

The form of the inversion formulas (7.17) and (7.18) naturally suggests evaluation by means of contour integration. Since the functions f under consideration are defined on the positive half-axis, the discussion of the previous section dictates use of a contour lying in the left half-plane. See Figure 7.36.

Combining the identifications made above with the discussion of contour integral inversion of the Fourier transform (Section 7.5), we may state the result as a theorem:

Theorem (Laplace Transform Inversion). *Suppose* $f : [0, \infty) \mapsto \mathbb{C}$ *is such that for some real* σ,

$$\int_0^\infty |f(t)\, e^{-\sigma t}|^2\, dt < \infty.$$

Let F denote the Laplace transform of f and suppose that:

1. $F(\sigma + Be^{i\theta}) \to 0$ as $B \to \infty$ uniformly in θ, for $\theta \in [\frac{\pi}{2}, \frac{3\pi}{2}]$, and

---

[19]In fact, this may be taken as the "usual form" of the inversion, with the implicit understanding that (7.17) is to be employed in the case of L^2 convergence.

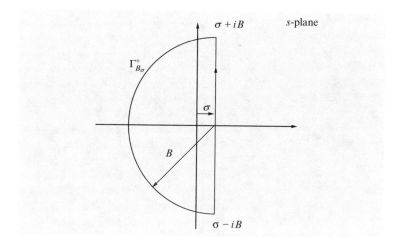

Fig. 7.36 The s-plane closed contour

2.

$$\lim_{B \to \infty} e^{-\sigma t} \frac{1}{2\pi i} \oint_{\Gamma_{B+}} F(s)\, e^{st}\, ds$$

exists in $L^2(0, \infty)$.

Then

$$f(t)\, e^{-\sigma t} = \lim_{B \to \infty} e^{-\sigma t} \frac{1}{2\pi i} \oint_{\Gamma_{B+}} F(s)\, e^{st}\, ds,$$

and hence

$$f(t) = e^{\sigma t} \left[\lim_{B \to \infty} e^{-\sigma t} \frac{1}{2\pi i} \oint_{\Gamma_{B+}} F(s)\, e^{st}\, ds \right].$$

The interpretation of this result is entirely similar to that of the corresponding Fourier transform inversion result. Assuming (the hard part) that the transform F vanishes in the limit along the augmenting arc of the contour, then the convergence of the contour integral calculations to a limit which is a candidate for the inversion result guarantees the validity of the calculation.

We have written these results as though the contour radius parameter B tends to infinity in a continuous fashion. This interpretation was sufficient for the Fourier transform examples presented in the previous section. All that is really required, however, is that the indicated conditions hold for a suitable sequence of radii $\{B_n\}$, with the restriction that $\lim_{n \to \infty} B_n = +\infty$. This freedom to judiciously place the

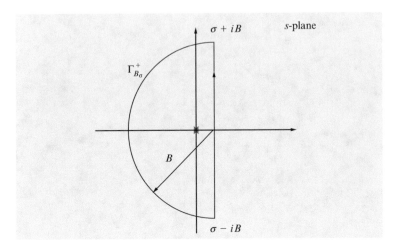

Fig. 7.37 Contour and poles

integration contour is essential in problems in which the function g (in the Fourier transform case) or the Laplace transform function F has an infinite number of singularities. This is illustrated in one of the examples below.

Example. We consider the inversion of the Laplace transform

$$F(s) = \frac{1}{s^n},$$

where $n > 0$ is some integer. The parameter σ may be chosen as any positive number, and the appropriate contour is as indicated in Figure 7.37.

We note first that

$$F(\sigma + Be^{i\theta}) = \frac{1}{(\sigma + Be^{i\theta})^n} \to 0$$

as $B \to \infty$ as required. Since the origin is the only singularity associated with F (where there is a pole of order n), we compute

$$\frac{1}{2\pi i} \oint_{\Gamma_{B_\sigma}^+} F(s)\, e^{st}\, ds = \text{Residue} \left(\frac{e^{st}}{s^n} \bigg| s = 0 \right)$$

$$= \frac{1}{(n-1)!} \left(\frac{d}{ds} \right)^{n-1} \left(\frac{s^n\, e^{st}}{s^n} \right) \bigg|_{s=0}$$

$$= \frac{t^{n-1}}{(n-1)!}.$$

Since for any $\sigma > 0$, we have that

$$e^{-\sigma t} \frac{t^{n-1}}{(n-1)!}$$

is square-integrable, we immediately conclude from the Laplace transform inversion theorem that

$$f(t) = \mathcal{L}^{-1} \left\{ \frac{1}{s^n} \right\} = \frac{t^{n-1}}{(n-1)!}.$$

Example. The inversion of rational Laplace transforms was treated earlier as an exercise in the partial fraction method. The form of the inversion may also be obtained directly from the Laplace inversion integral.

We suppose that F is defined as a rational function

$$F(s) = \frac{q(s)}{p(s)},$$

with $\deg(p) > \deg(q)$. (This restriction is required if F is to represent the transform of a function of the sort discussed above.)

Assuming that the denominator polynomial p factors as

$$p(s) = (s - z_1)_1^n (s - z_2)_2^n \ldots (s - z_N)_N^n$$

(and that F is expressed in lowest terms) we see that F has a pole at each of the zeroes of p. Since the Laplace transform must represent an analytic function for $\mathrm{Re}(s) > \sigma$, an appropriate contour for the inversion is constructed by fixing σ greater than the maximum real part of the zeroes of p.

The appropriate pole pattern and contour is illustrated in Figure 7.38.

In order to apply the Laplace Inversion Theorem, we verify in passing that F vanishes uniformly along the circular arc portion of $\Gamma_{B_\sigma^+}$. Since σ is chosen to the right of all of the poles, $\Gamma_{B_\sigma^+}$ for sufficiently large B encloses all of the poles of F. Hence

$$\frac{1}{2\pi i} \oint_{\Gamma_{B_\sigma^+}} F(s) e^{st} \, ds = \sum_{z_k} \mathrm{Residue} \left(e^{st} F(s) \Big|_{s \,=\, z_k} \right)$$

$$= \sum_{z_k} \frac{1}{(n_k - 1)!} \left(\frac{d}{ds} \right)^{n_k - 1} \left((s - z_k)^{n_k} F(s) e^{st} \right) \Bigg|_{s = z_k}.$$

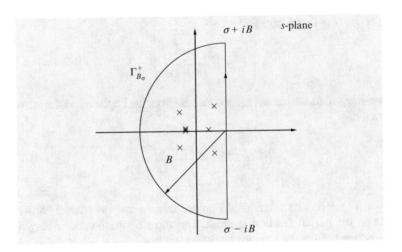

Fig. 7.38 Rational transform poles and contour

Since we have $\mathrm{Re}(z_k) < \sigma$, for $k = 1, \ldots, N$, it follows that

$$\lim_{B \to \infty} e^{-\sigma t} \frac{1}{2\pi i} \oint_{\Gamma_{B_\sigma}^+} F(s)\, e^{st}\, ds$$

(a sum of polynomials in t multiplied by exponential functions of negative real exponent) exists in $L^2[0, \infty)$.

This easily verifies the hypotheses of the Laplace Inversion Theorem, and we conclude that

$$f(t) = \sum_{z_k} \frac{1}{(n_k - 1)!} \left(\frac{d}{ds}\right)^{n_k - 1} \left((s - z_k)^{n_k} F(s)\, e^{st}\right)\Bigg|_{s = z_k}.$$

This result should be compared with the corresponding Fourier transform inversion result. It should be verified that the results are in agreement, provided that $\mathrm{Re}(z_k) < 0$, $k = 1, \ldots, N$. In this case, the function f (vanishing for negative arguments) has both a Laplace and a Fourier transform. The Laplace transform exists for purely imaginary arguments, and

$$F(i\omega) = \hat{f}(\omega).$$

In the case that $\mathrm{Re}(z_k) > 0$ for some pole of F, the results are considerably different. It is clear that terms of the inversion arising from such poles display (constant or) unbounded behavior as $t \to \infty$. As a result, they have no L^2 Fourier transform. Such zeroes (with $\mathrm{Re}(z_k) > 0$) would correspond in the context of Fourier transforms to functions decreasing as $t \to -\infty$, and vanishing for $t > 0$. Since the Laplace transform is defined only for functions $f : [0, \infty) \mapsto C$, the difference in the results is inevitable.

This distinction can be seen directly by a comparison of the contours associated with the inversion integrals. See Figures 7.38 and 7.29.

Example. A more challenging inversion arises from problems involving the Laplace transform of a periodic function f. In this case the contour integral inversion method leads to the rediscovery of the appropriate Fourier series expansion formula.

We consider the solution of an initial value problem for an ordinary differential equation. While it is a simple matter to consider an arbitrary state variable style problem formulation, a simple model suffices to illustrate the problems of the Laplace transform inversion. The problem is given by

$$p(\frac{d}{dt}) x(t) = e(t),$$

with p a polynomial of degree n. We assume for simplicity that

$$p(s) = (s - z_1)^{n_1} (s - z_2)^{n_2} \ldots (s - z_N)^{n_N}$$

with $\text{Re}(z_k) < 0$, $k = 1, \ldots, N$.

The forcing function e is periodic of period T,

$$e(t + T) = e(t), \ t > 0,$$

and of finite energy

$$\int_0^T |e(t)|^2 \, dt < \infty. \tag{7.19}$$

The incorporation of nonzero initial conditions in the differential equation leads to a problem of the sort considered above in Chapter 5. For this reason, we assume also that

$$x(0) = \frac{dx}{dt}(0) = \cdots = \frac{d^{n-1}x}{dt^{n-1}}(0) = 0,$$

and so consider only the forced response. The solution of the differential equation by Laplace transform methods leads to

$$X(s) = \frac{1}{p(s)} E(s),$$

where (for $\text{Re}(s) > 0$)

$$E(s) = \mathcal{L}\{e(t)\}$$

$$= \int_0^\infty e(t) \, e^{-st} \, dt$$

$$= \sum_{n=0}^{\infty} \int_{nT}^{(n+1)T} e(t)\, e^{-st}\, dt$$

$$= \frac{\int_0^T e(\tau)\, e^{-s\tau}\, d\tau}{1 - e^{-sT}}. \quad 20$$

The transform of the solution in general has a pole at each of the zeroes of the polynomial p. With our assumption that these zeroes all lie in the left half-plane, the parameter σ may be chosen as any positive number.

The inversion contours $\{\Gamma_{B_\sigma^+}\}$ for sufficiently large B enclose all of the poles arising from p.

We next consider the singularities arising from the transform of the forcing function. Considered as a function of the complex variable s, the numerator of the forcing function transform is an entire function. The denominator vanishes whenever

$$1 - e^{-sT} = 0,$$

that is, for

$$sT = 2n\pi i, \; n = 0, \pm 1, \pm 2. \ldots$$

$$s = \frac{2n\pi i}{T} = in\omega_0.$$

Since the zeroes of the denominator are simple, the transform of the solution in general has a simple pole at each of the points given by $s = in\omega_0$. (However, in a particular case any of these potential singularities may be removable.)

The closed inversion contour $\{\Gamma_{B_\sigma^+}\}$ must obviously be chosen to avoid each of these singularities.

We select a sequence of such contours adjusted to pass halfway between these points of singularity. This is accomplished (see Figure 7.39) by choosing the radius sequence $\{B_N\}$ defined by (with $\sigma > 0$ fixed, and $\omega_0 = \frac{2\pi}{T}$)

$$B_N^2 = \sigma^2 + (N + \frac{1}{2})^2 \omega_0^2.$$

Figure 7.39 also illustrates the pattern of the poles associated with the transform X. This clearly indicates that the sequence of inversion contours $\{\Gamma_{B_\sigma^+}\}$

---

[20]The integrals encountered are convergent since

$$\left| \int_0^T e(\tau)\, e^{-s\tau}\, d\tau \right| \le \left(\int_0^T |e(\tau)|^2\, d\tau \right)^{\frac{1}{2}} \left(\int_0^T |e^{-s\tau}|^2\, d\tau \right)^{\frac{1}{2}}$$

while $e^{-s\tau} < 1$ for $\mathrm{Re}(s) > 0$.

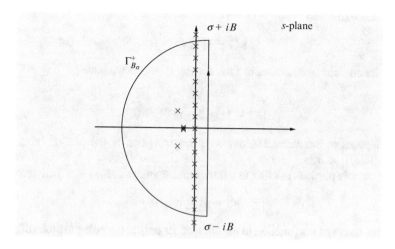

Fig. 7.39 An example with an infinite number of poles

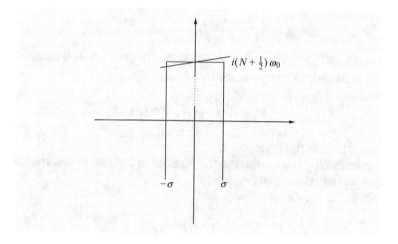

Fig. 7.40 Path of the inversion contours

encloses an increasing sequence of poles. As a result, the use of our inversion theorem is at least a slightly more delicate issue than in the case of the examples discussed above (Figure 7.40).

In order to use the inversion theorem, we must first verify that the transform X vanishes uniformly along the circular portion of the contour $\Gamma_{B_\sigma^+}$ as $B_N \to \infty$.

Since $\mathrm{Re}(s)$ is (predominantly) negative in the region of interest, $E(s)$ as written appears as the ratio of two "large" quantities. It is therefore more useful to rewrite the formula in the more tractable form

$$E(s) = \frac{\int_0^T e(\tau)\, e^{s(T-\tau)}\, d\tau}{e^{sT} - 1}.$$

We show first that

$$|e^{sT} - 1| \geq M > 0$$

along the contour in question. In fact, for $\mathrm{Re}(s) < -\sigma$ we have $|e^{sT}| \leq e^{-\sigma T} < 1$, and

$$|e^{sT} - 1| \geq 1 - e^{-\sigma T} > 0.$$

This disposes of the entire contour with the exception of the portion lying in the region $|\mathrm{Re}(s)| < \sigma$.

But for this portion (tending to a horizontal segment as $B_N \to \infty$) we have

$$|e^{sT} - 1| \geq \epsilon > 0$$

since the zeroes of the function in question occur only at the pole positions discussed above. Since the function $(e^{sT} - 1)$ is periodic in the direction parallel to the imaginary axis, this establishes the desired inequality. We note that the quantity

$$\frac{1}{p(s)}$$

tends to zero uniformly along $\Gamma_{B_N^+}$ as $B_n \to \infty$. To draw our desired conclusion, then, it remains only to verify that the transform numerator

$$N(s) = \int_0^T e(\tau) \, e^{s(T-\tau)} \, d\tau$$

is uniformly bounded on the contour.

The quantity $N(s)$ (for fixed s) may be regarded as the inner product between the function e and an exponential function.

Using the Cauchy–Schwarz inequality,

$$|N(s)| \leq \left(\int_0^T |e(\tau)|^2 \, d\tau \right)^{\frac{1}{2}} \left(\int_0^T |e^{s(T-\tau)}|^2 \, d\tau \right)^{\frac{1}{2}} \tag{7.20}$$

$$= \left(\int_0^T |e(\tau)|^2 \, d\tau \right)^{\frac{1}{2}} \left(\frac{e^{2\mathrm{Re}(s)T} - 1}{2\mathrm{Re}(s)} \right)^{\frac{1}{2}}.$$

(The last factor above should be interpreted as $\sqrt{T}$ for $\mathrm{Re}(s) = 0$.) From this we conclude

$$|N(s)| \leq \left(\int_0^T |e(\tau)|^2 \, d\tau \right)^{\frac{1}{2}} \left(\frac{e^{2\sigma T} - 1}{2\sigma} \right)^{\frac{1}{2}}$$

for $\mathrm{Re}(s) < \sigma$. This may be combined with the previous results to show that the transform $E(s)$ is uniformly bounded on $\Gamma_{B_N^+}$ as required.

These considerations verify the basic hypothesis of the inversion theorem. We next evaluate the contour integral

$$\frac{1}{2\pi i} \oint_{\Gamma_{B_N^+}} \frac{\int_0^T e(\tau) e^{-s\tau} \, d\tau}{p(s)(1 - e^{-sT})} \, ds.$$

The contour $\Gamma_{B_N^+}$ encloses the poles associated with p, as well as $2N + 1$ (simple) poles located at

$$s = in\omega_0, \ n = 0 \pm 1, \pm 2, \ldots, \pm N.$$

Evaluation of the contour integral by residues gives

$$\frac{1}{2\pi i} \oint_{\Gamma_{B_N^+}} \frac{\int_0^T e(\tau) e^{-s\tau} \, d\tau}{p(s)(1 - e^{-sT})} \, ds$$

$$= \sum_{z_k} \frac{1}{(n_k - 1)!} \left(\frac{d}{ds}\right)^{(n_k - 1)} \left(\frac{(s - z_k)^{n_k} \left[\int_0^T e(\tau) e^{-s\tau} \, d\tau\right] e^{st}}{p(s)(1 - e^{-st})}\right)\Bigg|_{s=z_k}$$

$$+ \sum_{n=-N}^{N} \lim_{s \to in\omega_0} \left(\frac{(s - in\omega_0) \int_0^T e(\tau) e^{-s\tau} \, d\tau \, e^{st}}{p(s)(1 - e^{-sT})}\right)$$

$$= x_1(t) + \sum_{n=-N}^{N} \frac{1}{p(in\omega_0)} \left(\frac{1}{T} \int_0^T e(\tau) e^{-in\omega_0 \tau} \, d\tau\right) e^{in\omega_0 t}.$$

Here x_1 is the result of the residue calculation at the zeroes of p. It has the form

$$x_1(t) = \sum_{k=1...M, \, j=0...n_M-1} \sum \alpha_{kj} t^j e^{z_k t}.$$

The term

$$x_2^N(t) = \sum_{n=-N}^{N} \frac{1}{p(in\omega_0)} \left(\frac{1}{T} \int_0^T e(\tau) e^{-in\omega_0 \tau} \, d\tau\right) e^{in\omega_0 t}$$

is readily recognized as the partial sum of the Fourier series for the periodic solution of the differential equation problem. (See Chapter 2, Section 2.8).

To complete the details of the inversion, we must verify that

$$e^{-\sigma t} \frac{1}{2\pi i} \oint_{\Gamma_{B_N^+}} \frac{\int_0^T e(\tau) e^{-s\tau} \, d\tau}{p(s)(1 - e^{-sT})} e^{st} \, ds$$

converges in $L^2(0, \infty)$. Since $\text{Re}(z_k) < 0$, and x_1 is independent of the contour $\Gamma_{B_N^+}$, we need only to verify that

$$\phi_N(t) = e^{-\sigma t} x_2^N(t)$$

defines an L^2 convergent sequence. However this is easy to show, because the (limiting) series

$$x_2(t) = \sum_{n=-\infty}^{\infty} \frac{1}{p(in\omega_0)} \left(\frac{1}{T} \int_0^T e(\tau) e^{-in\omega_0 \tau} \, d\tau \right) e^{in\omega_0 t}$$

is in fact uniformly convergent. If

$$e_n = \frac{1}{T} \int_0^T e(\tau) e^{-in\omega_0 \tau} \, d\tau$$

is the nth Fourier coefficient of the forcing function, then by the Cauchy–Schwarz inequality,

$$\sum_{n=-\infty}^{\infty} \left| \frac{1}{p(in\omega)} \right| |e_n| \le \left(\sum_{n=-\infty}^{\infty} \left| \frac{1}{p(in\omega)} \right|^2 \right)^{\frac{1}{2}} \left(\sum_{n=-\infty}^{\infty} |e_n|^2 \right)^{\frac{1}{2}} < \infty.$$

This inequality is sufficient (by the M-test) to guarantee that x_2 is uniformly convergent. That is,

$$\left| \sum_{|n|>N} \frac{1}{p(in\omega_0)} e_n e^{in\omega_0 t} \right| \le \epsilon(N),$$

where $\epsilon(N) \to 0$ as $N \to \infty$. Then (calculating the difference between the contour calculation and expected limit) we obtain

$$e^{-\sigma t} \left[x_1(t) + x_2(t) - \frac{1}{2\pi i} \oint_{\Gamma_{B_N^+}} X(s) e^{st} \, ds \right] = e^{-\sigma t} \sum_{|n|>N} \frac{1}{p(in\omega_0)} e_n e^{in\omega_0 t}.$$

Since the L^2 norm of this error is bounded by

$$\int_0^\infty \left| \sum_{|n|>N} \frac{1}{p(in\omega_0)} e_n e^{in\omega_0 t} \right|^2 dt \le \int_0^\infty \epsilon^2(N) e^{-2\sigma t} \, dt = \frac{\epsilon^2(N)}{2\sigma}$$

we conclude that the required convergence holds, and finally that

$$x(t) = x_1(t) + \sum_{n=-\infty}^{\infty} \frac{1}{p(in\omega_0)} \left(\frac{1}{T} \int_0^T e(\tau) e^{-in\omega_0 \tau} \, d\tau \right) e^{in\omega_0 t}.$$

It should be noted that this result is identical to that obtained by the formal summation of the infinite number of residues associated with the integrand of the inversion integral.

The transient term x_1 may be connected with our previous considerations of the same problem. This aspect of the solution is explored in the problems below.

Problems 7.7.

1. Suppose that F is a proper rational function

$$F(s) = \frac{q(s)}{p(s)}$$

with degree(p)> degree(q). Prove that

$$\lim_{B \to \infty} F(\sigma + B e^{i\theta}) = 0$$

uniformly in the angular variable θ.

2. Compute

$$\mathcal{L}^{-1} \left\{ \frac{1}{(s + 1)(s + 2)(s + 3)} \right\}$$

by contour integral methods.

3. Verify by contour integration that

$$\mathcal{L}^{-1} \left\{ \frac{s}{s^2 + a^2} \right\} = \cos at.$$

4. Calculate

$$\mathcal{L}^{-1} \left\{ \frac{s}{s^4 - a^4} \right\}$$

by contour integration.

5. Define f by

$$f(t) = \begin{cases} t & 0 \le t < 1, \\ 1 & t \ge 1 \end{cases}$$

and verify that

$$\mathcal{L}\{f(t)\} = \frac{1}{s^2}[1 - e^{-s}].$$

Recover f by evaluating the Laplace inversion integral.

Hint: This is clearly closely related to examples and problems of the previous section. Also, the inversion is carried out much more easily by use of the shift formula if such is allowed. It isn't.

6. Consider the Laplace transform solution of the periodically forced differential equation

$$p(\frac{d}{dt})x(t) = f(t),$$

where $f(T + \tau) = f(\tau)$, f has finite energy, and $x(0)$, $\frac{dx}{dt}(0), \ldots, \frac{d^{n-1}}{dt^{n-1}}(0)$ are all prescribed.

Describe the solutions in the case that p has some zeroes with positive real part (exclude the case of zero real-part zeroes).

7. Consider the Problem 6, in the case where p is allowed to have zeroes which are purely imaginary. Prove by use of complex inversion methods that the existence of zeroes of p of the form $ik\omega_0$ where k is an integer gives the possibility that the solution x contains terms of the form of a polynomial times a sinusoid at the frequency $k\omega_0$.

8. Consider Problem 6 and again assume that the real parts of all zeroes of p are negative. For simplicity assume that the zeroes of p are all distinct.

By use of complex inversion methods, find the initial conditions which guarantee that the solution is entirely periodic. Show that these initial conditions are exactly those associated with the value at time zero of the periodic solution.

7.8 An Introduction to Generalized Functions

In the previous sections we have discussed some aspects of the theory of the Fourier transform as it applies to certain classes of functions. For the solution of many problems, these results and methods are a sufficient tool.

On the other hand, there are various problems whose formulation leads in a natural fashion to considerations of "objects" beyond the scope of our previously defined transform.

Such an example has been previously encountered in the context of our discussion of the impulse response of an ordinary differential equation (Section 6.6). There it was found convenient to consider the notion of an "ideal impulse" (called a delta function) with a (Laplace) transform of unity.

This object is a prototype of what is known as a distribution. Other distributions arise naturally in other problems, and they are particularly useful in the formulation of models for partial differential equations with "singular" forcing terms.

The study of the theory of distributions is part of the broad area of functional analysis. As such, we cannot give a complete account here. Instead, we give an introduction to the basic notions and definitions below.

Our treatment of the Fourier transform in the previous sections has emphasized the vector space approach to the topic. The theory of distributions is again based on vector space notions. While our previous discussions have exploited most heavily the notions of a norm (distance) function, the theory of distributions is based on the idea of a dual vector space associated with a given vector space.

This notion is usually first encountered in connection with finite-dimensional linear algebra. There one is presented with a (finite-dimensional) vector space V — the dual vector space V' is defined as the set of all linear mappings from V to the field over which V is defined.[21] (In what follows we assume the field to be the complex number field C.)

The definition of V' in this fashion leads to the introduction of a bilinear form $< \cdot, \cdot >$ mapping $V \times V'$ to C. This is defined as

$$< v, v' > = v'(v). \tag{7.21}$$

(This definition makes sense since v' is a function on V. It is bilinear since the elements of V' are linear functions.) This construction plays the role of a replacement for an inner product in the ensuing theory. Requiring that (7.21) behaves formally in the manner which would be expected of the usual inner product of functions leads directly to the definitions of derivative and Fourier transform for distributions.

In the case of an infinite-dimensional vector space V, the definition of V' as all linear mappings made above is too broad to be tractable. It is necessary to restrict V' to those linear functions on V which are continuous in the variable $v \in V$.[22]

A concrete example of the pairing introduced above has already been encountered with the example of the (Hilbert) space $V = L^2(-\infty, \infty)$. For any fixed function $g \in L^2(-\infty, \infty)$, the function

$$v'(f) = \int_{-\infty}^{\infty} f(x)\, g(x)\, dx$$

is linear and (by the Cauchy–Schwarz inequality) continuous in f. A fundamental result concerning this example states that every element of $\left(L^2(-\infty, \infty)\right)'$ arises in

---

[21] In the finite-dimensional case, it is usual to show that V' is isomorphic to V. A common reaction to this result is to wonder why the subject was mentioned at all. However, for infinite-dimensional V, the V' is typically quite distinct from V. This distinction is the source of distribution theory.

[22] This means that a notion of closeness of elements of V is defined. In short, V has a *topology*, and is a *topological vector space*.

this fashion. In this case (and essentially for the last time) V' may again be identified with the original vector space V.

In the theory of distributions the inner product expression is taken as a motivating model and source of notation. The theory is constructed by simply defining a suitable vector space V, and declaring a distribution to be an element of the associated V'. The notion of a distribution is therefore entirely dependent on the vector space V chosen.

The elements of V are typically referred to as the space of *test functions*. A useful class of such is given by

$$S = \left\{ f(\cdot) \,\middle|\, f \text{ infinitely differentiable, and such that} \right.$$

$$\left. (1 + x^2)^k \frac{d^p f}{dx^p} \to 0 \text{ as } |x| \to \infty, \text{ for all } k, p \geq 0 \right\}.$$

S consists of functions which, together with all derivatives tend to zero at infinity at faster than polynomial rate. It is easy to see that such functions are closed under addition and scalar multiplication, and so form a vector space.

The delta function may now be defined as that element of S' which assigns to a function in S its value at the point $x = 0$. By definition, δ is the distribution with the property that

$$\delta(\phi) = <\phi, \delta> = \phi(0) \tag{7.22}$$

for all $\phi \in S$[23]. In view of the pairing that occurs in the Hilbert space case, it is a useful notation to symbolically express (7.22) in the form

$$\delta(\phi) = \int_{-\infty}^{\infty} \phi(t)\, \delta(t)\, dt = \phi(0).$$

It should be emphasized that logically the role of the integral expression involving δ is that of a beguiling notational device. The official declaration of δ is contained in (7.22), and there is no implication that the "integral" represents the evaluation of an integral in any of the usual senses.

On the other hand, the symbolic representation as an integral is an indispensable guide for the appropriate definitions of a calculus of distributions. As an example of such a definition, we consider the problem of defining the idea of derivative of a distribution. For functions, "slope" provides the motivation for the definition, but for distributions hints must be found elsewhere.

For two test functions $f,\ g \in S$, the usual formula for integration by parts provides

---

[23] We leave the definition of a topology on S and verification of continuity of δ for a more comprehensive treatment of the subject.

$$\int_{-\infty}^{\infty} f(t)\, g'(t)\, dt = f(t)g(t)\Big|_{-\infty}^{\infty} - \int_{-\infty}^{\infty} f'(t)\, g(t)\, dt$$

$$= -\int_{-\infty}^{\infty} f'(t)\, g(t)\, dt$$

(since both f and g tend to zero at infinity). If g in the above is replaced by the distribution δ the formalism gives (since $f' \in S$)

$$\int_{-\infty}^{\infty} f(t)\, \delta'(t)\, dt = -\int_{-\infty}^{\infty} f'(t)\, \delta(t)\, dt$$

$$= -f'(0).$$

This formal calculation leads us to define the distribution δ' using the formally derived expression above. The distribution δ' acta on a test function by:

$$\delta'(\phi) = <\phi, \delta'> = -<\phi', \delta> = -\delta(\phi') = -\phi'(0). \tag{7.23}$$

More generally, the same formalism may be used to define the derivative of an arbitrary distribution d. The definition is simply that

$$d'(\phi) = <\phi, d'> = -<\phi', d> = -d(\phi') \tag{7.24}$$

for all $\phi \in S$. This process may be repeated to define higher derivatives of the distribution. The second derivative is defined by

$$d'' = (d')',$$

so that

$$d''(\phi) = -d'(\phi') = +d(\phi'').$$

Example. The definition of the space of test functions S means that essentially any function which grows at infinity less rapidly than a polynomial may be used to define a distribution. Given such a function, we define its action on a test function by the obvious integral

$$g(\phi) = <\phi, g> = \int_{-\infty}^{\infty} \phi(t)\, g(t)\, dt.$$

Such a function may be differentiated (in the sense of a distribution) by applying the differentiation rule above. The function g defined by

$$g(t) = \begin{cases} -\frac{1}{2} & t < 0, \\ \frac{1}{2} & t \geq 0 \end{cases}$$

is clearly not differentiable in the classical sense. It does, however, have a derivative in the sense of a distribution. By the definition of derivative of a distribution, for all test functions ϕ we have

$$
\begin{aligned}
g'(\phi) &= \int_{-\infty}^{\infty} \phi(t)\, g'(t)\, dt = -\int_{-\infty}^{\infty} \phi'(t)\, g(t)\, dt \\
&= -\left[\int_{-\infty}^{0} -\frac{1}{2} \phi'(t)\, dt + \int_{0}^{\infty} \frac{1}{2} \phi'(t)\, dt \right] \\
&= -\left[-\frac{1}{2}\phi(0) + \frac{1}{2}\phi(-\infty) + \frac{1}{2}\phi(\infty) - \frac{1}{2}\phi(0) \right] \\
&= \phi(0) \\
&= <\phi, \delta> .
\end{aligned}
$$

Since this holds for all test functions ϕ, we conclude that

$$
g' = \delta.
$$

Considerations similar to the above allow the definition of "shifted" distributions, and certain "change of variables" compositions of functions with distributions. Some of these notions are included in the problems below.

The interpretation of the action of a distribution as a formal integral of a product also leads to the definition of the Fourier transform of a distribution.

In view of the definition of the class of test functions S, it follows readily that the Fourier transform

$$
\hat{\phi}(\omega) = \mathcal{F}\{\phi\}(\omega) = \int_{-\infty}^{\infty} \phi(t)\, e^{-i\omega t}\, dt
$$

is well defined for all test functions. In fact, the hypothesis that $\phi \in S$ is sufficiently strong to conclude that $\hat{\phi}$ is infinitely differentiable as a function of ω and rapidly decreasing at infinity. In short, $\hat{\phi}$ belongs to S (as a function of ω).

This observation that test functions in both the "original" and "transform" variable are naturally available leads us to consider two classes of test functions, $S(T)$ and $S(\Omega)$) in order to distinguish the variables involved.

Since (as the transform of a test function ϕ) $\hat{\phi} \in S(\Omega)$, we may consider the action of an element of $S'(\Omega)$ on $\hat{\phi}$. If one considers as a special case an element of $S'(\Omega)$ which arises through integration against a test function $f \in S(\Omega)$, then the defining relation takes the form

$$
\int_{-\infty}^{\infty} \left(\int_{-\infty}^{\infty} \phi(t)\, e^{-i\omega t}\, dt \right) f(\omega)\, d\omega = <\mathcal{F}\phi, f> .
$$

Since f and ϕ are both test functions, the order of integration in this expression may be interchanged to give

$$< \mathcal{F}\phi, f > = \int_{-\infty}^{\infty} f(\omega) \left(\int_{-\infty}^{\infty} \phi(t) e^{-i\omega t} \, dt \right) d\omega$$

$$= \int_{-\infty}^{\infty} \phi(t) \left(\int_{-\infty}^{\infty} f(\omega) e^{-i\omega t} \, d\omega \right) dt$$

$$=< \phi, \mathcal{F}f > .$$

This calculation result is now appropriated for the purpose of defining the transform of a distribution. If the function f in the expression is to be replaced by a distribution, then the form of this expression dictates that the Fourier transform operation $\mathcal{F}$ should be regarded as mapping $S'(\Omega)$ (i.e., distributions or generalized functions in the transform variable) to $S'(T)$. This leads to the definition of $\mathcal{F}$ for a distribution. For $d \in S'(\Omega)$, $\mathcal{F}d$ is that element of $S'(T)$ such that

$$< \phi, \mathcal{F}d >=< \mathcal{F}\phi, d >=< \hat{\phi}, d > \tag{7.25}$$

for all test functions ϕ. An evident consequence of this indirect definition is that Fourier transforms of distributions exist. It remains only to verify that it is possible to compute transforms from the definition.

Example. We compute the transform of a delta function this way. By definition, for all test functions ϕ

$$< \phi, \mathcal{F}\delta > =< \mathcal{F}\phi, \delta >$$

$$=< \hat{\phi}, \delta >$$

$$= \hat{\phi}(0)$$

$$= \int_{-\infty}^{\infty} \phi(t) e^{i0} \, dt$$

$$=< \phi, 1 > .$$

We therefore conclude that $\mathcal{F}\delta = 1$.

Remark. The *formal calculation*

$$\int_{-\infty}^{\infty} e^{-i\omega t} \delta(t) \, dt \text{ `` } ='' \ e^{i\omega 0} = 1$$

gives a result coinciding with the above. However, the exponential function is not rapidly decreasing at infinity, and so does not belong to the test function space S. This means that the definitions outlined above do not sanction the formalism, since δ does not apply to $e^{i\omega t}$.

Example. In keeping with the Fourier transform pair game played earlier with transform calculations, the next calculation to consider is $\mathcal{F}1$. Again using the definition of the transform of a distribution, we compute

$$
< \phi, \mathcal{F}1 > \ = \ < \mathcal{F}\phi, 1 > = \int_{-\infty}^{\infty} \hat{\phi}(\omega)\, 1 \, d\omega
$$

$$
= 2\pi \left(\frac{1}{2\pi} \int_{-\infty}^{\infty} \hat{\phi}(\omega)\, d\omega \right)
$$

$$
= 2\pi\, \phi(0)
$$

$$
= < \phi, 2\pi\, \delta > .
$$

This establishes that

$$
\mathcal{F}1 = 2\pi\, \delta.
$$

Remark. In this case the corresponding (very doubtful) formal calculation

$$
\int_{-\infty}^{\infty} e^{-i\omega t}\, 1 \, d\omega = 2\pi\, \delta
$$

must be regarded as ambiguous as an integral evaluation at best.

Example. The transform of the derivative of a delta function is computed as

$$
< \phi, \mathcal{F}\delta' > \ = \ < \mathcal{F}\phi, \delta' >
$$

$$
= -\hat{\phi}'(0)
$$

$$
= -\frac{d}{d\omega} \int_{-\infty}^{\infty} \phi(t) e^{-i\omega t}\, dt \ \Big|_{\omega = 0}
$$

$$
= \int_{-\infty}^{\infty} it\, \phi(t)\, dt.
$$

We therefore conclude that $\mathcal{F}\delta' = it$.

To complete the pair, we compute $\mathcal{F}\omega$ (the transform of the distribution defined by integration against the transform variable).

$$
< \phi, \mathcal{F}\omega > \ = \ < \hat{\phi}, \omega >
$$

$$
= \int_{-\infty}^{\infty} \hat{\phi}(\omega)\, \omega \, d\omega
$$

$$= \frac{2\pi}{i} \left(\frac{d}{dt} \frac{1}{2\pi} \int_{-\infty}^{\infty} \hat{\phi}(\omega)\, e^{i\omega t}\, d\omega \right) \Bigg|_{t=0}$$

$$= \frac{2\pi}{i}\, \phi'(0)$$

$$= -\frac{2\pi}{i} <\phi, \delta'>$$

$$= <\phi, 2\pi i\, \delta'> .$$

Hence $\mathcal{F}\omega = 2\pi i\, \delta'$.

Example. The calculations of the previous examples suggest the validity of the differentiation law for Fourier transforms in the distribution context. That this holds may be shown rather easily.[24] We let d denote a distribution (on $S(\Omega)$), and compute the transform of d by direct application of the definitions (7.25), (7.24):

$$<\phi, \mathcal{F}d'> = <\mathcal{F}\phi, d'>$$

$$= - <(\mathcal{F}\phi)', d> .$$

But

$$-(\mathcal{F}\phi)' = -\frac{d}{d\omega} \int_{-\infty}^{\infty} \phi(t)\, e^{-i\omega t}\, dt$$

$$= \mathcal{F}\{it\,\phi\} .$$

Hence

$$<\phi, \mathcal{F}d'> = <\mathcal{F}\{it\,\phi\}, d>$$

$$= <it\,\phi, \mathcal{F}d>$$

$$= <\phi, it\,\mathcal{F}d> .$$

(This last equality relies on the natural definition of multiplication of a distribution by a function). Since ϕ is arbitrary, we conclude that

$$\mathcal{F}d' = it\,\mathcal{F}d.$$

In short, this is just the usual rule for the transform of a derivative extended to this more general case.

---

[24]The apparent simplicity at this stage is due to the effort (alluded to above) involved in careful analysis of the vector space S.

This section has so far been devoted entirely to the explanation of some underlying ideas and basic formalisms associated with distributions (generalized functions).

A consequence of the differentiation and transform properties outlined above is that it is possible to use distributions as "forcing functions" in differential equations, and to seek solutions which are themselves distributions.

This possibility is widely exploited, especially in the investigation of partial differential equations. A gentle introduction to the basic ideas can be obtained from ordinary differential equation examples.

To discuss this topic, we require the introduction of a notion of convolution applicable to generalized functions. The convolution of two functions f and g is given by the formula

$$(f \star g)(\tau) = \int_{-\infty}^{\infty} f(\tau - t) g(t) \, dt.$$

In view of our heuristic identification of the function-distribution pairing with the integral of a product, it is natural to use the convolution formula to define the notion of the convolution of a function with a distribution.

The definition forced by the convolution formula is simply

$$(f \star d)(\tau) = \, <\tilde{f}_\tau, d>$$

where $\tilde{f}_\tau(t) = f(\tau - t)$ is the function (of t) defined by reversing and translating by τ the original function f. As long as f is a test function $\in S$, and d a distribution $\in S'$, the expression defines a function of τ.

Example. The simplest distribution is the delta function. For this example,

$$(f \star \delta)(\tau) = \, <\tilde{f}_\tau, \delta>$$

$$= f(\tau - t) \Big|_{t=0}$$

$$= f(\tau).$$

(This calculation is the source of the approximate delta function terminology employed earlier.)

Example. The derivatives of the delta function also have similar simple properties under convolution. The distribution δ' acts according to

$$\left(f \star \delta'\right)(\tau) = \, <\tilde{f}_\tau, \delta'>$$

$$= - \left(\tilde{f}_\tau\right)'(0)$$

$$= -\frac{d}{dt'}f(\tau - t)\Big|_{t=0}$$

$$= f'(\tau).$$

In a similar fashion,

$$(f \star \delta^n)(\tau) = <\tilde{f}_\tau, \delta^n>$$

$$= (-1)^n \left(\frac{d}{dt}\right)^n f(\tau - t)\Big|_{t=0}$$

$$= f^n(\tau).$$

In short, convolution with the appropriate delta function derivative is equivalent to differentiation of the indicated order.

The relevance of this idea to the solution of differential equations arises as follows.

We suppose that a solution of

$$p(\frac{d}{dt})x(t) = f(t), \quad -\infty < t < \infty, \tag{7.26}$$

is sought, where f is a test function. Here p is a polynomial of degree n, so that

$$p(\frac{d}{dt})x(t) = \left(p_n \frac{d^n x}{dt^n} + p_{n-1}\frac{d^{n-1}x}{dt^{n-1}} + \cdots + p_0 x\right)(t)$$

represents a constant coefficient differential operator. Rather than solve (7.26) directly, we consider the equation

$$p(\frac{d}{dt})X(t) = \delta(t), \tag{7.27}$$

to be solved for a distribution $X \in S'$. Assuming that such an X can be found, the solution of the original (7.26) follows from convolution. Specifically, we compute the convolution of both sides of (7.27) with f. Then

$$\left(f \star p(\frac{d}{dt})X(t)\right) = (f \star \delta)(\tau)$$

$$= f(\tau).$$

But the left side of this equality can also be evaluated using the definition of the derivative of a distribution. In the first place

$$\left(f \star p(\frac{d}{dt}) X(t)\right) = <\tilde{f}_\tau, \sum_{j=0}^{n} p_j \left(\frac{d}{dt}\right)^j X>$$

$$= < \sum_{j=0}^{n} p_j (-1)^j \left(\frac{d}{dt}\right)^j \tilde{f}_\tau, X > .$$

But since $\tilde{f}_\tau(t) = f(\tau - t)$

$$(-1)^j \left(\frac{d}{dt}\right)^j \tilde{f}_\tau(t) = \left(\frac{d}{d\tau}\right)^j \tilde{f}_\tau(t),$$

and the convolution takes the form

$$\left(f \star p(\frac{d}{dt}) X(t)\right) = < \sum_{j=0}^{n} p_j \left(\frac{d}{d\tau}\right)^j \tilde{f}_\tau, X >$$

$$= p(\frac{d}{d\tau}) <\tilde{f}_\tau, X >\ ^{25}$$

$$= p(\frac{d}{d\tau}) (f \star X)(\tau).$$

Combining these calculation,

$$p(\frac{d}{d\tau}) (f \star X)(\tau) = f(\tau)$$

so that

$$x(t) = (f \star X)(t)$$

is evidently the required solution of the original (function) differential equation problem (7.26).

The object X (in principle a distribution, although usually in practice representable as a function) is variously referred to as either the impulse response or Green's function associated with equation (7.26).

The problem in equation (7.26) is also discussed in the following section on the basis of transforms of functions. We defer examples to that point. For discussions of the use of distributions in problems of the form of equation (7.26) but with variable coefficients (as well as applications in boundary value problems) we refer the reader to other references.

---

[25] Strictly speaking, the step moving derivatives outside of the pairing, although intuitively clear, requires more justification than we have provided.

Problems 7.8.

1. A formal "integral" calculation

$$\int_{-\infty}^{\infty} f(t)\, \delta(t - \tau)\, dt = f(\tau)$$

provides a means of defining a "shifted" delta function. If δ_τ denotes the "delayed by τ" delta function, provide a formal definition of δ_τ in terms of test functions. More generally, define the "τ-delayed" version of an arbitrary distribution d.

2. Use the formal calculation

$$\int_{-\infty}^{\infty} f(t)\, g(at)\, dt = \int_{-\infty}^{\infty} \frac{1}{a} f(\frac{\tau}{a})\, g(\tau)\, d\tau$$

as a guide to the definition of the "time-scaled" delta function $d = \delta(at)$.

3. Define a distribution g as integration against the function g,

$$g(t) = \begin{cases} t^2 - 1 & |t| \leq 1, \\ 0 & |t| > 1. \end{cases}$$

Compute (as distributions) both

$$\frac{dg}{dt}$$

and

$$\frac{d^2 g}{dt^2}.$$

4. Compute the Fourier transform $\mathcal{F}\{t^2\}$.
5. Compute $\mathcal{F}\{\delta''\}$.
6. Compute $\mathcal{F}\{t^n\}$.
7. Compute $\mathcal{F}\{\delta^{(n)}\}$.
8. Prove a shift theorem for the Fourier transform of a distribution, using the shift definition of Problem 1.
9. Since distributions have derivatives and Fourier transforms, if a solution of the equation (for the distribution X)

$$\frac{dX}{dt} + X = \delta(t)$$

exists, the Fourier transform should discover it. Assuming such existence, and that Fourier transforms may be uniquely inverted in the context of distributions, solve the above equation.

7.9   Fourier Transforms, Differential Equations, and Circuits

The existence of the simple relationship between differentiation and the Fourier transform makes it possible to use the transform as a vehicle for the solution of differential equations. The situation is entirely similar to the use of Laplace transform methods for the same purpose.

The difference between the use of Fourier and Laplace transform methods lies entirely in the formulation of the problem to be solved. Laplace transforms are applicable to functions defined on a half-axis and are therefore the natural tool for the solution of initial value problems in differential equations. On the other hand, Fourier transforms are applicable to (suitably restricted classes of) functions defined on a whole line. In the context of ordinary differential equations, problems of the calculation of forced responses (input output problems) fall naturally into the purview of Fourier transform methods.[26]

Example. Suppose that it is required to solve the ordinary differential equation

$$\frac{dx}{dt} + x(t) = e^{-|t|}, \quad -\infty < t < \infty.$$

In order to apply Fourier transform methods to this problem, we *assume* the existence of a Fourier transformable solution of the equation, with the properties that justify use of the usual differentiation theorem.[27] Transforming gives

$$(i\omega + 1)\hat{x}(\omega) = \mathcal{F}\left\{e^{-|t|}\right\} = \frac{4}{\omega^2 + 4}.$$

Hence

$$\hat{x}(\omega) = \frac{4}{(i\omega + 1)(\omega^2 + 4)}.$$

Out of the inversion methods available for $\hat{x}$, we select the partial fractions method.

$$\hat{x}(\omega) = \frac{A}{i\omega + 1} + \frac{B}{-i\omega + 2} + \frac{C}{i\omega + 2},$$

where

---

[26]Certain initial value problems may also be treated (with some care) by including distributions (generalized functions) among the admissible forcing functions.

[27]The procedure is essentially similar to the process of finding periodic solutions by use of Fourier series. If a solution of the required properties exists, the method discovers it.

$$A = (i\omega + 1)\hat{x}(\omega) \bigg|_{i\omega=-1} = \frac{4}{3},$$

$$B = (-i\omega + 2)\hat{x}(\omega) \bigg|_{i\omega=+2} = \frac{1}{3},$$

$$C = (i\omega + 2)\hat{x}(\omega) \bigg|_{i\omega=-2} = -1.$$

Inverting the transform,

$$x(t) = \frac{4}{3}U(t)\,e^{-t} + \frac{1}{3}U(-t)\,e^{2t} - 1\,U(t)e^{-2t}.$$

As an alternative inversion method, x may be calculated by use of the convolution theorem. This gives

$$x(t) = \left[e^{-2|t|} \star U(t)\,e^{-t}\right]$$

$$= \int_{-\infty}^{\infty} U(t-\tau)e^{-(t-\tau)}e^{-2|\tau|}\,d\tau$$

$$= \int_{-\infty}^{t} e^{-(t-\tau)}e^{-2|\tau|}\,d\tau.$$

Evaluation of the integral again gives the result.

The representation of the solution obtained above in terms of a convolution integral is typical of the solution of forced differential equations by Fourier transform methods. The typical n^{th}-order equation version of such a problem is usually written in the form

$$p(\frac{d}{dt})x(t) = u(t), \quad -\infty < t < \infty, \tag{7.28}$$

where p is a polynomial of degree n. Assuming the existence of a Fourier transformable solution, that u is transformable, and that the solution has a sufficient degree of differentiability to justify use of the differentiation law, we obtain

$$p(i\omega)\,\hat{x}(\omega) = \mathcal{F}\{u(t)\} = \hat{u}(\omega),$$

and the formal result that

$$\hat{x}(\omega) = \frac{1}{p(i\omega)}\,\hat{u}(\omega).$$

Inverse transforming by means of the convolution theorem,

$$x(t) = \left[\mathcal{F}^{-1} \left\{ \frac{1}{p(i\omega)} \right\} \star u(t) \right](t) \tag{7.29}$$

$$= \int_{-\infty}^{\infty} g(t - \tau) \, u(\tau) \, d\tau$$

is produced as the obvious candidate for the solution.

We refer to this as a *candidate* for the solution, since the result is derived under the tentative hypothesis of the existence of a solution with certain specified properties. If the original equation has no such solution, then we should not be surprised to find that (7.29) makes little sense.

Whether or not this situation occurs is dependent on the polynomial p (that is, the characteristic equation associated with the differential equation).

The calculation of the inverse transform

$$\mathcal{F}^{-1} \left\{ \frac{1}{p(i\omega)} \right\}$$

is carried out on the basis of determination of the zeroes of the polynomial p, using either a partial fractions or contour integration method. As long as p has no zero which is a purely imaginary number, then

$$\mathcal{F}^{-1} \left\{ \frac{1}{p(i\omega)} \right\}$$

is a function g such that

$$\int_{-\infty}^{\infty} |g(t)| \, dt < \infty. \tag{7.30}$$

As a consequence (7.29) is well defined for any transformable function u, and the solution makes sense.

On the other hand, if $p(i\Omega_0) = 0$ for some real frequency "Ω_0," then an attempt to evaluate the inverse transform by, say, partial fractions eventually leads one to the dilemma

$$\mathcal{F}^{-1} \left\{ \frac{1}{i\omega - i\Omega_0} \right\} = ?$$

This term in braces does not represent the transform of a function of the desired sort, and the attempt to evaluate (7.29) breaks down. The underlying reason for this problem is that a solution of the differential equation with the assumed properties (Fourier transformability, etc.) *need not exist* in the case that p has imaginary zeroes. This point is illustrated in an example below.

The solution form (7.29) obviously has a close connection with the solution of the corresponding Laplace transform problem discussed in Section 6.6. This problem has the form

$$p\left(\frac{d}{dt}\right) x(t) = u(t), \ t \geq 0, \tag{7.31}$$

$$x(0) = \frac{dx}{dt}(0) = \cdots = \frac{d^{n-1}x}{dt^{n-1}}(0) = 0.$$

It differs from (7.28) in that the range of the independent variable is $(0, \infty)$, and that the required initial conditions are specified. We have previously seen that the initial value problem has the unique solution

$$x(t) = \int_0^t \tilde{g}(t - \tau) u(\tau) \, d\tau, \ t \geq 0,$$

where the impulse response (or weighting pattern) function $\tilde{g}$ is determined as

$$\tilde{g}(t) = \mathcal{L}^{-1}\left\{\frac{1}{p(s)}\right\}, \ t \geq 0.$$

Since the transform involved is the Laplace transform, the solution formula may be readily evaluated with *no restriction* on the location of the zeroes of the polynomial p.

To show the connection between the two results, it is necessary to consider a problem for which both Laplace and Fourier methods are (potentially) applicable. This common ground is obtained by first restricting the forcing functions u in (7.28) to vanish identically for negative values of t, as well as to be Fourier transformable. Assuming that the polynomial p has no imaginary zeroes, the solution (7.29) obtained by Fourier transform methods takes the form

$$x(t) = \int_{-\infty}^{\infty} g(t - \tau) u(\tau) \, d\tau$$

$$= \int_0^{\infty} g(t - \tau) u(\tau) \, d\tau, \ -\infty < t < \infty$$

since $u(\tau) = 0$ for $\tau < 0$.

The natural question to ask is whether or not this result coincides with the Laplace transform version. If the function g,

$$g(t) = \mathcal{F}^{-1}\left\{\frac{1}{p(i\omega)}\right\},$$

is such that $g(t) = 0$ for $t < 0$, then the Fourier solution reduces to the form

$$x(t) = \begin{cases} 0 & t < 0, \\ \int_0^t g(t - \tau) u(\tau) \, d\tau & t \geq 0, \end{cases}$$

and the solutions computed by both methods then coincide for $t > 0$ as one might expect.

On the other hand, if $g(t)$ does not vanish for negative t, the Fourier solution takes the form

$$x(t) = \begin{cases} \int_0^\infty g(t - \tau) u(\tau) \, d\tau \neq 0 & t < 0, \\ \int_0^\infty g(t - \tau) u(\tau) \, d\tau & t \geq 0, \end{cases}$$

and the two results *differ* by an amount

$$\int_t^\infty g(t - \tau) u(\tau) \, d\tau$$

for $t > 0$. In order that the solutions should coincide[28] for $t > 0$ it is necessary and sufficient that

$$\mathcal{F}^{-1} \left\{ \frac{1}{p(i\omega)} \right\} = g(t) = 0 \text{ for } t < 0.$$

But this requirement is easily expressible in terms of the polynomial p. Considering the inversion of rational functions (Sections 7.2, 7.5 and problems below) we see that the requirement holds if and only if the zeroes of the polynomial p all have negative real parts. It is only in this case that the computation of the forced response due to the forcing function restricted to $(0, \infty)$ by use of Fourier and Laplace methods will produce the same result.

This result may appear paradoxical, but the reason for it is apparent. In the case that the polynomial p has zeroes of positive real part, the solution of the initial value problem (7.31) in general displays exponential growth with increasing t. It therefore has no Fourier transform. The calculation of the "solution" by means of the Fourier transform is based on the *assumption* that the solution being sought is Fourier transformable. It should not therefore be surprising that the results differ when that assumption is invalid.

Entirely similar arguments may be produced to explain the situation in the case that p has an imaginary zero. These points are illustrated further in the problems below.

---

[28] Recall that we assume $p(i\beta) \neq 0$, so that both exist.

It should be clear that the other Laplace transform differential equation topics discussed in Chapter 5 have counterparts in the context of Fourier transforms.

The system of first-order vector equations

$$\frac{d\mathbf{x}}{dt} = \mathbf{A}\mathbf{x} + \mathbf{B}\mathbf{u}, \quad -\infty < t < \infty, \tag{7.32}$$

may be formally solved by methods self evident from the above treatment of the single equation case. Transforming (7.32) produces successively

$$i\omega \,\hat{\mathbf{x}}(\omega) = \mathbf{A}\,\hat{\mathbf{x}}(\omega) + \mathbf{B}\,\hat{\mathbf{u}}(\omega)$$

$$\hat{\mathbf{x}}(\omega) = (i\omega\,\mathbf{I} - \mathbf{A})^{-1}\,\mathbf{B}\,\hat{\mathbf{u}}(\omega)$$

$$\mathbf{x}(t) = \mathcal{F}^{-1}\left\{(i\omega\,\mathbf{I} - \mathbf{A})^{-1}\,\mathbf{B}\,\hat{\mathbf{u}}(\omega)\right\}$$

$$= \int_{-\infty}^{\infty} \mathbf{G}(t - \tau)\,\mathbf{u}(\tau)\,d\tau,$$

where

$$\mathbf{G}(t) = \mathcal{F}^{-1}\left\{(i\omega\,\mathbf{I} - \mathbf{A})^{-1}\,\mathbf{B}\right\}(t).$$

The result may be discussed exactly as the (single equation) result (7.29) obtained above. In this case, the eigenvalues of the matrix $\mathbf{A}$ play the same role as the zeroes of the polynomial p above. These results may be directly compared with the corresponding Laplace transform calculations of Section 6.7 in a similar fashion.

The Fourier transform also finds application as a tool for the analysis of linear RLC circuits. In fact, the mechanics of this procedure follow directly from our discussion in Section 6.8.

The basic circuit relations

$$e(t) = R\,i(t) \tag{7.33}$$

$$e(t) = L\,\frac{di}{dt}$$

$$i(t) = C\,\frac{de}{dt}$$

are assumed to hold for all time, $-\infty < t < \infty$, and it is further assumed that the circuit elements are connected in a network-source configuration such that the functions of (7.33) are all Fourier transformable. The differentiation law for Fourier transforms then gives (the impedance relations).

$$\hat{e}(\omega) = R\,\hat{i}(\omega), \tag{7.34}$$

$$\hat{e}(\omega) = i\omega\,L\,\hat{i}(\omega),$$

$$\hat{i}(\omega) = i\omega\,C\,\hat{e}(\omega).$$

The impedance relation associated with (7.34) is illustrated in Figure 7.41.

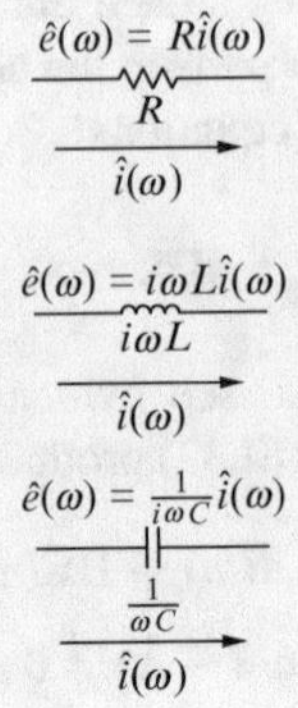

$$\underset{\displaystyle \overset{\hat{i}(\omega)}{\longrightarrow}}{\overset{\hat{e}(\omega) = R\hat{i}(\omega)}{\underset{R}{\wedge\!\!\wedge\!\!\wedge}}}$$

$$\underset{\displaystyle \overset{\hat{i}(\omega)}{\longrightarrow}}{\overset{\hat{e}(\omega) = i\omega L\hat{i}(\omega)}{\underset{i\omega L}{\text{mmm}}}}$$

$$\underset{\displaystyle \overset{\hat{i}(\omega)}{\longrightarrow}}{\overset{\hat{e}(\omega) = \frac{1}{i\omega C}\hat{i}(\omega)}{\underset{\frac{1}{\omega C}}{-\!|\,|\!-}}}$$

Fig. 7.41 Fourier domain impedances

This should be compared with the corresponding Laplace transform domain impedance diagram Figure 6.21, and the results obtained in the context of periodic steady state solutions of a circuit model, Section 2.8.

As in the case of transient analysis (Section 6.8), we conclude that Kirchoff's laws are preserved in Fourier transformed form, and that the desired solution may be determined on the same algebraic basis as previously discussed.

On an algebraic basis, the solution of the network equations is identical to that which occurs in the periodic solution case, with the sole exception that "$i\omega$" replaces "$in\omega_0$" throughout, while the Fourier transform of the source terms replaces the Fourier coefficient of the sources appearing in the periodic solution case.

The algebraic solution may also be identified in terms of the transient analysis (Laplace domain) solution of Section 6.8. The current situation differs only in that the initial condition source terms of Section 6.8 are absent, while "$i\omega$" (and Fourier transforms) replace "s" (and Laplace transforms) throughout. The solution therefore takes the form

$$\begin{bmatrix} \hat{\mathbf{i}}(\omega) \\ \hat{\mathbf{e}}(\omega) \end{bmatrix} = \mathbf{W}(i\omega)\,\hat{\mathbf{f}}(\omega),$$

expressing the network current and voltage transforms (arranged in vector form) in terms of the (vector of) transforms of the forcing function.[29]

---

[29]In the case that the network in question has only Fourier transformable solutions in response to transformable inputs, $W(i\omega)$ is referred to as the *frequency response* of the network. In the contrary case, Fourier transforms are inapplicable, and the transient analysis of Section 6.8 should be employed.

Problems 7.9.

1. Show that the function

$$x(t) = \frac{4}{3} U(t) e^{-t} + \frac{1}{3} U(-t) e^{2t} - U(t) e^{-2t}$$

actually is a solution of

$$\frac{dx}{dt} + x(t) = e^{-2|t|}, \quad -\infty < t < \infty.$$

That is, verify that x is everywhere differentiable, and that the governing equation is satisfied for each value of t, $-\infty < t < \infty$.

2. Show by explicit evaluation that the convolution integral

$$x(t) = \int_{-\infty}^{\infty} U(t - \tau) e^{-(t-\tau)} e^{-2|\tau|} \, d\tau$$

gives the same solution as in (1) above to the equation

$$\frac{dx}{dt} + x(t) = e^{-2|t|}, \quad -\infty < t < \infty.$$

3. Find (by means of Laplace transforms or more classical methods) the solution to the initial value problem

$$\frac{d^2x}{dt^2} + x(t) = f(t),$$

$$x(0) = \frac{dx}{dt}(0) = 0,$$

$$f(t) = \begin{cases} 1 & 0 \leq t < 1, \\ 0 & \text{otherwise.} \end{cases}$$

Verify that the solution of this problem is not a Fourier transformable function.

4. Solve the equation

$$\frac{d^2x}{dt^2} + 3\frac{dx}{dt} + 2x(t) = U(-t) e^{4t}, \quad -\infty < t < \infty,$$

by Fourier transform methods.

5. Suppose that u is defined by

$$u(t) = \begin{cases} 1 & 0 \leq t < 1, \\ 0 & \text{otherwise}. \end{cases}$$

Solve the initial value problems

$$\frac{dx_1}{dt} + x_1(t) = u(t), \ x_1(0) = 0,$$

$$\frac{dx_2}{dt} - x_2(t) = u(t), \ x_2(0) = 0,$$

and the "forced system" problems

$$\frac{dx_3}{dt} + x_3(t) = u(t), \ -\infty < t < \infty,$$

$$\frac{dx_4}{dt} - x_4(t) = u(t), \ -\infty < t < \infty,$$

the latter by means of Fourier transform methods.

In what cases do the solutions coincide for $t > 0$? Explain this result.

6. Consider the solution of the first-order system

$$\frac{d\mathbf{x}}{dt} = \mathbf{A}\,\mathbf{x}(t) + \mathbf{B}\,\mathbf{u}(t), \ -\infty < t < \infty,$$

by means of Fourier transforms, assuming that the forcing function $\mathbf{u}$ is Fourier transformable. Show that a solution exists under the hypothesis that no eigenvalue of $\mathbf{A}$ has a zero real part.

Show by means of an example, that such a system may have solutions which are not Fourier transformable functions in case the coefficient matrix $\mathbf{A}$ has an eigenvalue with zero real part.

7. Given the circuit of Figure 7.42 (with no stored energy), and supposing that

$$e(t) = U(t)\,e^{-t},$$

find (for all time) the capacitor voltage and inductor current in the circuit by means of (Fourier) impedance methods.

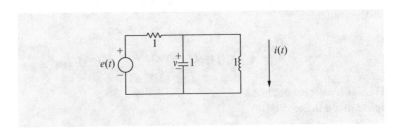

Fig. 7.42 Parallel RLC circuit

7.10 Transform Solutions of Boundary Value Problems

In previous sections we have exploited the "transform-differentiation" relations to find solutions of ordinary differential equations. It is also the case that these rules can be used to find solutions of certain partial differential equations (boundary value problems).

This is a subject which may be approached at various levels of rigor. There exist extensive theoretical treatments on the use of Fourier transforms in the solution of partial differential equations, particularly in the area of constant coefficient problems. There also exists a complete treatment of the use of Laplace transforms in connection with equations of evolution (initial value problems).

As the analytical level of these treatments is well beyond what we assume here, we treat the use of transforms below as simply a computational device. As in the case of our earlier consideration of boundary value problems, explicit solution formulas are obtained. In principle, any desired analytical properties of the solution may be verified after the fact from these explicit calculations.

In the examples below, we concentrate on boundary value problems with only two independent variables (although the basic methods readily extend to higher dimensions). The basic technique is to use a transform to "remove" the derivatives with respect to one of the variables, solve the resulting equations, and then invert the transform. The idea is entirely similar to the case encountered in ordinary differential equations, with the exception that the reduction is to a differential equation with fewer independent variables, rather than to an algebraic equation.

A further analogy with the ordinary differential case arises in the issue of the choice of an appropriate transform for a given problem. Again, an intuitive guide is found in the observation that Fourier transforms are associated with functions defined on a full line, while Laplace transforms are defined for functions whose domain is a half-line.

Example. We consider a problem of "heat conduction in a bar of infinite extent."[30] The formal problem statement is

$$\frac{\partial u}{\partial t} = \frac{\partial^2 u}{\partial x^2}, \quad t > 0, \ -\infty < x < \infty, \tag{7.35}$$

with an initial temperature distribution

$$u(x, 0) = f(x), \quad -\infty < x < \infty$$

prescribed.

---

[30]The physical interpretation of this model is that the physical extent is so large that "end effects" are negligible.

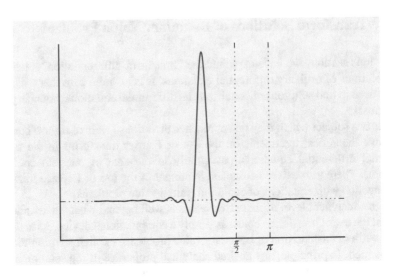

Fig. 7.43 Spatial dependence at time t_1

The solution u defines a function of two variables; for fixed $t = t_1$, the solution provides a function of the remaining variable (x) interpreted as the spatial temperature distribution at time t_1 (Figure 7.43).

Assuming that this function is Fourier transformable, we define

$$\hat{u}(\omega, t_1) = \mathcal{F}_x\{u(x, t_1)\} = \int_{-\infty}^{\infty} u(x, t_1)\, e^{-i\omega x}\, dx.$$

Applying the Fourier transform to (7.35),

$$\int_{-\infty}^{\infty} \frac{\partial u}{\partial t}(x, t)\, e^{-i\omega x}\, dx = \int_{-\infty}^{\infty} \frac{\partial^2 u}{\partial x^2}(x, t)\, e^{-i\omega x}\, dx,$$

or

$$\frac{\partial}{\partial t}\hat{u}(\omega, t) = -\omega^2\, \hat{u}(\omega, t).^{31}$$

The transformed equation is now regarded as an ordinary differential equation with respect to the time variable t. (ω is treated as a parameter.) The initial condition

---

[31]These manipulations assume sufficient smoothness of the solution sought. We make such assumptions without further apology, since they are required to carry out the formal calculations of the solution process.

statement from the original problem statement provides an initial value for the transformed problem. From the initial condition

$$u(x, 0) = f(x),$$

we obtain

$$\mathcal{F}_x \{u(x, 0)\} = \hat{u}(\omega, 0) = \mathcal{F}_x \{f(x)\} = \hat{f}(\omega).$$

The solution of the transformed equation is given by

$$\hat{u}(\omega, t) = e^{-\omega^2 t} \hat{u}(\omega, 0)$$

$$= e^{-\omega^2 t} \hat{f}(\omega).$$

The solution of the original problem is given by Fourier inversion of this expression. Since the solution transform is expressed as the product of two transforms, the inversion can be expressed as a convolution.

$$u(x, t) = \mathcal{F}_x^{-1} \left\{ e^{-\omega^2 t} \hat{f}(\omega) \right\} \tag{7.36}$$

$$= \mathcal{F}_x^{-1} \left\{ e^{-\omega^2 t} \right\} \star \mathcal{F}_x^{-1} \left\{ \hat{f}(\omega) \right\}$$

$$= \int_{-\infty}^{\infty} \mathcal{F}_x^{-1} \left\{ e^{-\omega^2 t} \right\} (x - y) f(y) \, dy.$$

The computation of $\mathcal{F}_x^{-1} \left\{ e^{-\omega^2 t} \right\}$ is essentially the computation of the transform of a Gaussian distribution function carried out explicitly in Section 7.4 above. From these results (or a suitable set of tables), we conclude that

$$\mathcal{F}_x^{-1} \left\{ e^{-\omega^2 t} \right\} = \frac{1}{2\sqrt{\pi t}} e^{-\frac{x^2}{4t}}.$$

The solution (7.36) therefore takes the form

$$u(x, t) = \int_{-\infty}^{\infty} \frac{1}{2\sqrt{\pi t}} e^{-\frac{(x-y)^2}{4t}} f(y) \, dy \tag{7.37}$$

in explicit terms.

The result may be interpreted as representing a "smearing" operation on the original temperature distribution f. The shape of the convolution kernel is illustrated in Figure 7.44. As $t \to 0^+$, the kernel tends to a delta function, and the original distribution is recovered. As t increases, the kernel flattens and spreads, leading to a smoothing and averaging of the initial distribution.

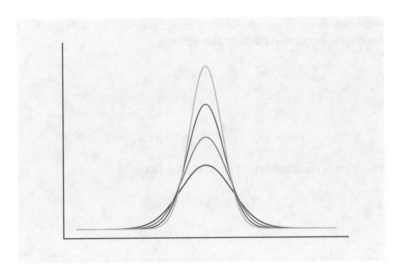

Fig. 7.44 Gaussian curves

Example. The wave equation

$$\frac{1}{c^2} \frac{\partial^2 u}{\partial t^2} = \frac{\partial^2 u}{\partial x^2}, \quad -\infty < x < \infty, \tag{7.38}$$

may be treated by a method parallel to that used above. In connection with this problem, we must also specify initial conditions for position and velocity. To simplify computations, we consider the case

$$u(x, 0) = f(x),$$

$$\frac{\partial u}{\partial t}(x, 0) = 0,$$

corresponding to zero initial velocity.

Introducing

$$\hat{u}(\omega, t) = \mathcal{F}_x \{u(x, t)\}$$

as above, we transform the governing equation to obtain

$$\int_{-\infty}^{\infty} \frac{1}{c^2} \frac{\partial^2 u}{\partial t^2}(x, t)\, e^{-i\omega x}\, dx = \int_{-\infty}^{\infty} \frac{\partial^2 u}{\partial x^2}(x, t)\, e^{-i\omega x}\, dx,$$

$$\frac{1}{c^2} \frac{\partial^2}{\partial t^2} \int_{-\infty}^{\infty} u(x, t)\, e^{-i\omega x}\, dx = -\omega^2 \int_{-\infty}^{\infty} u(x, t)\, e^{-i\omega x}\, dx.$$

This is the (ordinary differential) equation

$$\frac{1}{c^2} \frac{\partial^2}{\partial t^2} \hat{u}(\omega, t) = -\omega^2 \hat{u}(\omega, t)$$

for the transform of the solution.

Since this is a second-order ordinary differential equation, we require two initial conditions. These are obtained by transforming the given initial conditions to provide

$$\frac{\partial}{\partial t} \hat{u}(\omega, 0) = 0,$$

$$\hat{u}(\omega, 0) = \mathcal{F}_x \{u(x, 0)\} = \mathcal{F}_x \{f(x)\} = \hat{f}(\omega).$$

The general solution of the differential equation may be written in the form

$$\hat{u}(\omega, t) = A e^{i\omega ct} + B e^{-i\omega ct}$$

$$= A(\omega) e^{i\omega ct} + B(\omega) e^{-i\omega ct}.$$

The "constants" $A(\omega), B(\omega)$ are determined from the transformed initial conditions. Since

$$\frac{\partial \hat{u}}{\partial t} = i\omega c A(\omega) e^{i\omega ct} - i\omega c B(\omega) e^{i\omega ct}$$

we have

$$\frac{\partial \hat{u}}{\partial t}(\omega, 0) = i\omega c [A(\omega) - B(\omega)] = 0,$$

as well as

$$\hat{u}(\omega, 0) = A(\omega) + B(\omega) = \hat{f}(\omega).$$

These give

$$A(\omega) = B(\omega) = \frac{\hat{f}(\omega)}{2},$$

and the transform in the form

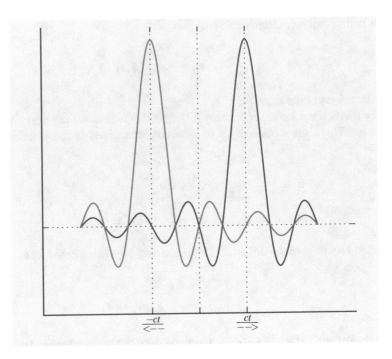

Fig. 7.45 Traveling wave propagation

$$\hat{u}(\omega, t) = \frac{\left[\hat{f}(\omega)\, e^{i\omega ct} + \hat{f}(\omega)\, e^{-i\omega ct}\right]}{2}.$$

Inverting this transform gives

$$u(x, t) = \frac{1}{2}f(x + ct) + \frac{1}{2}f(x - ct)$$

(by using the shift formula). The interpretation of this result is that the original displacement f divides into a left-traveling and right-traveling copy of itself. Geometrical interpretations are available for several other standard wave equation examples. Some of these are considered in the problems (Figure 7.45).

Example. As an example of the solution of a boundary value problem of the third standard type, we consider the problem of Laplace's equation in a half-plane. The problem statement is

$$\frac{\partial^2 \phi}{\partial x^2} + \frac{\partial^2 \phi}{\partial y^2} = 0, \ y > 0, \ -\infty < x < \infty, \tag{7.39}$$

with $\phi(x, 0) = \psi(x)$ given.

This problem may be interpreted as modeling the temperature distribution in a laterally insulated flat plate, with temperature prescribed along one edge. An

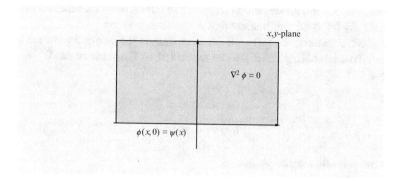

Fig. 7.46 Laplace's equation in a half-plane

alternative interpretation is that of the potential distribution in an electrostatics problem (Figure 7.46). In either case, we expect on a physical basis that the condition

$$\lim_{y\to\infty} \phi(x, y) = 0$$

holds.

This problem may be treated by the same method employed in the last two examples. Fourier transforming (7.39) with respect to x, we obtain[32]

$$\frac{\partial^2}{\partial y^2}\hat{\phi}(\omega, y) = \omega^2\,\hat{\phi}(\omega, y)$$

and

$$\hat{\phi}(\omega, 0) = \mathcal{F}_x\{\psi(x)\} = \hat{\psi}(\omega).$$

The general solution of this ordinary differential equation is

$$\hat{\phi}(\omega, y) = A(\omega)\,e^{\omega y} + B(\omega)\,e^{-\omega y}.$$

If this solution is to represent a Fourier transform (for each fixed $y > 0$), then this imposes conditions on $A(\cdot)$ and $B(\cdot)$. If $A(\omega) \neq 0$ for $\omega > 0$, then the solution

---

[32]One might attempt to use a Laplace transform with respect to the y variable, but in view of the ultimate answer, that procedure leads to unfamiliar transforms. In principle it would work.

grows as $y \to \infty$, and hence cannot represent the transform of a function satisfying the physically imposed requirement that it vanish as $y \to \infty$.

In a similar fashion, if $B(\omega) \neq 0$ for $\omega < 0$, then the solution grows as $e^{|\omega| y}$ as $y \to \infty$. This possibility must also be discarded on the same basis. We conclude then that

$$A(\omega) = 0, \ \omega > 0,$$
$$B(\omega) = 0, \ \omega < 0,$$

so that the solution may be written as

$$\hat{\phi}(\omega, y) = C(\omega) e^{-|\omega| y},$$

where $C(\cdot)$ is chosen to assume the value of either $A(\cdot)$ or $B(\cdot)$ according to the sign of ω.

The function C is determined by imposing the boundary condition

$$\hat{\phi}(\omega, 0) = C(\omega) e^0 = \hat{\psi}(\omega),$$

so that the solution takes the form

$$\hat{\phi}(\omega, y) = \hat{\psi}(\omega) e^{-|\omega| y}.$$

The representation may be inverse transformed by use of the convolution theorem to give

$$\phi(x, y) = \mathcal{F}_x^{-1} \left\{ \hat{\psi} \, e^{-|\omega| y} \right\}$$
$$= \mathcal{F}_x^{-1} \left\{ \hat{\psi} \right\} \star \mathcal{F}_x^{-1} \left\{ e^{-|\omega| y} \right\}$$
$$= \psi(x) \star \frac{y}{\pi (x^2 + y^2)}.$$

Explicitly writing out the convolution, we find that the solution takes the form

$$\phi(x, y) = \int_{-\infty}^{\infty} \frac{y}{\pi \left((x - \zeta)^2 + y^2 \right)} \psi(\zeta) \, d\zeta. \tag{7.40}$$

The solution (7.40) is a representation of the (potential) ϕ in terms of its boundary value ψ. This is similar to some results obtained earlier on examples of finite range boundary value problems. Such representations are widely used in connection with systematic treatments of related problems.

Example. The integral transform method lends itself readily to the solution of boundary value problems with forcing functions. As an example, we modify the diffusion equation problem treated above to include a source term. The model is

$$\frac{\partial u}{\partial t} = \kappa \frac{\partial^2 u}{\partial x^2} + g(x, t) \tag{7.41}$$

$$u(x, 0) = f(x).$$

Transforming with respect to x (as usual) provides

$$\frac{\partial \hat{u}(\omega, t)}{\partial t} = -\omega^2 \hat{u}(\omega, t) + \mathcal{F}_x\{g(x, t)\}$$

$$= -\omega^2 \hat{u}(\omega, t) + \hat{g}(\omega, t),$$

$$\hat{u}(\omega, 0) = \hat{f}(\omega).$$

The system is a (forced) first-order ordinary differential equation with an initial condition at time $t = 0$. The solution takes the form[33]

$$\hat{u}(\omega, t) = e^{-\omega^2 t} \hat{f}(\omega) + \int_0^t e^{-\omega^2 (t-\tau)} \hat{g}(\omega, \tau) \, d\tau.$$

The solution of the original problem follows by inverse transforming this expression.
We have

$$u(x, t) = \mathcal{F}_x^{-1}\left\{e^{-\omega^2 t} \hat{f}(\omega)\right\} + \mathcal{F}_x^{-1}\left\{\int_0^t e^{-\omega^2 (t-\tau)} \hat{g}(\omega, \tau) \, d\tau\right\}$$

$$= \int_{-\infty}^{\infty} \frac{1}{2\sqrt{\pi t}} e^{-\frac{(x-\zeta)^2}{4t}} f(\zeta) \, d\zeta + \mathcal{F}_x^{-1}\left\{\int_0^t e^{-\omega^2 (t-\tau)} \hat{g}(\omega, \tau) \, d\tau\right\},$$

where we invoke the inversion result from our previous example to evaluate the first term. To evaluate the second inversion, we express the inversion in terms of an integral and interchange the order of integration. This gives in succession

$$\mathcal{F}_x^{-1}\left\{\int_0^t e^{-\omega^2 (t-\tau)} \hat{g}(\omega, \tau) \, d\tau\right\}$$

$$= \frac{1}{2\pi} \int_{-\infty}^{\infty} e^{i\omega x}\left\{\int_0^t e^{-\omega^2 (t-\tau)} \hat{g}(\omega, \tau) \, d\tau\right\} d\omega$$

$$= \int_0^t \frac{1}{2\pi} \int_{-\infty}^{\infty} e^{i\omega x}\left\{e^{-\omega^2 (t-\tau)} \hat{g}(\omega, \tau)\right\} d\omega \, d\tau$$

---

[33]This is a special case of the linear, constant coefficient problems of Section 6.6.

$$= \int_0^t \mathcal{F}_x^{-1}\left\{e^{-\omega^2(t-\tau)}\,\hat{g}(\omega,\tau)\right\}\,d\tau$$

$$= \int_0^t \left[\mathcal{F}_x^{-1}\left\{e^{-\omega^2(t-\tau)}\right\} \star \mathcal{F}_x^{-1}\{\hat{g}(\omega,\tau)\}\right]\,d\tau$$

$$= \int_0^t \int_{-\infty}^{\infty} \frac{1}{2\sqrt{\pi(t-\tau)}}\,e^{-\frac{(x-\zeta)^2}{4(t-\tau)}}\,g(\zeta,\tau)\,d\zeta\,d\tau.$$

The solution then takes the form

$$u(x,t) = \int_{-\infty}^{\infty} \frac{1}{2\sqrt{\pi t}}\,e^{-\frac{(x-\zeta)^2}{4t}}\,f(\zeta)\,d\zeta$$

$$+ \int_0^t \int_{-\infty}^{\infty} \frac{1}{2\sqrt{\pi(t-\tau)}}\,e^{-\frac{(x-\zeta)^2}{4(t-\tau)}}\,g(\zeta,\tau)\,d\zeta\,d\tau, \qquad (7.42)$$

expressing the solution as a linear combination of an initial condition response and a weighted sum of the past spatial distribution of the forcing function.

Example. It was suggested above that Laplace transform methods are a natural candidate for the solution of boundary value problems with an evolution equation interpretation. We illustrate this by producing the solution of a standard heat equation by use of a Laplace transform technique. This approach produces the result directly, without any recourse to the separation of variables methods of Chapters 3 and 4. The price associated with this service is that the required computations are somewhat tedious.

The problem in question is given by

$$\frac{\partial u}{\partial t} = \frac{\partial^2 u}{\partial x^2}, \quad t > 0,\ 0 < x < 1 \qquad (7.43)$$

with boundary and initial conditions

$$u(0,t) = u(1,t) = 0,$$

$$u(x,0) = f(x).$$

We use a Laplace transform with respect to t to convert (7.43) to an ordinary differential equation. Using

$$U(x,s) = \mathcal{L}_t\{u(x,t)\} = \int_0^{\infty} u(x,t)\,e^{-st}\,dt,$$

the problem becomes

$$s\,U(x, s) - u(x, 0) = \frac{\partial^2}{\partial x^2}\,U(x, s),$$

or,

$$\frac{\partial^2}{\partial x^2}\,U(x, s) - s\,U(x, s) = -f(x),$$

introducing the initial condition. The equation is a second-order ordinary differential equation (with respect to x), with $-f$ as forcing function, and s as a parameter. The solution of the differential equation must be found subject to the (Laplace transformed version of the) boundary conditions

$$U(0, s) = \mathcal{L}_t\{u(0, t)\} = \mathcal{L}_t\{u(1, t)\} = 0.$$

The equation may be solved by any of a variety of the methods for ordinary differential equations. We elect to use a Laplace transform with respect to x. We use λ as the transform variable, and the notation

$$\mathcal{U}(\lambda, s) = \mathcal{L}_x\{U(x, s)\} = \int_0^\infty U(x, s)\,e^{-\lambda x}\,dx^{34}$$

for the twice transformed solution. Application of the transform gives

$$\lambda^2\,\mathcal{U}(\lambda, s) - \lambda U(0, s) - U'(0, s) - s\,\mathcal{U}(\lambda, s) = -\mathcal{L}_x\{f(x)\}$$

and

$$\mathcal{U}(\lambda, s) = \frac{U'(0, s)}{\lambda^2 - s} - \frac{\mathcal{L}_x\{f(x)\}}{\lambda^2 - s}.$$

The unknown initial value $U'(0, s)$ is determined by inverse transforming this representation, and applying the remaining boundary condition at $x = 1$ to the result. Using the fact that

---

[34]Even though we seek U only on the interval $[0, 1]$, it is convenient to extend the solution beyond $x = 1$ in order to use the transform.

$$\mathcal{L}_x^{-1}\left\{\frac{1}{\lambda^2-s}\right\} = \frac{1}{\sqrt{s}}\sinh\sqrt{s}\,x, \,^{35}$$

inverting produces

$$U(x,s) = U'(0,s)\frac{\sinh\sqrt{s}x}{\sqrt{s}} - \int_0^x \frac{\sinh\sqrt{s}(x-\zeta)}{\sqrt{s}}f(\zeta)\,d\zeta.$$

Since $U(1,s) = 0$, this can be solved for $U'(0,s)$. It provides

$$U'(0,s) = \int_0^1 \frac{\sinh\sqrt{s}(1-\zeta)}{\sinh\sqrt{s}}f(\zeta)\,d\zeta,$$

and gives the transform U in the explicit form

$$U(x,s)$$

$$= \frac{\sinh\sqrt{s}x}{\sqrt{s}\,\sinh\sqrt{s}}\int_0^1 \sinh\sqrt{s}\,(1-\zeta)f(\zeta)\,d\zeta - \sinh\sqrt{s}\int_0^x \frac{\sinh\sqrt{s}(x-\zeta)}{\sqrt{s}\,\sinh\sqrt{s}}f(\zeta)\,d\zeta$$

$$= \frac{\sinh\sqrt{s}x}{\sqrt{s}\,\sinh\sqrt{s}}\int_x^1 \sinh\sqrt{s}\,(1-\zeta)f(\zeta)\,d\zeta$$

$$+ \int_0^x \frac{\sinh\sqrt{s}x\,\sinh\sqrt{s}(1-\zeta) - \sinh\sqrt{s}\,\sinh\sqrt{s}(x-\zeta)}{\sqrt{s}\,\sinh\sqrt{s}}f(\zeta)\,d\zeta.$$

(This latter form is useful for subsequent calculations. The reason for this manip-ulation becomes evident when the inversion of the result by use of the complex inversion integral is contemplated.)

We complete the solution of the problem by calculating the inverse transform.

The complex form of the Laplace inversion integral takes the form

$$u(x,t) = \frac{1}{2\pi}\int_{\sigma-i\infty}^{\sigma+i\infty} U(x,s)\,e^{st}\,ds,$$

where σ is chosen so that the integration is carried out with the transform $U(x,s)$ analytic on and to the right of the path of integration.

Anticipating an inversion by means of residue calculations, we consider the singularities of the transform $U(x,s)$. In view of the formula, there is a possibility of a singularity at $s = 0$. However, in a neighborhood of $s = 0$, calculations of the form

---

[35]This expression apparently contains a branch cut, when regarded as a function of the complex variable s. Consideration of the power series shows, however, that the result is actually an entire function of s with no branch cut required.

$$\frac{\sinh \sqrt{s}x \sinh \sqrt{s}(1 - \zeta)}{\sqrt{s} \sinh \sqrt{s}} = \frac{\left(x + \frac{x^3 s}{3!} + \cdots\right)\left((1 - \zeta) + (1 - \zeta)^3 \frac{s}{3!} + \cdots\right)}{1 + \frac{s}{3!} + \cdots}$$

show that $U(x, s)$ is actually analytic at $s = 0$.

The other candidates for singularities of $U(x, s)$ arise from values of s such that

$$\sinh \sqrt{s} = 0,$$

that is for

$$\sqrt{s} = in\pi, \; n = \pm1, \pm2, \dots, \tag{7.44}$$

$$s = -\left(n^2\pi^2\right), \; n = 1, 2, 3, \dots.$$

In view of the fact that the zeroes of the complex sine function are simple, and considering the form of the expansion for the $s = 0$ case, we suspect that the singularities of the transform $U(x, s)$ at the points given by (7.44) are simple first-order poles. To verify this hypothesis, we attempt to compute the residue associated with $U(x, s)$ at each of the points $s = -n^2\pi^2$ by evaluating

$$\lim_{s \to -n^2\pi^2} \left\{\left(s + n^2\pi^2\right) U(x, s)\right\}.$$

In order to evaluate this, it suffices to compute the less menacing limit

$$\lim_{s \to -n^2\pi^2} \frac{s + n^2\pi^2}{\sqrt{s} \sinh \sqrt{s}}$$

since the other factors involved in $U(x, s)$ are entire functions of s, in particular analytic at $s = -n^2\pi^2$.

The limit is readily evaluated by means of L'Hopital's rule.[36] We have

$$\frac{\frac{d}{ds}(s + n^2\pi^2)}{\frac{d}{ds}\sqrt{s} \sinh \sqrt{s}} = \frac{1}{\frac{1}{2\sqrt{s}} \sinh \sqrt{s} + \sqrt{s}\cosh \sqrt{s}\frac{1}{2\sqrt{s}}},$$

so that

$$\lim_{s \to -n^2\pi^2} \frac{s + n^2\pi^2}{\sqrt{s} \sinh \sqrt{s}} = \frac{1}{\frac{\cosh in\pi}{2}} = 2(-1)^n.$$

---

[36] Strictly speaking, in order to apply L'Hopital's rule, we define $\sqrt{s}$ as an analytic function by installing a branch cut in the s-plane, avoiding the negative real axis.

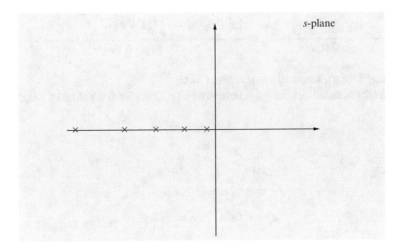

Fig. 7.47 Heat equation poles

This computation verifies that the singularities of $U(x, s)$ are in fact simple poles at the points $s = -(n\pi)^2$, $n = 1, 2, 3, \ldots$. The locations of these singularities are illustrated in Figure 7.47.

The fact that the transform $U(x, s)$ has no singularity at $s = 0$ makes it possible to take $\sigma = 0$ in the complex inversion integral, and to calculate the solution as

$$U(x, t) = \lim_{B \to \infty} \int_{-iB}^{iB} U(x, s) \, e^{st} \, ds.$$

This is evaluated by means of the usual contour integral methods. The line integral is replaced by a closed contour created by the addition of a semicircular arc in the left half-plane.

We consider a sequence of such contours $\{\Gamma_{B_N}\}$ located so that the arc passes half way between the points $s_N = -(N^2\pi^2)$ and $s_{N+1} = -\big((N+1)^2\pi^2\big)$. See Figure 7.47.

Since $U(x, s)$ has simple poles at each of the points $s = -(n\pi)^2$, we have

$$\frac{1}{2\pi i} \oint_{\Gamma_{B_N}} U(x, s) \, e^{st} \, ds = \sum_{n=1}^{N} \text{Residue} \left(U(x, s) \, e^{st} \Big|_{s \, = \, -(n\pi)^2} \right)$$

by an application of the residue theorem.

Assuming that the calculation of the inversion may be replaced by a limit of the contour integrals, we obtain[37]

$$u(x, t) = \lim_{N \to \infty} \sum_{n=1}^{N} \text{Residue}\left(U(x, s)\, e^{st} \Big|_{s = -(n\pi)^2} \right).$$

To complete the problem, it remains to evaluate the residues in this expression, noting that the essence of the required calculation was made above. Using our previous observation that the hyperbolic sine terms in the numerators of the transform provide entire functions of s, a similar observation that e^{st} is entire, and the preliminary calculation, we compute

$$\text{Residue}\left(U(x, s)\, e^{st} \Big|_{s = -(n\pi)^2} \right)$$

$$= e^{-(n\pi)^2 t} 2(-1)^n \int_x^1 \{\sinh(in\pi\, x)\, \sinh(in\pi\, (1 - \zeta)) f(\zeta)\}\, d\zeta$$

$$+ e^{-(n\pi)^2 t} 2(-1)^n \int_0^x \{\sinh(in\pi\, x)\, \sinh(in\pi\, (1 - \zeta))$$

$$- \sinh(in\pi)\, \sinh(in\pi\, (x - \zeta))\}\, f(\zeta)\, d\zeta$$

$$= e^{-(n\pi)^2 t} 2(-1)^n \sin(n\pi\, x) \int_0^1 i^2 \cos(n\pi)\, \sin(-n\pi\zeta) f(\zeta)\, d\zeta$$

$$= e^{-(n\pi)^2 t} \sin(n\pi\, x) \left\{ 2 \int_0^1 \sin(n\pi\zeta)\, d(\zeta)\, d\zeta \right\}.$$

The solution therefore takes the form

$$u(x, t) = \sum_{n=1}^{\infty} \left\{ 2 \int_0^1 \sin(n\pi\zeta) f(\zeta)\, d\zeta \right\} e^{-(n\pi)^2 t} \sin(n\pi\, x). \tag{7.45}$$

This result is readily recognized as the usual form obtained by separation of variable methods.

Example. In view of the tedious nature of some of the calculations encountered in the previous example, the reader may view the Laplace transform method as an overly masochistic approach to the problem in question. It should be borne in mind that the process involved is at least predictably mechanical, if a bit dirty in spots.

---

[37]The validity of the procedure relies on considerations of the sort used in Section 7.6, substantially complicated by the "x" dependence of the integrand. As the computations of this section are being made on a formal basis, we casually sidestep this issue.

In Section 4.6 some boundary value problems were discussed which involved the presence of forcing terms in the governing equations and boundary conditions. A prototype problem of this sort is

$$\frac{\partial u}{\partial t} = \kappa \frac{\partial^2 u}{\partial x^2} + g(x, t), \ 0 < x < 1,$$

$$u(0, t) = f_0(t),$$

$$u(l, t) = f_1(t),$$

$$u(x, 0) = u_0(x).$$

This problem can be treated in principle in the same fashion as the previous example. Laplace transformation of the equations gives

$$s\, U(x, s) - u_0(x) = \frac{\partial^2}{\partial x^2} U(x, s) + G(x, s),$$

$$U(0, s) = F_0(s),$$

$$U(1, s) = F_1(s).$$

This system differs from the corresponding equation of the previous example only in the presence of an additional forcing term and inhomogeneous boundary conditions. The mechanical aspects of the solution of such a problem are not essentially more complicated than the previous case.

We ask the reader to verify his understanding of the previous example by becoming convinced that the results computed in Chapter 4 follow from a relentless pursuit of the problem by means of Laplace transforms.

Problems 7.10.

1. A model for diffusion in a moving medium is given by

$$\frac{\partial u}{\partial t} = \kappa \frac{\partial^2 u}{\partial x^2} - V_0 \frac{\partial u}{\partial x}, \ -\infty < x < \infty,$$

$$u(x, 0) = u_0(x), \ \text{given}.$$

Find a solution of this problem.

2. A wave equation in a region of infinite extent, and for which an initial velocity (but no initial displacement) is prescribed has the model

$$\frac{1}{c^2} \frac{\partial^2 u}{\partial t^2} = \frac{\partial^2 u}{\partial x^2}, \ -\infty < x < \infty,$$

$$u(x, 0) = 0,$$

$$\frac{\partial u}{\partial t}(x, 0) = g(x).$$

Find the solution of this problem, and by superimposing the solution obtained in the text, a solution of

$$\frac{1}{c^2} \frac{\partial^2 u}{\partial t^2} = \frac{\partial^2 u}{\partial x^2}, -\infty < x < \infty,$$

$$u(x, 0) = f(x),$$

$$\frac{\partial u}{\partial t}(x, 0) = g(x).$$

3. Consider the problem of heat conduction in a semi-infinite bar, insulated at $x = 0$. The governing equation is

$$\frac{\partial u}{\partial t} = \kappa \frac{\partial^2 u}{\partial x^2}, x > 0, t > 0,$$

$$\frac{\partial u}{\partial x}(0, t) = 0,$$

$$u(x, 0) = f(x), x > 0.$$

Hint: This is solved most readily by employing a symmetry argument (analogous to the half-range expansions encountered in finite range boundary value problems) in conjunction with the already derived full line solution.
4. Solve the problem of heat conduction in a semi-infinite bar, under the conditions that

$$u(0, t) = 0.$$

Hint: See Problem 3 above.
5. Use the symmetry method of Problems 3 and 4 above to solve the following two semi-infinite wave equations.

a.

$$\frac{1}{c^2} \frac{\partial^2 u}{\partial t^2} = \frac{\partial^2 u}{\partial x^2}, x > 0, t > 0,$$

$$u(0, t) = 0,$$

$$u(x, 0) = f(x),$$

$$\frac{\partial u}{\partial t}(x, 0) = 0.$$

b.

$$\frac{1}{c^2} \frac{\partial^2 u}{\partial t^2} = \frac{\partial^2 u}{\partial x^2}, x > 0, t > 0,$$

$$\frac{\partial u}{\partial x}(0, t) = 0,$$

$$u(x, 0) = f(x),$$

$$\frac{\partial u}{\partial t}(x, 0) = 0.$$

6. Provide a geometrical interpretation of the solution of

$$\frac{1}{c^2}\frac{\partial^2 u}{\partial t^2} = \frac{\partial^2 u}{\partial x^2}, x > 0, t > 0,$$

$$\frac{\partial u}{\partial x}(0, t) = 0,$$

$$u(x, 0) = f(x),$$

$$\frac{\partial u}{\partial t}(x, 0) = 0.$$

in terms of traveling and reflected waves.

7. A standard heat equation problem is given by

$$\frac{\partial u}{\partial t} = \kappa \frac{\partial^2 u}{\partial x^2}, 0 < x < 1, t > 0,$$

$$u(x, 0) = f(x),$$

$$u(0, t) = 0,$$

$$\frac{\partial u}{\partial x}(1, t) = 0.$$

Find the solution of this problem by means of Laplace transformation with respect to the time variable.

8. Use Laplace transformation with respect to time to solve the inhomogeneous heat equation problem

$$\frac{\partial u}{\partial t} = \kappa \frac{\partial^2 u}{\partial x^2}, 0 < x < 1, t > 0,$$

$$u(x, 0) = f(x),$$

$$u(0, t) = T_0,$$

$$u(1, t) = T_1.$$

Hint: This is a minor modification of the text example.

7.11   Band-limited Functions and Communications

Concepts and methods from Fourier transform theory play a central role in various problems of communication theory.

The basic problem of communications is to transmit a message across some distance to an intended receiver. The usual outline of the process is given in Figure 7.48.

A typical message to be sent is encoded on a transmitted signal by a process known as modulation. This is necessary in order that the transmitted signal should have the physical characteristics required in order that it should propagate through the medium (referred to as the "channel") connecting the originator of the message with the intended recipient.

At the receiving end, a received message is produced by processing the received signal through a demodulator, constructed to extract (an estimate of) the message from that part of the transmitted signal appearing at the receiving terminal.[38]

As indicated above, the applications of Fourier analysis in communications are numerous, and occur at essentially all levels of sophistication associated with the subject. Out of all of these, we select for discussion a topic whose discussion is self-contained, but which finds wide application.

This topic arises in the modeling of the message (i.e., leftmost) portion of the basic diagram 7.48.

The message typically consists of a speech waveform, or some other data arising sequentially in time (as the output of some measuring instrument). Assuming that the system 7.48 is to operate continuously in time, the messages may be identified with functions f defined on the (time) interval $(-\infty, \infty)$. On the basis of considerations of the physical processes generating the message, it is natural to assume that f has finite power. That is

$$\int_{-\infty}^{\infty} |f(t)|^2 \, dt < \infty.$$

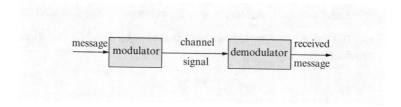

Fig. 7.48 A communication system

---

[38]This simple model is in practice augmented by models for stochastic noise (interference) and channel distortion. In such models, the messages are also modeled as stochastic (rather than deterministic) functions.

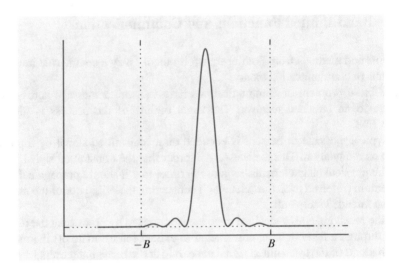

Fig. 7.49 Band-limited spectrum

These considerations immediately identify "messages" with functions to which the Fourier transform theory may be applied.

Out of all the potential messages restricted by the finite energy constraint, one class of messages of particular interest are the "band-limited messages." These are motivated by an appeal to Parseval's Theorem. We recall that Parseval's equality

$$\int_{-\infty}^{\infty} |f(t)|^2 \, dt = \frac{1}{2\pi} \int_{-\infty}^{\infty} |\hat{f}(\omega)|^2 \, d\omega,$$

has the intuitive interpretation that the magnitude squared of the Fourier transform provides the frequency distribution of the power in the message.

Physical sources have the property that their power is largely concentrated in the lower frequencies (e.g., speech and music are essentially concentrated in a band below 20 KHz.). See Figure 7.49.

A useful idealization of this observation is the concept of a band-limited function.

Definition. The square-integrable function f is said to be *band-limited* to the band $[-B, B]$, provided that

$$\hat{f}(\omega) \equiv 0 \text{ for } |\omega| > B.$$

That is, the power in f is entirely concentrated in the (frequency) range $|\omega| < B$.

The utility of this idealization is illustrated by the example of an amplitude modulation scheme. In such a system, a band-limited message f is modulated for transmission by multiplication by a sinusoid at the carrier frequency, ω_c. That is, the transmitted signal is given by

$$s_f(t) = \cos(\omega_c t) f(t).^{39}$$

The Fourier transform of the transmitted signal is

$$\hat{s}_f(\omega) = \mathcal{F}\left\{ \frac{e^{i\omega_c t} + e^{-i\omega_c t}}{2} f(t) \right\}$$

$$= \frac{1}{2} \left\{ \hat{f}(\omega - \omega_c) + \hat{f}(\omega + \omega_c) \right\}.$$

Assuming that the carrier frequency ω_c greatly exceeds the message bandwidth B, the power density of the transmitted signal is as illustrated in Figure 7.50.

If a second band-limited message g is to be transmitted over the same channel, it may be modulated as the original signal, but at a different carrier frequency Ω_c. Then the second transmitted signal takes the form

$$s_g(t) = \cos(\Omega_c t) g(t),$$

$$\hat{s}_g(\omega) = \frac{1}{2} [\hat{g}(\omega - \Omega_c) + \hat{g}(\omega + \Omega_c)].$$

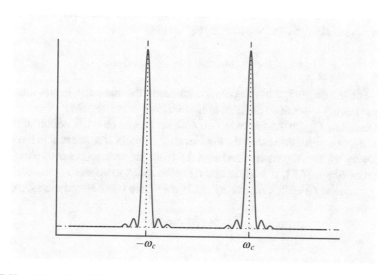

Fig. 7.50 AM modulated signal

---

[39]This is double sideband suppressed carrier amplitude modulation (DSBSC-AM). Commercial AM transmits $(1 + mf(t)) \cos(\omega_c t)$, and so contains a term at the carrier frequency.

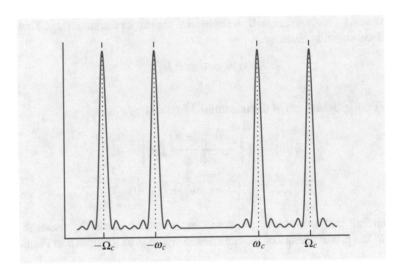

Fig. 7.51 Received signal power density

The receiver accepts the sum of these f and g transmissions in the form

$$r(t) = s_f(t) + s_g(t),$$
$$\hat{r}(\omega) = \hat{s}_f(\omega) + \hat{s}_g(\omega).$$

The power density of r takes the form of Figure 7.51. provided that the carrier frequencies satisfy the band separation requirement

$$|\Omega_c - \omega_c| > 2B,$$

that is, the carrier frequency separation exceeds the message bandwidth. (This argument readily extends to the case of more than two messages.)

The result of Figure 7.51 is based on nothing deeper than the notion of a band-limited function, combined with the Fourier shift formula. These are also the keys to the recovery of, say, the transmitted signal f from the received transmission. From the form of Figure 7.51, it is clear that f can be made to appear at the receiver by shifting frequency distribution of r by ω_c. If we define (the heterodyned signal) by

$$h(t) = r(t) \cos \omega_c t$$

then

$$h(t) = \cos^2(\omega_c t) f(t) + \cos(\omega_c t) \cos(\Omega_c t) g(t)$$
$$= \frac{1}{2} f(t) + \frac{1}{2} \cos(2\omega_c t) f(t) + \cos(\omega_c t) \cos(\Omega_c t) g(t)$$

$$\hat{h}(\omega) = \frac{1}{2}\hat{f}(\omega) + \frac{1}{4}\hat{f}(\omega + 2\,\omega_c) + \frac{1}{4}\hat{f}(\omega - 2\,\omega_c)$$

$$+ \frac{1}{2}\,[\hat{g}(\omega + \Omega_c + \omega_c) + \hat{g}(\omega - \Omega_c - \omega_c)$$

$$+ \hat{g}(\omega - \Omega_c + \omega_c) + \hat{g}(\omega + \Omega_c - \omega_c)]$$

We recall that a time domain convolution corresponds to a frequency domain multiplication. One half of the desired f may be recovered from the heterodyned signal by multiplying the transform by a function $\hat{w}(\omega)$, defined by

$$\hat{w}(\omega) = \begin{cases} 1 & |\omega| < B, \\ 0 & |\omega| > B. \end{cases}$$

Then

$$\hat{w}(\omega)\,\hat{h}(\omega) = \frac{\hat{w}(\omega)}{2}\,\hat{f}(\omega)$$

$$= \frac{1}{2}\hat{f}(\omega).$$

This means that

$$f(t) = 2\,\mathcal{F}^{-1}\left\{\hat{w}(\omega)\,\hat{h}(\omega)\right\} \tag{7.46}$$

$$= 2 \int_{-\infty}^{\infty} w(t - \tau)\,h(\tau)\,d\tau, \tag{7.47}$$

where $w(t) = \mathcal{F}^{-1}\{\hat{w}(\omega)\} = \frac{1}{\pi}\frac{\sin(Bt)}{t}$. [40]

A "block diagram" of the overall scheme is given in Figure 7.52.

The example discussed above illustrates the utility of band-limited functions in the description of certain modulation schemes. These functions also play a key conceptual role in the design of various digital communication systems. [41]

The central result underlying such schemes is known as the *Sampling Theorem* (for band-limited functions). This theorem may be readily derived on the basis of elementary notions of Fourier integrals and series.

The intuitive notion of a band-limited function is that it is composed entirely of low frequency components. Accordingly, a band-limited function is expected to vary

---

[40] The function w is referred to as the impulse response of an "ideal low pass filter." Since (7.46) as it stands requires future values of the received signal r to compute the present message value, the processing in practice is replaced by a physically realizable approximation to the ideal low pass filter.

[41] Examples include digital telephony, voice compression, and digital media recordings.

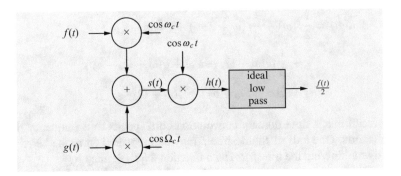

Fig. 7.52 Two AM signals and receiver

slowly, and to contain (transmit) considerably less information than is contained in an arbitrary square-integrable function. The quantitative demonstration of the truth of this intuition is the content of the Sampling Theorem.

To derive this result, consider a square-integrable function f, band-limited to the band $[-B, B]$. Then, by Parseval's equality, it follows that the integral

$$\int_{-B}^{B} |\hat{f}(\omega)|^2 \, d\omega < \infty$$

is convergent as indicated. In view of the finite energy of the band-limited signal, it is possible to represent f over the interval $[-B, B]$ by means of a Fourier series of the form

$$\hat{f}(\omega) = \sum_{n=-\infty}^{\infty} c_n e^{inT\omega}, \quad -B < \omega < B. \tag{7.48}$$

Here T denotes the "fundamental frequency" associated with the expansion interval $[-B, B]$. Since the expansion variable ω has the interpretation as a frequency, T has the dimension "time," motivating the choice of notation.

The expansion holds in the mean-square sense over the interval $[-B, B]$. Since f by hypothesis vanishes outside of $[-B, B]$, the expansion can be extended to all values of ω by defining a collection of functions $\{u_n\}$, where

$$u_n(\omega) = \begin{cases} e^{inT\omega} & |\omega| \leq B, \\ 0 & |\omega| > B. \end{cases}$$

Then we have the expansion

$$\hat{f}(\omega) = \sum_{n=-\infty}^{\infty} c_n u_n(\omega), \tag{7.49}$$

valid in the mean-square sense over $-\infty < t < \infty$. That is,

$$\lim_{N \to \infty} \int_{-\infty}^{\infty} \left| \hat{f}(\omega) - \sum_{n=-N}^{N} c_n u_n(\omega) \right|^2 d\omega = 0. \tag{7.50}$$

The expansion coefficients $\{c_n\}$ are computed from the standard Fourier series coefficient formula appropriate to the range $[-B, B]$.

$$c_n = \frac{1}{2B} \int_{-B}^{B} \hat{f}(\omega) e^{-in T \omega} d\omega.$$

This computation is readily identifiable in terms of the Fourier inversion integral. Since

$$f(t) = \frac{1}{2\pi} \int_{-\infty}^{\infty} \hat{f}(\omega) e^{i\omega t} d\omega,$$

$$= \frac{1}{2\pi} \int_{-B}^{B} \hat{f}(\omega) e^{i\omega t} d\omega,$$

(since $\hat{f}$ vanishes elsewhere) we have

$$c_n = \frac{\pi}{B} f(-n T). \tag{7.51}$$

The expansion therefore takes the form

$$\hat{f}(\omega) = \sum_{n=-\infty}^{\infty} \frac{\pi}{B} f(-n T) e^{in T \omega}$$

$$= \sum_{k=-\infty}^{\infty} \frac{\pi}{B} f(k T) e^{-ik T \omega}, \quad -B < \omega < B.$$

The corresponding whole axis expansion (7.49) is therefore

$$\hat{f}(\omega) = \sum_{k=-\infty}^{\infty} \frac{\pi}{B} f(k T) u_{-k}(\omega), \quad -\infty < \omega < \infty.$$

The sampling theorem now follows from a term-by-term inversion of this expansion. Such a process is justified by an appeal to Parseval's Theorem, and the mutual orthogonality of the functions $\{u_n\}$ as a function of ω. Since (7.49) is a mean-square convergent orthogonal expansion, the inverse transform of a partial sum of the series defined by

$$f_N(t) = \mathscr{F}^{-1}\left\{ \sum_{k=-N}^{N} \frac{\pi}{B} f(kT)\, u_{-k}(\omega) \right\}$$

converges to f in the mean-square sense as $N \to \infty$.

But by explicit calculation of the inversion integral we have

$$\mathscr{F}^{-1}\left\{ u_{-k}(\omega) \right\}(t) = \frac{1}{2\pi} \int_{-B}^{B} e^{-ikT\omega}\, e^{i\omega t}\, d\omega$$

$$= \frac{1}{\pi} \frac{\sin B(t - kT)}{(t - kT)},$$

and so

$$f_N(t) = \sum_{k=-N}^{N} f(kT) \frac{\sin B\,(t - kT)}{B\,(t - kT)}.$$

Since (in the mean-square sense)

$$f = \lim_{N\to\infty} f_N,$$

we obtain

$$f(t) = \sum_{k=-\infty}^{\infty} f(kT) \frac{\sin B\,(t - kT)}{B\,(t - kT)}. \qquad (7.52)$$

Theorem (Sampling Theorem). *Suppose that f is square-integrable and band-limited to the band $[-B, B]$. Then f is completely determined by its values at the equally spaced set of points $\{kT\}$. The function f is reconstructed from these sample values by the expansion*

$$f(t) = \sum_{k=-\infty}^{\infty} f(kT) \frac{\sin B\,(t - kT)}{B\,(t - kT)}.$$

This expansion is mean square convergent in the sense that

$$\lim_{N\to\infty} \int_{-\infty}^{\infty} \left| f(t) - \sum_{k=-N}^{N} f(kT) \frac{\sin B\,(t - kT)}{B\,(t - kT)} \right|^2 dt = 0.$$

It is of interest to consider the form of the functions appearing in the expansion (7.52). These are illustrated in Figure 7.53. Since the sampling time is

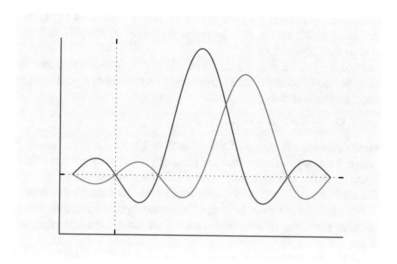

Fig. 7.53 Sampling expansion functions

determined from the usual "fundamental frequency" relation for Fourier series
expansions as

$$T = \frac{2\pi}{2B} = \frac{\pi}{B},$$ (7.53)

we note that for integral n

$$\frac{\sin B(nT - kT)}{B(nT - kT)} \equiv 0, \; n \neq k.$$

At the sample points, therefore, the expansion (7.52) reduces to a single term, with
a coefficient of 1, the limiting value of the sampling function $\frac{\sin B(t - nT)}{B(t - nT)}$ as $t \to nT$.

It is also conventional to express the sampling time information in terms of the
frequency of sampling required. Since B above is computed in terms of radian
measure, B is expressible as

$$B = 2\pi f$$

where f denotes the associated frequency in Hz. The relation is therefore

$$T = \frac{1}{2f}, \; \frac{1}{T} = 2f,$$

and the sampling rate is twice the highest frequency present in the band-limited signal f. The sampling rate is also the reciprocal of the bandwidth (in Hz). This is referred to as the "Nyquist rate." .

In practice, this result plays the role of a lower limit to the required sampling rate. The expansion also in practice is replaced by a realizable approximation rather than a strict implementation.

Problems 7.11.

1. The description given above of the process in Figure 7.54 for recovery of the transmitted message assures that the transmitter and receiver have access to a common clock (so that each may generate the function $\cos \omega_c t$).

    The scheme is referred to as synchronous demodulation for this reason.

    If there is a timing error between the transmitter and receiver, the receiver implements the above scheme with a shifted sinusoid in the demodulator shown in Figure 7.55.

    Show that in this case the recovered message is diminished by a factor of $\cos(\alpha)$ over the result in the ideal case.
2. Suppose m is a periodic function of period T (fundamental frequency ω_c) with finite average power

$$< m >^2 = \frac{1}{T} \int_0^T |m(t)|^2 \, dt.$$

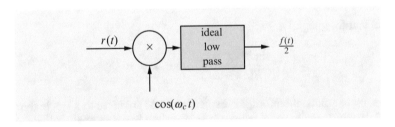

Fig. 7.54 Phase locked AM demodulator

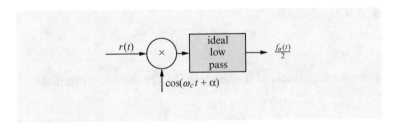

Fig. 7.55 A phase shifted demodulator

Let f be a band-limited function, limited to the band $[-B, B]$, and suppose that the total energy in f is

$$E_f^2 = \int_{-\infty}^{\infty} |f(t)|^2 \, dt.$$

Assume finally that $\omega_c > 2B$.

Use Parseval's Theorems (both the Fourier series and Fourier transform versions) to prove that the total energy in the function mf is given by

$$E_{mf}^2 = \int_{-\infty}^{\infty} |m(t)f(t)|^2 \, dt = <m>^2 E_f^2.$$

3. The calculation of Problem 2 may be used to calculate the effect of "imperfect modulation" in a DSBSC-AM scheme.

The ideal transmitted signal is

$$\cos \omega_c t f(t),$$

and has the spectral distribution of Figure 7.50. If the transmitter generates less than a perfect sinusoid, the transmitted signal takes the form

$$m(t)f(t),$$

where m is periodic with fundamental frequency ω_c.

Show that if e denotes the modulation error, and assuming that

$$e(t) = m(t) - \cos \omega_c t$$

where e contains no energy at the fundamental frequency, then the energy in the transmitted signal outside of a $2B$ band about $\omega = \pm \omega_c$ is given by

$$R_{\text{out of band}}^2 = <e>^2 E_f^2.$$

4. The previous problem considers the problem of "imperfect modulation." A problem related to "imperfect demodulation" is given by the following.

Suppose that (instead of the desired $\cos \omega_c t$ function) the receiver generates a function $\tilde{d}$ of period $T = \frac{2\pi}{\omega_c}$, and of finite average power,

$$<\tilde{d}>^2 = \frac{1}{T} \int_0^T |\tilde{d}(t)|^2 \, dt < \infty.$$

The received signal is (in the case of a single transmission)

$$r(t) = \cos \omega_c t f(t),$$

with message f band limited to $[-B, B]$, and $2B < \omega_c$, as usual.

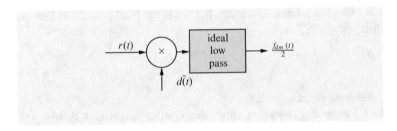

Fig. 7.56 Demodulating with $\overline{d(t)}$

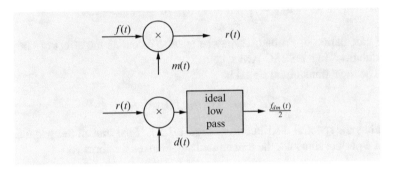

Fig. 7.57 Periodically modulated and demodulated signals

Show that the scheme of Figure 7.56 for suitable choice of the demodulator phase α results in an f_α which is a multiple of the intended message f, provided that d contains energy at the fundamental frequency ω_c.

5. A case of "imperfect modulation and demodulation" arises from the (transmitted and received) signal

$$r(t) = m(t)f(t),$$

which is demodulated according to the scheme of Figure 7.57. Assume that

a. m and d are both periodic of period $T = \frac{2\pi}{\omega_c}$,

b. the product function $m\,d$ if (as well as m and d separately) has finite power

$$< md >^2 = \frac{1}{T} \int_0^T |m(t)\,d(t)|^2 \, dt < \infty,$$

and

c. the message f is band limited to $[-B, B]$, $2B < \omega_c$.

Then

a. show that $f_{dm}(t)$ is a multiple of the intended message provided that (the inner product of m and d)

$$\frac{1}{T}\int_0^T m(t)\,\overline{d}(t)\,dt \neq 0,$$

b. assuming that m is fixed, and d is at our disposal, find the function d which maximizes the demodulated energy

$$E_{f_{dm}}^2 = \int_{-\infty}^{\infty} |f_{dm}(t)|^2\,dt,$$

subject to the constraint that

$$\frac{1}{T}\int_0^T |d(t)|^2\,dt = 1,$$

(preventing the incorporation of arbitrary large amplification in the demodulator).

6. Use Parseval's Theorem to show that

$$\int_{-\infty}^{\infty} \frac{\sin B(t-nT)}{B(t-nT)}\frac{\sin B(t-kT)}{B(t-kT)}\,dt$$

vanishes for $k \neq j$, so that these functions are orthogonal.

7. Suppose that f is band limited to $[-B, B]$, and that T is the associated sampling interval. Show that

$$\int_{-\infty}^{\infty} |f(t)|^2\,dt = \text{Constant} \sum_{n=-\infty}^{\infty} |f(kT)|^2$$

and evaluate the constant in question.

8. Show that if f is band limited to $[-B, B]$, then

$$f(nT) = K \int_{-\infty}^{\infty} f(t)\,\frac{\sin B(t-nT)}{B(t-nT)}\,dt$$

for some K. Also, find K.

Further Reading

A.B. Carlson, *Communication Systems: An Introduction to Signals and Noise in Electrical Communication*, 2nd edn. (McGraw-Hill, New York, 1975)

I.N. Gelfand, C.B. Shilov, *Generalized Functions*, vol. 1 (Academic, New York, 1964)

I.N. Gelfand, C.B. Shilov, *Generalized Functions*, vol. 2 (Academic, New York, 1968)

R.R. Goldberg, *Fourier Transforms* (Cambridge University Press, Cambridge, 1962)

J. Horvarth, *Topological Vector Spaces and Distributions* (Addison-Wesley, Reading, 1966)

E.M.J. Lighthill, *Introduction to Fourier Analysis and Generalized Functions* (Cambridge University Press, Cambridge, 1964)

J.F. Marsden, *Basic Complex Analysis* (W.H. Freeman and Company, San Francisco, 1975)

A. Papoulis, *The Fourier Integral and Its Applications* (McGraw-Hill, New York, 1962)

J.G. Proakis, *Digital Communications*, 3rd edn. (McGraw-Hill, New York, 1995)

H. Schwartz, W.R. Bennett, S. Stein, *Communication Systems and Techniques* (McGraw-Hill, New York, 1966)

Chapter 8
Discrete Variable Transforms

The models and methods of analysis discussed in previous chapters deal largely with problems in which the independent variable takes continuous values. The Fourier and Laplace transforms are specifically designed for application to functions of a real valued variable.

Such functions by no means exhaust the list of mathematical structures available for the construction of models of phenomena of various sorts. A rich variety of models may be based on functions whose natural domain is a discrete set.

It will be seen below that functions defined on the integers have an associated theory of transform methods which is completely parallel to the (by now familiar) theory of Laplace and Fourier transforms. The theory in question is also closely related to Fourier series. This observation greatly simplifies the required exposition, since the analytical heart of the theory is familiar from the treatment of Chapter 2 above.

8.1 Some Discrete Variable Models

In the following examples we discuss a variety of problems which lead naturally to models involving functions defined on the integers. In the following sections we treat the transform theories associated with such problems and examples of typical applications.

Example. The reader has no doubt noticed that economic information is commonly reported on a quarterly (or annual, semiannual) basis. As a result, the data takes the form of a set of numbers indexed by the quarter of occurrence. A formal notation for this practice arises from defining a function $x(\cdot)$: $x(k)$ is the value at quarter number k (counting from some origin) of the economic variable in question.

© Springer International Publishing Switzerland 2016
J.H. Davis, *Methods of Applied Mathematics with a Software Overview,*
Applied and Numerical Harmonic Analysis, DOI 10.1007/978-3-319-43370-7_8

The construction of certain economic models proceeds from the hypothesis that the value of the variable in question at time k is dependent on (at least) its values at (usually a finite number of) previous times. In the event that the value at time k is completely determined by the values at N previous times, such a hypothesis is simply expressed in the functional form

$$x(k) = f(x(k-1), x(k-2), \ldots, x(k-N), k).$$

The admission that disturbances (as well as previous values of the variable) may have an effect leads to a model of the form

$$x(k) = g(x(k-1), \ldots, x(k-N), k, u(k))$$

where $u(k)$ is the disturbance at time k. Such models are obviously capable of extension to include multiple disturbances, as well as interaction between more than one economic variable

A particular case of this model results from the hypothesis that the functional relation is linear (and independent of the time variable k).

In this case the equation takes the form

$$x(k) = -\alpha_1 x(k-1) - \alpha_2 x(k-2) - \cdots - \alpha_N x(k-N) + \beta u(k).$$

This is a typical example of a linear, time-invariant difference equation. We show below that such an equation may be analyzed in a fashion analogous to that employed in the case of linear time-invariant differential equations.

Example. Computational treatment of functions of a continuous variable requires (on the basis of storage limitations, if nothing else) that only function values at a discrete set of points be considered.

If the continuous interval $[0, 1]$ is divided into subintervals of length Δ, associated with each (continuous) function $f(\cdot)$ on $[0, 1]$ is a function defined (on the integers) as the value of $f(\cdot)$ at the corresponding grid point

$$x(k) = f(k\Delta).$$

An example of this process arises in the numerical solution of ordinary differential equations.

Among a variety of numerical schemes which might be used to approximate the solution of the equation

$$\frac{d^2 f}{dt^2} + \omega^2 f(t) = 0,$$

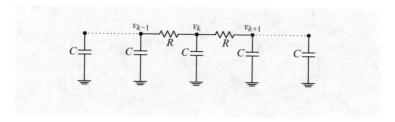

Fig. 8.1 RC network

one might be based on the "second difference approximation"

$$\frac{d^2f}{dt^2} \approx \frac{1}{\Delta^2} \{f(t+\Delta) - 2f(t) + f(t-\Delta)\}.$$

Evaluating the differential equation at the points $t = k\Delta$, replacing the second derivative by the second difference approximation gives

$$\frac{1}{\Delta^2} \{x(k+1) - 2x(k) + x(k-1)\} + \omega^2 x(k) = 0$$

as a means of approximating the original differential equation.

The equation again is a linear constant coefficient difference equation of the same type encountered above.

Example. Examples of discrete models for boundary value problems were discussed in Section 3.5 above. A discrete model for the heat equation is illustrated in Figure 8.1.

The governing equation for this system (at least away from the boundaries) is

$$RC \frac{dv_j}{dt} = v_{j+1} - 2v_j + v_{j-1}. \tag{8.1}$$

This equation involves a function of two variables (time and location), and might be described as a partial differential difference equation.

Example. Difference equations arise in the analysis of the so-called random walk problems (with discrete time parameter).

These problems arise as a (probabilistic) model for a process of diffusion. In the one-dimensional discrete-position case a particle is considered to be located along a line, with integer location n (Figure 8.2).

At regular time intervals, the particle moves (in a probabilistic fashion). This motion is assumed independent of the present location and of the past history of moves.

A simple rule for the (incremental) motion is given by the following:

- With probability .5, the particle remains at the present location.

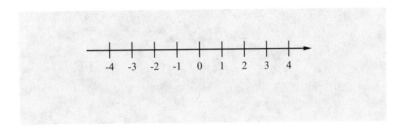

Fig. 8.2 Random walk locations

- With probability .25, the particle moves one location in a rightward direction.
- With probability .25, the particle moves one location in a leftward direction.

A concrete description of the process may be given by requiring the particle (marker) to move (at the sound of a bell) in accordance with the results of two flips of a fair coin carried out in response to the bell signal.

The location of the particle at any fixed time is a random variable. The distribution of this (integer valued) random value is the quantity of interest.

We define a function of two discrete (integer valued) variables by

$$p(n, k) = \text{probability that the particle is at location } n \text{ at time } k.$$

It is clear that if the initial distribution

$$p(n, 0), -\infty < n < \infty$$

is given, then the above description of the "unfolding of the process" should determine $p(n, k)$ for all values of $k > 0$.

We seek to determine $p(n, k + 1)$.

If the particle is at location n at time $k + 1$, then it follows from the transition rule above that it must have been at one of locations $n - 1, n,$ or $n + 1$ at time k.

From the rules above (using either formally Bayes' rule or an appeal to common sense), we obtain the relation

$$p(n, k + 1) = \frac{1}{2} p(n, k) + \frac{1}{4} p(n - 1, k) + \frac{1}{4} p(n + 1, k), \quad -\infty < n < \infty, \, k > 0. \tag{8.2}$$

The model might be referred to as a partial difference equation for the particle probability function.

It should be noted that the model has a close relationship to the diffusion equation

$$\frac{\partial u}{\partial t} = \frac{\partial^2 u}{\partial x^2}.$$

To see the relationship, consider replacing the derivatives in the diffusion equation by differences according to

$$\frac{\partial u}{\partial t} \approx \frac{u(x, t + \Delta) - u(x, t)}{\Delta},$$

$$\frac{\partial^2 u}{\partial x^2} \approx \frac{u(x + h, t) - 2\,u(x, t) + u(x - h, t)}{h^2}.$$

Evaluating at the grid points $t = k\,\Delta,\ x = n\,h$, we are led to consider

$$\tilde{u}(n, k) - \tilde{u}(n, k) = \frac{\Delta}{h^2}\,[\tilde{u}(n + 1, k) - 2\,\tilde{u}(n, k) + \tilde{u}(n - 1, k)].$$

In the case that the grid spacings are chosen to satisfy

$$\frac{\Delta}{h^2} = \frac{1}{4},$$

the system is essentially identical to (8.2) above.

Example. It is also possible to construct random walk problems with a continuous location variable. In this situation, the location at time $k + 1$ is obtained from the location at time k by the addition of an independent random variable.

One example arises as a model for price behavior in a stock market. The hypothesis is that the logarithm of the price of the stock is a homogeneous Gaussian random walk. This means that (if $\pi(k)$ denotes the price of the stock on day k) the quantity

$$\log \pi(k + 1) - \log \pi(k)$$

is (for each k) an independent Gaussian random variable with a fixed mean and variance (independent of k). We have then

$$\log \pi(k + 1) = \log \pi(k) + n_k$$

where n_k is the independent increment (jump), distributed as $N(m, \sigma^2)$.

If $f_k(\cdot)$ is the probability density of $\log \pi(k)$, then (in analogy with the previous example) we seek an iterative equation for $f_k(\cdot)$.

Using the governing equation, and the fact that the probability density of the sum of two independent random variables is the convolution of the densities of the components, we obtain

$$f_{k+1}(x) = \int_{-\infty}^{\infty} \frac{1}{\sqrt{2\pi}\,\sigma}\, e^{-\frac{(x - \zeta - m)^2}{2\sigma^2}}\, f_k(\zeta)\, d\zeta, \quad k > 0,\ -\infty < x < \infty. \tag{8.3}$$

This relation may be called a partial integral difference equation.

In the following sections we treat some of the transform methods available for the solution of difference equations, and provide solutions for some of the problems posed above.

Problems 8.1.

1. Consider a linear, constant coefficient vector differential equation of the form

$$\frac{d\mathbf{x}}{dt} = \mathbf{A}\,\mathbf{x}(t), \ \mathbf{x}_0 = \mathbf{x}_0.$$

Use the Euler approximation

$$\frac{d\mathbf{x}}{dt} \approx \frac{1}{\Delta}\,[\mathbf{x}(t+\Delta) - \mathbf{x}(t)]$$

to formulate a vector difference equation approximating the solution of the vector differential equation.

2. Repeat Problem 1 utilizing the implicit approximation

$$\frac{d\mathbf{x}}{dt} \approx \frac{1}{2\,\Delta}\,[\mathbf{x}(t+\Delta) - \mathbf{x}(t-\Delta)]$$

for the derivative.

3. Repeat Problem 1 using the Runge-Kutta method RK4 to take the numerical step.

4. Consider the discrete-location random walk problem, and suppose that at each time the particle jumps a distance of j units with probability π_j. Suppose that $\pi_j = 0$, $j > N$, $j < -M$, so that the particle is constrained to move at most M units leftward, N units rightward at each time. Show that the probability distribution $p(n, k)$ for the position satisfies a relation of the form

$$p(n, k+1) = \sum_{-M}^{N} g_j\, p(n-j, k).$$

5. A continuous location random walk problem on a circle may be described as follows. A particle is located at each time k at some (angular) point θ on a circle as in Figure 8.3.

   At time $k + 1$ the particle advances to a location $\theta + \phi$ (modulo 2π). ϕ is a "random angle" (generated independently at each stage) with a given probability density $p_\phi(\phi)$. Such a density is illustrated in Figure 8.4.

   Suppose that at time $k = 0$ the particle location has a given probability density f_0. Show that the probability density of the location at time k satisfies the integral difference equation

$$f_{k+1}(\theta) = \int_0^{2\pi} \tilde{p}_\phi(\theta - \zeta) f_k(\zeta)\, d\zeta,$$

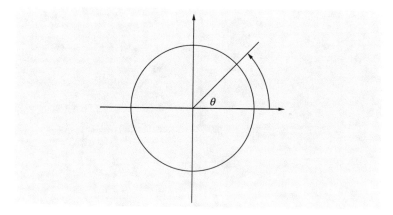

Fig. 8.3 Continuous circular random walk

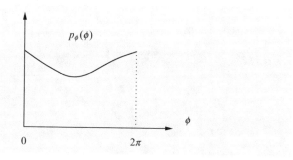

Fig. 8.4 Transfer angle probability density

where $\tilde{p}_\phi(\theta)$ is the 2π periodic extension of $p_\phi(\theta)$.

6. A discrete-discrete random walk on a circle may be constructed by locating N equidistant points on a circle, and postulating that the particle in question jumps a certain number of locations around the circle at each time step. The locations are illustrated in Figure 8.5

Suppose that the jumps at each time are independent, and that the probabilities of jumps (in a counter clockwise direction) of length $0, 1, 2, \ldots, N-1$ are specified as $p_0, p_1, p_2, \ldots, p_{N-1}$.

If $f(n, k)$ denotes the probability of finding the particle in location n at time k, find a recursive formula for $f(n, k)$ analogous to the result of Problem 5 above.

Write the result in terms of the "probability vector"

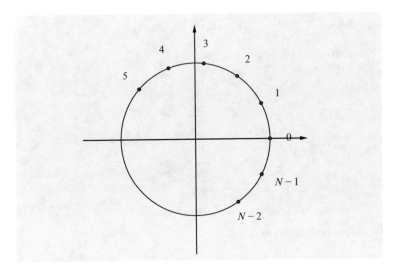

Fig. 8.5 Circular random walk

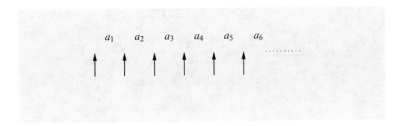

Fig. 8.6 Permutations

$$\mathbf{p}(k) = \begin{bmatrix} f(0,k) \\ f(1,k) \\ f(2,k) \\ \vdots \\ f(N-1,k) \end{bmatrix}.$$

7. The problem of discovering the formula for the number of arrangements in order of n distinct objects (and in fact various other combinatorial problems) may be formulated as a recursive function definition (Figure 8.6). Show that if p_n denotes the number of such arrangements of n objects, the observation that the $n + 1^{st}$ may be placed either between two of the present positions, or at one of the ends leads to

$$p_{n+1} = (n+1)p_n.$$

8. A homogeneous constant coefficient linear difference equation takes the form

$$x(k) + \alpha_1 x(k-1) + \alpha_2 2x(k-2) + \cdots + \alpha_N x(k-N) = 0.$$

Show that it is possible to find solutions to such an equation in the form

$$x(k) = \lambda^k,$$

for suitably chosen λ.

8.2 Z-Transforms

The z-transform may be concisely defined as the "discrete-time version" of the Laplace transform.

The transform finds application in the solution of difference equations of the sort encountered in the previous section, in the cases in which the functions in question are defined on the nonnegative integers. The typical examples may be described as initial value problems for difference equations. These are simply the discrete variable analogs of the initial value problems for differential equations treated earlier by Laplace transform methods.

To formally define the z-transform, we consider a function x defined on the nonnegative integers $\mathbb{Z}^+ = \{0, 1, 2, 3, \ldots\}$ and taking complex values. We assume initially that x has at most exponential (geometric) growth, so that

$$|x(n)| \leq M \rho^n, \ n \geq 0$$

for some constants M, ρ.

We define a function of the complex variable z, called the z-transform of x, by the formula

$$\mathbb{Z}\{x(n)\} = X(z) = \sum_{n=0}^{\infty} x(n) z^{-n}. \tag{8.4}$$

The transform function X is defined for values of z for which the infinite series (8.4) converges.

In the case that x satisfies the growth restriction above, standard power series considerations show that

1. $X(z)$ converges absolutely for $|z| > \rho$, and
2. $X(z)$ defines an analytic function of the complex variable z for $|z| > \rho$ (since it is defined by an absolutely convergent power series in this region).

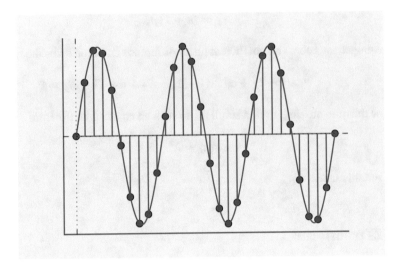

Fig. 8.7 Discrete sinusoid

The transform definition (8.4) should be regarded as an exponentially (geometrically) weighted sum of the values of the original function. In this way it is exactly analogous to the continuous-time Laplace transform.

Example. Suppose that $x(n) = (-1)^n$, $n \geq 0$. Then

$$X(z) = \sum_{n=0}^{\infty} (-1)^n z^{-n}$$

$$= \frac{1}{1 - \left(\frac{-1}{z}\right)} = \frac{z}{z+1}, \quad |z| > 1.$$

Example. If $x(n) = a^n$, $n \geq 0$, then (as above)

$$X(z) = \frac{z}{z-a}, \quad |z| > a.$$

Example. Consider the function x defined by $x(n) = \sin(\frac{2\pi}{K} n)$ (Figure 8.7). Then

$$X(z) = \sum_{n=0}^{\infty} \sin(\frac{2\pi}{K} n) z^{-n}$$

$$= \sum_{n=0}^{\infty} \frac{e^{i\frac{2\pi}{K} n} - e^{-i\frac{2\pi}{K} n}}{2i} z^{-n}$$

$$= \frac{1}{2i} \left[\frac{1}{1 - \dfrac{e^{\frac{2\pi i}{K}}}{z}} - \frac{1}{1 - \dfrac{e^{\frac{-2\pi i}{K}}}{z}} \right]$$

$$= \frac{z}{2i} \left[\frac{e^{\frac{2\pi i}{K}} - e^{\frac{-2\pi i}{K}}}{z^2 - z\left(e^{\frac{2\pi i}{K}} + e^{\frac{-2\pi i}{K}} \right) + 1} \right]$$

$$= \frac{z \sin \frac{2\pi}{K}}{z^2 - 2z \cos \frac{2\pi}{K} + 1}.$$

The geometric growth condition is a useful restriction for an initial definition of the z-transform. However, in view of the treatment given earlier of the Fourier and Laplace inversion integrals, it may be expected that there are natural classes of sequences for which the z-transform may be defined. These are simply the discrete variable analogues of the classes employed in the continuous-time case.

We recall that in the case of Laplace transforms, appropriate functions f are those for which either

$$\int_0^\infty |f(t)| \, e^{-\sigma t} \, dt < \infty,$$

or

$$\int_0^\infty |f(t)|^2 \, e^{-2\sigma t} \, dt < \infty$$

for some real σ. These restrictions correspond to those appropriate to the L^1 and L^2 Fourier transform theories respectively.

In the case of a function x defined on $\mathbb{Z}^+$, the analogue of the first class is the requirement that the sequence satisfy

$$\sum_{n=0}^{\infty} |x(n)| \, \rho^{-n} < \infty \tag{8.5}$$

for some real ρ. The discrete-time L^2 condition takes the form

$$\sum_{n=0}^{\infty} |x(n)|^2 \, \rho^{-2n} < \infty. \tag{8.6}$$

It is a simple matter to see that the assertions made above with regard to the domain of definition and analyticity of the z-transform function also hold under the assumption of (8.5) or (8.6).

In the case of the hypothesis (8.6), we consider

$$X(z) = \sum_{n=0}^{\infty} x(n)\, z^{-n} = \sum_{n=0}^{\infty} [x(n)\, \rho^{-n}]\, \rho^n\, z^{-n}.$$

Since $|z| > \rho$, we have

$$\sum_{n=0}^{\infty} \left|\frac{\rho}{z}\right|^{2n} < \infty.$$

Viewing the transform sum as an inner product and applying the Cauchy–Schwarz inequality,

$$|X(z)| \le \left(\sum_{n=0}^{\infty} |x(n)\, \rho^{-n}|^2\right)^{\frac{1}{2}} \left(\sum_{n=0}^{\infty} \left|\frac{\rho}{z}\right|^{2n}\right)^{\frac{1}{2}}.$$

We conclude that (8.4) indeed converges for all z with $|z| > \rho$. Similar considerations verify that the function so defined is analytic in this region.

The mechanical properties of the Laplace transform described in Section 6.3 essentially all have z-transform versions. For completeness, we note first that the z-transformation is a linear operation. We introduce the operational symbol $\mathcal{Z}$ defined by

$$\mathcal{Z}\{x(n)\} = X(z) = \sum_{n=0}^{\infty} x(n)\, z^{-n}.$$

This notation simplifies the algebraic description of the formal properties of the transform operation.

8.3 Z-Transform Properties

The analogies between z and Laplace transforms become evident when the formal manipulation properties are considered.

n Multiplication Law

Suppose that $x(n) = n\,a^n$. From the fact that

$$\mathcal{Z}\{a^n\} = \frac{z}{z-a} = \sum_{n=0}^{\infty} a^n\,z^{-n},$$

we note that

$$\frac{d}{dz}\sum_{n=0}^{\infty} a^n\,z^{-n} = \sum_{n=0}^{\infty} (-n)\,a^n\,z^{-(n+1)}.$$

Hence

$$\mathcal{Z}\{n\,a^n\} = -z\frac{d}{dz}\mathcal{Z}\{a^n\} = -z\frac{d}{dz}\frac{z}{z-a}$$

$$= \frac{a\,z}{(z-a)^2}.$$

The technique employed in this example is evidently generally applicable. This observation provides the following n-multiplication law.
    If

$$X(z) = \mathcal{Z}\{x(n)\},$$

then

$$\mathcal{Z}\{n\,x(n)\} = -z\frac{d}{dz}\mathcal{Z}\{x(n)\}.$$

This result may be compared with the Laplace transform formula

$$\mathcal{L}\{t\,x(t)\} = -\frac{d}{ds}\mathcal{L}\{x(t)\}.$$

As in the Laplace transform case, shift formulas appear both in exponential and time delay form.

Exponential Shift Law

Suppose that

$$\mathcal{Z}\{x(n)\} = X(z);$$

then (for $a \neq 0$)

$$\mathcal{Z}\{a^n x(n)\} = \sum_{n=0}^{\infty} x(n)\, a^n\, z^{-n}$$

$$= \sum_{\frac{z}{a}=0}^{\infty} x(n) \left(\frac{z}{a}\right)^{-n}$$

$$= X\left(\frac{z}{a}\right),\ \left|\frac{z}{a}\right| > \rho.$$

Convolution Law

A convolution theorem in the z-transform context may also be established. In the case of continuous-time problems, we have proceeded by defining the convolution operation at the outset, and then calculating the required transform. This procedure may also be followed in the discrete-time case. However, it is instructive to approach the problem from the opposite direction.

We suppose that f and g are z-transformable functions defined on $\mathcal{Z}^+$. Then (from our considerations above)

$$\mathcal{Z}\{f(n)\} = \sum_{n=0}^{\infty} f(n)\, z^{-n} = F(z),$$

and

$$\mathcal{Z}\{g(n)\} = \sum_{n=0}^{\infty} g(n)\, z^{-n} = G(z)$$

define analytic functions of the variable z for z sufficiently large. We compute the product of these transforms as

$$H(z) = \left(\sum_{i=0}^{\infty} f(i)\, z^{-i}\right)\left(\sum_{j=0}^{\infty} g(j)\, z^{-j}\right)$$

$$= \sum_{i=0}^{\infty}\sum_{j=0}^{\infty} f(i)\, g(j)\, z^{-(i+j)}.$$

Identifying the coefficient of the typical power of z in the above, H is expressed in the form

$$H(z) = \sum_{k=0}^{\infty} h(k) \, z^{-k}$$

where

$$h(k) = \sum_{l=0}^{k} f(k-l)g(l)$$

is identified as the "sum of all products associated with the exponent k." This is, of course, just the usual rule that multiplication of absolutely convergent power series is evaluated by the convolution of their coefficients.

We define the function h computed above as the convolution of $f(\cdot)$ and $g(\cdot)$,

$$h(k) = [f \star g](k)$$

$$= \sum_{l=0}^{k} f(k-l)\, g(l).$$

This definition and the above computation give the discrete form of the convolution theorem.

Suppose that

$$\mathcal{Z}\{f(n)\} = F(z),$$
$$\mathcal{Z}\{g(n)\} = G(z),$$

and that h is defined as

$$h(k) = \sum_{l=0}^{k} f(k-l)g(l).$$

Then

$$\mathcal{Z}\{h(n)\} = H(z) = F(z)\, G(z).$$

Strictly speaking, the derivation of the above result should include a demonstration that h, defined as a convolution, is a z-transformable sequence. This gap is covered by the problems at the end of this section.

As might be expected, the major use for the convolution theorem is in the analysis of models containing forcing functions. We reserve such examples until more z-transform tools are at our disposal.

Example. Suppose that h is defined by

$$h(k) = \sum_{l=0}^{k} (k-l)\, a^{k-l}\, b^l.$$

Using the convolution theorem, we compute

$$\mathcal{Z}\{h(n)\} = \mathcal{Z}\{n\, a^n\}\, \mathcal{Z}\{b^n\}$$

$$= \frac{az}{(z-a)^2}\, \frac{z}{z-b}.$$

An explicit formula for $h(k)$ could be computed by inverse transforming that expression.

Delay Law

To further pursue the analogy with the continuous-time case, we compute the z-transform of a delayed function.

Suppose that

$$\mathcal{Z}\{x(n)\} = X(z).$$

Define the "N-step delayed version of x" by

$$g(n) = \begin{cases} 0 & 0 \le n \le N-1, \\ x(n-N) & n \ge N. \end{cases}$$

(See Figure 8.8.)

Then we compute

$$\mathcal{Z}\{g(n)\} = \sum_{n=0}^{\infty} g(n)\, z^{-n}$$

$$= \sum_{n=N}^{\infty} g(n)\, z^{-n}$$

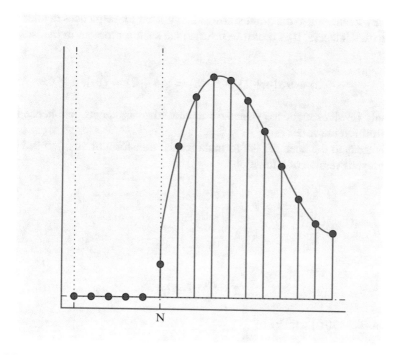

Fig. 8.8 Delayed sequence

$$= \sum_{n=N}^{\infty} x(n-N)\, z^{-n}$$

$$= z^{-N} \sum_{n=N}^{\infty} x(n-N)\, z^{-(n-N)}$$

$$= z^{-N} \sum_{j=0}^{\infty} x(j)\, z^{-j}$$

$$= z^{-N} \, \mathcal{Z}\{x(n)\}.$$

In short, this shows that a factor of z^{-N} (in the transform) corresponds to an N-step delay in time.

Linear difference equations are often initially formulated in terms involving delayed arguments, as in the model

$$x(n) + \alpha\, x(n-1) + \beta\, x(n-2) = g(n),\ n > 0.$$

If this is regarded as an initial value problem (providing $x(0)$ from $x(-1)$, $x(-2)$ and so forth) it evidently involves values of the sequence in question for negative values of the argument. Since the z-transform is formulated for sequences defined on the nonnegative integers, it is useful to reformulate such a problem in the "advanced form"

$$x(n+2) + \alpha\, x(n+1) + \beta\, x(n) = g(n+2) = f(n), \; n \geq 0.$$

This only involves sequence values for nonnegative arguments, and hence fits the z-transform framework.

This leads to the need for the formula for the transform of an advanced (rather than delayed) version of a function.

Shift Law

Suppose that

$$\mathcal{Z}\{x(n)\} = X(z)$$

and consider $\{h(n)\}$ defined by

$$h(n) = x(n+N), \; n > 0.$$

(See Figure 8.9.)
  Then

$$\mathcal{Z}\{h(n)\} = \sum_{n=0}^{\infty} h(n)\, z^{-n}$$

$$= \sum_{n=0}^{\infty} x(n+N)\, z^{-n}$$

$$= z^{N} \sum_{n=0}^{\infty} x(n+N)\, z^{-(n+N)}$$

$$= z^{N} \sum_{j=N}^{\infty} x(j)\, z^{-j}$$

$$= z^{N} \left\{ \sum_{j=0}^{\infty} x(j)\, z^{-j} - \sum_{j=0}^{N-1} x(j)\, z^{-j} \right\}$$

$$= z^{N} \left\{ \mathcal{Z}\{x(n)\} - \sum_{j=0}^{N-1} x(j)\, z^{-j} \right\}.$$

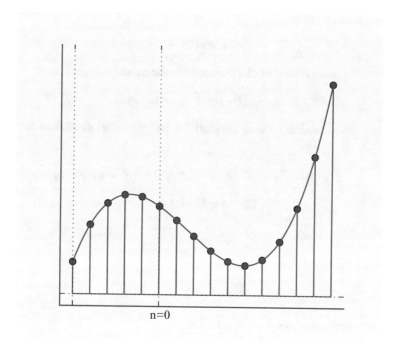

n=0

Fig. 8.9 Advance shift

This result should be considered as the operational equivalent of the differentiation law

$$\mathcal{L}\left\{\frac{dx}{dt}\right\} = s\,\mathcal{L}\{x(t)\} - x(0^+)$$

appearing in the theory of Laplace transforms.

Example. One of the more notorious sequences of integers is the so-called Fibonacci sequence. This is the sequence

$$\{1, 1, 2, 3, 5, 8, 13, 21, 34, \ldots\}.$$

Recognizing this sequence as a likely constituent of an intelligence test, we seek a general formula for the nth term of the sequence.

We first note that the sequence is generated by adding the two preceding members of the sequence. The process is started by taking the first two entries to be one. In short, the sequence satisfies the (second-order) difference equation

$$x(n + 2) = x(n + 1) + x(n), \ n \geq 0 \tag{8.7}$$

with the initial conditions

$$x(1) = x(0) = 1.$$

To find the general term of (8.7), we rewrite the equation in the form

$$x(n + 2) - x(n + 1) - x(n) = 0,$$

and apply the z-transform. Using the shift law and the initial conditions above, we find

$$z^2 X(z) - z^2 x(0) - z x(1) - z X(z) + z x(0) - X(z) = 0,$$

$$(z^2 - z - 1) X(z) = z^2,$$

and

$$X(z) = \frac{z^2}{z^2 - z - 1}.$$

We next invert the transform to obtain $x(\cdot)$.

Since we have not yet discussed the inversion integral for z-transforms, our options for inverting $X(z)$ are essentially limited to partial fractions techniques and use of tables.

In view of our previous computation that

$$\mathcal{Z}\{a^n\} = \frac{1}{1 - \frac{a}{z}} = \frac{z}{z - a},$$

it is convenient to treat the partial fractions problem in terms of the variable $\frac{1}{z}$. We have

$$X(z) = \frac{1}{1 - \left(\frac{1}{z}\right) - \left(\frac{1}{z^2}\right)}$$

$$= \frac{-1}{\left[\frac{1}{z} - r_1\right]\left[\frac{1}{z} - r_2\right]},$$

where

$$r_{1,2} = \frac{-1 \pm \sqrt{5}}{2}$$

are zeroes of the polynomial p,

$$p(\lambda) = \lambda^2 + \lambda - 1.$$

By the usual partial fractions method,

$$\frac{-1}{\left(\frac{1}{z} - r_1\right)\left(\frac{1}{z} - r_2\right)} = -\left[\frac{\frac{1}{r_1 - r_2}}{\frac{1}{z} - r_1} + \frac{\frac{1}{r_2 - r_1}}{\frac{1}{z} - r_2}\right]$$

$$= \frac{-1}{r_1 - r_2}\left[\frac{1}{\frac{1}{z} - r_1} - \frac{1}{\frac{1}{z} - r_2}\right]$$

$$= \frac{-1}{r_1 \, r_2 \, (r_1 - r_2)}\left[\frac{z \, r_2}{\frac{1}{r_1} - z} - \frac{z \, r_1}{\frac{1}{r_2} - z}\right].$$

Inverse transforming, we compute

$$x(n) = \frac{-1}{r_1 \, r_2 \, (r_1 - r_2)}\left[-r_2\left(\frac{1}{r_1}\right)^n + r_1\left(\frac{1}{r_2}\right)^n\right] \tag{8.8}$$

$$= \frac{1}{(r_2 - r_1)}\left[\left(\frac{1}{r_2}\right)^{n+1} - \left(\frac{1}{r_1}\right)^{n+1}\right].$$

We note that the above procedure implicitly assumes that the calculation of the inverse z-transform is a well-defined operation. This is in fact the case. We place the burden of this on the discussion of the z-transform inversion integral below.

8.4 Z-Transform Inversion Integral

It is clear that the partial fractions inversion carried out above requires a certain amount of algebraic foresight. The integral inversion method is less demanding in this respect.

The z-transform inversion integral may be approached as a simple exercise in Fourier series.

We know from our previous discussion that (with the natural restrictions on the sequence $x(\cdot)$ the z-transform function

$$X(z) = \sum_{n=0}^{\infty} x(n) \, z^{-n} \tag{8.9}$$

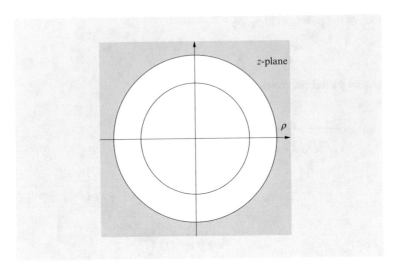

Fig. 8.10 Z-transform convergence region

defines an analytic function of z for $|z| > \rho$, with ρ (dependent on x) chosen sufficiently large. It is conventional to illustrate this region of convergence as the exterior of a disk in the z-plane (Figure 8.10).

Our object is to recover the function values $\{x(n)\}$ from the transform expression. If the complex variable z is expressed in polar form,

$$z = |z|\, e^{i \arg(z)}$$
$$= r\, e^{i\theta},$$

then we have

$$x(r\, e^{i\theta}) = \sum_{n=0}^{\infty} x(n) \left(r\, e^{i\theta}\right)^{-n}$$
$$= \sum_{n=0}^{\infty} \frac{x(n)}{r^n} e^{-in\theta}.$$

For fixed $r > \rho$, the series is absolutely convergent, and defines a function (of the angle θ) by

$$f(\theta) = \sum_{n=0}^{\infty} \frac{x(n)}{r^n} e^{-in\theta}, \; 0 \le \theta \le 2\pi. \tag{8.10}$$

Since the series is absolutely convergent, the function f defined by (8.10) is continuous, periodic with period 2π, and square-integrable (of finite power)

$$\int_0^{2\pi} |f(\theta)|^2 \, d\theta < \infty.$$

We immediately recognize (8.10) as the Fourier series expansion of the function f. The function f just happens to have the special property that half of its Fourier coefficients vanish identically.

The nonvanishing Fourier coefficients may be read from (8.10) as

$$c_{-n} = \frac{x(n)}{r^n}, \quad n \geq 0.$$

Invoking the usual Fourier coefficient formula,

$$\frac{x(n)}{r^n} = \frac{1}{2\pi} \int_0^{2\pi} f(\theta) \, e^{in\theta} \, d\theta.$$

Identifying f in terms of the z-transform function X, we obtain the (real form of the) z-transform inversion integral,

$$\frac{x(n)}{r^n} = \frac{1}{2\pi} \int_0^{2\pi} X(r e^{i\theta}) \, e^{in\theta} \, d\theta \qquad (8.11)$$

$$x(n) = \frac{1}{2\pi} \int_0^{2\pi} X(r e^{i\theta}) \left(r e^{i\theta} \right)^n d\theta.$$

A formal statement of this result, incorporating the relevant assumptions, is provided in the following theorem.

Theorem (z-Transform Inversion Theorem). *Suppose that $x(\cdot)$ is a complex valued function defined on the nonnegative integers. Suppose further that there exists a (radius of convergence) ρ such that one of the following conditions holds:*[1]

- $|x(n)| \leq M \rho^n$, *for some M,*
- $\sum_{n=0}^{\infty} |x(n)| \, \rho^{-n} < \infty$,
- $\sum_{n=0}^{\infty} |x(n)|^2 \, \rho^{-n} < \infty$.

  Then

$$X(z) = \sum_{n=0}^{\infty} x(n) \, z^{-n}$$

---

[1] These conditions are not independent, but define classes of sequences analogous to the continuous-time case.

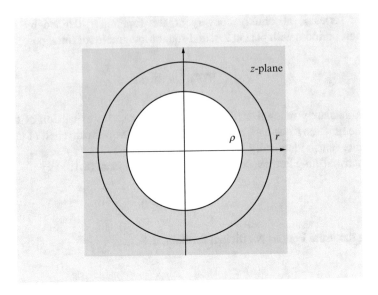

Fig. 8.11 Inversion contour location

defines an analytic function of z for $|z| > \rho$, and the sequence (function) x may be recovered from the transform function X by the inversion integral

$$x(n) = \frac{1}{2\pi} \int_0^{2\pi} X(r e^{i\theta}) \left(r e^{i\theta} \right)^n d\theta,$$

where r is any real number such that $r > \rho$.

The inversion integral may be visualized by modifying Figure 8.11 to include the circle of radius r associated with the inversion.

This diagram immediately suggests identifying the inversion integral

$$\frac{1}{2\pi} \int_0^{2\pi} X(r e^{i\theta}) \left(r e^{i\theta} \right)^n d\theta$$

as the parameterization of a contour integral in the z-plane. Using the polar form

$$z = r e^{i\theta}$$

we calculate

$$dz = i r e^{i\theta} d\theta$$

and write the inversion theorem in complex form.

Theorem (Complex z-Transform Inversion Theorem).
  *Let $x(\cdot)$ satisfy the conditions of the real form of the inversion theorem. Then x
may be recovered from the z-transform X by evaluation of the contour integral*

$$x(n) = \frac{1}{2\pi i} \oint_{|z|=r} X(z)\, z^n \, \frac{dz}{z}$$

$$= \frac{1}{2\pi i} \oint_{|z|=r} X(z)\, z^{n-1} \, dz$$

  The complex form of the inversion leads directly to the evaluation of the inverse
transform of a rational z-transform by means of residue calculations. This is in fact
the major method for inverse transform calculations, since the forms of the integrals
encountered in the real form inversion integral are typically intractable by use of
real variable methods.

  The presence of both a real and complex variable form of an inversion theorem
is another instance of the parallels between Laplace and z-transforms. It might
be noted that the discussion in the z-transform case requires much less involved
analytical justification than the continuous-time case. This is essentially due to the
fact that the inversion integral is over a compact domain (finite interval) in the z-
transform case. This removes the necessity for dealing with what in the continuous
case are essentially improper integrals in the classical sense.

  For the sake of completeness, we note that the inversion results above provide a
form of uniqueness theorem for the z-transform, and hence justify the use of "tricks
and tables" methods of z-transform inversion. The justification for this remark lies
in the observation that Fourier series expansions of square-integrable functions are
unique.

Example. We earlier computed

$$\mathcal{Z}\{a^n\} = \frac{z}{z-a}.$$

This computation is also valid for complex values of a, as long as $|z| > |a|$.
The complex form of the inversion theorem is

$$x(n) = \frac{1}{2\pi i} \oint_{|z|=r>a} \frac{z}{z-a}\, z^n \, \frac{dz}{z}.$$

  The contour of this integral, together with the single pole of the integrand at the
point $z = a$, is illustrated in Figure 8.12.
  Using the residue theorem, we compute

$$x(n) = \text{Residue}\left(\frac{z^n}{z-a}\bigg|z = a\right) = a^n.$$

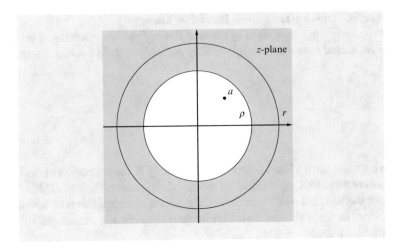

Fig. 8.12 Single pole z-transform inversion

Example. The Fibonacci sequence introduced above has z-transform

$$X(z) = \frac{z^2}{z^2 - z - 1}.$$

This transform has simple poles at the points

$$z = r_{1,2} = \frac{1 \pm \sqrt{5}}{2},$$

determined as solutions of

$$z^2 - z - 1 = 0.$$

The inversion integral takes the form (Figure 8.13)

$$x(n) = \frac{1}{2\pi i} \oint_{|z|=r} \frac{z^2}{(z - r_1)(z - r_2)} z^{n-1} \, dz$$

$$= \frac{1}{2\pi i} \oint_{|z|=r} \frac{z^{n+1}}{(z - r_1)(z - r_2)} \, dz.$$

Evaluating the integral by residues,

$$x(n) = \frac{r_1^{n+1}}{r_1 - r_2} + \frac{r_2^{n+1}}{r_2 - r_1}.$$

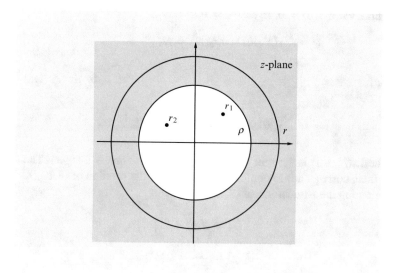

Fig. 8.13 Two pole inversion diagram

This is of course the same result as that obtained by partial fractions above. It should be noted that the calculation proceeds virtually by inspection in the case of contour integral inversion.

Example. An example of a problem with an algebraic flavor which can be solved by use of z-transforms is the following. We seek a formula for the sum of the kth powers of the first N integers.

For $k = 0, 1, 2$, the results are well known:

$$f_0(N) = \sum_{j=1}^{N} 1 = N,$$

$$f_1(N) = \sum_{j=1}^{N} j = 1 + 2 + \cdots + N = \frac{N(N+1)}{2},$$

$$f_2(N) = \sum_{j=1}^{N} j^2 = 1^2 + 2^2 + \cdots + N^2 = \frac{N(N+1)(2N+1)}{6}.$$

To cast this problem in the framework of difference equations, we observe that the difference between two successive values of the function in question is simply the last k^{th} power added.

This leads to consideration of the difference equation

$$x(n+1) = x(n) + n^k, \tag{8.12}$$

with initial value $x(0) = 0$. In case $k = 0$,

$$x(N + 1) = f_0(N) = N,$$

for k = 1,

$$x(N + 1) = f_1(N) = \frac{N(N + 1)}{2},$$

in general $x(N + 1)$ is the sum of the kth powers of the first N integers. The desired result, then, corresponds to obtaining a formula for the solution of (8.12).

Computing the z-transform of (8.12), we obtain

$$z X(z) = X(z) + \mathcal{Z}\{n^k\}$$

$$X(z) = \frac{1}{z - 1} \mathcal{Z}\{n^k\}.$$

To compute the z-transform of n^k, we use the multiplication law

$$\mathcal{Z}\{n x(n)\} = -z \frac{d}{dz} X(z).$$

From the fact that

$$\mathcal{Z}\{1\} = \frac{z}{z - 1},$$

we compute

$$\mathcal{Z}\{n\} = -z \frac{d}{dz} \frac{z}{z - 1},$$

and find that the desired transform is representable in the form

$$\mathcal{Z}\{n^k\} = \left(-z \frac{d}{dz}\right)^k \frac{z}{z - 1}.$$

The transform of the desired sequence is therefore

$$X(z) = \frac{1}{z - 1} \left(-z \frac{d}{dz}\right)^k \frac{z}{z - 1}.$$

This is a rational function whose only singularity is a pole of order $(k + 2)$ at the point $z = 1$.

The inversion integral therefore takes the form

$$x(n) = \frac{1}{2\pi i} \oint_{|z|=r>1} \frac{z^{n-1}}{z-1} \left(-z \frac{d}{dz}\right)^k \frac{z}{z-1} \, dz.$$

The desired sum is therefore

$$\sum_{j=1}^{N} j^k = x(N+1) = \frac{1}{2\pi i} \oint_{|z|=r>1} \frac{z^N}{z-1} \left(-z \frac{d}{dz}\right)^k \frac{z}{z-1} \, dz.$$

The integrand is everywhere analytic, with the single exception of the pole of order $(k+2)$ at $z = 1$. The integral may be evaluated as the residue of the integrand at this pole. This gives the (more or less explicit) formula

$$\sum_{j=1}^{N} j^k = \frac{1}{(k+1)!} \frac{d}{dz}^{k+1} \left[(z-1)^{k+2} \frac{z^N}{z-1} \left(-z \frac{d}{dz}\right)^k \frac{z}{z-1} \right] \Bigg|_{z=1}$$

$$= \frac{1}{(k+1)!} \frac{d}{dz}^{k+1} \left[(z-1)^{k+1} z^N \left(-z \frac{d}{dz}\right)^k \frac{z}{z-1} \right] \Bigg|_{z=1}$$

This result readily reproduces the results of the simple cases, and the result for any specific value of k may be computed in principle from the above.

Example. There exists a discrete-time analog of the Fourier series expansion of a periodic function. The topic in question is referred to as the finite Fourier transform and is discussed in some detail in the following section.

Here we introduce the basic expansion result as an exercise in the application of z-transforms.

We consider a function f defined on $\mathcal{Z}^+$ and periodic of period N,

$$f(k+N) = f(k), k = 0, 1, 2, \ldots. \tag{8.13}$$

Since such an f is of necessity bounded, we compute its z-transform as

$$F(z) = \sum_{k=0}^{\infty} f(k) z^{-k}$$

$$= \sum_{l=0}^{\infty} \sum_{k=0}^{N-1} f(lN + k) z^{-(lN+k)}$$

$$= \sum_{l=0}^{\infty} \sum_{k=0}^{N-1} f(k) z^{-(lN)} z^{-k}$$

$$= \frac{z^N}{z^N - 1} \sum_{k=0}^{N-1} f(k) z^{-k}, \ |z| > 1.$$

Using the complex form of the inversion theorem, we represent $f(n)$ in the form

$$f(n) = \frac{1}{2\pi i} \oint_{|z|=r>1} \frac{z^N}{z^N - 1} \left(\sum_{k=0}^{N-1} f(k) z^{-k} \right) z^{n-1} \, dz.$$

We evaluate this by residues. The integrand has (by virtue of the Nth-order zero z^N) no pole at the origin. The only poles occur at the zeroes of the polynomial

$$p(z) = z^N - 1.$$

These occur at each of the N roots of unity. Using $\omega_j = \left(e^{\frac{2\pi i}{N}} \right)^j$, we have the factorization

$$z^N - 1 = (z - \omega_0)(z - \omega_1) \cdots (z - \omega_{N-1}). \tag{8.14}$$

The associated pole locations are equally distributed about the unit circle in the complex plane, as illustrated in Figure 8.14.

Since the poles of the integrand are all simple, we evaluate the integral as the sum of the residues at each of the poles $\{\omega_j\}$.

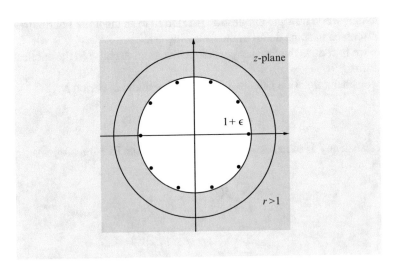

Fig. 8.14 Roots of unity

Then

$$f(n) = \sum_{j=0}^{N-1} \text{Residue} \left\{ \frac{z^N}{z^N - 1} \left(\sum_{k=0}^{N-1} f(k) z^{-k} \right) z^{n-1} \right\}. \tag{8.15}$$

To evaluate each of the residues, we use the factorization of $z^N - 1$ to express the required residue in the form

$$\frac{\left(\omega_j\right)^N \left(\omega_j\right)^{n-1} \sum_{k=0}^{N-1} f(k) \left(\omega_j\right)^{-k}}{(\omega_j - \omega_0)(\omega_j - \omega_1) \cdots (\omega_j - \omega_{j-1}) \cdot (\omega_j - \omega_{j+1}) \cdots (\omega_j - \omega_{N-1})}$$

$$= \frac{\left(\omega_j\right)^n \sum_{k=0}^{N-1} f(k) \left(\omega_j\right)^{-k}}{\left(1 - \frac{\omega_0}{\omega_j}\right)\left(1 - \frac{\omega_1}{\omega_j}\right) \cdots \left(1 - \frac{\omega_{j-1}}{\omega_j}\right) \cdot \left(1 - \frac{\omega_{j+1}}{\omega_j} \cdots \right)\left(1 - \frac{\omega_{N-1}}{\omega_j}\right)}.$$

To obtain the desired result, it remains only to simplify and interpret this expression.

The most imposing aspect of the formula is the denominator

$$D = \left(1 - \frac{\omega_0}{\omega_j}\right)\left(1 - \frac{\omega_1}{\omega_j}\right) \cdots \left(1 - \frac{\omega_{j-1}}{\omega_j}\right) \cdot \left(1 - \frac{\omega_{j+1}}{\omega_j}\right) \cdots \left(1 - \frac{\omega_{N-1}}{\omega_j}\right).$$

However, this term in fact may be explicitly evaluated. In the first place, we note that each of the terms $\frac{\omega_k}{\omega_j}$ is *itself* an Nth root of unity. Moreover, the collection $\{\frac{\omega_k}{\omega_j}\}$ consists exactly of the $N - 1$ roots of unity distinct from the root $\omega_0 = 1$. Therefore, we can write the denominator D in the form

$$D = (1 - \omega_1)(1 - \omega_2)(1 - \omega_3) \cdots (1 - \omega_{N-1}).$$

We note that this result is independent of which of the N residues is being evaluated.[2]

The evaluation of D now follows from the factorization (8.14). We have

$$(z - \omega_1)(z - \omega_2)(z - \omega_3) \cdots (z - \omega_{N-1}) = \frac{z^N - 1}{z - \omega_0}$$

$$= \frac{z^N - 1}{z - 1}$$

$$= z^{N-1} + z^{N-2} + z^{N-3} + \cdots + z + 1.$$

---

[2] We have written the formulas as though an intermediate ω_j is under consideration, although it is easy to see that the result also holds for the "extreme" roots ω_0 and ω_{N-1}.

Hence, we evaluate the common denominator as

$$D = (1 - \omega_1)(1 - \omega_2)(1 - \omega_3) \cdots (1 - \omega_{N-1})$$
$$= 1^{N-1} + 1^{N-2} + 1^{N-3} + \cdots + 1 + 1$$
$$= N.$$

Using this result in the residue calculation, we express the periodic f in the form

$$f(n) = \sum_{j=0}^{N-1} (\omega_j)^n \left(\frac{1}{N} \sum_{k=0}^{N-1} f(k) (\omega_j)^{-k} \right) \tag{8.16}$$

$$= \sum_{j=0}^{N-1} \left(e^{\frac{2\pi i}{N} jn} \right) \left(\frac{1}{N} \sum_{k=0}^{N-1} f(k) \left(e^{-\frac{2\pi i}{N} jk} \right) \right).$$

This result should be compared with the Fourier series expansion

$$f(t) = \sum_{j=0}^{\infty} \left(e^{\frac{2\pi i}{T} jt} \right) \left(\frac{1}{T} \int_0^T f(\tau) \left(e^{-\frac{2\pi i}{T} j\tau} \right) d\tau \right) \tag{8.17}$$

appropriate to a periodic function f defined on the real axis. The functions $\{u_j\}$ (defined on $\mathbb{Z}^+$) by

$$u_j(n) = e^{\frac{2\pi i}{N} jn}$$

play the same role in (8.16) that the complex exponential functions play in the familiar expansion (8.17).

Expansions of this sort are discussed systematically in Section 8.5 below.

Problems 8.4.

1. Find the z-transforms of the sequences defined by

   a. $x(n) = \cos n\omega_0$,

   b. $x(n) = \begin{cases} 1 & 0 \le n \le N - 1, \\ 0 & n \ge N. \end{cases}$

   c. $x(n) = n^3$,

   d. $x(n) = \alpha^n \sin n\omega_0$.

2. Find the z-transforms of the sequences defined by

   a. $x(n) = \begin{cases} 0 & 0 \le 0 \le N - 1, \\ \cos(n - N)\omega_0 & N, \le n. \end{cases}$

b. $x(n) = \begin{cases} 1 & k \le n \le j, \\ 0 & \text{otherwise.} \end{cases}$

3. Compute the z-transforms of the following sequences:

   a. $x(n) = n \cos n\, \omega_0$,
   b. $x(n) = n^3 \alpha^n$.

4. Define a sequence $x(\cdot)$ by

$$x(n) = \begin{cases} \sin \frac{n\pi}{N}, & 0 \le n \le N, \\ 0 & n > N. \end{cases}$$

   By use of the shift theorem and the graphical techniques of Section 6.4, compute $\mathcal{Z}\{x(n)\}$.

5. Using the methods of Problem 3 compute the z-transforms of the following functions:

   a. $x(n) = \begin{cases} n & 0 \le n \le N - 1, \\ 0 & n \ge N, \end{cases}$

   b. $x(n) = \begin{cases} n & 0 \le n \le N - 1, \\ 2N - n & N \le n \le 2N, \\ 0 & n > 2N, \end{cases}$

   c. $x(n) = \begin{cases} n^2 & 0 \le n \le N, \\ 0 & n > N. \end{cases}$

6. Write out the real form of the z-transform inversion integral for the transforms computed in Problems 1a,b.

7. By use of the residue theorem and the complex z-transform inversion integral, invert each of the transforms of Problem 1.

8. Use the complex z-transform inversion method to invert the transforms of Problem 2.

9. Consider the general second-order difference equation

$$\alpha\, x(k + 2) + \beta\, x(k + 1) + \gamma\, x(k) = 0,$$

   with $x(0), x(1)$ prescribed.
   Assume that $\alpha \neq 0$, and find an expression for the general solution of such a recurrence relation. (Treat all special cases which arise.)

10. A time-invariant vector difference equation is a relation of the form

$$\mathbf{x}_{k+1} = \mathbf{A}\, \mathbf{x}_k, \quad \mathbf{x}_0 \text{ prescribed.}$$

Here $\mathbf{x}$ is an n-vector, and $\mathbf{A}$ represents an $n \times n$ (real or complex) matrix. Use the z-transform and inversion integral to find a representation of the solution in the form of a contour integral.

11. Use the result of Problem 10 above to prove that

$$\mathbf{A}^n = \frac{1}{2\pi i} \oint_\Gamma (\mathbf{I} z - \mathbf{A})^{-1} z^n \, dz$$

for any $n > 0$, and any complex (square) matrix $\mathbf{A}$. Explicitly describe the restrictions required on the simple closed contour Γ so that the above result holds.

12. Consider the scalar difference equation (recurrence relation)

$$x(k + N) + p_{N-1} x(k + N - 1) + \cdots + p_1 x(k + 1) + p_0 x(k) = 0.$$

Show that by defining a vector sequence

$$\mathbf{x}(k) = \begin{bmatrix} x(k) \\ x(k + 1) \\ \vdots \\ x(k + N - 1) \end{bmatrix}$$

it is possible to express the scalar recurrence relation in the equivalent form

$$\tilde{\mathbf{x}}_{k+1} = \mathbf{A} \, \tilde{\mathbf{x}}_k.$$

What is the form of the matrix $\mathbf{A}$ for this case?

13. Show that the governing equation for the random walk on a circle described in Problem 6, Section 8.1 may be written in the form

$$\mathbf{x}_{k+1} = \mathbf{A} \, \mathbf{x}_k,$$

where $\mathbf{x}_k$ is the vector whose j^{th} component is the probability that the particle is at location j at time k. Show that the entries of $\mathbf{A}$ are such that

$$a_{ij} = \alpha_{i-j}$$

where

$$\alpha_{j+N} = \alpha_j.$$

(Such a matrix $\mathbf{A}$ is referred to as a circulant matrix of order N).

14. Find the solution of the difference equation

$$x(k + 1) + x(k) = k, \ k > 0,$$
$$x(0) = 1.$$

15. Define the convolution of two z-transformable sequences $\{f(n)\}$, $\{g(n)\}$ by the sequence $\{h(k)\}$

$$h(k) = \sum_{n=0}^{k} f(k - n) g(n).$$

Show first that for $\rho \neq 0$ we have

$$h(k) \rho^{-k} = \sum_{n=0}^{k} f(k - n) \rho^{-(k)} g(n)$$

$$= \sum_{n=0}^{k} f(k - n) \rho^{-(k-n)} g(n) \rho^{-n}.$$

Regarding this last expression (for fixed k) as an inner product, of $(k + 1)$-dimensional vectors, use the Cauchy–Schwarz inequality to show that

$$|h(k) \rho^{-k}| \leq \left(\sum_{j=0}^{k} |f(j)|^2 \rho^{-2j} \right)^{\frac{1}{2}} \left(\sum_{j=0}^{k} |g(j)|^2 \rho^{-2j} \right)^{\frac{1}{2}}.$$

Use this result to prove that the convolution sequence is z-transformable, assuming that for some ρ, we have

$$\sum_{n=0}^{\infty} |f(n)|^2 \rho^{-2n} < \infty,$$

$$\sum_{n=0}^{\infty} |g(n)|^2 \rho^{-2n} < \infty.$$

16. Consider the linear homogeneous difference equation

$$x(k + N) + p_{N-1} x(k + N - 1) + \cdots + p_1 x(k + 1) + p_0 x(k) = 0, \ k \geq 0$$

with $x(0), x(1), \ldots x(N - 1)$ prescribed. Prove that the solution of this equation may be expressed in the form

$$x(k) = \sum_{j=1}^{m} \left(\alpha_{j0} + \alpha_{j1}\, k + \alpha_{j2}\, k^2 + \cdots + \alpha_{jm_j}\, k^{m_j} \right) \left(\beta_j \right)^k,$$

(that is, as a sum of polynomials in k multiplied by exponential functions of k).

17. A linear N^{th}-order difference equation with a forcing term takes the form

$$x(k+N) + p_{N-1}\, x(k+N-1) + \cdots + p_1\, x(k+1) + p_0\, x(k) = f(k), \ k \geq 0,$$

with $x(0), \ldots, x(N-1)$ prescribed.

Show that the solution of this equation takes the form of a sum of terms of the form of 16 above, together with a convolution sum

$$\sum_{n=0}^{k} g(k-n) f(n).$$

What is the weighting sequence g? Show that $g(0) = \cdots = g(N-1) = 0$ in this case.

18. Find an explicit expression for

$$\sum_{n=0}^{N} n^3.$$

19. Prove that the set of complex numbers $\{ \frac{\omega_k}{\omega_j} \}_{k \neq j}$ consists exactly of the $N-1$ out of the Nth roots of unity distinct from 1.

20. By use of z-transform methods, find an expression for

$$\sum_{n=1,3,5,\ldots}^{2N-1} n^2,$$

the sum of the squares of the first N odd integers.

8.5   Discrete Fourier Transforms

On the basis of the pervasive analogies between z-transforms and Laplace transforms exploited in the previous section, we must expect the existence of a transform theory appropriate to functions defined on the set of the integers, $\mathbb{Z}$.

Given such a function $f : \mathbb{Z} \mapsto \mathbb{C}$ it is possible to guess on this basis that an appropriate transform definition may be[3]

$$\hat{f}(\theta) = \sum_{n=-\infty}^{\infty} f(n) e^{-i\theta n}. \tag{8.18}$$

As usual, it is necessary to identify the class of functions for which the transform is defined. In view of the fact that the transform is evidently nothing more than a completely familiar Fourier series, the appropriate identifications follow readily from our discussions of the convergence properties of Fourier series (as well as from analogy to the similar considerations involved in the theory of Fourier transforms on the real line).

Definition. We say that the sequence $\{f(n)\}$ belongs to the class l^1 (or is absolutely summable) provided that

$$\sum_{n=-\infty}^{\infty} |f(n)| < \infty.$$

The sequence $\{g(n)\}$ is said to belong to the class l^2 (or to be square-summable) provided that

$$\sum_{n=-\infty}^{\infty} |g(n)|^2 < \infty.$$

These definitions delineate classes of functions (sequences) for which (8.18) is naturally defined.

Definition. Suppose that $f \in l^1$. Then we define the discrete Fourier transform $\hat{f}(\cdot)$ of f by

$$\hat{f}(\theta) = \sum_{n=-\infty}^{\infty} f(n) e^{-i\theta n}.$$

With the assumption that $f \in l^1$, it follows that the series defining $\hat{f}(\cdot)$ is absolutely convergent, and that $\hat{f}(\cdot)$ is a uniformly continuous function (regarded either as a function defined on the unit circle of the complex plane, or as a 2π-periodic function defined on the real numbers).

---

[3] Among the possible excuses for this guess, we cite the observation that points on the unit circle are associated with "sinusoidal" discrete-time functions, coupled with an incipient desire to make a connection with the z-transform of the previous section.

Definition. Suppose that $g \in l^2$. Then the *discrete Fourier transform* $\hat{g}$ of g is defined as

$$\hat{g}(\theta) = \sum_{n=-\infty}^{\infty} g(n) e^{-i\theta n}.$$

We recall from our consideration of Fourier series convergence that the condition that g belong to l^2 is the necessary and sufficient condition that the partial sums

$$\hat{g}_N(\theta) = \sum_{n=-N}^{N} g(n) e^{-i\theta n}$$

define a mean-square convergent sequence of functions on the interval $[0, 2\pi]$. The transform $\hat{g}$ is defined as the unique mean-square limit of this function sequence.

From these definitions, it is evident that the theory of discrete Fourier transforms is substantially identical to that of Fourier series. The difference (more properly distinction) between the two theories lies entirely in the point of view.

In our previous discussion of Fourier series, attention was focused on the functions defined on the interval $[0, 2\pi]$. The Fourier coefficient sequence may be regarded as the transform of such a function. From the present point of view, sequences (functions defined on the integers) have the conceptual status of the initial objects. The functions defined on $[0, 2\pi]$ then play the role of the transformed objects.[4]

One consequence of the identification of the discrete Fourier transform with the standard Fourier series is that many of the properties of this transform are "already known" from previous considerations.

Theorem (Discrete Fourier Inversion Theorem).
   Suppose that the sequence
$\{f(n)\}$ belongs to either l^1 or l^2. Then the sequence may be recovered from its transform according to the inversion integral

$$f(n) = \frac{1}{2\pi} \int_0^{2\pi} \hat{f}(\theta) e^{in\theta} \, d\theta.$$

The above theorem of course amounts to a restatement of the usual Fourier coefficient formula, adjusted for the fact that the coefficients of (8.18) are indexed in the reverse of the usual manner.[5]

---

[4]The systematic theory of such transforms is based on the theory of groups and is known as abstract harmonic analysis. A further example of this transform theory is discussed below as the finite Fourier transform.

[5]This choice is made for the sake of the convenience of the connection with the z-transform.

Example. Suppose that $f(n) = a^{|n|}$, where $|a| < 1$, ensuring that $f(\cdot)$ belongs to both l^1 and l^2. Then

$$\hat{f}(\theta) = \sum_{n=-\infty}^{\infty} a^{|n|} e^{-i\theta n}$$

$$= \sum_{n=-\infty}^{-1} a^{-n} e^{-i\theta n} + 1 + \sum_{n=1}^{\infty} a^n e^{-i\theta n}$$

$$= \sum_{k=1}^{\infty} a^k e^{i\theta k} + 1 + \sum_{n=1}^{\infty} a^n e^{-i\theta n}$$

$$= \frac{ae^{i\theta}}{1 - ae^{i\theta}} + 1 + \frac{ae^{-i\theta}}{1 - ae^{-i\theta}}$$

$$= \frac{1 - a^2}{1 - 2a \cos\theta + a^2}.$$

The inversion theorem then asserts that

$$a^{|n|} = \frac{1}{2\pi} \int_0^{2\pi} \left(\frac{1 - a^2}{1 - 2a \cos\theta + a^2} \right) e^{in\theta} d\theta.$$

We note that the inversion integral is not an example of what is commonly regarded as an elementary integral. A standard method for the evaluation of such integrals is the use of complex variable methods (this should be entirely predictable on the basis of the examples of the usual Fourier, Laplace, and z-transforms).

The usual form of the inversion formula is

$$f(n) = \frac{1}{2\pi} \int_0^{2\pi} \hat{f}(\theta) e^{in\theta} d\theta. \tag{8.19}$$

In order to evaluate this by complex variable methods, we must identify this expression as a parameterization of a contour integral. If the transform $\hat{f}$ is identifiable as the restriction to the unit circle (in the complex plane) of a function of a complex variable, then the desired identification is immediate.

We therefore assume that there exists a function g, of a complex variable ζ analytic in a neighborhood of $|\zeta| = 1$, and such that

$$\hat{f}(\theta) = g(e^{i\theta}), \ 0 \le \theta \le 2\pi.$$

The inversion integral (8.19) then takes the form of the contour integral

$$f(n) = \frac{1}{2\pi i} \oint_{|\zeta|=1} g(\zeta) \zeta^n \frac{d\zeta}{\zeta}. \tag{8.20}$$

Example. For the transform computed above, we may write

$$\hat{f}(\theta) = \frac{1 - a^2}{1 - a\left(e^{i\theta} + e^{-i\theta}\right) + a^2}.$$

The natural identification of g is

$$g(\zeta) = \frac{1 - a^2}{1 - a\left(\zeta + \frac{1}{\zeta}\right) + a^2}$$

$$= \frac{\zeta\left(1 - a^2\right)}{-a\zeta^2 + (1 + a^2)\,\zeta - a}.$$

The zeroes of the denominator of the integrand occur at

$$\zeta = \frac{(1 + a^2) \pm \sqrt{(1 + a^2)^2 - 4a^2}}{2a}$$

$$= \frac{(1 + a^2) \pm (1 - a^2)}{2a}$$

$$= \frac{1}{a},\ a.$$

Hence $g(\zeta)$ in factored form is given by

$$g(\zeta) = \frac{(a^2 - 1)\,\zeta}{(a\zeta - 1)(\zeta - a)}.$$

The contour inversion integral to be evaluated is

$$f(n) = \frac{1}{2\pi i} \oint_{|\zeta|=1} \frac{(a^2 - 1)\,\zeta}{(a\zeta - 1)(\zeta - a)}\, \zeta^n\, \frac{d\zeta}{\zeta}.$$

The poles associated with the integrand are illustrated in Figure 8.15.

Poles are located at $\zeta = a$, $\zeta = \frac{1}{a}$. In the case $n > 0$ these are the only poles of the integrand. In the case $n < 0$, there is also a pole of order $|n|$ at the origin.

For $n \geq 0$ we evaluate the integral as the residue at the pole $\zeta = a$. This gives

$$f(n) = \frac{(a^2 - 1)}{(a\,a - 1)}\, a^n = a^n,\ n \geq 0.$$

For the case that $n < 0$, the integral may be evaluated by regarding the contour as encircling the pole at $\zeta = \frac{1}{a}$ (on the Riemann sphere) in the negative sense.

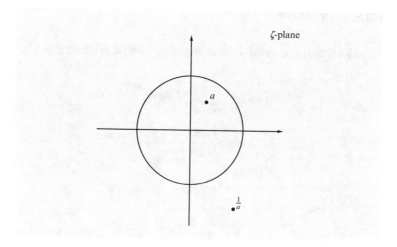

Fig. 8.15 Integrand pole locations

The evaluation of the integral by the residue at this single pole is then

$$f(n) = \frac{-(a^2 - 1)}{a\left(\frac{1}{a} - a\right)} \left(\frac{1}{a}\right)^n = \left(\frac{1}{a}\right)^n$$

$$= a^{|n|}, \; n < 0.$$

This second case may also be treated by changing the parameterization in the original contour integral conversion (see the problems below), or by evaluating the integral by including the residues at the pole at the origin with the calculations of the previous case.

In order to discuss the mechanical properties of the discrete Fourier transform, it is useful to have available an operational notation for the transform analogous to the $\mathcal{L}\{\}$, $\mathcal{F}\{\}$ and $\mathcal{Z}\{\}$ notations employed above. Our notation is given by

$$F\{f(n)\} = \hat{f}(\theta) = \sum_{n=-\infty}^{\infty} f(n)\, e^{-i\theta n}. \tag{8.21}$$

8.6   Discrete Fourier Transform Properties

The transform is clearly a linear operation, and beyond this the formal properties follow the pattern previously encountered.

n-Multiplication Law

Suppose that $\{nf(n)\}$ is a transformable sequence (that is, in either l^1 or l^2). Then

$$F\{nf(n)\} = \sum_{n=-\infty}^{\infty} nf(n)\, e^{-i\theta n}$$

$$= \frac{1}{-i}\frac{d}{d\theta} \sum_{n=-\infty}^{\infty} f(n)\, e^{-i\theta n}$$

$$= i\frac{d}{d\theta}\hat{f}(\theta)$$

$$= i\frac{d}{d\theta} F\{f(n)\}.$$

Example. If $f(n) = a^{|n|}$, $|a| < 1$, then

$$F\left\{n\, a^{|n|}\right\} = i\frac{d}{d\theta}\left(\frac{1-a^2}{1-2a\cos\theta + a^2}\right)$$

$$= \frac{i\,(1-a^2)\,2a\,\sin\theta}{(1-2a\cos\theta + a^2)^2}.$$

Exponential Shift Law

If $F\{f(n)\} = \hat{f}(\theta)$, then

$$F\left\{e^{in\Omega} f(n)\right\} = \sum_{n=-\infty}^{\infty} f(n)\, e^{in\Omega}\, e^{-i\theta n}$$

$$= \sum_{n=-\infty}^{\infty} f(n)\, e^{in(\Omega-\theta)}$$

$$= \hat{f}(\theta - \Omega).$$

This shift formula therefore takes the same form as the continuous-time transform version. This is also the case for the relation between a delay operation and the transform.

Delay Law

Suppose that

$$F\{f(n)\} = \hat{f}(\theta),$$

then

$$F\{f(n-N)\} = \sum_{n=-\infty}^{\infty} f(n-N)\,e^{-i\theta n}$$

$$= \sum_{k=-\infty}^{\infty} f(k)\,e^{-i\theta(k+N)}$$

$$= e^{-iN\theta}\,\hat{f}(\theta).$$

The use of this shift law allows the treatment of difference equations defined over all the integers. The connections between these problems and the (positive integer) difference equations of the previous section are completely parallel to the situation encountered earlier in the Fourier/Laplace transform treatments of differential equations. We reserve the derivation of these results for the problems below.

Convolution Law

The convolution theorems for discrete Fourier transforms take the expected form. The convolution of two sequences is defined by

$$h(n) = [g \star f](n)$$

$$= \sum_{k=-\infty}^{\infty} g(n-k)f(k).$$

In order that the sequence $\{h(n)\}$ should be transformable in the sense used above, we require that at least one of the convolved sequences $\{g(n)\}$, $\{f(n)\}$ be absolutely summable. This hypothesis is sufficient to guarantee that $\{h(n)\}$ is transformable.[6]

---

[6]See the problems below.

Theorem. *Suppose that one of* $\{g(n)\}$, $\{f(n)\}$ *belongs to* l^1 *and that the other sequence is transformable. Then*

$$F\{h(n)\} = F\{[g \star f]\}$$
$$= F\{g(n)\}\, F\{f(n)\}.$$

This formula is simply the usual one associated with multiplication of absolutely convergent series in the case that both of the convolved sequences are absolutely summable. In the case that one of the sequences is merely square-summable, an additional limiting argument is involved in the detailed proof.

In addition to the time domain convolution above, there is a frequency domain version of the convolution theorem.

Theorem. *Suppose that* $\{f(n)\}$, $\{g(n)\}$ *are both transformable, with discrete Fourier transforms*

$$\hat{f}(\theta) = F\{f(n)\},$$
$$\hat{g}(\theta) = F\{g(n)\}.$$

Define a function $\hat{h}$ *(periodic of period* 2π*) as the convolution of the transforms* $\hat{f}$ *and* $\hat{g}$.

$$\hat{h}(\theta) = \int_0^{2\pi} \hat{g}(\theta - \phi)\hat{f}(\phi)\, d\phi.$$

Then $\hat{h}$ *is the discrete Fourier transform of a transformable sequence* $\{h(n)\}$ *where*

$$h(n) = 2\pi\, g(n)f(n).$$

Proof. The demonstration that the convolution defines $\hat{h}$ as a square-integrable function of θ (sufficient to identify it as the transform of a square-summable sequence) is yet another exercise in the Cauchy–Schwarz inequality (see Problem 12 below.)

The sequence $\{h(n)\}$ is determined by inverse transforming the expression.

$$h(n) = \frac{1}{2\pi} \int_0^{2\pi} e^{in\theta}\, \hat{h}(\theta)\, d\theta$$
$$= \frac{1}{2\pi} \int_0^{2\pi} e^{in\theta} \left\{ \int_0^{2\pi} \hat{g}(\theta - \phi)\hat{f}(\phi)\, d\phi \right\} d\theta.$$

Our hypotheses guarantee that the double integral

$$\int_0^{2\pi} \int_0^{2\pi} |\hat{g}(\theta - \phi)\hat{f}(\phi)| \, d\phi \, d\theta$$

is convergent. We evaluate the integral by interchanging the order of integration. This gives successively

$$h(n) = \frac{1}{2\pi} \int_0^{2\pi} \left\{ \int_0^{2\pi} \hat{g}(\theta - \phi)\hat{f}(\phi) e^{in\theta} \, d\theta \right\} d\phi$$

$$= \frac{1}{2\pi} \int_0^{2\pi} \left\{ \int_{-\phi}^{2\pi-\phi} \hat{g}(\psi)\hat{f}(\phi) e^{in(\phi+\psi)} \, d\psi \right\} d\phi$$

$$= \frac{1}{2\pi} \int_0^{2\pi} \int_0^{2\pi} \hat{g}(\psi)\hat{f}(\phi) e^{in\phi} e^{in\psi} \, d\psi \, d\phi,$$

(using the fact that $g(\psi) e^{in\psi}$ is periodic of period 2π)

$$= (2\pi) \left(\frac{1}{2\pi} \int_0^{2\pi} \hat{g}(\psi) e^{in\psi} \, d\psi \right) \left(\frac{1}{2\pi} \int_0^{2\pi} \hat{f}(\phi) e^{in\phi} \, d\phi \right)$$

$$= (2\pi) g(n) f(n).$$

The verbal description of the above results is that a time convolution corresponds to a multiplication in the transform (frequency) domain, a frequency domain convolution corresponds to (2π times) a multiplication in the time domain.

Software

Octave has a convolution operation appropriately named `conv`. It is usually introduced (early in the Octave documentation) as polynomial multiplication, with coefficients represented as Octave arrays. A little experimentation shows that it respects delays (one would think a polynomial multiplier might throw away superfluous zero coefficients). It corresponds to the convolution of finite, one-sided sequences, with a transform product corresponding to

$$\left(a_0 + a_1 e^{-i\theta} + a_2 e^{-i2\theta} + a_n e^{-in\theta}\right) \left(b_0 + b_1 e^{-i\theta} + b_2 e^{-i2\theta} + b_m e^{-im\theta}\right).$$

OCTAVE

```
a = [1 2 3 4 5];
b = [0 1 0];
conv(a, b)

ans =

   0    1    2    3    4    5    0

b = [0 0 2];
conv(a, b)

ans =

   0    0    2    4    6    8   10

b = [1 -2 1];
conv(a, b)

ans =

   1    0    0    0    0   -6    5
```

The last computation is a hint of edge effects in digital signal processing. The variable a is a linear progression, so one would naturally expect the second difference to vanish. It does, except where the differencing overlaps the boundaries of the data.

These effects also occur with a segment of sinusoid and data generated by uniform sampling.

OCTAVE

```
fs = 1000;
t = 0: 1/fs: 6*pi;
s = sin(t);
plot(s)

d = fs*[-.5 0 .5];
diff = conv(s, d);
plot(diff)

diffsq = conv(diff, d);
```

A plot of the first difference is given in Figure 8.16.

A plot of the second difference is dominated by the edge effects. This disappears if the boundary data is omitted.

OCTAVE

```
plot(diffsq(5:end-5));

print -deps diffsq55.eps
```

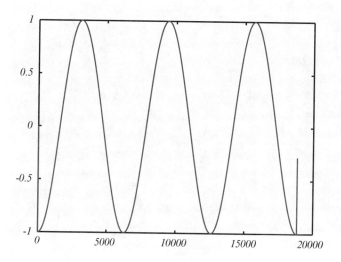

Fig. 8.16 Numerical symmetric difference by convolution.

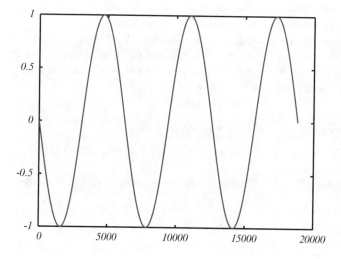

Fig. 8.17 Second difference omitting five end boundary samples.

The results are shown in Figures 8.17 and 8.18. Figure 8.18 shows the result of computing a slope between an augmented 0 and an original step edge from the data boundary. The resulting large value determines the vertical scale in the plot.

Some applications of discrete transform methods are described in the following section, where certain difference equations are treated. However, it is appropriate

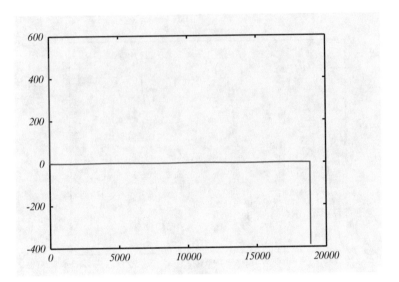

Fig. 8.18 Second difference boundary edge effects.

to make separate mention of the notion of a discrete delta function, and the role of convolutions in the solution of forced difference equations.

Such topics can be discussed in the framework of vector-matrix difference equations. However (as in the case of ordinary differential equations) the basic ideas may be discovered through the example of a single N^{th}-order equation. This is written in the form $(-\infty < n < \infty)$

$$x(n + N) + p_{N-1} x(n + N - 1) + p_{N-2} x(n + N - 2)$$
$$+ \cdots + p_1 x(n + 1) + p_0 x(n) = f(n). \qquad (8.22)$$

The close analogy between this equation and the ordinary differential equation

$$p(\frac{d}{dt}) x(t) = f(t)$$

may be brought out by defining a shift operator E, defined (together with its iterates) by

$$E x(n) = x(n + 1),$$

$$E^2 x(n) = E(E x(n)) = x(n + 2),$$

$$\vdots$$

$$E^N x(n) = x(n + N).$$

With the added convention that

$$E^0 x(n) = x(n),$$

the difference equation may be compactly written as

$$p(E) x(n) = f(n), \tag{8.23}$$

where

$$p(E) = E^N + p_{N-1} E^{N-1} + \cdots + p_1 E + p_0 E^0.$$

Equation (8.23) may be treated by means of the discrete Fourier transform, provided that one makes the explicit a priori assumptions that $\{f(n)\}$ is transformable, and that a transformable solution sequence $\{x(n)\}$ exists.[7]

Making such assumptions, we may transform the constant coefficient difference equation (using the delay law) to obtain

$$F\{p(E) x(n)\} = p(e^{i\theta}) F\{x(n)\}$$
$$= p(e^{i\theta}) \hat{x}(\theta)$$
$$= F\{f(n)\}$$
$$= \hat{f}(\theta),$$

and finally

$$\hat{x}(\theta) = \frac{1}{p(e^{i\theta})} \hat{f}(\theta).$$

Assuming that

$$\frac{1}{p(e^{i\theta})}$$

is the discrete Fourier transform of a summable sequence, the solution transform may be inverted by use of the convolution law to obtain

$$x(n) = \sum_{k=-\infty}^{\infty} g(n-k) f(k) \tag{8.24}$$

---

[7]The condition that this be the case is the same as the condition required to guarantee that (8.23) has a transformable solution for arbitrary forcing functions f. This is simply that $p(e^{i\theta}) \neq 0$, $0 \leq \theta < 2\pi$.

where

$$g(n) = F^{-1}\left\{\frac{1}{p(e^{i\theta})}\right\} = \frac{1}{2\pi}\int_0^{2\pi}\frac{1}{p(e^{i\theta})}e^{in\theta}\,d\theta. \tag{8.25}$$

In the case of the transform solution of the ordinary differential equation, the function

$$g(t) = \mathcal{L}^{-1}\left\{\frac{1}{p(s)}\right\}$$

was referred to as the impulse response (weighting pattern) associated with the differential equation. The same terminology is applied to the function (sequence) defined by (8.25).

A source for this terminology is provided by consideration of the discrete delta function (delta function sequence) δ defined by

$$\delta(n) = \begin{cases} 1 & n = 0, \\ 0 & n \neq 0. \end{cases}$$

This sequence has the properties that

$$\sum_{n=-\infty}^{\infty} f(n)\,\delta(n) = f(0),$$

$$\sum_{n=-\infty}^{\infty} f(n-k)\,\delta(k) = \sum_{n=-\infty}^{\infty} \delta(n-k)f(k) = f(n),$$

and

$$F\{\delta(n)\} = \sum_{n=-\infty}^{\infty} \delta(n)\,e^{in\theta} = 1.$$

These properties are formal analogues of the properties of the continuous-time delta function. The additional theory associated with the discrete-time version of the delta function is essentially nonexistent, as the object is simply a distinguished discrete-time function.

Consideration of the special case of (8.23) given by

$$p(E)x(n) = \delta(n)$$

provides the solution

$$\hat{x}(\theta) = \frac{1}{p(e^{i\theta})} F\{\delta(n)\}$$

$$= \frac{1}{p(e^{i\theta})},$$

$$x(n) = F^{-1}\left\{\frac{1}{p(e^{i\theta})}\right\} = g(n).$$

This calculation justifies the use of the term impulse response for g of (8.25). The general solution may also be written in the form

$$x(n) = \sum_{k=-\infty}^{\infty} g(n-k)f(k),$$

giving the solution in the form of a weighted sum of (delayed) impulse responses. This interpretation is amplified in the problems below.

Parseval's Theorem

The Parseval Theorem for continuous-time transforms takes the form

$$\int_{-\infty}^{\infty} |f(t)|^2 \, dt = \frac{1}{2\pi} \int_{-\infty}^{\infty} |\hat{f}(\omega)|^2 \, d\omega.$$

Combined with the inversion integral

$$f(t) = \frac{1}{2\pi} \int_{-\infty}^{\infty} \hat{f}(\omega) e^{i\omega t} \, d\omega$$

this provides a powerful source of intuition in the use of transform methods. This use is particularly apparent in various applications in communications problems.

This frequency domain point of view also plays a prominent role in certain applications of discrete-time transforms, so that it is useful to consider the discrete transform version of Parseval's theorem in this light.[8]

Periodic discrete functions were encountered in the previous section, where a discrete-time analogue of the complex form of Fourier series was derived by z-transform methods. This procedure led to the functions $(j = 0, 1, \ldots, N-1)$

---

[8]The "bare formula" is of course already familiar from encounters with Fourier series. In effect, the new interpretation is the only point at issue.

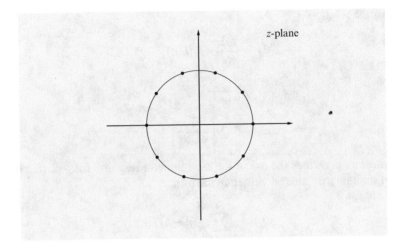

Fig. 8.19 Roots of unity

$$u_j(n) = e^{\frac{2\pi i}{N} jn}, \tag{8.26}$$

periodic of period N. These functions (since they are in fact simply powers of roots of unity) are naturally associated with points on the unit circle of the complex plane (See Figure 8.19).

In fact, for any point on the unit circle in the complex plane (identified with an angle Ω, $-\pi < \Omega \le \pi$), with the property that $\frac{\Omega}{2\pi}$ is a rational number, there is associated a periodic function

$$u_\Omega(n) = e^{i\Omega n}.$$

If the ratio $\frac{\Omega}{2\pi}$ is not rational, then the expression still defines what might be called an oscillatory (but not periodic) function of the integer valued variable n. Functions of this form (with arbitrary real Ω) are intrinsically identified with the interval $(-\pi, \pi]$ since (for integers k)

$$u_{\Omega+2\pi k}(n) = e^{i(\Omega+2\pi k)n}$$
$$= e^{i\Omega n} e^{i(2\pi k)n}$$
$$= e^{i\Omega n}$$
$$= u_\Omega(n).$$

This shows that an arbitrary Ω is equivalent (as far as the values of the sequence are concerned) to some Ω' for $-\pi < \Omega' \le \pi$. This equivalence provides the desired identification.

The inversion integral for the discrete Fourier transform takes the form

$$f(n) = \frac{1}{2\pi} \int_0^{2\pi} \hat{f}(\theta) e^{in\theta} \, d\theta. \tag{8.27}$$

This may be interpreted as the representation of the (arbitrary) transformable sequence $\{f(n)\}$ as a continuous weighted sum of the oscillatory functions $u_\theta(n)$.[9]

Since discrete-time Fourier transforms are "really" conventional Fourier series, the appropriate form of Parseval's theorem is just that appropriate to Fourier series.

Theorem (Parseval's Theorem).

Suppose that $\{f(n)\}$ is a square-summable sequence, and that $\hat{f}$ denotes its discrete Fourier transform. Then

$$\sum_{n=-\infty}^{\infty} |f(n)|^2 = \frac{1}{2\pi} \int_{-\pi}^{\pi} |\hat{f}(\theta)|^2 \, d\theta.$$

The present interpretation of this relation is that it represents the partition of the energy in the sequence $\{f(n)\}$ into the sum of incremental power contributions at the frequencies θ, for $-\pi < \theta < \pi$.[10]

In this section we have emphasized the basic mechanics of discrete Fourier transform theory. Applications of these techniques are included in the problems and in the following section.

Problems 8.6.

1. a. Define the sequence $f(\cdot)$ by

$$f(n) = \begin{cases} 0 & n < 0, \\ \beta^n & n \geq 0, \ |\beta| < 1. \end{cases}$$

   Compute $F\{f(n)\}$, and explicitly recover $f(\cdot)$ from the inversion integral by evaluating the required expression.
   b. Repeat 1a for the sequence g defined by

$$g(n) = \begin{cases} \alpha^{|n|} & n \leq 0, \\ 0 & n > 0. \end{cases}$$

---

[9]We note that the frequency range is $|\theta| < \pi$, so that the range of frequencies is limited. Intuitive support for this idea may be found in digital systems, in which frequencies are derived by dividing down the clock frequency. The clock (highest frequency) corresponds to $\theta = \pi$, since $e^{in\pi} = (-1)^n$.

[10]The reader may ponder the fact that the labels "time" and "frequency" are in the present situation exactly reversed from the interpretation in Chapter 2. This is another aspect of the duality evident throughout transform theory.

2. Evaluate the inversion integral ($|a| < 1$)

$$f(n) = \frac{1}{2\pi i} \oint_{|\zeta|=1} \frac{(a^2 - 1)\,\zeta}{(a\zeta - 1)(\zeta - a)}\, \zeta^{n-1}\, d\zeta$$

for the case $n < 0$ by explicitly evaluating the residues of the integrand at the poles inside of the contour $|\zeta| = 1$.

3. The contour integral of Problem 2 arises from the identification of

$$\frac{1}{2\pi} \int_0^{2\pi} \frac{(a^2 - 1)}{1 - 2a\cos\theta + a^2}\, e^{in\theta}\, d\theta$$

as a contour integral by means of the parameterization

$$\zeta = e^{i\theta},$$

$$d\zeta = i\,e^{i\theta}\, d\theta$$

and the identifications

$$e^{in\theta} = \zeta^n,$$

$$d\theta = \frac{d\zeta}{\zeta},$$

$$\frac{(a^2 - 1)}{1 - 2a\cos\theta + a^2} = \frac{(a^2 - 1)\,\zeta}{(a\zeta - 1)(\zeta - a)} \bigg|_{\zeta = e^{i\theta}}.$$

Show that the integral may also be parameterized by

$$\zeta' = e^{-i\theta},$$

$$d\zeta' = -i\,e^{-i\theta}\, d\theta$$

with the identification

$$e^{in\theta} = \left(\zeta'\right)^{-n}.$$

Use this parameterization to evaluate the inversion integral in the case that $n < 0$.

4. Compute the transform of the sequence

$$x(n) = \begin{cases} 1 & |n| \le N, \\ 0 & \text{otherwise.} \end{cases}$$

Explicitly evaluate the inversion integral.
5. Compute the convolution

$$[x * x] (n)$$

where $x(n)$ is as in Problem 4 above.
6. Suppose that $R(\cdot)$ is a rational function,

$$R(e^{i\theta}) = \frac{q(e^{i\theta})}{p(e^{i\theta})}.$$

Show that $R(e^{i})$ is the discrete Fourier transform of a transformable sequence (as defined above) if and only if the polynomial p has no zero of unit magnitude. (Assume that R is expressed in lowest terms. Why is it not necessary to assume that R is "proper"?)

7. Suppose that in Problem 6 above, the polynomial p has distinct zeroes (none of unit magnitude). By use of partial fractions methods, provide an expression for the inverse transform of

$$\hat{f}(\theta) = R(e^{i\theta}).$$

8. Give a graphical interpretation of the convolution calculation

$$h(n) = \sum_{k=-\infty}^{\infty} g(n-k)f(k).$$

9. Suppose that the sequence f vanishes for arguments greater than N in magnitude, while g vanishes for arguments of magnitude greater than M. Prove that the convolution of f and g vanishes for arguments greater than $M + N$ in magnitude.

10. Suppose that $f \in l^1$ and $g \in l^1$. Show that $h = [f * g] \in l^1$, so that h is transformable.
   Hint: $|h(n)| < \sum_{k=-\infty}^{\infty} |f(n-k)| \, |g(k)|$.

11. Suppose that $f \in l^1$, and $g \in l^2$. Define (the truncated sequence) $\{g_N\}$ by

$$g_N(n) = \begin{cases} g(n) & |n| \leq N, \\ 0 & \text{otherwise.} \end{cases}$$

Show that (using the delay law)

$$F\{[f \star g_N]\} = \hat{f}(\theta) \, \hat{g}_N(\theta).$$

Show that the sequence of square-integrable functions

$$\hat{h}_N(\theta) = \hat{f}(\theta)\,\hat{g}_N(\theta)$$

converges in the mean-square sense to

$$\hat{h}(\theta) = \hat{f}(\theta)\,\hat{g}(\theta).$$

Use Parseval's theorem to conclude that

$$\sum_{k=-\infty}^{\infty} f(n-k)\,g(k)$$

defines a sequence $h \in l^2$, with transform $\hat{h} = \hat{f}\,\hat{g}$.

12. Suppose that $f \in l^2$, $g \in l^2$. Use the Cauchy–Schwarz inequality to show that

$$\hat{h}(\theta) = \frac{1}{2\pi} \int_0^{2\pi} \hat{f}(\theta - \phi)\,\hat{g}(\phi)\,d\phi$$

defines $\hat{h}$ as a square-integrable function, and hence (by Parseval's equality) as the discrete Fourier transform of a square-summable sequence.

13. Show that the difference equation

$$p(E)\,x(n) = f(n),\ -\infty < n < \infty,$$

has a transformable solution x for any transformable forcing sequence $\{f(n)\}$, provided that

$$p(e^{i\theta}) \neq 0,\ 0 \leq \theta < 2\pi.$$

14. Show by means of an example that the conclusion of Problem 13 does not necessarily follow if for some θ_0,

$$p(e^{i\theta_0}) = 0.$$

   Hint: Consider a problem solvable by z-transform methods.

15. Consider the difference equation problems

   (P1): $p(E)\,x(n) = f(n),\ -\infty < n < \infty,$
   (P2): $p(E)\,\tilde{x}(n) = f(n),\ n \geq 0,$ with the initial conditions $\tilde{x}(0) = \tilde{x}(1) = \dots = \tilde{x}(N-1) = 0.$

   Suppose that f is a discrete Fourier transformable sequence such that $f(n) = 0$, $n < 0$. Show that both (P1) and (P2) have solutions, and

$$\tilde{x}(n) = x(n), n \geq 0,$$

if and only if the polynomial $p(\cdot)$,

$$p(z) = z^N + p_{N-1} z^{N-1} + \cdots + p_0$$

has all of its zeroes strictly inside the unit circle of the complex plane.
Hint: Consider Problems 13 and 14 above.

16. A forced vector difference equation takes the form

$$\mathbf{x}_{k+1} = \mathbf{A}\,\mathbf{x}_k + \mathbf{B}\,\mathbf{f}_k, \quad -\infty < k < \infty,$$

where $\{f_k\}$ is a transformable vector-valued sequence. Show that this model has a solution provided that the matrix $\mathbf{A}$ has no eigenvalues of unit magnitude.

17. Compute the transform of a delayed unit impulse sequence

$$\delta(n - l).$$

Show that any sequence $\{f_N(n)\}$ vanishing outside of the interval $[-N, N]$ is expressible as a weighted sum of delayed unit impulse sequences.

$$f_N(n) = \sum_{l=-N}^{N} \alpha_l\, \delta(n - l).$$

Assuming $p(e^{i\theta}) \neq 0$, for $0 < \theta \leq 2\pi$, interpret the solution x to

$$p(E)\,x(n) = f_N(n)$$

as a weighted sum of impulse responses.

18. State and prove a version of Parseval's Theorem applicable to the z-transform.

8.7   Some Applications of Discrete Transform Methods

In this section we treat some examples of problems involving difference equations, partial difference equations, and partial differential difference equations by means of discrete transform methods.

The problems treated arise naturally from some of our earlier discussions. The methods used may also be regarded as natural adaptations of the discrete transform techniques to the problems posed.

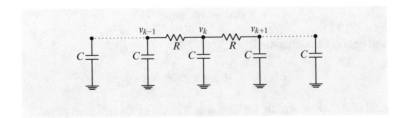

Fig. 8.20 RC ladder network

Example. In Section 3.5 the *RC* ladder network of Figure 8.20 was introduced as an analogue of the diffusion equation

$$\frac{\partial u}{\partial t} = \kappa \frac{\partial^2 u}{\partial x^2}, \; 0 < x < 1.$$

The governing equation of the ladder network takes the form

$$\frac{dv_j}{dt} = \frac{1}{RC} \left[v_{j+1} - 2\, v_j + v_{j-1} \right]. \tag{8.28}$$

In the case of the standard diffusion equation, the model is supplemented by an initial condition, together with the boundary conditions

$$u(0, t) = u(1, t) = 0.$$

With the identification of $v_j(t)$ in (8.28) with (an approximation to) $u(\frac{j}{N}, t)$ of the partial differential equation, the system (8.28) is augmented with the "boundary terminations"

$$v_0 = v_N = 0.$$

With these conditions, (8.28) may be written in the vector-matrix form

$$\frac{d}{dt} \begin{bmatrix} v_1 \\ v_2 \\ \vdots \\ v_{N-1} \end{bmatrix} = \frac{1}{RC} \begin{bmatrix} -2 & 1 & & & \\ 1 & -2 & 1 & & \\ & 1 & \ddots & \ddots & \\ & & & -2 & 1 \\ & & & 1 & -2 \end{bmatrix} \begin{bmatrix} v_1 \\ v_2 \\ \vdots \\ v_{N-1} \end{bmatrix}.$$

From Section 3.5 we recall that the component value choices satisfying

$$\frac{1}{RC} = \kappa N^2$$

provide the desired identification between the models. We therefore consider the
model

$$
\frac{d}{dt}
\begin{bmatrix} v_1 \\ v_2 \\ \vdots \\ v_{N-1} \end{bmatrix}
= \kappa \, N^2
\begin{bmatrix}
-2 & 1 & & & \\
1 & -2 & 1 & & \\
& 1 & \ddots & \ddots & \\
& & & -2 & 1 \\
& & & 1 & -2
\end{bmatrix}
\begin{bmatrix} v_1 \\ v_2 \\ \vdots \\ v_{N-1} \end{bmatrix}
\tag{8.29}
$$

with the intent of comparing solutions of this analogue system with those of the
boundary value problem.

The discrete model is a system of first-order ordinary differential equations
which can be solved by various means. A general procedure outlined in Chapter 6
involves the use of Laplace transforms. This procedure may in fact be employed
here, although the details are somewhat tedious.

The Laplace transform method applies (in principle) to an arbitrary first-order
system. Special simplifications occur in the case in which the coefficient matrix
may be diagonalized.

Example. The coefficient matrix of (8.29) is in fact self-adjoint (this observation
should be expected on the basis of Chapter 4), so that the system of equations can
be diagonalized by means of an orthogonal change of basis. The solution of the
initial value problem may be expressed in terms of the eigenvalues and eigenvectors
of the coefficient matrix. To simplify subsequent notation, we consider the matrix $\mathbf{L}$
(of dimension $(N-1) \times ((N-1))$ given by

$$
\mathbf{L} =
\begin{bmatrix}
-2 & 1 & & & \\
1 & -2 & 1 & & \\
& 1 & \ddots & \ddots & \\
& & & -2 & 1 \\
& & & 1 & -2
\end{bmatrix}.
\tag{8.30}
$$

It is easy to show (arguing as in Chapter 4) that $\mathbf{L}$ is negative definite. We use
the notation $\{-\lambda_i^2\}_{i=1}^{N-1}$ for the eigenvalues of $\mathbf{L}$, and $\{\mathbf{X}_i\}_{i=1}^{N-1}$ for the corresponding
eigenvector set.

Then

$$
\mathbf{L}\,\mathbf{X}_i = -\lambda_i^2\,\mathbf{X}_i, \ i = 1 \ldots (N-1),
$$

and (see Chapter 4) the matrix $\mathbf{L}$ has the expansion

$$
\mathbf{L} = \sum_{i=1}^{N-1} -\lambda_i^2\,\mathbf{X}_i\,\mathbf{X}_i^*.
$$

The solution of (8.29) satisfying the initial condition

$$
\begin{bmatrix} v_1 \\ v_2 \\ \vdots \\ v_{N-1} \end{bmatrix}(0) = \begin{bmatrix} v_1(0^+) \\ v_2(0^+) \\ \vdots \\ v_{N-1}(0^+) \end{bmatrix}
$$

is given by (recall Section 3.6, equation (3.21))

$$
\begin{bmatrix} v_1 \\ v_2 \\ \vdots \\ v_{N-1} \end{bmatrix}(t) = \sum_{i=1}^{N-1} e^{-\kappa N^2 \lambda_i^2 t} \mathbf{X}_i \left(\begin{bmatrix} v_1(0^+) \\ v_2(0^+) \\ \vdots \\ v_{N-1}(0^+) \end{bmatrix}, \mathbf{X}_i \right). \tag{8.31}
$$

This form is of course strongly reminiscent of the usual solution of the boundary value problem in the form

$$
u(x,t) = \sum_{n=1}^{\infty} e^{-\kappa(n\pi)^2 t} X_n(x) \, (u(x,0), X_n(x)) \tag{8.32}
$$

$$
= \sum_{n=1}^{\infty} e^{-\kappa(n\pi)^2 t} \sin n\pi x \left\{ 2 \int_0^1 u(\zeta,0) \sin n\pi \zeta \, d\zeta \right\}.
$$

The similarity at this stage rests entirely on the self-adjointness of the coefficient matrix in (8.29). In order to verify the extent to which the solution (8.32) may be considered to be simulated (approximated) by (8.31), it is necessary to investigate the relation of the eigenvalues and eigenvectors of (8.30) to those associated with the boundary value problem. It is at this stage that the difference equations and transform methods of the previous sections come into play.

The eigenvalue-eigenvector equation takes the form

$$
\mathbf{L} \begin{bmatrix} x_1 \\ x_2 \\ \vdots \\ x_{N-1} \end{bmatrix} = -\lambda^2 \begin{bmatrix} x_1 \\ x_2 \\ \vdots \\ x_{N-1} \end{bmatrix}. \tag{8.33}
$$

Writing out the component equations gives

$$
x_2 - 2x_1 = -\lambda^2 x_1,
$$

$$
x_{k+2} - 2xk + 1 + x_k = -\lambda^2 x_{k+1}, \; 1 \le k \le N-3,
$$

$$
-2x_{N-1} + x_{N-2} = -\lambda^2 x_{N-1}.
$$

Solving this system of equations is equivalent to finding a solution of the difference equation

$$x_{k+2} - 2x_{k+1} + x_k = -\lambda^2 x_{k+1} \tag{8.34}$$

satisfying $x_0 = 0, x_N = 0$. With these boundary conditions the solution of the difference equation also includes the "end cases" of the matrix-vector component equations.

This problem is a two-point boundary value problem for the difference equation (8.34), since values of the solution are specified at both ends of the interval.

The difference equation may be solved by use of z-transforms. Introducing the z-transform

$$X(z) = \mathcal{Z}\{x_k\}$$

we obtain

$$z^2 X(z) - z^2 x_0 - z x_1 - (2 - \lambda^2)[z X(x) - z x_0] + X(z) = 0.$$

The required initial condition is $x_0 = 0$. x_1 may be taken as unity for convenience, since it amounts to an eigenvector scaling factor. Then

$$\left[z^2 - (2 - \lambda^2)z + 1\right] X(z) = z,$$

and the z-transform inversion theorem provides

$$x_k = \frac{1}{2\pi i} \oint \frac{z}{z^2 - (2 - \lambda^2)z + 1} z^{k-1}\, dz,$$

where the contour is chosen sufficiently large to include the poles of the integrand. These poles occur at the zeroes of the polynomial

$$z^2 - (2 - \lambda^2)z + 1.$$

If these zeroes are denoted by z_1, z_2, then

$$z^2 - (2 - \lambda^2)z + 1 = (z - z_1)(z - z_2) = z^2 - (z_1 + z_2)z + z_1 z_2. \tag{8.35}$$

This factorization provides the information that

$$z_1 z_2 = 1$$

and

$$z_2 = \frac{1}{z_1}.$$

In particular, $z_1 \neq 0$, and so may be represented as an exponential, say

$$z_1 = e^{\theta},$$

where $\theta = \alpha + i\beta$ is the principal logarithm of the (complex) number z_1.

In terms of θ, the factorization of the characteristic polynomial takes the form

$$z^2 - (2 - \lambda^2)z + 1 = (z - z_1)(z - z_2)$$
$$= (z - e^{\theta})(z - e^{-\theta})$$
$$= z^2 - 2\cosh\theta\, z + 1.$$

From this, the relationship between the eigenvalue $-\lambda^2$ and the parameter θ is given by

$$(2 - \lambda^2) = 2\cosh\theta. \tag{8.36}$$

The motivation behind the introduction of the parameter θ is that it facilitates the application of the boundary condition at $k = N$. Evaluating the inversion integral by use of the residues at the poles $z = e^{\theta}$, $z = e^{-\theta}$ we obtain

$$x_k = \frac{1}{2\pi i} \oint \frac{z^k}{(z - z_1)(z - z_2)} \tag{8.37}$$
$$= \frac{z_1^k}{(z_1 - z_2)} + \frac{z_2^k}{(z_2 - z_1)}$$
$$= \frac{e^{k\theta} - e^{-k\theta}}{e^{\theta} - e^{-\theta}}$$
$$= \frac{\sinh k\theta}{\sinh \theta}.$$

We now apply the boundary condition $x_N = 0$ to obtain

$$x_N = \frac{\sinh N\theta}{\sinh \theta} = 0.$$

This condition is met provided that θ takes one of the values

$$\theta = \frac{j\pi i}{N}, j = 1, 2, 3, \ldots, N - 1$$

(the case $j = 0$ provides a trivial solution). Corresponding to these values of θ are the critical eigenvalues determined from (8.36). These are given by ($j = 1, 2, \ldots$)

$$\lambda_j^2 = 2 \left(1 - \cosh \frac{j\pi}{N} i \right) \tag{8.38}$$

$$= 2 \left(1 - \cos \frac{j\pi}{N} \right)$$

$$= 4 \left(\sin^2 \frac{j\pi}{N} \right).$$

Since the matrix $\mathbf{L}$ is of dimension $(N-1) \times (N-1)$, we must determine $N-1$ eigenvalues from these candidates. In fact, $(N-1)$ distinct values are provided by the values $j = 1, 2, \ldots, (N-1)$ in (8.38). Integer multiples of these indices provide the same eigenvalues. For the sake of completeness, we show that it is possible to eliminate from (8.38) consideration of values of j which are integer multiples of N. In the case that $j = KN$, we have

$$\theta = K\pi i,$$

and the solution is apparently an indeterminate form. This observation leads us to reconsider the solution of the difference equation in the case that $\theta = K\pi i$ holds. In this case, we have

$$z_1 = e^{iK\pi} = \pm 1,$$

so that the characteristic equation (8.35) of the difference equation has a double zero. The solution form (8.37) is inappropriate, since it is based on a residue calculation assuming distinct roots. We leave as an exercise the calculation of the appropriate form, and the verification that this solution is incapable of meeting the boundary conditions in a nontrivial fashion.[11]

We next consider the eigenvectors associated with these eigenvalues. These are given by (from (8.37))

$$x_k^j = \frac{\sinh k\,\theta_j}{\sinh \theta_j}$$

$$= \text{constant } \sin j \frac{k\pi}{N}.$$

These "mode shapes" are illustrated in Figure 8.21.

It should be observed that these eigenvectors happen to coincide exactly with the sampled values of the half-range sine functions appropriate to the interval $[0, 1]$. These are the (continuous) eigenfunctions associated with the boundary value problem.

---

[11]This argument is an analogue of the $\lambda^2 = 0$ cases encountered in boundary value problems.

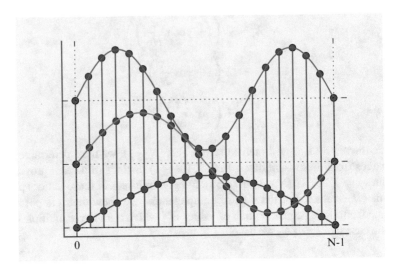

Fig. 8.21 Discrete mode shapes

The connection with the eigenvalues of the continuous problem is made by noting that for large values of N (appropriate for simulation) and moderate values of j, the eigenvalues (8.38) may be approximated by a Taylor series calculation. This gives

$$\lambda_j^2 = 2\left(1 - \cos\frac{j\pi}{N}\right)$$

$$= 2\left(1 - \left(1 - \frac{1}{2}\left(\frac{j\pi}{N}\right)^2 + \text{terms of order }\left(\frac{j\pi}{N}\right)^4\right)\right)$$

$$= \frac{(j\pi)^2}{N^2} + \text{terms of order }(\frac{j\pi}{N})^4.$$

This computation shows that the eigenvalues in fact coincide to within the indicated error. The use of the results in the overall solution form (8.31) shows that the analogue is capable of remarkably accurate simulations of the boundary value problem, at least for initial conditions whose energy is concentrated among the low frequency components.

Example. The model (partial difference equation)

$$p(n, k+1) = \frac{1}{2}p(n, k) + \frac{1}{4}p(n-l, k) + \frac{1}{4}p(n+1, k) \tag{8.39}$$

was introduced above in connection with the discrete variable random walk problem (and also as a computational model for the heat equation).

In the model (8.39), k (the time variable) ranges over the nonnegative integers, while n (the spatial location variable) ranges over the whole set of integer coordinates. It follows from the equation that

$$\sum_{n=-\infty}^{\infty} p(n, k+1) = \sum_{n=-\infty}^{\infty} p(n, k),$$

so that (as is obvious) the total probability of finding the particle anywhere is conserved. Since p is a probability density it is nonnegative, and this calculation guarantees that the discrete Fourier transform of this density exists for all k provided that the probability density is chosen to sum to 1 at the initial time.

Using the discrete Fourier transform shift formula,

$$\hat{p}(\theta, k+1) = F\{p(n, k+1)\}$$

$$= \left(\frac{1}{2} + \frac{1}{4} e^{i\theta} + \frac{1}{4} e^{-i\theta}\right) \hat{p}(\theta, k).$$

The solution of this simple difference equation is evidently given by

$$\hat{p}(\theta, k) = \left(\frac{1}{2} + \frac{1}{4} e^{i\theta} + \frac{1}{4} e^{-i\theta}\right)^k \hat{p}(\theta, 0).$$

This result may be obtained systematically by use of z-transforms. If

$$\hat{P}(\theta, z) = \mathcal{Z}\{p(\theta, k)\},$$

then

$$z\hat{P}(\theta, z) - z\hat{P}(\theta, 0) = \left(\frac{1}{2} + \frac{1}{4} e^{i\theta} + \frac{1}{4} e^{-i\theta}\right) \hat{P}(\theta, z)$$

and

$$\hat{p}(\theta, k) = \frac{1}{2\pi i} \oint \frac{z\hat{p}(\theta, 0)}{z - \left(\frac{1}{2} + \frac{1}{4} e^{i\theta} + \frac{1}{4} e^{-i\theta}\right)} z^{k-1} \, dz,$$

leading to the solution by a residue calculation.

The probability density is determined by computing the inverse Fourier transform. Since the form is a product of transforms, the convolution theorem provides

$$p(n, k) = [p(n, 0) \star g(n, k)] \qquad\qquad (8.40)$$

$$= \sum_{m=-\infty}^{\infty} p(n-m,0)\,g(m,k) \qquad (8.41)$$

$$= \sum_{m=-\infty}^{\infty} g(n-m,k)\,p(m,0).$$

where

$$g(n,k) = F^{-1}\left\{\left(\frac{e^{i\theta}+2+e^{-i\theta}}{4}\right)^{k}\right\}(n).$$

This inverse transform can virtually be obtained by inspection (using the binomial theorem). In deference to more challenging problems we describe the result through the inversion integral. This gives

$$g(n,k) = \frac{1}{2\pi}\int_{0}^{2\pi}\left(\frac{1}{2}+\frac{1}{4}e^{i\theta}+\frac{1}{4}e^{-i\theta}\right)^{k} e^{in\theta}\,d\theta$$

$$= \frac{1}{2\pi}\int_{0}^{2\pi} e^{-ik\theta}\,\frac{\left(e^{i\theta}+1\right)^{2k}}{4^{k}}\,e^{in\theta}\,d\theta.$$

The integrand is in fact a finite sum of powers of $e^{i\theta}$. To evaluate the integral, it is necessary only to identify the coefficients of terms corresponding to a vanishing exponent in the overall expansion.

Now since (by the binomial theorem)

$$\left(e^{i\theta}+1\right)^{2k} = \sum_{l=0}^{2k} e^{il\theta}\,\frac{(2k)!}{(2k-l)!\,l!},$$

we have

$$g(n,k) = \frac{1}{2\pi}\int_{0}^{2\pi}\sum_{l=0}^{2k} e^{il\theta}\,\frac{(2k)!}{(2k-l)!\,l!}\,e^{i(n-k)\theta}\,d\theta.$$

In this calculation, only terms for which

$$l + n - k = 0$$

contribute. Further, since $0 \le l \le 2k$, only values of n such that $-k \le n \le k$ result in a nonvanishing value. For such values of n, the term $l = k - n$ contributes

$$g(n,k) = \frac{1}{4^{k}}\,\frac{(2k)!}{(2k-(k-n))!(k-n)!}$$

$$= \frac{1}{4^k} \frac{(2k)!}{(k+n)!(k-n)!}.$$

We conclude, therefore, that

$$g(n,k) = \begin{cases} \frac{1}{4^k} \frac{(2k)!}{(k+n)!(k-n)!} & |n| \le k, \\ 0 & |n| > k. \end{cases} \qquad (8.42)$$

This result may be incorporated in (8.40) to provide a complete solution. A special case arises in the case that the particle in question is located at the origin (with probability 1) at time zero. The result (8.42) is then the probability of finding the particle at location n at time k. The fact that the density vanishes for $|n| > k$ in this case is an obvious consequence of the constraint that the particle moves only one location (at most) at each time stage.

Example. In the previous chapter, the frequency interpretation of the continuous-time Fourier transform was exploited in the discussion of several basic topics arising in communications problems.

Similar applications arise in the context of discrete-time Fourier transforms. The results find ready application in problems of digital signal processing. We illustrate the analogies involved with the derivation of a discrete-time version of the sampling theorem discussed in the previous chapter.

We recall from previous discussions that the frequency domain associated with the discrete-time functions is naturally identified with the interval $[-\pi, \pi]$ (even more naturally with the circumference of the unit circle in the complex plane).

The inversion representation

$$f(n) = \frac{1}{2\pi} \int_{-\pi}^{\pi} \hat{f}(\theta) \, d\theta$$

represents the function f as a (continuous) superposition of the oscillatory functions (of frequency θ)

$$u_\theta(n) = e^{in\theta}.$$

Based on this intuitive description, it is natural to refer to a sequence which has the property that its Fourier transform vanishes outside of some subinterval of $[-\pi, \pi]$ as band limited.

Definition (Band-Limited Sequence).
Suppose that $f \in l_2$, and that for some B, $0 < B < \pi$, we have

$$|\hat{f}(\theta)|^2 = 0, \ |\theta| > B.$$

Then f is said to be band limited to the band $[-B, B]$.

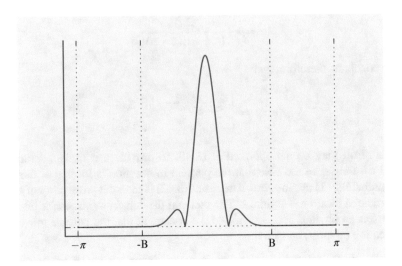

Fig. 8.22 Band limited discrete spectrum

Associated with this definition is the corresponding power spectrum plot of Figure 8.22.

In the above plot, of course, the point "π" is identified with $-\pi$. This is a consequence of the fact that the transform is periodic of period 2π (and so naturally defined on the unit circle).

To derive a discrete-time form of a sampling theorem, we consider the case of a function $f(\cdot)$ band-limited to the band $[-\frac{\pi}{2}, \frac{\pi}{2}]$ On an intuitive basis, such a function (sequence) is a superposition of oscillatory components limited to half of the maximum possible frequency. Accordingly, we expect that in some sense half of the function values must contain redundant information. To make this statement precise, we consider the inversion formula. Evidently

$$f(n) = \frac{1}{2\pi} \int_{-\pi}^{\pi} \hat{f}(\theta) \, d\theta$$

$$= \frac{1}{2\pi} \int_{-\frac{\pi}{2}}^{\frac{\pi}{2}} \hat{f}(\theta) \, d\theta,$$

since f is band-limited to $[-\frac{\pi}{2}, \frac{\pi}{2}]$.

On the interval $[-\frac{\pi}{2}, \frac{\pi}{2}]$ the transform f may be represented by a Fourier series. The fundamental frequency appropriate is

$$\omega_0 = \frac{2\pi}{\pi} = 2,$$

so that

$$\hat{f}(\theta) = \sum_{n=-\infty}^{\infty} d_n e^{-i2n\theta}, \quad -\frac{\pi}{2} \le \theta \le \frac{\pi}{2}.$$

The expansion coefficients are given by

$$d_n = \frac{1}{\pi} \int_{-\frac{\pi}{2}}^{\frac{\pi}{2}} \hat{f}(\theta) e^{2in\theta} \, d\theta$$

$$= 2 \frac{1}{2\pi} \int_{-\frac{\pi}{2}}^{\frac{\pi}{2}} \hat{f}(\theta) e^{2in\theta} \, d\theta$$

$$= 2f(2n)$$

(by virtue of the inversion formula above). This provides the expansion

$$\hat{f}(\theta) = \sum_{n=-\infty}^{\infty} 2f(2n) e^{-i2n\theta}, \tag{8.43}$$

showing that the transform $\hat{f}$ (and hence the sequence $\{f(n)\}$) is completely determined by knowledge of only the even argument values of the sequence. This observation verifies the intuition that "half" of the function values are redundant.

An explicit representation of the sequence in terms of the even argument samples is provided by term-by-term inversion of (8.43), justified by the observation that (8.43) is an orthonormal expansion. Compare the corresponding argument of the preceding chapter.

To compute $f(m)$ in the form

$$f(m) = \frac{1}{2\pi} \int_{-\frac{\pi}{2}}^{\frac{\pi}{2}} \hat{f}(\theta) e^{im\theta} \, d\theta,$$

we first compute

$$\frac{1}{2\pi} \int_{-\frac{\pi}{2}}^{\frac{\pi}{2}} e^{-2in\theta} e^{im\theta} \, d\theta = \frac{e^{i(m-2n)\theta}}{2\pi i(m-2n)} \Big|_{-\frac{\pi}{2}}^{\frac{\pi}{2}}$$

$$= \frac{1}{\pi} \frac{\sin(m-2n)\frac{\pi}{2}}{(m-2n)}, \quad m \ne 2n$$

$$= \frac{1}{2}, \quad m = 2n.$$

Carrying out the term-by-term inversion, we obtain

$$f(m) = \sum_{n=-\infty}^{\infty} f(2n) \frac{\sin(m-2n)\frac{\pi}{2}}{(m-2n)\frac{\pi}{2}} \tag{8.44}$$

(with the convention that the indeterminate form arising for $m = 2n$ is to be replaced by its limiting value of unity).[12]

It is a simple matter to extend the discussion above to the case of a function band limited to the band $|\theta| < \frac{\pi}{N}, N = 3, 4, 5, \dots$. This is left as an exercise below.

In view of the fact that there exists a discrete-time analogue of the frequency-translation law for Fourier transforms, it is possible to consider amplitude modulation problems in the framework of discrete-time functions. Such results are in fact of rather limited interest in the discrete-time case. Multiple-message/single channel transmission is in practice more naturally handled in digital systems by interleaving messages (time-division multiplexing) in the natural way, or coding the data streams on mutually orthogonal bases.

Problems 8.7.

1. Show that the second difference matrix **L** given by (8.30) is negative definite.
2. Show that in case $\lambda^2 = 0$, it is not possible to find a nontrivial solution of

$$x_{k+2} - (2 - \lambda^2) x_{k+1} + x_k = 0$$

   satisfying the boundary conditions $x_0 = x_N = 0$, $(N > 0)$.
3. In Chapter 3, the spring-mass model

$$\frac{M}{\kappa} \frac{d^2 x_k}{dt^2} = x_{k+1} - 2 x_k + x_{k-1},$$

$$x_0 = x_N = 0$$

   was proposed as a wave equation analogue. Find the solution to such a model, and compare with the solution of the corresponding wave equation (Figure 8.23).
4. The finite difference equation

$$\left[v_{k+1,j} - 2 v_{k,j} + v_{k-1,j} \right] + \frac{M^2}{N^2} \left[v_{k,j+1} - 2 v_{k,j} + v_{k,j-1} \right] = 0$$

---

[12]This result is closely connected to the continuous-time sampling theorem. It is even possible to derive (8.44) by use of the continuous-time theorem by associating continuous-time functions (via the sampling expansion) to the sequences under consideration. The approach taken here avoids this diversion and tends to emphasize the "frequency domain" interpretations of discrete Fourier transforms.

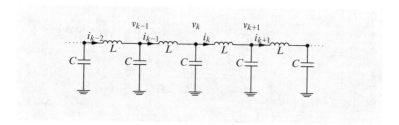

Fig. 8.23 Wave equation analogue

is employed for numerical solutions of the Laplace equation. Find explicit solutions of this equation on the rectangular grid $0 \le k \le N,\, 0 \le j \le M$, given the boundary conditions

$$v_{0,j} = v_{n,j} = 0,\ 0 \le j \le M,$$

$$v_{k,0} = 0,\ 0 \le k \le N,$$

$$v_{k,M} = f(k).$$

Hint: Would you believe separation of variables?

5. Formulate and solve the following discrete random walk problem: at each time the particle moves rightward only — one step with probability $\frac{3}{4}$, or two with probability $\frac{1}{4}$.

6. Provide an integral form of solution to the random walk model of Problem 4, Section 8.1.

7. Suppose that the sequence $\{f(n)\}$ is band-limited to the band $[-\frac{\pi}{3}, \frac{\pi}{3}]$. Find a representation of such a sequence in terms of one-third of its samples.

8. Repeat Problem 7 for the case of a sequence band-limited to the band $[-\frac{\pi}{M}, \frac{\pi}{M}]$, where M is an integer.

9. Describe the relationship between the square-summable sequence $\{f(n)\}$ with transform $\hat{f}(\theta)$, and the function of a real variable g defined as band limited to the band $[-B, B]$, with Fourier transform

$$\hat{g}(\omega) = \hat{f}(\pi \frac{\omega}{B}).$$

10. Use the results of Problem 9 to deduce the results of equations (8.43) and (8.44) by an appeal to the usual continuous-time version of the sampling theorem.

11. Consider a rectangular array of one-ohm resistors as shown in Figure 8.24. Suppose that a current of value f_{kj} is drawn from each node kj.

   a. Show that the node voltages satisfy a discrete version of Poisson's equation.
   b. Find the solution of the above problem.

   Hint: Compare Section 4.6, and Problem 4 in Section 8.4 above.

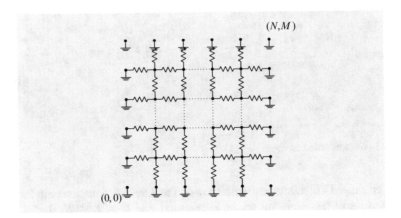

Fig. 8.24 Planar resistor grid

12. Suppose that the entire plane is tiled with one-ohm resistors (so that the pattern of 8.24 extends over the whole plane). Find the equivalent resistance between the node $(k + 1, j)$ and the node $(k, j + 1)$.

Hint: Recall that $V = IR$, and proceed as in problem (10) as far as the setup is concerned.

Can you derive an expression for the resistance between any two nodes?

8.8 Finite and Fast Fourier Transforms

The transform methods treated in previous chapters are applicable to various problems in which the independent variable takes values in an "infinite" set. The variables have been either continuous (in the case of Fourier and Laplace transforms) or identifiable with the integers (in the case of the discrete Fourier and z-transforms).

Another class of problems arises in the case in which the independent variable is discrete and *finite*. Certain of these problems arise in an intrinsic fashion in that a natural problem variable has a finite and discrete character. Other problems of this sort (probably in fact a large majority of the total) arise from the attempt to make numerical computations in connection with many of the problems treated earlier by infinite extent transform methods. The nature of digital computer algorithms requires that functions be represented (approximated) by discrete sample values. The inherent limits on storage availability and computation time guarantee that only a finite number of such samples can be retained in any practical computation.

The transform theory associated with functions defined on a finite discrete set is referred to as the finite Fourier transform. It will be shown below that this theory shares many of the properties of the previous transform theories. The theory is of

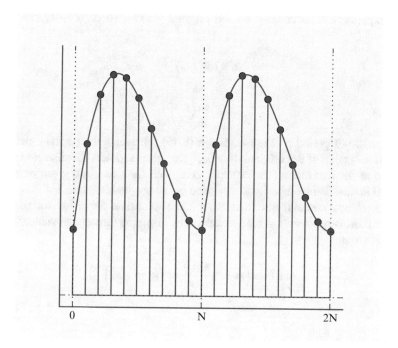

Fig. 8.25 Finite discrete function

interest largely because of the connections with the computational problems alluded to above. Interest increases as a result of the existence of computationally efficient algorithms for the calculation of the finite Fourier transform. Those algorithms (described below) have the capability to dramatically reduce the computation time for the finite Fourier transform and are referred to as *Fast Fourier Transform (FFT)* algorithms.

The functions of finite Fourier transform theory are defined on a finite set. The set in question may (initially) be thought of as consisting of the integers $\{0, 1, 2, \ldots, N-1\}$.

A function defined on such a set may be represented in a form analogous to that previously adapted for functions defined on the integers (Figure 8.25).

In Section 8.2 above, functions obtained by extending a given function (defined on $\{0, \ldots, N-1\}$) to be periodic of period N were analyzed by means of the z-transform technique. The result of this computation was that such a function is representable in the form of a weighted sum of powers of the Nth roots of unity. The explicit form of this representation is (recall (8.16), Section 8.2)

$$f(n) = \frac{1}{N} \sum_{j=0}^{N-1} (\omega_j)^n \left(\sum_{k=0}^{N-1} f(k) (\omega_j)^{-k} \right). \tag{8.45}$$

This expression reproduces f on the original interval $\{0, \ldots N - 1\}$. Since the functions

$$u_j(n) = \left(\omega_j\right)^n \tag{8.46}$$

$$= \left(e^{\frac{2\pi ij}{N}}\right)^n$$

are periodic of period N, the expression (8.45) represents the periodic extension of f for n outside of the original domain. (This observation is the discrete variable version of the fact that a Fourier series computed for a function given on a finite interval in turn defines a periodic extension of the original function.)

The relation contains simultaneously the appropriate definition for the finite Fourier transform, as well as the corresponding inversion formula. In view of (8.46), we may write

$$\left(\omega_j\right)^n = \left(\omega = e^{\frac{2\pi ij}{N}}\right)^n = \left(\omega = e^{\frac{2\pi i}{N}}\right)^{jn} \tag{8.47}$$

$$= \left(\omega\right)^{jn}$$

where

$$\omega = e^{\frac{2\pi i}{N}} \tag{8.48}$$

denotes the principal N^{th} root of unity. In these terms, we define (from (8.45)) the finite Fourier transform by

$$\hat{f}(j) = \sum_{k=0}^{N-1} f(k) \left(\omega\right)^{-kj}. \tag{8.49}$$

The corresponding inversion formula is (also from (8.45))

$$f(n) = \frac{1}{N} \sum_{j=0}^{N-1} \hat{f}(j) \left(\omega\right)^{jn}. \tag{8.50}$$

We have already noted above that $f(\cdot)$ regarded as a function of the integer valued variable n is periodic of period N. In view of the form of the transform formula, it follows that $\hat{f}(\cdot)$ (regarded as a function of the integer j) is also periodic of period N.[13]

---

[13] One consequence of this periodicity is that both the original function and the transform may be regarded as functions whose domain is the group of the integers mod N. This set is isomorphic to the group of powers of the Nth roots of unity, and is conventionally identified with a set of equally

8.9 Finite Fourier Properties

The finite Fourier transform has, of course, the usual formal properties.

Shift Laws

We compute in turn the finite transform of a delayed function (Figure 8.26), and the transform of an exponentially multiplied function.

The delay law is computed as

$$\{f(\hat{n} - l)\}(j) = \sum_{n=0}^{N-1} f(n-l)\,(\omega)^{-nj}$$

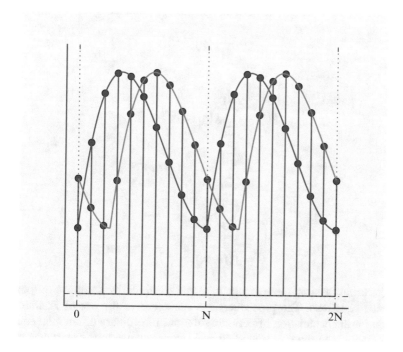

Fig. 8.26 Delayed finite discrete function

---

spaced points on the unit circle of the complex plane. This identification corresponds to the practice of regarding continuous Fourier series as being defined on the unit circle.

$$= \sum_{k=-l}^{N-1-l} f(k)(\omega)^{-(l+k)j}$$

$$= (\omega)^{-lj} \sum_{k=-l}^{N-1-l} f(k)(\omega)^{-kj}$$

$$= (\omega)^{-lj} \hat{f}(j).$$

(This last equality follows from the observation that the summand is periodic of period N. The sum over the period is independent of the starting index.)

Exponential Shift

The appropriate "exponential function multipliers" are those which fit the periodicity constraint. We consider the periodic function $g(\cdot)$ defined by

$$g(n) = (\omega)^{mn} f(n).$$

Then

$$\hat{g}(j) = \sum_{n=0}^{N-1} g(n)(\omega)^{-nj}$$

$$= \sum_{n=0}^{N-1} (\omega)^{mn} f(n)(\omega)^{-nj}$$

$$= \sum_{n=0}^{N-1} f(n)(\omega)^{-n(j-m)}$$

$$= \hat{f}(j-m).$$

In the usual fashion, these results may be used to derive solutions of (periodic) difference equations. The procedures are analogous to the use of Fourier series methods for the calculation of periodic solutions of ordinary differential equations (and are subject to the corresponding restrictions relative to the existence of such periodic solutions).

Example. Consider the difference equation

$$x(n) + \frac{1}{4}x(n-2) = f(n)$$

and suppose that $f(n + N) = f(n)$. A periodic solution of the difference equation is sought. The forcing function $f(\cdot)$ has an expansion of the form

$$f(n) = \frac{1}{N} \sum_{k=0}^{N-1} \hat{f}(k) (\omega)^{kn},$$

where $\omega = e^{\frac{2\pi i}{N}}$. We seek $x(\cdot)$ in the form of a finite Fourier series

$$x(n) = \frac{1}{N} \sum_{k=0}^{N-1} \hat{x}(k) (\omega)^{kn}.$$

Then (substituting the assumed solution form in the equation), using the delay law

$$\frac{1}{N} \sum_{k=0}^{N-1} \left\{ \hat{x}(k) (\omega)^{kn} + \frac{1}{4} \hat{x}(k) (\omega)^{-2k} (\omega)^{kn} \right\} = \frac{1}{N} \sum_{k=0}^{N-1} \hat{f}(k) (\omega)^{kn}.$$

By the uniqueness of the finite Fourier transform expansion (i.e., equating Fourier coefficients) we obtain

$$\left(1 + \frac{1}{4} (\omega)^{-2k}\right) \hat{x}(k) = \hat{f}(k),$$

$$\hat{x}(k) = \frac{\hat{f}(k)}{1 + \frac{1}{4} (\omega)^{-2k}}.$$

The desired solution is then given by

$$x(n) = \frac{1}{N} \sum_{k=0}^{N-1} \frac{\hat{f}(k)}{1 + \frac{1}{4} (\omega)^{-2k}} (\omega)^{kn}.$$

A general discussion of periodic solution problems can be presented. This topic is treated in the problems below.

Convolution Laws

The convolution of two periodic sequences may be defined as

$$[f \star g] (n) = \sum_{k=0}^{N-1} f(n - k) g(k). \tag{8.51}$$

It follows readily that $[f \star g]$ is itself periodic of period N. The finite Fourier transform of the convolution may be computed as

$$[f \,\hat{\star}\, g](j) = \sum_{n=0}^{N-1} (\omega)^{-nj} \sum_{k=0}^{N-1} f(n-k)\, g(k)$$

$$= \sum_{k=0}^{N-1}\sum_{n=0}^{N-1} f(n-k)\, g(k)\, (\omega)^{-nj}$$

$$= \sum_{k=0}^{N-1}\sum_{l=-k}^{N-1-k} f(l)\, g(k)\, (\omega)^{-(k+l)j}$$

$$= \left(\sum_{k=0}^{N-1} g(k)\, (\omega)^{-kj}\right)\left(\sum_{m=0}^{N-1} f(m)\, (\omega)^{-mj}\right)$$

$$= \hat{f}(j)\, \hat{g}(j).$$

In view of the fact that the discrete Fourier transform may be interpreted as a sequence of period N, the dual convolution result follows essentially immediately from the above. If the transform convolution is defined as

$$\hat{h}(j) = \frac{1}{N}\sum_{k=0}^{N-1} \hat{f}(j-k)\, \hat{g}(k),$$

then the corresponding inverse transform is given by

$$h(n) = f(n)\, g(n).^{14}$$

Finite Fourier Calculations with Octave

Octave is often said to "compute Fourier transforms." Without qualification this is a little misleading, because a finite Fourier transform is what is being computed (with a Fast Fourier Transform algorithm). If the data involved is a segment of sample of some continuous signal, then concerns of sampling (approximately) band-limited functions are involved. Even this scenario results in data which theoretically are in the domain of the discrete Fourier transform, with the "time" indices ranging over

---

[14]It might be noted that these results can also be considered as exercises in the multiplication of polynomials in the Nth root of unity.

the entire integers. By the time the data is transformed it has been truncated in time, and will require zero padding, or windowing to avoid introducing spurious edge effects in the transform calculations.

We assume these considerations as a given, and describe here some tools for the finite transform stage.

A shift procedure is useful for signal processing calculations. This version shifts in both directions.

```
OCTAVE
% shift (m)
% circular  shift  of  vector  as  a  column

function  y  =  shift (a,  n)

  x= a (:);

  N  =  size  (x);

  y  =  [x(1  +  mod(N-n,  N):N);  x(1:  mod(N-n,N))];
endfunction
```

The circular convolution represents a linear operation, and so can be represented as a matrix, called a circulant. Because of the way matrix multiplication is defined, the entries must be shifted along the rows of the matrix.

```
OCTAVE
% circulant (a)
% make  a  circulant  matrix  out  of  row  or  column
% (or  strung  out  columns)

function  y  =  circulant (a)

  x  =  a (:) ';

  for  k=0:length  (x)-1
    y(k+1,  :)  =  (shift (x,  k)) ';
  endfor
endfunction
```

For large data lengths, representing a convolution as a matrix multiplication may not be feasible. The usual computing insight is that you can trade storage space for computing time, so this function uses no more storage than is required for the variables used and generated.

OCTAVE

```
% circonv(a, b)
% circular convolution of vectors a, b
% treats desired length as max of a, b  lengths

function c = circonv(a, b)

   x = a(:);
   y = b(:);

   modulus = max(length(x), length(y));

   if (length(x) == modulus)
     c = zeros(size(x));
     len = length(y);
     for k = 1:1:modulus
        temp = [x(k: modulus); x(1: k-1)];
        c(k) = sum(temp(1:len).* y);
     endfor
   else
     c = zeros(size(y));
     len = length(x);
     for k = 1:1:modulus
        temp = [y(k: modulus); y(1: k-1)];
        c(k) = sum(temp(1:len).* x);
     endfor
   endif
endfunction
```

We illustrate the circular convolution calculation by calculating finite difference derivatives of a sinusoid. The data length would make the circulant calculation involve rather large matrices, so we trade space for time and use the procedural version of circular convolution.

OCTAVE
```
t = 0: pi/fs: 6*pi-(pi/fs);
s = sin(t);
d = (fs/pi)*[-.5 0 .5];

circdiff = circonv(s, d);
plot(circdiff)

circdiffsq = circonv(circdiff, d);
plot(circdiffsq)
```

The plot of the first derivative calculated with a symmetric difference circular convolution is shown in Figure 8.27.

The numerical second derivative calculated in the same way is shown in Figure 8.28. Comparing this with the computations made earlier in Section 8.5, it should be evident that "edge effects" are an issue that must be dealt with as part

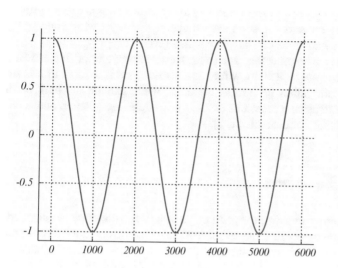

Fig. 8.27   Circular convolution derivative

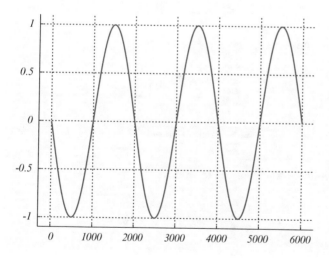

Fig. 8.28   Circular convolution second derivative

of digital signal processing. In particular, the presence of discontinuities at the data extremes easily results in unwanted artifacts in the processed data.

One way to handle the problem is to use a circular convolution as done above. Of course, the data above was artificially generated with the sampling interval chosen to cause a "smooth" transition at the wraparound point. If the data arises from measurements, then adjustments must be made before processing begins. If the use of circular processing is desired, one way to accomplish this is to double the original data length by reflecting it as an even function across the wraparound point. This entails concatenating the original data with a reversed copy of itself.

If the data is to be processed as a (finite) discrete convolution, it is more common to multiply the data by a window function whose effect is to more or less smoothly change the data values to zero at the ends, while leaving the bulk of the data unaffected. Extensive discussions of these procedures are given in books on digital signal processing, for example [1, 2].

Parseval's Equation

The finite Fourier transform version of Parseval's relation may be obtained on the basis of a linear algebraic interpretation of the transform operation (see the problems below). A derivation based on the convolution theorem (analogous to that used in the continuous-time Fourier transform case, Chapter 7 above) may also be constructed.

We require an evaluation of the expression (total energy)

$$E = \sum_{k=0}^{N-1} |f(k)|^2.$$

In terms of a convolution,

$$E = \sum_{k=0}^{N-1} f(k)\overline{f(k-n)}\bigg|_{n=0}$$

$$= \sum_{k=0}^{N-1} f(k)\overline{f(-(n-k))}\bigg|_{n=0}.$$

Using the convolution and inversion theorems

$$E = \frac{1}{N} \sum_{j=0}^{N-1} \{f(\hat{k})\}(j)\ \overline{\{f(\hat{-k})\}(j)}\,(\omega)^0$$

$$= \frac{1}{N} \sum_{j=0}^{N-1} \hat{f}(j)\overline{\hat{f}(j)}$$

$$= \frac{1}{N} \sum_{j=0}^{N-1} |\hat{f}(j)|^2.$$

This calculation establishes the Parseval relation

$$\sum_{k=0}^{N-1} |f(k)|^2 = \frac{1}{N} \sum_{j=0}^{N-1} |\hat{f}(j)|^2 \tag{8.52}$$

for the finite Fourier transform.

8.10   Fast Finite Transform Algorithm

The discussion above shows that the finite Fourier transform has the usual formal structure associated with a transform. While the finite Fourier transform is of some interest for this reason alone, it finds increasing use in problems arising from approximations to the usual (Fourier) transform calculations. This is particularly the case because of the existence of rapid computational algorithms for the computation of the finite Fourier transform. Such algorithms are referred to as *Fast Fourier Transform* algorithms, and their existence has led to a rapid increase in the use of Fourier methods in computational problems.

Example. In view of the fact that FFT algorithms perform calculations on finite Fourier transforms, the use of these algorithms in a given situation requires a reduction of the given problem to one to which finite Fourier transforms may be applied. This reduction typically involves approximation procedures, whose errors must be analyzed in order to justify computations based on the finite Fourier transform.

Various data processing schemes (arising, for example, in geophysical problems) require computation of convolutions as part of the processing. If the data is compiled on a discrete (sampled) basis, then the required computation takes the usual form

$$y(n) = \sum_{k=-\infty}^{\infty} g(n-k)\,x(k). \tag{8.53}$$

Here $\{x(n)\}$ represents a (sampled) data sequence, $\{g(k)\}$ is the weighting pattern associated with a smoothing or filtering operation used to produce the sequence $\{y(n)\}$. In conventional models the sequences $\{x(n)\}$, $\{g(n)\}$ are considered infinite in extent. The relation (8.53) then formally requires an infinite number of multiplications and additions in order to compute (a single value of) $y(n)$.

The infeasibility inherent in this situation is overcome by truncating the sequences $\{x(n)\}$, $\{g(n)\}$ (and hence $\{y(n)\}$) to an approximation of finite extent. As a practical matter, the data sequence $\{x(n)\}$ is of finite extent (due to storage limitations). The weighting sequence $\{g(n)\}$ may be truncated as well; this introduces an error into the computed values of $\{y(n)\}$ which may be estimated on an a priori basis.

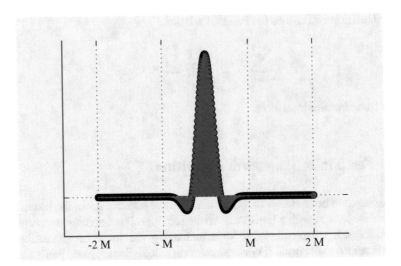

Fig. 8.29 Convolution sequence

We may therefore assume that the sequences $\{x(n)\}$ and $\{g(n)\}$ are of limited duration. In order to simplify the notation, we assume that

$$g(n) \equiv 0, \ |n| > M,$$
$$x(n) \equiv 0, \ |n| > M$$

for some integer M. It then follows that

$$y(n) = \sum_{k=-M}^{M} g(n-k)\, x(k).$$

In view of the interpretation of the convolution as a "shift, multiply, and add" operation (Figure 8.29) we see that $\{y(n)\}$ vanishes identically for $|k| \geq 2M+1$. The value $y(k)$ is nonvanishing only for $|k| \leq 2M$. The sequence $\{y(k)\}$ is completely determined by these $4M + 1$ values.

Our objective is to interpret the computation in the context appropriate to the finite Fourier transform. In view of the above observation, it follows that a (minimal) sequence length of $4M + 1$ is appropriate.

One way to accomplish this (and at the same time preserve the arguments of the original sequence) is to proceed as follows. With the original sequence $\{x(n)\}$ we associate the sequence $\{\tilde{x}(n)\}$ defined by appending M zeroes before and after the original, and extending periodically.

$$\tilde{x}(n) = x(n), \; |n| \leq M$$
$$\tilde{x}(n) = 0, \; M + 1 \leq n \leq 2M$$
$$\tilde{x}(n) = 0, \; -2M \leq n \leq -M - 1$$
$$\tilde{x}(n) = \tilde{x}(n + (4M + 1)).$$

The sequence $\{\tilde{g}(n)\}$ is defined in an analogous fashion. If the sequence $\{\tilde{y}(n)\}$ (periodic of period $4M + l$) is defined by

$$\tilde{y}(n) = \sum_{k=-2M}^{2M} \tilde{g}(n - k)\,\tilde{x}(k),$$

it is easily seen that $\{\tilde{y}(n)\}$ is the periodic extension of the desired sequence $\{y(n)\}$.

The above example shows that the discrete-time convolution may be computed using a finite Fourier transform convolution (i.e., by convolution of a periodic discrete sequence) by the device of embedding the given sequences in periodic sequences of twice the original length. In a similar fashion, an approximation to a continuous-time convolution

$$\int_{-\infty}^{\infty} g(t - \tau) f(\tau)\,d\tau$$

may be calculated as a finite Fourier convolution by just approximating the continuous convolution as a discrete sum, and then embedding the discrete problem as described above.

We next consider the problem of the computation of such finite discrete convolutions, in particular, the time required for such a computation. In such considerations, it is conventional to account for only the time associated with multiplication operations (on the ground that multiplications are inherently slower than additions in conventional implementations). For the computation of the convolution of two N-length periodic sequences by use of the defining relation

$$y(n) = \sum_{k=0}^{N-1} g(n - k)\,x(k),$$

evidently N multiplications are required to compute each value $y(n)$. Computation of the entire $y(\cdot)$ sequence directly therefore involves N^2 operations.

An alternative method for the evaluation of the convolution makes use of the convolution theorem. Evidently

$$\hat{f}(j) = \hat{g}(j)\,\hat{x}(j),$$

so that evaluation of $\hat{f}(\cdot)$ from $\hat{g}(\cdot)$ and $\hat{x}(\cdot)$ requires N multiplications.[15] A computation of $\hat{x}(\cdot)$ from

$$\hat{x}(j) = \sum_{k=0}^{N-1} x(k) \, (\omega)^{-jk}$$

requires formally N^2 operations (assuming that the required roots of unity are stored). The same number would be required to invert $\hat{x}(\cdot)$ to obtain $x(\cdot)$, for a total of

$$N^2 + N + N^2 = 2N^2 + N$$

operations. At first appraisal, the use of the convolution theorem appears inferior to a straightforward evaluation of the convolution.

In fact, the estimate of N^2 for the transform calculations is overly pessimistic. The computation (at least when N is a power of 2) may be accomplished with an algorithm requiring a number of operations proportional to $N \log N$. With the use of this (fast) algorithm for the Fourier transform calculation, the total for the convolution computation using the transform method is

$$2 \, \alpha \, N \log N + N = (2 \, \alpha \log N + 1) \, N.$$

The ratio of this to the result for a direct calculation is

$$\frac{(2 \, \alpha \log N + 1) \, N}{2N^2 + N} = \frac{(2 \, \alpha \log N + 1)}{2N + 1}. \tag{8.54}$$

As $N \to \infty$, this ratio tends to zero, establishing the superiority of the fast algorithms for the convolution computation.

FFT Algorithm Derivation

To show that the finite Fourier transform may be evaluated with the efficiency claimed above, we may argue as follows.

The transform appropriate to a sequence of length N is given by

$$\hat{x}(j) = \sum_{n=0}^{N-1} x(n) \, (\Omega_N)^{-jn}. \tag{8.55}$$

---

[15]For the purposes of this order of magnitude accounting, we do not distinguish between real- and complex-valued operations.

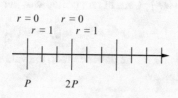

Fig. 8.30 Modular indexing

Here we use the notation

$$\Omega_N = e^{\frac{2\pi i}{N}}$$

for the root of unity appropriate to sequences of length N. The additional notation is required in order to distinguish transforms of sequences of various lengths.

We assume that N is a composite integer, so that N factors as

$$N = PQ.$$

The integers of summation in (8.55) may be represented in terms of a "reduction modulo P." For each n we construct the representation

$$n = lP + r$$

where $0 \leq r \leq P - 1$, and $0 \leq l \leq Q - 1$ (Figure 8.30).

The summation (8.55) may be replaced by

$$\hat{x}(j) = \sum_{r=0}^{P-1}\sum_{l=0}^{Q-1} x(lP + r)\,(\Omega_N)^{-(lP+r)j}$$

$$= \sum_{r=0}^{P-1}(\Omega_N)^{-rj}\sum_{l=0}^{Q-1} x(lP + r)\,(\Omega_N)^{-lPj}$$

In the inner sum, we have

$$(\Omega_N)^{-lPj} = \left(e^{\frac{2\pi i}{N}}\right)^{-lPj} \tag{8.56}$$

$$= \left(e^{\frac{2\pi i P}{N}}\right)^{-lj}$$

$$= \left(e^{\frac{2\pi i}{Q}}\right)^{-lj}$$

$$= (\Omega_Q)^{-lj}.$$

The functions $u_l(j) = (\Omega_Q)^{-lj}$ are in fact periodic of period Q,

$$u_l(j + mQ) = e^{2\pi i\, lm}\, u_l(j) = u_l(j).$$

The expression (8.56) is therefore dependent on the value of the index j evaluated modulo Q. If we denote this by

$$j \mod Q = [j],$$

then the transform (8.55) may be written in the form

$$\hat{x}(j) = \sum_{r=0}^{P-1} (\Omega_N)^{-rj} \left(\sum_{l=0}^{Q-1} x(lP + r)\, (\Omega_Q)^{-l[j]} \right). \tag{8.57}$$

The inner summation of this may (for fixed r) be identified as the finite Fourier transform of a sequence of length Q. The transformed sequence is defined by taking every P^{th} element of the original sequence, starting with the rth element. This expression of the desired transform in terms of lower-dimensional transforms is the basic reduction loading to the Fast Fourier Transform algorithm. As long as the integer Q of the inner summation is composite, the reduction may be carried out in turn for the computation of the Q-length transforms.

In order to evaluate the number of (complex) multiplication operations associated with such a reduction procedure, we define

$$\#[N]$$

as the number of operations required for the computation of the finite Fourier transform by means of the algorithm described below.

As the algorithm is based on factorization of the integer N, the case of prime N requires special treatment.

For prime N, a direct evaluation of the transform by (8.55) would require N multiplications; in the summation, the terms corresponding to $n = 0$ require no multiplication since the exponential factor is 1. This gives a total count of $N^2 - N = N(N - 1)$ multiplications. The FFT algorithm may now be described as follows:

FFT

The description of the algorithm is:

1. If N is prime, then the transform is computed as described above, giving $\#[N] = N(N - 1)$, and the algorithm terminates.
2. If N is not prime, then (8.57) is employed recursively. $\#[N]$ is the count resulting from this procedure (taking the factors of N in, say, decreasing order.)

(In the above and in what follows, we ignore a possible marginal reduction from the case $j = 0$.)

In the case that N is not prime, the count associated with recursive use of (8.57) must be determined.

Assuming that the inner summations of (8.57) have been evaluated, computation of $\hat{x}(\cdot)$ requires a multiplication for each of $P - 1$ nonzero values of r, and N values of j, for a subtotal of $(P - 1)N$. The inner summation we have identified with (for each of P values of r) the computation of a transform of length $Q = \frac{N}{P}$. This gives in total the recursive count

$$\#[N] = (P - 1)\,N + P\,\#[Q]$$

$$= (P - 1)\,N + P\,\#[\frac{N}{P}].$$

If the reduced order Q is in turn composite, say R divides Q, then by the same argument

$$\#[Q] = (R - 1)\,Q + R\,\#[\frac{Q}{R}],$$

$$\#[\frac{N}{P}] = (R - 1)\,\frac{N}{P} + R\,\#[\frac{N}{PR}],$$

so that the count for the overall computation is expressed as

$$\#[N] = (P - 1)\,N + P\,\#[\frac{N}{P}]$$

$$= (P - 1)\,N + (R - 1)\,N + RP\,\#[\frac{N}{RP}].$$

If $\frac{N}{RP}$ is divisible by S, then

$$\#[N] = (P - 1)\,N + (R - 1)\,N + (S - 1)\,N + RPS\#[\frac{N}{RPS}].$$

Proceeding in this fashion, it is easy to see that, given a factorization of N in the form

$$N = P_1 P_2 \cdots P_m$$

the recursive algorithm produces an operation count of

$$\#[N] = (P_1 - 1)\,N + (P_2 - 1)\,N + \cdots + (P_{m-1} - 1)\,N + P_1 P_2 \cdots P_{m-1}\#[P_m].$$

Assuming that the factorization of N is prime (or that the calculation of the order P_m transforms is to be made with the straightforward "prime" version of algorithm

FFT), the total count is

$$\#[N]$$

(8.58)

$$= (P_1 - 1) N + (P_2 - 1) N + \cdots + (P_{m-1} - 1) N + P_1 P_2 \cdots P_{m-1} \#[P_m]$$

(8.59)

$$= (P_1 - 1) N + (P_2 - 1) N + \cdots + (P_{m-1} - 1) N + P_1 P_2 \cdots P_{m-1} (P_m(P_m - 1))$$

$$= (P_1 - 1) N + (P_2 - 1) N + \cdots + (P_{m-1} - 1) N + (P_m - 1) N$$

$$= N \sum_{j-1}^{m} (P_j - 1).$$

From the description of the algorithm, it is clear that the major reduction occurs in the case of highly composite N. Of particular interest (to the virtual exclusion of other cases) is the case in which N is a power of 2, say

$$N = 2^m.$$

In that case

$$P_j = 2, j = 1, \ldots, m,$$

and the count is

$$\#[N] = N \sum_{j=1}^{m} 1 = N m$$

$$= N \log_2 N.$$

This establishes the basic computational estimate for the fast fourier transform algorithm, using the basic reduction of algorithm FFT. In the case $N = 2^m$, detailed consideration of symmetries involved allows a further reduction by a factor of 2 beyond the estimate above. For details of this, and information on storage requirements, hardware implementations, and other problems associated with FFT algorithms, we refer the reader to other sources.

8.11   Computing The FFT

The interest in the computational speed of the FFT calculation arises from the need to perform real-time computations of the FFT on measured-data values. The machines doing the calculation might be divided into two classes: digital signal processor (DSP) architectures and general purpose (workstation or desktop) machines.

DSP Applications

The FFT algorithm (especially when the data length is a power of 2) has an associated pattern of data access referred to as a "butterfly." This is a different pattern of memory access than that encountered when a program iterates through an array of floating point numbers, or searches for a character match in a string.

These last events are so common that general purpose processors include addressing modes (assembly language instruction formats) that provide sequential access at full hardware speed.

Digital signal processors are designed for use as embedded processors in communication or media processing devices. For marketing reasons, their performance on such applications must exceed that of general purpose CPU devices. One of the ways this is accomplished is to include addressing modes adapted to computation of the FFT. Since FFT performance is one of the benchmarks for DSP devices, examples of FFT codes written in assembly language are usually included in the application notes and user guides provided by the manufacturer [16].

Desktop/Workstation Code

For data analysis and simulation purposes it is useful to incorporate FFT routines in your own projects. While there is nothing that clarifies the understanding of an algorithm the way attempting to code a routine does, for production use one should seek tested and tuned codes.

One widely used set of routines is provided in the "Fastest Fourier Transform in the West" library. This library contains general purpose routines as well as versions tuned to a variety of common CPU architectures.

---

[16]DSP performance changes rapidly. The web sites of manufacturers such as *Texas Instruments, Motorola* and *Analog Devices* should be consulted for current data sheets, user guides, and application notes.

The Octave FFT

Although we associate the FFT algorithm with signal processing, the Octave `fft` routine is provided with the standard configuration as well as the signal processing package installation.

The Octave version of `fft` computes the finite Fourier transform of a finite sequence of (real or complex) values. The order of the returned values is the same as that in the standard finite transform definition[17].

The magnitude of the finite Fourier transform of

$$f(n) = \begin{cases} n/256 & \text{for } 0 \le n \le 255, \\ \text{undefined} & \text{otherwise} \end{cases}$$

is plotted in Figure 8.31.

It can be noted that the energy in that signal is concentrated at the 0 frequency neighborhood. This is more easily seen by generating a plot over (in our notation) the symmetric frequency range $[-\pi, \pi]$. For many applications of the FFT the interest (for analyzing vibrations, for example) lies in picking out frequency components. This seems easier with a centered frequency plot, since that corresponds to the usual Fourier transform illustration.

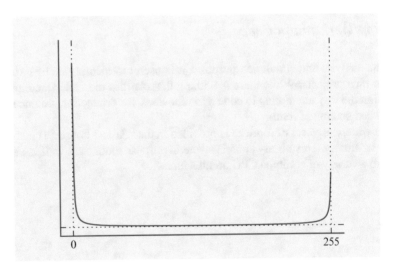

Fig. 8.31 Original FFT data

---

[17]This is true except for the origin of the subscript count. The FORTRAN heritage of Octave shows through in the convention of numbering subscripts starting from 1.

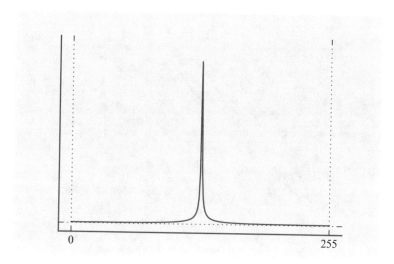

Fig. 8.32 Shifted FFT data

Octave includes the `fftshift` routine for reordering the FFT components into a symmetric array, with the 0 frequency component at the "center." The shifted plot corresponding to Figure 8.31 is in Figure 8.32.

The finite Fourier versions of transform and the convolution theorems can be illustrated with Octave. The example below is based on a Fourier domain bandpass filter characteristic. Construct a function which is defined as 1 in a band symmetric about the zero frequency, and inverse transform it to the time domain version. The calculations illustrate both the convolution theorem and frequency response ideas in the finite Fourier transform domain.

The `real` call removes some imaginary roundoff values in the inversion calculation. The function g represents the convolution square in the time domain.

OCTAVE

```
passband = [ones([1  33]) zeros([1  256 - 33])];
passband = shift(passband, -16);

f = real(ifft(passband));
Mf = circulant(f);

g = Mf* f;

plot([1:256], g, [1:256], f);
```

The generated plot is shown in Figure 8.33.

The fact that only one curve is evident follows from the convolution theorem. Because $\hat{f}$ only assumes the values 0 and 1, we have $\hat{f}^2 = \hat{f}$. Convince yourself that

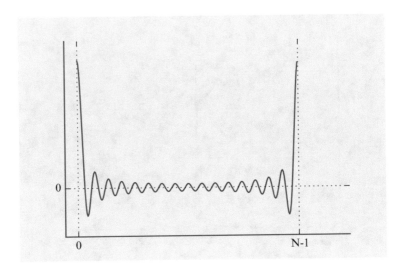

Fig. 8.33 Time domain f *and* convolution $f * f$

this bandpass example is one out of exactly 2^{256} that could have been chosen with this property.

Problems 8.11.

1. Since the periodic sequences $f(\cdot)$ of this section are determined by their values on the points $\{0, 1, \ldots, N-1\}$, they may be represented by an N-dimensional (column) vector

$$\mathbf{x} = \begin{bmatrix} f(0) \\ f(1) \\ \vdots \\ f(N-1) \end{bmatrix}.$$

The finite Fourier transform of $f(\cdot)$ may be represented in a similar fashion by

$$\hat{\mathbf{x}} = \begin{bmatrix} \hat{f}(0) \\ \hat{f}(1) \\ \vdots \\ \hat{f}(N-1) \end{bmatrix}.$$

The finite Fourier transform is a linear transformation from $f(\cdot)$ to $\hat{f}(\cdot)$. Show that in terms of the column vectors above this transformation is represented by the matrix

$$
T_N = \begin{bmatrix}
1 & 1 & \cdots & 1 \\
\left(\omega^1\right)^{-0} & \left(\omega^1\right)^{-1} & \cdots & \left(\omega^1\right)^{-(N-1)} \\
\left(\omega^2\right)^{-0} & \left(\omega^2\right)^{-1} & \cdots & \left(\omega^2\right)^{-(N-1)} \\
\left(\omega^3\right)^{-0} & \left(\omega^3\right)^{-1} & \cdots & \left(\omega^3\right)^{-(N-1)} \\
\vdots & \vdots & \vdots & \vdots \\
\left(\omega^{N-1}\right)^{-0} & \left(\omega^{N-1}\right)^{-1} & \cdots & \left(\omega^{N-1}\right)^{-(N-1)}
\end{bmatrix}
$$

$$
= \left[(\omega)^{-ij}\right], \; i,j = 0, 1, \ldots, N-1.
$$

(This matrix is a matrix of the Vandermonde type, with variables ω^{-i}, $i = 1, \ldots, N-1$).

2. By considering the inverse finite Fourier transform, show that

$$
(T_N)^* \, (T_N) = N \, I,
$$

so that the matrix

$$
U_N = \frac{1}{\sqrt{N}} \, T_N
$$

is a unitary transformation.

3. Use Problem 2 above to establish the Parseval formula for the finite Fourier transform.

4. The discussion of the section above includes no explicit calculation of a finite Fourier transform. Show that explicit calculations of finite Fourier transforms may be made by computing the z-transform of the sequence

$$
\tilde{f}(n) = \begin{cases} f(n) & 0 \le n \le N-1, \\ 0 & n \ge N, \end{cases}
$$

and using $\hat{f}(j) = \mathcal{Z}\left\{\tilde{f}(n)\right\}(\omega^j)$.

5. Use the result of Problem 4 (and standard z-transform methods) to compute finite Fourier transforms for

   a. $f(n) = a^n$, $0 \le n \le N-1$,
   b. $f(n) = n$, $0 \le n \le N-1$.

6. For the examples of Problem 5 compute and plot the finite Fourier transform using the Octave fft routine.

7. For the previous problem display the frequency distribution of the signal power using the Octave fftshift command along with fft.

8. Find the formal (see Problems 9 and 10 below) expression for the periodic solution of the difference equation

$$p(E) x(n) = f(n),$$
$$f(n + N) = f(n).$$

9. Consider the Problem of 8 above, and show that a periodic solution exists (for any periodic forcing function $f(\cdot)$) provided that $p(\omega^j) \neq 0$, for $j = 0, 1, \ldots, N-1$. Show further that even in the absence of this condition, periodic solutions may exist for suitably restricted forcing functions $f(\cdot)$. Can you interpret the required condition in a "physical" sense?

10. Consider the difference equation

$$p(E) x(n) = f(n), \; n \geq 0,$$

with $f(n + N) = f(n)$, and $x(0) \ldots x(N - 1)$ prescribed. Prove that the solution tends to the formal periodic solution of problem (8), provided that all zeroes of the polynomial

$$p(z) = z^M + p_{M-1} z^{M-1} + \cdots + p_0$$

lie inside the unit circle of the complex plane.

11. Show that the solutions of the circular resistor-capacitor network of Figure 8.34 may be explicitly found. Comment on the use of this network as an analogue device for the solution of the heat transfer in a thin ring boundary value problem.

12. It is possible to associate with each periodic sequence of period N an $N \times N$ circulant matrix. This matrix is defined by taking the given values of the sequence as the first row of the matrix, and filling in the remaining rows by means of a cyclic shift of the first row. The matrix associated with f is

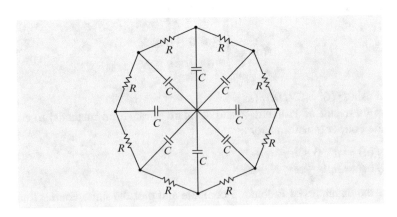

Fig. 8.34 Circular RC network

$$\mathbf{M}_f = \begin{bmatrix} f(0) & f(1) & f(2) \ldots f(N-1) \\ f(N-1) & f(0) & f(1) \ldots f(N-2) \\ f(N-2) & f(N-1) & f(0) \ldots f(N-3) \\ \vdots & \ddots & \ddots \ddots & \vdots \\ f(1) & f(2) & f(3) \ldots & f(0) \end{bmatrix}.$$

Prove that $\mathbf{M}_f \cdot \mathbf{M}_g = \mathbf{M}_{[f \star g]}$ and $c_1 \mathbf{M}_f + c_2 \mathbf{M}_g = \mathbf{M}_{[c_1 f + c_2 g]}$, showing that the mapping from periodic sequences to circulant matrices is a *ring homomorphism*.

13. Prove that the matrix $\mathbf{M}_f$ of Problem 12 is diagonalized by the unitary transformation $\mathbf{U}_N$ of Problem 2 above.
14. What sequence has

$$\frac{1}{N} \sum_{k=0}^{N-1} \hat{f}(j-k)\,\hat{g}(k)$$

as finite Fourier transform? ($\hat{f}, \hat{g}$ here are finite transforms of period N).
15. The sample auto-correlation function associated with a data sequence

$$\{x(n)\}_{n=-N}^{N}$$

is defined as

$$R_{xx}(k) = \frac{1}{2N+1} \sum_{n=-N}^{N} x(k+n)\,x(n).$$

Estimate the time required to compute the discrete Fourier transform of $\{R_{xx}(k)\}$.
16. The results of Problem 1 indicate that the finite Fourier transform has a representation as a matrix multiplication operation. Find a matrix representation of the basic Fast Fourier transform reduction algorithm equation (8.57). This entails representing the matrix $\mathbf{T}_N$ of Problem 1 in terms of a permutation matrix (to reorder the original sequence) and matrix factors of the form

$$\begin{bmatrix} \mathbf{T}_Q & & & & \mathbf{0} \\ & \mathbf{T}_Q & & & \\ & & \mathbf{T}_Q & & \\ & & & \ddots & \\ \mathbf{0} & & & & \mathbf{T}_Q \end{bmatrix}$$

corresponding to P (parallel) computations of the Q-dimensional transforms.
17. What is the form of the representation obtained in Problem 16 for the case $P = 2$?

References

1. A.V. Oppenheim, R.W. Schafer, *Digital Signal Processing* (Prentice-Hall, Englewood Cliffs, 1975)
2. L.R. Rabiner, C.M. Rader (eds.), *Digital Signal Processing* (IEEE Press, New York, 1972)

Further Reading

P. Cootner (ed.) *The Random Character of Stock Market Prices* (MIT Press, Cambridge, 1964)
J.R. Johnson, *Introduction to Digital Signal Processing* (Prentice-Hall, Englewood Cliffs, 1989)
E.I. Jury, *Theory and Application of the Z-Transform Method* (Wiley, New York, 1964)
Mathworks Incorporated, *Signal Processing Toolbox User's Guide* (Mathworks Incorporated, Natick, 2000)
W. Rudin, *Fourier Analysis on Groups* (Interscience, New York, 1962)

Chapter 9
Additional Topics

Fourier methods broadly construed have applications beyond the problems discussed in previous chapters. All of these consist, in a sense, of different decompositions for functions. The motivation for the decomposition varies from a need for efficient storage, shifted point of view, to geometrically motivated adaptations of standard transforms.

Some of these topics are of relatively recent origin, and actually arose and were developed according to the needs of digital signal processing. The geometrically based transforms are more closely connected to boundary value problems and models of mathematical physics.

9.1 Local Waveform Analysis

Conventional Fourier analysis is well adapted to analysis of time-invariant systems. There are function analysis and expansion problems, however, that have applications beyond the solution of differential and integral equations. Applications arise in identification, data compression, noise reduction, and modeling where the economy of representation and presentation is a significant aspect of the problem.

It is not the case that descriptions in terms of, for example, the Fourier transform and the associated power spectrum are the appropriate tools for all situations.

The problem is that time resolution and frequency resolution of signals are in a precise sense complementary and mutually exclusive properties of a signal. Some hint of this has already been encountered in the discussion of the Fourier transform inversion integrals. There inversion was carried out (in the L^1 case) by using "approximate delta functions" to regularize the inversion integrals. These entities have the property that the more concentrated they become in the time domain, the more spread out they are in the frequency domain. The extreme example of this

© Springer International Publishing Switzerland 2016
J.H. Davis, *Methods of Applied Mathematics with a Software Overview*,
Applied and Numerical Harmonic Analysis, DOI 10.1007/978-3-319-43370-7_9

(a distribution rather than a function) is the delta function, whose Fourier transform (appropriately defined) is actually a constant and not in the least bit concentrated in the frequency domain.

In the context of "functions," there is actually a numerical inequality which prevents a given function from being simultaneously localized in the time and frequency domains. The result is referred to as the *uncertainty principle*, a name that is borrowed from the application of the result in the physics context of quantum mechanics.[1]

9.2   Uncertainty Principle

For square-integrable functions (i.e., functions in L^2, of finite energy) Parseval's Theorem provides the relation that

$$\int_{-\infty}^{\infty} |f(t)|^2 \, dt = \frac{1}{2\pi} \int_{-\infty}^{\infty} |\hat{f}(\omega)|^2 \, d\omega,$$

so that assuming that $f(\cdot)$ is a unit vector, we have an interpretation of a probability density in both the "time domain" and "frequency domain." Explicitly, the "densities" are

$$p(t) = |f(t)|^2,$$

$$\hat{p}(\omega) = \frac{1}{2\pi} |\hat{f}(\omega)|^2.$$

The density interpretation arises solely from the fact that the functions are nonnegative, and integrate to 1 (assuming a unit vector) over the respective domain.

In statistical terms, the uncertainty principle derivation assumes that each of those probability distributions has a finite variance. That is,

$$\int_{-\infty}^{\infty} |t|^2 \, |f(t)|^2 \, dt < \infty,$$

$$\frac{1}{2\pi} \int_{-\infty}^{\infty} |\omega|^2 \, |\hat{f}(\omega)|^2 \, d\omega < \infty.$$

Because of the fact that

$$\mathcal{F}\{f'(t)\} = i\omega \, \mathcal{F}\{f(t)\},$$

---

[1] The author was stunned to learn that the uncertainty principle was a manifestation of the Cauchy–Schwarz inequality rather than a mystical property of subatomic particles.

the second assumption can be read as assuming that $f' \in L^2$. The assumption that the "time distribution" has finite variance is the same as assuming that $tf(t) \in L^2$. The uncertainty principle arises from applying the Cauchy–Schwarz inequality to the inner product

$$< tf(t), f'(t) > = \int_{-\infty}^{\infty} tf(t)\overline{f'(t)}\, dt.$$

To derive the inequality, consider the integration-by-parts identity

$$\int_{-B}^{B} t \frac{d}{dt} |f(t)|^2\, dt = t\, |f(t)|^2 \Big|_{-B}^{B} - \int_{B}^{B} |f(t)|^2\, dt.$$

If it can be shown (as we do below) that the "boundary limit" terms vanish, then this identity gives

$$\int_{-\infty}^{\infty} t \frac{d}{dt} |f(t)|^2\, dt = \int_{-\infty}^{\infty} t\overline{f(t)} \frac{d}{dt} f(t)\, dt + \int_{-\infty}^{\infty} tf(t) \frac{d}{dt} \overline{f(t)}\, dt$$

$$= < tf(t), \frac{d}{dt} f(t) > + < \frac{d}{dt} f(t), tf(t) >$$

$$= 2\, \mathrm{Re} \left(< tf(t), \frac{d}{dt} f(t) > \right)$$

$$= - \int_{-\infty}^{\infty} |f(t)|^2\, dt.$$

Assuming that $\|f(t)\| = 1$, so that f is normalized, this implies that

$$\left| \mathrm{Re} \left(< tf(t), \frac{d}{dt} f(t) > \right) \right| = \frac{1}{2}.$$

But then by the Cauchy–Schwarz inequality

$$\frac{1}{2} = \left| \mathrm{Re} \left(< tf(t), \frac{d}{dt} f(t) > \right) \right| \leq \left| < tf(t), \frac{d}{dt} f(t) > \right| \leq \|tf(t)\| \, \|\frac{d}{dt} f(t)\|.$$

Written out, this is the inequality

$$\frac{1}{4} \leq \int_{-\infty}^{\infty} t^2\, |f(t)|^2\, dt \int_{-\infty}^{\infty} \frac{1}{2\pi} \omega^2\, |\hat{f}(\omega)|^2\, d\omega,$$

which in words says that the product of the "time variance" and the "frequency variance" of the signal exceeds $\frac{1}{4}$. This inequality is known as the uncertainty principle. [2]

The fact that the boundary terms $t|f(t)|^2 \to 0$ as $t \to \pm\infty$ actually follows from our original assumptions about $f(t)$. It is very nearly a consequence of the integration by parts argument above, and actually follows by a "one-sided" version of that.

Consider

$$\int_0^B t \frac{d}{dt} |f(t)|^2 \, dt = t |f(t)|^2 \Big|_0^B - \int_0^B |f(t)|^2 \, dt.$$

The left side of this equality tends to the limit

$$2 \operatorname{Re} \left(< U(t) \, t f(t), \frac{d}{dt} f(t) > \right)$$

as $B \to \infty$. The last term

$$\int_0^B |f(t)|^2 \, dt$$

has a limit since (by assumption) $f \in L^2$. As a result (by the two out of three limit theorem) we conclude that

$$\lim_{t \to \infty} t |f(t)|^2 = L$$

exists. We expect that actually $L = 0$. If not, for t sufficiently large we have the inequality

$$t |f(t)|^2 \geq (L - \epsilon),$$

$$|f(t)|^2 \geq \frac{(L - \epsilon)}{t},$$

and this would make $\int_0^\infty |f(t)|^2 \, dt$ divergent, in contradiction to the assumption that $f \in L^2$. Clearly we can similarly show that $\lim_{t \to -\infty} t |f(t)|^2$ exists and is 0. These details complete the proof of the uncertainty principle, assuming only that the function has finite energy, and that the variance quantities involved in the inequality are well defined.

---

[2] The connection of this with the quantum mechanical version comes from identifying $f(t)$ with the *wave function*. Then t corresponds to the position variable, and the time variance is the "spatial uncertainty." The quantum mechanical momentum is identified as $\frac{h}{2\pi i} \frac{d}{dt} f(t)$, and our frequency variance differs by factors of Planck's constant and 2π from the "momentum uncertainty."

Sheep May Safely Graze
J. S. Bach

Fig. 9.1 A frequency time diagram

The uncertainty principle result might be thought of as demonstrating the impossibility of simultaneous time and frequency localization. A more accurate description is that it demonstrates that considering only the original function and its Fourier transform provides "inadequate information"[3]. The issue is actually one of point of view and appropriate presentation of information.

There are applications for which the idea of the "time variation of the frequency content" is central. A common instance of this idea appears in standard musical notation, as in Figure 9.1. Each of the notes corresponds to a specific frequency, but has a limited duration and definite initiation point.

This timing information is completely lost when the "sound signal" associated with the measures is Fourier transformed. This can be verified with an Octave computation of the fft of an audio waveform.

Musical Scores

Musical notes are conventionally modeled as sinusoids modulated by a finite duration envelope function. This idea is in fact part of the MIDI standard employed in synthesizer keyboards. The note envelope consists of a rapid rise ("attack") followed by a constant period ("sustain") and a more gradual decay ("release") to extinction. A typical note waveform is shown in Figure 9.2. This diagram is created with a python script:

---

[3]The statement clearly is quite imprecise as it stands, since the time function (or the Fourier transform, for that matter) contains all there is to know about the function.

PYTHON

```python
import numpy as np
import matplotlib.pyplot as plt
from sheep_graze import qnote, sheep_graze
from setup_plots import setup_default




def Epsfile(plot_dict, var, filename, fun_list, range_list):
    """ write out a pdf file containing plots
    other figure content may be set before calling epsfile
    """

    for kk in range(len(fun_list)):
        exec(range_list[kk])
        yy = eval(fun_list[kk])
        tem = eval(var)
        plt.plot(tem, yy)


        # label the axes

    plt.xticks(plot_dict['xtick'], plot_dict['xticklabel'])
    plt.yticks(plot_dict['ytick'], plot_dict['yticklabel'])

    s = .05

    # get [xmin, xmax, ymin, ymax]

    limits = plt.axis()

    xlims = [limits[0], limits[1]]
    ylims = [limits[2], limits[3]]

    lrange = xlims[1] - xlims[0]
    xlims[0] = xlims[0] - s * lrange
    xlims[1] = xlims[1] + s * lrange

    lrange = ylims[1] - ylims[0]
    ylims[0] = ylims[0] - s * lrange
    ylims[1] = ylims[1] + s * lrange

    plt.axis([xlims[0], xlims[1], ylims[0], ylims[1]])

    # cut down to simple x -y axes

    fig_axis = plt.gca()
```

```
# hide top and right spines

fig_axis.spines['right'].set_color('none')
fig_axis.spines['top'].set_color('none')

# grid on wheat='#F5DEB3' goldenrod= '#DAA520',

plt.grid(color='#404040')

# add space around the plot in the figure

plt.margins(0.25)

setup_default()

plt.savefig(filename, dpi=300)

plt.close('all')
#######


plot_dict = {}
#
fig_axis = plt.gca()
fig = plt.gcf()
#
setup_default()
#
plot_dict['ytick'] = [0, 2.5]
plot_dict['yticklabel'] = ["0", "0"]
plot_dict['xtick'] = [0, 2*np.pi]
plot_dict['xticklabel'] = ['0', 'T']
#
#

Epsfile(plot_dict, 'x', 'qnote_wave.pdf', \
        ['2.5 + (qnote(x, 2 * np.pi) * np.sin(20*x))',\
         'qnote(x, 2 * np.pi)'], \
        ['x = np.linspace(0, 2*np.pi, 1000)',\
         'x = np.linspace(0, 2*np.pi, 1000)'])
```

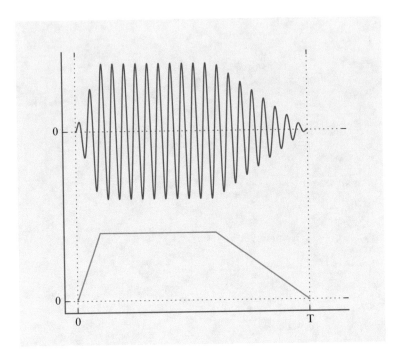

Fig. 9.2 A quarter note waveform

The envelope is implemented with a combination of step functions and straight lines. Note that calling the waveform a quarter note is a matter of convention, since the function incorporates a T duration parameter in the definition. It is convenient to think of it as a quarter note since that is a middle duration in most musical scores. The other length notes are created by scaling the T parameter with powers of 2.

The frequency data for the musical notes is stored in a python dictionary: the dictionary keys are the octave number, and the stored data is a list of frequencies. The python multiple assignment syntax allows assigning integer indices with the usual note names. The frequency is then accessed as $note[octave][note - name]$.

The musical fragment is the opening measure of Bach's "Sheep may safely graze." The score is in Figure 9.1:

The definition of the waveform was created by editing in the bass line as a sum of delayed notes, and then going back and filling in the treble notes based on the bass line timing. One is tempted to add generation of the waveform expression to one of the music score editing programs, since it is an entirely mechanical transcription.

PYTHON

```python
# note-names.py + Sheep may safely graze intro measure
#

import numpy as np

def heaviside(t):

    return(.5*(1  + np.sign(t)))

def qnote(t, T):
    """
    quarter note envelope
    note duration T seconds
    """

    y = (heaviside(t)* (1/(.1 *T))*t \
    - heaviside(t - .1*T)*(1/(.1*T))*(t - .1*T)\
    - heaviside(t - .6*T) *(1/(.4*T))*(t-.6*T) )\
    * (heaviside(t) - heaviside(t- T))

    return(y)

# index names  for notes dictionary

C, Csharp ,D, Eflat ,E, F, Fsharp ,G, Gsharp ,A, Bflat ,B = range(12)

# dictionary of A = 440Hz equally tempered scales
# arranged as index above


notes = {\
    0: [16.35, 17.32, 18.35, 19.45, 20.60, 21.83,\
        23.12, 24.50, 25.96, 27.50, 29.14, 30.87],\
    1: [32.70, 34.65, 36.71, 38.89, 41.20, 43.65,\
        46.25, 49.00, 51.91, 55.00, 58.27, 61.74],\
    2: [65.41, 69.30, 73.42, 77.78, 82.41, 87.31,\
        92.50, 98.00, 103.8, 110.0, 116.5, 123.5],\
    3: [130.8, 138.6, 146.8, 155.6, 164.8, 174.6,\
        185.0, 196.0, 207.7, 220.0, 233.1, 246.9],\
    4: [261.6, 277.2, 293.7, 311.1, 329.6, 349.2,\
        370.0, 392.0, 415.3, 440.0, 466.2, 493.9],\
    5: [523.3, 554.4, 587.3, 622.3, 659.3, 698.5,\
        740.0, 784.0, 830.6, 880.0, 932.3, 987.8],\
```

```
6: [1047,  1109,  1175,  1245,  1319,  1397,\
    1480,  1568,  1661,  1760,  1865,  1976],\
7: [2093,  2217,  2349,  2489,  2637,  2794,\
    2960,  3136,  3322,  3520,  3729,  3951],\
8: [4186,  4435,  4699,  4978,  5274,  5588,\
    5920,  6272,  6645,  7040,  7459,  7902]}

# first measure of "Sheep may safely graze"

def sheep_graze(t, T):
    """
    waveform definition
    """
    waveform = \
        (qnote(t, .5*T) * np.sin(2*np.pi*notes[2][G]*t)\
        + qnote(t-.5*T, .5*T)\
        * np.sin(2*np.pi*notes[2][G]*(t - .5*T)))\
        + (qnote(t, .5*T) * (np.sin(2*np.pi*notes[5][D]*t)\
        + np.sin(2*np.pi*notes[4][B]*t)))\
        + (qnote(t-.5*T, .25*T)\
        * (np.sin(2*np.pi*notes[5][D]*(t-.5*T))\
        + np.sin(2*np.pi*notes[4][B]*(t-.5*T))))\
        + (qnote(t-.75*T, .25*T)\
        * (np.sin(2*np.pi*notes[5][D]*(t-.75*T))\
        + np.sin(2*np.pi*notes[4][B]*(t-.75*T))))\
        + (qnote(t-T, .5*T)\
        * np.sin(2*np.pi*notes[2][G]*(t-T))\
        + qnote(t-1.5*T, .5*T)\
        * np.sin(2*np.pi*notes[2][G]*(t -1.5*T)))\
        + (qnote(t-T, .5*T)\
        * (np.sin(2*np.pi*notes[5][D]*(t-T))\
        + np.sin(2*np.pi*notes[4][B]*(t-T))))\
        + (qnote(t-1.5*T, .25*T)\
        * (np.sin(2*np.pi*notes[5][D]*(t-1.5*T))\
        + np.sin(2*np.pi*notes[4][B]*(t-1.5*T))))\
        + (qnote(t-1.75*T, .25*T)\
        * (np.sin(2*np.pi*notes[5][D]*(t-1.75*T))\
        + np.sin(2*np.pi*notes[4][B]*(t-1.75*T))))\
        + (qnote(t - 2*T, .5*T)\
        * np.sin(2*np.pi*notes[2][G]*(t-2*T))\
        + qnote(t-2.5*T, .5*T)\
        * np.sin(2*np.pi*notes[2][G]*(t - 2.5*T)))\
        + (qnote(t-2*T, .5*T)\
        * (np.sin(2*np.pi*notes[5][D]*(t-2*T))\
        + np.sin(2*np.pi*notes[4][B]*(t-2*T))))\
        + (qnote(t-2.5*T, .5*T)\
        * (np.sin(2*np.pi*notes[5][E]*(t-2.5*T))\
        + np.sin(2*np.pi*notes[5][C]*(t-2.5*T))))\
        + (qnote(t-3*T, .5*T)\
        * np.sin(2*np.pi*notes[2][G]*(t-3*T))\
        + qnote(t-3.5*T, .5*T)\
        * np.sin(2*np.pi*notes[2][G]*(t - 3.5*T)))\
        + (qnote(t-3*T, .5*T)\
        * (np.sin(2*np.pi*notes[5][E]*(t-3*T))\
```

```
   +  np.sin(2*np.pi*notes[5][C]*(t-3*T))))\
   +  (qnote(t-3.5*T, .5*T)\
   *  (np.sin(2*np.pi*notes[5][E]*(t-3.5*T))\
   +  np.sin(2*np.pi*notes[5][G]*(t-3.5*T))))
```

return waveform

The other sections of this chapter are concerned with topics that might be thought of as bridging the gap between the time and Fourier domain descriptions of a signal. The time domain might be thought of as providing perfect time resolution, but no information about frequency content. The Fourier transform computes the frequency domain decomposition of the signal, but at the expense of obscuring the time resolution characteristics of the signal.

The topic of short-time Fourier Transform attempts to join the views of the signal by computing the Fourier resolution of time-localized portions of the signal.

9.3   Short-Time Fourier Transforms

The short-time Fourier transform is actually not a uniquely defined operation. The lack of uniqueness lies in the use of different *window functions* to localize the transformed function in time. Some of the most common window waveforms are illustrated in Figure 9.3.

In terms of a given window function $w(\cdot)$, the short-time Fourier transform (STFT) is a function of two variables τ, ω (time and frequency) defined by

$$F(\tau, \omega) = \mathcal{SF}\{f(t)\} = \int_{-\infty}^{\infty} w(t - \tau) f(t) e^{-i\omega t} \, dt.$$

The window function w is typically symmetric about the origin and of finite duration. The short-time transform then corresponds to the ordinary Fourier transform applied to that portion of the signal around time τ. A plot of the magnitude of the STFT (or magnitude squared) gives information on the frequency power distribution in the signal around time τ.

Fig. 9.3 Sample window
waveforms

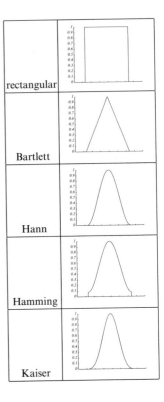

The SciPy signal package provides a variety of signal processing procedures, including the usual window functions. These can be investigated through the python help system.

PYTHON

```
>>> import scipy.signal as sig

>>> dir(sig)
['BadCoefficients', 'StateSpace', 'Tester', 'TransferFunction',
 'ZerosPolesGain', '__all__', '__builtins__', '__doc__',
 '__file__', '__name__', '__package__', '__path__',
 '_arraytools', '_max_len_seq','_max_len_seq_inner',
 '_peak_finding', '_savitzky_golay', '_spectral',
```

```
'abcd_normalize', 'absolute_import', 'argrelextrema',
'argrelmax', 'argrelmin', 'band_stop_obj', 'barthann',
'bartlett', 'bench', 'bessel', 'besselap', 'bilinear',
'blackman', 'blackmanharris', 'bode', 'bohman', 'boxcar',
'bspline', 'bsplines', 'buttap', 'butter','buttord',
'cascade', 'cheb1ap', 'cheb1ord', 'cheb2ap','cheb2ord',
'chebwin', 'cheby1', 'cheby2', 'chirp', 'cmplx_sort',
'coherence', 'cont2discrete', 'convolve', 'convolve2d',
'correlate', 'correlate2d', 'cosine', 'csd', 'cspline1d',
'cspline1d_eval', 'cspline2d', 'cubic', 'cwt', 'daub',
'decimate', 'deconvolve', 'detrend', 'dimpulse', 'division',
'dlsim', 'dltisys', 'dstep', 'ellip', 'ellipap', 'ellipord',
'exponential', 'fftconvolve','filter_design', 'filtfilt',
'find_peaks_cwt', 'findfreqs', 'fir_filter_design', 'firwin',
'firwin2','flattop', 'freqresp', 'freqs', 'freqz',
'gauss_spline', 'gaussian', 'gausspulse', 'general_gaussian',
'get_window','group_delay', 'hamming', 'hann', 'hanning',
'hilbert', 'hilbert2', 'iirdesign', 'iirfilter', 'impulse',
'impulse2','invres', 'invresz', 'kaiser', 'kaiser_atten',
'kaiser_beta', 'kaiserord', 'lfilter', 'lfilter_zi',
'lfiltic', 'lombscargle', 'lp2bp', 'lp2bs', 'lp2hp', 'lp2lp',
'lsim', 'lsim2', 'lti', 'ltisys', 'max_len_seq', 'medfilt',
'medfilt2d', 'morlet', 'normalize', 'nuttall', 'order_filter',
'parzen', 'periodogram', 'place_poles', 'print_function', 'qmf',
'qspline1d', 'qspline1d_eval', 'qspline2d', 'quadratic',
'remez', 'resample', 'residue', 'residuez','ricker', 's',
'savgol_coeffs', 'savgol_filter', 'sawtooth', 'sepfir2d',
'signaltools', 'sigtools', 'slepian', 'sos2tf', 'sos2zpk',
'sosfilt', 'sosfilt_zi', 'spectral', 'spectrogram', 'spline',
'spline_filter', 'square', 'ss2tf', 'ss2zpk', 'step', 'step2',
'sweep_poly','symiirorder1','symiirorder2', 'test', 'tf2sos',
'tf2ss','tf2zpk', 'triang', 'tukey', 'unique_roots',
'vectorstrength','waveforms', 'wavelets', 'welch', 'wiener',
'windows', 'zpk2sos', 'zpk2ss', 'zpk2tf']
```

```
>>> help(sig.windows)
```

The help for *signal.windows* contains a lot of information. Along with the formulas for the standard window shapes, there is actually example python code that will plot the windows along with their frequency response information. The python help displays text in a pager program, typically called "less" in deference to an original pager called "more."

Less is keystroke driven, *q* quits and returns to the terminal prompt. However, "s" will save the viewed text under a chosen name: "sig.win.hlp" works. This is a large file, with the included python code lines all prefaced by the python interactive prompt ">>." Extracting the python code from the help file is the purpose of the unix-based command line text processing tools [5]. First, grep extracts lines with the python prompt.

BASH

```
$ grep ">>>" sig.win.hlp

>>> from scipy import signal
>>> from scipy.fftpack import fft, fftshift
>>> import matplotlib.pyplot as plt
>>> window = signal.barthann(51)
>>> plt.plot(window)
>>> plt.title("Bartlett-Hann window")
>>> plt.ylabel("Amplitude")
>>> plt.xlabel("Sample")
>>> plt.figure()
>>> A = fft(window, 2048) / (len(window)/2.0)
>>> freq = np.linspace(-0.5, 0.5, len(A))
    . . . . . . . . . . . . . .
```

The spurious ">>" prompts need to be removed. The "grep" output could be caught in a file, and the prompts removed with a global search and replace. That is tedious and not necessary because the "stream editor" sed can snip out the prompts on the way to saving the file; The pipeline that does that is

BASH

```
grep ">>>" sig.win.hlp | sed 's/>>>//g' > sig_win_plot.py
```

The inline sed command string substitutes the prompt with an empty string (globally). This gives a file of clean python code, the head of which appears as

PYTHON

```
from scipy import signal
from scipy.fftpack import fft, fftshift
import matplotlib.pyplot as plt
window = signal.barthann(51)
plt.plot(window)
plt.title("Bartlett-Hann window")
plt.ylabel("Amplitude")
plt.xlabel("Sample")
plt.figure()
A = fft(window, 2048) / (len(window)/2.0)
freq = np.linspace(-0.5, 0.5, len(A))
response = 20 * np.log10(np.abs(fftshift(A / abs(A).max())))
plt.plot(freq, response)
plt.axis([-0.5, 0.5, -120, 0])
plt.title("Frequency response of the Bartlett-Hann window")
plt.ylabel("Normalized magnitude [dB]")
plt.xlabel("Normalized frequency [cycles per sample]")
    . . . . . . . .
```

That fragment represents about 95% of the code to display the information for a Bartlett-Hann window. The code is missing an import of numerical python as "np," and does not display on the screen or save an image to a file even though it generates the matplotlib figures for that.

PYTHON

```python
from scipy import signal
from scipy.fftpack import fft, fftshift
import matplotlib.pyplot as plt

import numpy as np

window = signal.barthann(51)
plt.plot(window)
plt.title("Bartlett-Hann window")
plt.ylabel("Amplitude")
plt.xlabel("Sample")
plt.figure()

plt.show()

A = fft(window, 2048) / (len(window)/2.0)
freq = np.linspace(-0.5, 0.5, len(A))
response = 20 * np.log10(np.abs(fftshift(A / abs(A).max())))
plt.plot(freq, response)
plt.axis([-0.5, 0.5, -120, 0])
plt.title("Frequency response of the Bartlett-Hann window")
plt.ylabel("Normalized magnitude [dB]")
plt.xlabel("Normalized frequency [cycles per sample]")

plt.show()
```

Because the factor of the windowing function "distorts" the original function, the STFT is not simply related to the Fourier transform of the original signal.[4] There actually are "inversion integrals" that apply to the short time Fourier transform (See for example [2, 6]) which seem primarily of interest for their connections to wavelets. Short-time Fourier transforms do not have derivative properties that make Fourier transforms so operationally useful in differential equation problems, but they are useful in the time-frequency context that motivated the definition.

The *spectrogram* of a signal $f(\cdot)$ is simply a plot of the magnitude of the short-time Fourier transform as a function of the two variables. Geometrically, this is a surface, but the information is usually not presented in three-dimensional form. Usually a contour plot, or a two-dimensional representation using a gray scale to suggest altitude is the format encountered.

---

[4]Of course, for particular windows relations can be found.

One of the principal uses of the short-time Fourier transform is in the area of speech recognition and analysis. There the topic is referred to as *formant analysis*, and the problem is to identify patterns in the spectrogram in terms of *phonemes*.[5] Spectrograms of speech waveforms display characteristic patterns for vowels, for example.

Calculating Spectrograms

The definition of the STFT given above is based on the continuous-time Fourier transform. The short-time transform can also be naturally defined for discrete time signals by simply computing the discrete Fourier transform (DFT) of a windowed signal. The windows used are effectively the same ones described above, but in a scaled and sampled version. The definition produces a function of a frequency variable, and delay parameter

$$\mathcal{SF}\{f(n)\} = \sum_{-\infty}^{\infty} f(n)\, w(n-k)\, e^{i n \omega}.$$

For the case of numerical computation, only a discrete set of frequency parameters can be handled, so we might as well take

$$\omega_j = \frac{2\pi j}{N}$$

for some N, to obtain equally spaced points in the discrete-time frequency domain. Since the weight function $w(n)$ is chosen to be nonzero only on a symmetric finite interval about 0, the formula becomes

$$\mathcal{SF}\{f(n)\} = \sum_{n-W}^{n+W} f(n)\, w(n-k)\, e^{i n \frac{2\pi j}{N}},$$

and as long as the window width $2W$ is at least as large as the sampling parameter N, the result identifies as a finite discrete Fourier transform. This means it can be computed using a FFT (fast Fourier transform) algorithm.

The Octave Signal Processing package [7] provides a spectrogram calculating routine. Like other routines in that package, output is formatted on the assumption that the data arises from sampling a continuous-time signal at a sampling rate

---

[5]Phonemes are a linguistic notion, corresponding to an individually identifiable speech sound segment. The spectrogram allows one to "see" the phonemes in terms of the time frequency distribution of the power in the sonic signal.

(denoted by *fs* in the code below). A spectogram procedure is also included in the scipy signal package, and that is used as an example below.

Other parameters control the number of time delay samples used in the output and the choice of sampling window for the computation. The documentation can be consulted for details of the parameter list specification, and doing so is the cost of using prepackaged routines.

See Problem 8 for a suggestion on writing your own spectrogram routine.

PYTHON

```python
# plots spectogram of the sheep_graze waveform
#

from scipy import signal
import matplotlib.pyplot as plt
import numpy as np
from sheep_graze import qnote, sheep_graze
from setup_plots import setup_default


#seconds per beat

T = 60.0/90.0

# sample frequency 2k

fs = 2048

# samples = 5 beats * T sec/beat * fs smmples/sec

t = np.linspace(-.5*T, 4.5*T, 5*T*fs )

x = sheep_graze(t, T)

freq, times, Sxx = signal.spectrogram(x, fs, nperseg=256)

setup_default()

plt.pcolormesh(times, freq, Sxx, cmap='YlGnBu')

plt.ylabel('Frequency')
plt.xlabel('Time')
plt.savefig('sheep_spectogram.pdf', dpi=300)
```

The plot generated by this instance of spectogram is shown in Figure 9.4. The note frequencies, and especially the time sequencing points, are very evident in the plot. The appearance of the plot is largely determined by the *fs* sampling rate parameter, and the frequency axis is scaled internally by the spectogram procedure based on the sampling theorem "twice the highest frequency present" requirement.

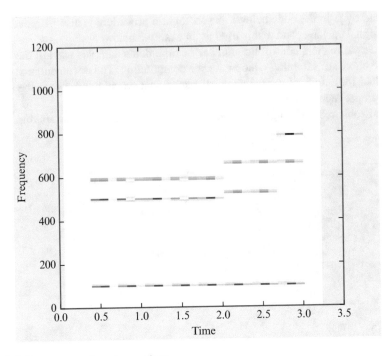

Fig. 9.4 Spectrogram of music waveform

The frequency content information of this spectrogram is quite simple, based on the single frequency note model in the tune definition. A more interesting spectrogram arises from incorporating note harmonics into the tune definition. See Problem 9 for this.

The application of STFT to linguistics and formant analysis is a wide subject area with a variety of mathematical models and associated problems. An introduction to these topics is provided in [8]. Identification of spoken phonemes through analysis of measured spectrogram data is a component of speech recognition software.

Problems 9.3.

1. Show that a function that achieves minimum joint uncertainty (that is, equality in the uncertainty principle inequality) must be a Gaussian waveform.
2. Evaluate the uncertainty principle expression for $f(t) = e^{-|t|}$.
3. Find the short-time Fourier transform of

$$f(t) = U(t)\, e^{-t}$$

using a rectangular window of width W.

4. Evaluate the short-time Fourier transform of

$$f(t) = \begin{cases} 1 & |t| < a, \\ 0 & \text{otherwise.} \end{cases}$$

   Use a Hamming window to evaluate the transform.
5. Derive a "delay law" for the short-time Fourier transform.
6. Calculate the short-time Fourier transform of

$$e^{i\Omega t} f(t)$$

   in terms of the short-time transform of f.
7. Some popular windows are constructed as (a segment of) combinations of cosine waveforms. Use Octave to plot graphs of

$$A - B \cos(2\pi t) + C \cos(4\pi t)$$

   over the interval $t = [0, 1]$, for various A, B, C. Try to adjust the parameters to give a wide, flat top to the waveform, as well as a value of zero at the limiting extents. Several of the standard windows are of this form.
8. Construct

OCTAVE

```
function drawspec(data, samples, window)
```

   to draw a three-dimensional spectrogram rendition.
   Hint: Plot the three-dimensional picture with

OCTAVE

```
surf(Omega, T, AbsoluteFFT);
```

   Generate the plotting grid with

OCTAVE

```
[Omega, T] = meshgrid(1:length(window)/2, 1:samples);
```

   The windowed data for a time shift k can be calculated with by selecting a subset of the data and multiplying elementwise with the window. This can be transformed, and the positive frequency portion can be placed in a surface plot array.

OCTAVE

```
W = length(window)/2;

for k = W+1 : length(data) - W
    y = data(k-W:k-W + length(window)-1) .* window;
    absfft = abs(fft(y));
    plotvalues = absfft(1: length(window)/2);

%  add plotvalues to AbsoluteFFT array
    ....
endfor
```

Obviously, the use of a *samples* parameter substantially complicates the management of the surface plot arrays. The maximum number of samples corresponds to fitting the window without overlapping the given *data* limits. The simplest implementation uses that observation. Elaborations that place the windows in order to end up with a given number of samples can be added later.

9. Define the basis of a guitar simulator by changing the quarter note MATLAB definition to include fundamental frequency and "a_over_L" parameters. The discussion of Section 3.7 computes the harmonic amplitudes for a plucked guitar string. Construct

OCTAVE

```
function y = string_qnote(t, fund\_freq, a\_over\_L)
    ....
    y = ....
```

to multiply the original quarter note envelope by a sum of sinusoids at harmonics of "fund_freq," and relative amplitudes determined by the "pluck position" parameter "a_over_L."

Adapt the tune waveform definition (that is, the original two-measure music notation) to be played with $a\_over\_L = .25$, and generate a spectrogram of the result.

10. What is the frequency domain condition that, for a given normalized $f \in L^2$, the translates by integer multiples of some fixed T are orthogonal. That is,

$$\int_{-\infty}^{\infty} T_{nT} f \, \overline{T_{mT} f} \, dt = \delta nm.$$

11. Show that a 2π periodic L^2 function α which satisfies

$$\alpha(\theta) + \alpha(\theta + \pi) = 0$$

has only odd terms in its Fourier series.

9.4 Two-Sided Transforms

In our treatment of Fourier and Laplace transforms, we have emphasized the connections between physical problems defined on a half axis and Laplace transforms on one hand, and whole-axis problems and Fourier transforms on the other hand. These associations are natural, and these transforms are appropriate for many such problems.

There also exists a formalism (i.e., a transform theory) referred to as the *Two-Sided Laplace Transform* which is sometimes encountered. This approach extends the definition of the Laplace transform to the case of certain functions defined on the whole axis. The effect is to avoid introduction of the Fourier transform for such problems. However, the resulting transform requires some technical care in use. The details are closely related to the process of Fourier transform inversion by complex variable methods. The fact that the formulas involved have "the look" of standard Laplace transforms sometimes leads to erroneous inversion procedures on the part of the unwary.

The definition of the two-sided transform is as follows. For functions f and values of the complex transform variable p such that the indicated integrals converge, define

$$\mathcal{L}_{II}\{f(t)\} = \int_{-\infty}^{\infty} e^{-pt} f(t)\, dt. \tag{9.1}$$

The expression may be regarded as being defined as the sum of two half-line (one-sided) Laplace transforms:

$$\mathcal{L}_{II}\{f^{+}(t)\} = \int_{0}^{\infty} e^{-pt} f(t)\, dt$$

and

$$\mathcal{L}_{II}\{f^{-}(t)\} = \int_{0}^{\infty} e^{-(-p)t} f(-t)\, dt.$$

Note that the transform of f^{+} is typically well defined for

$$\mathrm{Re}(p) > \sigma_1$$

while that of f^{-} exists for

$$\mathrm{Re}(-p) > \sigma_2.$$

The two-sided transform (9.1) is well defined only when there is a region of overlap between the two regions. (See Figure 9.5).

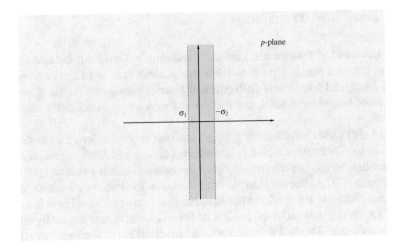

Fig. 9.5 Overlap region

Beyond this, when the required overlap exists, the transform is defined only for values of p satisfying

$$\sigma_1 < \text{Re}(p) < (-\sigma_2).$$

This behavior is in contrast to the case of the conventional Laplace transform, which is well defined in a half-plane.

Example. Let $f(t) = e^{-a|t|}$, $a > 0$. Then

$$\mathcal{L}_{II}\{f(t)\} = \int_{-\infty}^{0} e^{at} e^{-pt} dt + \int_{0}^{\infty} e^{-at} e^{-pt} dt.$$

The first integral exists for $\text{Re}(a - p) > 0$ or

$$\text{Re}(p) < a.$$

The second requires that $\text{Re}(a + p) > 0$ or

$$\text{Re}(p) > -a.$$

Therefore,

$$\mathcal{L}_{II}\{f(t)\} = \frac{1}{a-p} + \frac{1}{a+p}, \quad |\text{Re}(p)| < a.$$

Example. Let $f(t) = 1$. Then

$$\int_{-\infty}^{0} 1\, e^{-pt}\, dt = -\frac{1}{p}, \ \mathrm{Re}(p) < 0,$$

and

$$\int_{0}^{\infty} 1\, e^{-pt}\, dt = \frac{1}{p}, \ \mathrm{Re}(p) > 0.$$

There is no region of overlap, so $\mathcal{L}_{II}\{1\}$ is undefined.

Note that if one fails to note the regions of convergence, the puzzling "calculation"

$$\mathcal{L}_{II}\{1\} \stackrel{?}{=} -\frac{1}{p} + \frac{1}{p} = 0$$

follows.

The last example emphasizes that regions of convergence matter when dealing with two-sided Laplace transforms. With one-sided transforms, the fact that (assuming transforms exist) taking the transform variable with sufficiently large real part guarantees simultaneous existence of transforms for all of several functions in a given problem allows the issue to be sidestepped until the transforms are inverted.

For the sake of symmetry, it should be mentioned that the definition of two-sided z-transforms is also possible. The motivations are analogous to those in the continuous variable case, and the costs of the use of such are again that attention must be paid to the region of convergence of the transform. The definition is simply

$$\mathcal{Z}_{II}\{f(n)\} = \sum_{n=-\infty}^{\infty} f(n)\, \zeta^{-n} \tag{9.2}$$

$$= \sum_{n=0}^{\infty} f(n)\, \zeta^{-n} + \sum_{k=1}^{\infty} f(-k) \left(\frac{1}{\zeta}\right)^{-k}.$$

This can be expected to be well defined for

$$|\zeta| > R, \ \left|\frac{1}{\zeta}\right| > \rho,$$

and so makes sense only if there is an annulus of convergence (Figure 9.6)

$$R < |\zeta| < \frac{1}{\rho}.$$

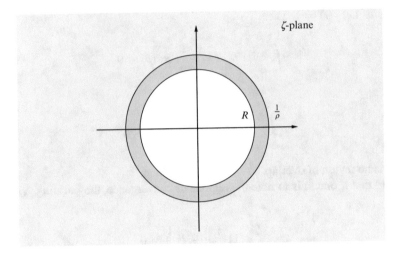

Fig. 9.6 Annulus of convergence

The properties of two-sided transforms are rather predictable, and are left for the exercises below.

Problems 9.4.

1. For each of the following functions, compute (if possible) the two-sided Laplace transform, and indicate the associated strip of convergence.

   a.

$$f(t) = \begin{cases} 0 & t < 0, \\ e^{-at} & t \geq 0. \end{cases}$$

   b.

$$f(t) = \begin{cases} e^{at} & t < 0, \\ 0 & t \geq 0. \end{cases}$$

   c.

$$f(t) = \begin{cases} 0 & t > 0, \\ e^{-at} & t \leq 0. \end{cases}$$

d.

$$f(t) = e^{a|t|}, \quad -\infty < t < \infty.$$

2. Suppose that f and f' both have two-sided Laplace transforms with a common strip of convergence. Compute

$$\mathcal{L}_{II}\left\{f'(t)\right\}.$$

3. For each of the following sequences compute (if possible) the two-sided z-transform, and in each case indicate the associated annulus of convergence.

a.

$$f(n) = \begin{cases} 0 & n < 0, \\ \alpha^n & n \geq 0, \ (|\alpha| < 1). \end{cases}$$

b.

$$f(n) = \begin{cases} \alpha^{-n} & n \leq 0, \ (|\alpha| < 1), \\ 0 & n > 0. \end{cases}$$

c.

$$f(n) = \begin{cases} \beta^n & n \geq 0, \ (|\beta| > 1), \\ 0 & n < 0. \end{cases}$$

d.

$$f(n) = \begin{cases} 0 & n > 0, \\ \beta^{-n} & n \leq 0 \ (|\beta| > 1). \end{cases}$$

e.

$$f(n) = \alpha^{|n|}, \quad -\infty < n < \infty, \ (|\alpha| < 1)$$

f.

$$f(n) = \beta^{|n|}, \quad -\infty < n < \infty, \ (|\beta| > 1)$$

4. Suppose that $\{f_n\}$ is a sequence with a two-sided z-transform. Compute

$$\mathcal{Z}_{II}\, \{f_{n+k}\}\, .$$

5. Solve the problem

$$\frac{dx}{dt} + x(t) = \begin{cases} e^t & t \le 0, \\ 0 & t > 0 \end{cases}$$

by two-sided Laplace transforms. Immediately afterward, solve the problem

$$\frac{dx}{dt} + x(t) = e^t, \; t \ge 0$$

subject to $x(0) = 1$, by means of one-sided Laplace transforms.

6. Suppose that f and g have two-sided Laplace transforms with a common strip of convergence. Compute

$$\mathcal{L}_{II}\left\{ \int_{-\infty}^{\infty} f(t-\tau)\, g(\tau)\, d\tau \right\}\, .$$

7. Establish a convolution theorem (in the time domain) for the two-sided z-transform. Make clear all assumptions about regions of convergence.

9.5   Walsh Functions

The theories of the Fourier series and transform are closely based on properties of the trigonometric functions

- $\sin x$
- $\cos x$
- e^{ix}

which have their roots in the basic relation

$$e^{i(x+y)} = e^{ix}\, e^{iy}.$$

This relation underlies the orthogonality properties enjoyed by these functions, as well as the intimate connections between the notion of time invariance (behavior under time shifts) and the appearance of the Fourier transform in one of its several guises.

These factors combine to make conventional Fourier analysis a powerful tool in applied mathematics, as should be evident from the previous chapters. However, the trigonometric system is not the only system of orthogonal functions of interest in applications. Other orthogonal systems (arising from Sturm–Liouville problems) have appeared in the solution of boundary value problems. Orthogonal expansions also find direct application in areas such as communication theory.

One example of such expansions is the theory of Walsh functions. These are an example of an orthonormal function set whose mathematical structure is rather more intricate (and "interesting") than that of the usual Fourier expansion set and whose properties may be argued to be more in harmony with the physical characteristics of digital hardware circuitry than are those of the trigonometric functions. These functions have been studied (see [4], for example) with the aim of developing a systems analysis theory (analogous to the usual Fourier-based theory) capable of efficient digital implementation, as well as for their intrinsic mathematical interest (in [3] for example).

From the point of view of an observer used to the orthonormal function sets arising from boundary value problems, the Walsh functions are striking, since they are defined to be discontinuous and to assume only two values. These latter properties provide the connection to digital hardware.

There are several approaches to the definition of Walsh functions. Various properties are more or less easily derived by means of the alternative definitions. A definition with certain graphical appeal uses a recursive description. The basic interval of definition is $[-\frac{1}{2}, \frac{1}{2}]$, and the 0th Walsh function is defined by

$$Wal(0, t) = \begin{cases} 1 & -\frac{1}{2} < t \le \frac{1}{2}, \\ 0 & \text{otherwise.} \end{cases} \tag{9.3}$$

The later functions in the sequence are defined by

$$Wal(2j + q, t)$$
$$= (-1)^{\lfloor j/2 \rfloor + q} \left(Wal(j, 2(t + \frac{1}{4})) + (-1)^{j+q} Wal(j, 2(t - \frac{1}{4})) \right) \tag{9.4}$$

for $q = 0, 1, j = 0, 1, 2, \ldots$. The graphical interpretation of (9.4) is that

$$Wal(j, 2(t + \frac{1}{4}))$$

represents $Wal(j, t)$ on a time scale squeezed by a factor of 2. The other term involves a time delay of this squeezed version (perhaps with sign reversal). The definition translated into a python function gives:

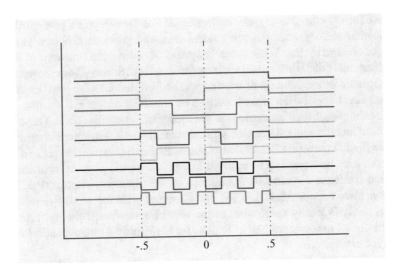

Fig. 9.7 Walsh functions

PYTHON

```python
def walsh(k, t):
    """ y = walsh(k,t) returns walsh functions
    """
    if k == 0:
        return heaviside(t + .5001) - heaviside(t - .5001)
    else:
        if (k % 2) == 0:
            q = 0
        else:
            q = 1

    j = (k - q)/2

    y = ((-1)**(j+q)*(walsh(j, 2*(t+ .25))\
        + (-1)**(j+q)*walsh(j, 2*(t- .25))))

    return y
```

The content of the definition is more apparent from the graphs of the resulting functions (see Figure 9.7).

In the context of Fourier analysis, the relation

$$e^{inx} e^{imx} = e^{i(m+n)x}$$

is a basic tool. For the Walsh functions (9.4), an analogous relation holds. This is

$$Wal(k, t) \cdot Wal(n, t) = Wal(n \oplus k, t) \tag{9.5}$$

where $n \oplus k$ is computed by componentwise modulo 2 addition of the binary digits of n and k. The relation (9.5) is an example of one for which a definition different from (9.4) provides a far easier derivation. Granting (9.5) (and the accuracy of the diagrams Figure 9.7) it is easy to see that the Walsh function set is orthogonal (with respect to the standard inner product) over the interval $[-\frac{1}{2}, \frac{1}{2}]$. For

$$\int_{-\frac{1}{2}}^{\frac{1}{2}} Wal(k, t) \cdot Wal(n, t) \, dt = \int_{-\frac{1}{2}}^{\frac{1}{2}} Wal(n \oplus k, t) \, dt$$

$$= \begin{cases} \int_{-\frac{1}{2}}^{\frac{1}{2}} 1 \, dt = 1 & n = k, \\ 0 & n \neq k, \end{cases}$$

this last result arising from the fact that each Walsh function beyond the 0th has average value 0. An analogue of the Riesz-Fischer Theorem also holds in the case of the Walsh functions: the set $\{Wal(n, t)\}$ is a complete orthonormal set in $L_2[-\frac{1}{2}, \frac{1}{2}]$. It follows immediately that a Parseval relation holds in this context. That is,

$$\int_{-\frac{1}{2}}^{\frac{1}{2}} |f(t)|^2 \, dt = \sum_{n=0}^{\infty} \left| \int_{-\frac{1}{2}}^{\frac{1}{2}} f(t) \, Wal(n, t) \, dt \right|^2.$$

In the case of Fourier analysis, the Parseval relation leads to the notion of the power spectrum of a function, providing distribution of energy according to frequency. The Parseval relation provides the decomposition of energy into frequency components. It remains to construct an extension of the notion of frequency which will naturally apply to Walsh functions.

The notion that arises is called "sequency." For trigonometric functions, the notion of frequency may be identified with "one-half the average number of zero-crossings per second." If the idea is formulated in this fashion, it is also applicable to non-trigonometric situations: *sequency* may be defined as this quantity, with the observation that it coincides with frequency for sinusoids (Figure 9.7).

It is easy to see (Figure 9.7 and equation (9.5)) that the sequency associated with the basic interval for $Wal(k, t)$ is just k.

The discussion above provides a brief introduction to Walsh functions and their properties. Most of the deeper properties of these functions have their mathematical basis in the relation between Walsh functions and the so-called *dyadic group*. This relation is the analog of the exponential addition formula.

The corresponding Walsh function identity is (for $0 \leq t, \tau < 1$)

$$Wal(n, t) \cdot Wal(n, \tau) = Wal(n, t \dotplus \tau)$$

where $t \dot{+} \tau$ is the result of the dyadic group operation: (crudely) expand t and τ into binary form

$$t = a_0 + a_1 \, 2^{-1} + a_2 \, 2^{-2} + a_3 \, 2^{-3} + \cdots$$

$$\tau = b_0 + b_1 \, 2^{-1} + b_2 \, 2^{-2} + b_3 \, 2^{-3} + \cdots$$

and compute the operation $\dot{+}$ by adding digits componentwise modulo 2. Pursuing this approach in detail is beyond the scope of this exposition, and we refer interested readers to the references [4] and [3] for an entry to the literature on the subject.

Problems 9.5.

1. Prove that Walsh functions of even number are even functions on $[-\frac{1}{2}, \frac{1}{2}]$, while Walsh functions of odd order are odd functions on the same interval.
2. Prove (using the recursive equation defining the Walsh functions) that the number of zero-crossings per unit time of $Wal(k, t)$ is in fact k.
3. In Chapter 7 we formally derived the Fourier transform from the Fourier series by rescaling the independent variable, and computing an appropriate limit. Can this be done with Walsh functions?
4. Generate the plots of Figure 9.7 from the defining relation (9.4).

9.6   Geometrically Based Transforms

In previous chapters the Fourier and Laplace transforms have been treated at some length. The heavy interest in these transforms arises from their deep connections to time and space invariant models, and the common occurrence of such models in applications.

It should also be evident (from Chapter 4, for example) that problems arising in mathematical physics often result in models involving the differential operator ∇^2. As a result, differential equations often appear in applications in a form determined by the expression for ∇^2 in terms of some coordinate system associated with the problem. Since such formulations are no more unnatural than (and indeed may be a reformulation of) a spatially invariant model, it is natural to seek a transform theory adapted to such problem formulations. We refer to such transform theories as "geometrically based," since it is ultimately the underlying geometry of the physical problem that determines the coordinate system and, in turn, the form of the governing equations in terms of the chosen coordinates. The resulting theories have an air of less general applicability than the Fourier and Laplace transforms. On the other hand, they are conspicuously effective for problems in the class for which the transforms were designed.

Our aim is to give below an introduction which explains the form and structure of some of these transforms. Rigorous technical discussions of the associated inversion theorems are avoided.

The transforms under discussion may be viewed as the (continuous) limit of the Sturm–Liouville expansion problems of Chapter 4. It should be recalled that the Laplace transform is effective in the solution of ordinary differential equations largely because of the relation

$$-\frac{d}{dt}e^{-st} = s\,e^{-st}.$$ (9.6)

The relation in effect identifies e^{-st} as an eigenfunction of the operator $-\frac{d}{dt}$. This observation raises the hope that the eigenvalue problems of Chapter 4 may be adapted to provide an operational calculus applicable to other problems. While the Laplace transform reduces an arbitrary constant coefficient differential operator to a multiplication operation, the transforms derived from Sturm–Liouville problems serve "only" to reduce the Sturm–Liouville operator in question to a multiplication in the transform domain.

We introduce these methods with the example of the sine transform. The associated Sturm–Liouville problem is

$$\frac{d^2}{dx^2}X = -\lambda^2 X,$$

$$X(0) = X(L) = 0.$$

The eigenfunctions are

$$X_n(x) = \sin\frac{n\pi x}{L},$$

and the expansion result is

$$f(x) = \sum_{n=1}^{\infty} b_n \frac{2}{L}\sin\frac{n\pi x}{L}$$

where

$$b_n = \int_0^L \sin\frac{n\pi x}{L} f(x)\, dx.$$

If $f \in L_2(0, \infty)$, and in addition vanishes for $x > M$, then for $L > M$, we can compute the expansion coefficient as

$$b_n = \frac{2}{L}\int_0^{\infty} \sin\frac{n\pi x}{L} f(x)\, dx,$$

and represent f in a mean-square convergent sine series over the interval $[0, L]$.

Introducing

$$\tilde{f}(p) = \int_0^\infty \sin px f(x)\, dx \tag{9.7}$$

we have

$$f(x) = \sum_{n=1}^\infty \tilde{f}(\frac{n\pi}{L}) \frac{2}{L} \sin(\frac{n\pi x}{L}). \tag{9.8}$$

With the definitions

$$\Delta p = \frac{\pi}{L}, \quad \frac{2}{L} = \frac{2\Delta p}{\pi}$$

the relation (9.8) takes the form of a Riemann sum

$$f(x) = \frac{2}{\pi} \sum_{n=1}^\infty \tilde{f}(n\,\Delta p) \sin(n\,\Delta p\,x)\, \Delta p.$$

As $L \to \infty$ the expansion above leads to the inversion formula

$$f(x) = \frac{2}{\pi} \int_0^\infty \tilde{f}(p) \sin px\, dp \tag{9.9}$$

for the sine transform (9.7).

The relations (9.7) and (9.9) can in fact be easily justified by means of the standard Fourier transform results, applied to functions with odd symmetry about the origin. The heuristic derivation of the inversion relation (9.9) may also be used as a guide in the case of other transforms, and so is of interest beyond the formulas for the sine transform.

The operational use of the sine transform is the removal of second derivatives from a differential expression. The formula arises simply by computing the sine transform of a second derivative.

$$\mathcal{S}\{f''\} = \int_0^\infty \sin px f''(x)\, dx$$

$$= \sin px f'(x)\Big|_0^\infty + p\int_0^\infty \cos px f'(x)\, dx$$

$$= \sin px f'(x)\Big|_0^\infty + p\cos px f(x)\Big|_0^\infty - p^2 \int_0^\infty \sin px f(x)\, dx.$$

Assuming f, f' tend to zero at infinity,

$$\mathcal{S}\left\{f''\right\} = -p^2\, \mathcal{S}\left\{f\right\} + pf(0). \tag{9.10}$$

This result plays an operational role exactly parallel to that of the formula for the Laplace transform of a derivative. It can be used to convert a second derivative operation to an algebraic one in terms of the transformed quantities.

Example. Consider the problem of heat conduction in a very long (i.e., infinite) laterally insulated bar, the near end of which is held at temperature T_0. The model is

$$\frac{\partial u}{\partial t} = \kappa \frac{\partial^2 u}{\partial x^2}, \; x > 0, \, t > 0,$$

$$u(0, t) = T_0,$$

$$u(x, 0) = 0, \;\; \text{for simplicity.}$$

Introduce

$$U(p) = \mathcal{S}_x\left\{u(x, t)\right\} = \int_0^\infty \sin px\, u(x, t)\, dx.$$

Then (sine transforming)

$$\frac{\partial}{\partial t} U(p) = \kappa\left(p\, T_0 - p^2\, U(p)\right), \tag{9.11}$$

providing the transform

$$U(p) = \frac{T_0}{p}\left(1 - e^{-\kappa p^2 t}\right)$$

and the solution via inversion

$$u(x, t) = \frac{2\, T_0}{\pi} \int_0^\infty \left[1 - e^{-\kappa p^2 t}\right] \frac{\sin px}{p}\, dp.$$

Using

$$\int_0^\infty \frac{\sin xp}{p}\, dp = \frac{\pi}{2}, ^6$$

the above may be evaluated as (see Problem 5)

---

[6]This integral was effectively considered in Section 7.6 as a Fourier inversion example.

$$u(x, t) = T_0 \left[1 - \frac{2}{\pi} \int_0^\infty e^{-\kappa p^2 t} \frac{\sin px}{p} \, dp \right]$$

$$= T_0 \, \mathrm{erfc}(\frac{x}{2\sqrt{\kappa t}}).$$

Among the most commonly encountered geometrically based transforms are those associated with cylindrical polar coordinates. These are the Mellin and Hankel transforms.

The Mellin transform is associated with the differential operator

$$\mathbf{L}[X] = x^2 X'' + x X'. \tag{9.12}$$

Since we have (the evident eigenvector)

$$\mathbf{L}[x^p] = x^2 (x^p)'' + x (x^p)' \tag{9.13}$$

$$= (p(p-1) + p) \, x^p$$

$$= p^2 x^p,$$

consideration of the likely transform

$$\int f(x) x^p \, dx$$

is suggested. In fact, it is more conventional and convenient to use $p - 1$ instead of p as the transform variable, and to define the Mellin transform as

$$\tilde{f}(p) = \mathcal{M} \{f\} = \int_0^\infty f(x) x^{p-1} \, dx. \tag{9.14}$$

To obtain an inversion formula for the Mellin transform, we recall that the Euler (or equidimensional) differential expression is reducible to a constant coefficient expression through a logarithmic change of variables. If in (9.14) we make the substitution

$$x = e^y,$$

then

$$\tilde{f}(p) = \int_{-\infty}^\infty f(e^y) e^{py} \, dy,$$

revealing the Mellin transform as a disguised form of a two-sided Laplace transform (taking p complex valued) or a Fourier transform (taking p purely imaginary).

In either case, an inversion formula is obtained from the standard Fourier inversion. The form which corresponds to the two-sided Laplace transform interpretation is

$$f(x) = \frac{1}{2\pi i} \int_{\gamma - i\infty}^{\gamma + i\infty} \tilde{f}(p) x^{-p} \, dp. \tag{9.15}$$

The path in this integral must of course lie within the strip of convergence of the corresponding two-sided transform.

In view of the Mellin-Laplace relation it is to be expected that most of the Laplace transform formulas have Mellin transform analogues. An example is

$$\mathcal{M}\{f'\}(p) = \int_0^\infty f'(x) x^{p-1} \, dx \tag{9.16}$$

$$= x^{p-1} f(x) \Big|_0^\infty - (p-1) \int_0^\infty f(x) x^{p-2} \, dx$$

$$= -(p-1) \mathcal{M}\{f\}(p-1),$$

provided that the indicated boundary terms vanish for p within the strip of convergence.

Example. Consider the problem of steady state heat conduction within a solid wedge of angle θ_0 with the temperature on the boundary specified as zero on the lower side, and a given function f of the radial coordinate on the upper side. The problem is then

$$r^2 \frac{\partial^2 u}{\partial r^2} + r \frac{\partial u}{\partial r} + \frac{\partial^2 u}{\partial \theta^2} = 0, \tag{9.17}$$

$$u(r, \theta_0) = f(r),$$

$$u(r, 0) = 0.$$

Applying the Mellin transform gives

$$\mathcal{M}\left\{ r^2 \frac{\partial^2 u}{\partial r^2} + r \frac{\partial u}{\partial r} \right\} = r^{p+1} \frac{\partial u}{\partial r} - p \, r^p u \Big|_0^\infty + p^2 \mathcal{M}\{u\}.$$

Assuming that u is such that the indicated boundary terms vanish for p within the strip of convergence, the transformed equation is

$$p^2 U(p) + \frac{d^2}{d\theta^2} U(p) = 0,$$

$$U(p, \theta_0) = F(p), \quad U(p, 0) = 0.$$

The solution of the transformed equation is simply

$$U(p) = F(p)\frac{\sin p\theta}{\sin p\theta_0}.$$

Inversion of this expression is most readily carried out by use of the Mellin version of the convolution theorem (see the problems below).

The Hankel transform is derived from the eigenfunctions associated with Bessel's equation. We recall that Bessel's equation of order n (encountered in solution of a heat equation in circular polar coordinates) takes the form

$$r^2 X'' + r X' + \left(\lambda^2 r^2 - n^2\right) X = 0 \tag{9.18}$$

or (in eigenvalue form)

$$X'' + \frac{1}{r} X' - \frac{n^2}{r^2} X = -\lambda^2 X.$$

If the pattern of the sine and Mellin transforms is to be repeated, it should be possible to define a transform by integrating against the (eigenfunction) solution of (9.18). The form of the integral should be that of the inner product associated with the Sturm–Liouville problem. For the case of (9.16) we choose the eigenfunction bounded at the origin, and define the *Hankel transform of order n* in terms of the corresponding Bessel function of the first kind. This gives

$$\tilde{f}(p) = \mathcal{H}_n \{f\} = \int_0^\infty r f(r) J_n(p\,r)\, dr \tag{9.19}$$

as the natural candidate for the transform.

One expects that this transform should reduce the differential operator associated with Bessel's equation to a multiplication in the transform domain (at least with suitable restrictions on the transformed function). To verify this, compute

$$\mathcal{H}_n \left\{ f'' + \frac{1}{r} f' - \frac{n^2}{r^2} f \right\} = \int_0^\infty r \left\{ f'' + \frac{1}{r} f' - \frac{n^2}{r^2} f \right\} J_n(pr)\, dr$$

$$= \int_0^\infty \left\{ \frac{d}{dr} \left(r f'(r) \right) J_n(pr) - \frac{n^2}{r^2} J_n(pr) \right\} dr$$

$$= r \left(f' J_n - f J_n' \right) \Big|_0^\infty + \int_0^\infty \left\{ f \frac{d}{dr} (r J_n') - \frac{n^2}{r} f J_n \right\} dr.$$

Assuming f such that the boundary terms vanish, the fact that J_n satisfies

$$r J_n''(pr) + J_n'(pr) - \frac{n^2}{r} J_n(pr) = -p^2 r J_n(pr)$$

gives the expected result

$$\mathcal{H}_n \left\{ f'' + \frac{1}{r} f' - \frac{n^2}{r^2} f \right\} = -p^2 \int_0^\infty r f(r) J_n(pr)\, dr = -p^2\, \mathcal{H}_n \{f\}.$$

A technical proof of the validity of the inversion formula for the Hankel transform is complicated (and beyond the scope of this introduction to the topic). However, an heuristic derivation of the inversion can be based on the two-dimensional Fourier transform and the integral representation for the Bessel function J_n. For the case of integral n, this latter formula is easy to derive, and of some independent interest.[7] The representation in question is

$$J_n(p) = \frac{1}{2\pi} \int_0^{2\pi} e^{-in\theta + ip\sin\theta}\, d\theta. \tag{9.20}$$

This may be derived by combining the power series for the Bessel function

$$J_n(x) = \sum_{m=0}^{\infty} \frac{(-1)^m \left(\frac{x}{2}\right)^{n+2m}}{m!\,(n+m)!} \tag{9.21}$$

with an integral representation for the reciprocal factorial expressions of the series. The series

$$e^z = \sum_{k=0}^{\infty} \frac{z^k}{k!}$$

produces the residue calculation

$$\frac{1}{2\pi i} \oint_C z^{-(m+n+1)} e^z\, dz = \frac{1}{(m+n)!} \tag{9.22}$$

where C is any positively oriented simple closed contour encircling the origin. The expression series for $J_n(x)$ then may be written

$$J_n(x) = \left(\frac{x}{2}\right)^n \sum_{m=0}^{\infty} \left(\frac{x}{2}\right)^{2m} \frac{(-1)^m}{m!} \frac{1}{2\pi i} \oint_C z^{-(m+n+1)} e^z\, dz \tag{9.23}$$

$$= \left(\frac{x}{2}\right)^n \frac{1}{2\pi i} \oint_C \sum_{m=0}^{\infty} \left(\frac{x^2}{4z}\right)^m \frac{(-1)^m}{m!} e^z\, z^{-(n+1)}\, dz$$

$$= \left(\frac{x}{2}\right)^n \frac{1}{2\pi i} \oint_C e^{(z - \frac{x^2}{4z})}\, z^{-(n+1)}\, dz.$$

---

[7]The formulas occur in communication problem calculations, especially in models involving phase modulation.

(The interchange of summation and integration is justified by the observation that the series converges uniformly on a closed domain containing C.)

If C is now deformed (the only singularity of the integrand is at the origin) to the circle $|z| = \frac{1}{2}|x|$ (assuming $x \neq 0$), and the variable of integration is changed according to

$$z = \frac{1}{2}x\zeta,$$

the integral takes the form

$$J_n(x) = \frac{1}{2\pi i} \oint_{|\zeta|=1} e^{\frac{1}{2}x(\zeta - \frac{1}{\zeta})} \zeta^{-(n+1)} \, d\zeta. \tag{9.24}$$

When this contour integral is evaluated with the parameterization $\zeta = e^{i\theta}$, the expression becomes

$$J_n(x) = \frac{1}{2\pi} \int_0^{2\pi} e^{-in\theta + ix\sin\theta} \, d\theta. \tag{9.25}$$

In order to derive a Hankel transform inversion formula, the two-dimensional inversion formulas for the Fourier transform

$$\hat{f}(\omega, v) = \int_{-\infty}^{\infty} \int_{-\infty}^{\infty} f(x, y) \, e^{-i\omega x} \, e^{-ivy} \, dx \, dy \tag{9.26}$$

$$f(x, y) = \frac{1}{(2\pi)^2} \int_{-\infty}^{\infty} \int_{-\infty}^{\infty} \hat{f}(\omega, v) e^{i\omega x} \, e^{ivy} \, d\omega \, dv$$

are expressed in terms of polar coordinates. Using

$$x = r\cos\theta, \; y = r\sin\theta,$$
$$\omega = \rho\cos\phi, \; v = \rho\sin\phi$$

in (9.26) we obtain (the transform)

$$F(\rho, \phi) = \int_0^{\infty} \int_0^{2\pi} \tilde{f}(r, \theta) e^{r\rho(\cos\theta\cos\phi + \sin\theta\sin\phi)} \, r \, dr \, d\theta \tag{9.27}$$

$$= \int_0^{\infty} \int_0^{2\pi} \tilde{f}(r, \theta) e^{-r\rho\cos(\phi - \theta)} \, r \, dr \, d\theta,$$

(and inversion)

$$\tilde{f}(r, \theta) = \frac{1}{(2\pi)^2} \int_0^{\infty} \int_0^{2\pi} F(\rho, \phi) \, e^{ir\rho\cos(\phi - \theta)} \, \rho \, d\rho \, d\phi. \tag{9.28}$$

The desired Hankel transform results arise from computing the (polar-Fourier) transform of the particular function

$$\tilde{f}(r, \theta) = e^{-in\theta} g(r).$$

We have for this particular case

$$F(\rho, \phi) = \int_0^\infty \int_0^{2\pi} g(r) e^{-in\theta} e^{-ir\rho} e^{-ir\rho \cos(\theta - \phi)} r \, dr \, d\theta.$$

Since

$$\cos(x + \frac{\pi}{2}) = -\sin x$$

the substitution

$$\theta - \phi = x + \frac{\pi}{2}$$

is appropriate in order to introduce the representation (9.25). Using the above substitution, we find

$$F(\rho, \phi) = e^{-in\phi} e^{-in\frac{\pi}{2}} 2\pi \int_0^\infty J_n(\rho r) g(r) r \, dr$$

$$= e^{-in\phi} e^{-in\frac{\pi}{2}} 2\pi \, \mathcal{H}_n \{g\}.$$

Using the polar form of the Fourier inversion gives

$$e^{-in\theta} g(r) = \frac{1}{(2\pi)^2} \int_0^\infty \int_0^{2\pi} F(\rho, \phi) e^{ir\rho \cos(\phi - \theta)} \rho \, d\rho \, d\phi$$

$$= \frac{1}{2\pi} \int_0^\infty \int_0^{2\pi} \mathcal{H}_n \{g\} e^{-in\phi} e^{-in\frac{\pi}{2}} e^{ir\rho \cos(\phi - \theta)} \rho \, d\rho \, d\phi.$$

With the substitution

$$\phi - \theta = y - \frac{\pi}{2}$$

we have (since $\cos(y - \frac{\pi}{2}) = \sin y$)

$$e^{-in\theta} g(r) = \frac{1}{2\pi} \int_0^\infty \int_0^{2\pi} \mathcal{H}_n \{g\} e^{-in(\theta + y - \frac{\pi}{2})} e^{-in\frac{\pi}{2}} e^{ir\rho \sin\phi} \rho \, d\rho \, dy$$

and in turn

$$g(r) = \int_0^\infty \mathcal{H}_n\{g\} \left(\frac{1}{2\pi}\int_0^{2\pi} e^{-iny+irp\sin y}\,dy\right) \rho\,d\rho$$

$$= \int_0^\infty \mathcal{H}_n\{g\}\, J_n(r\rho)\,\rho\,d\rho.$$

The Hankel transform-inversion pair relation is therefore

$$\mathcal{H}_n\{g\} = \int_0^\infty g(r)\, J_n(pr)\, r\,dr, \tag{9.29}$$

$$g(r) = \int_0^\infty \mathcal{H}_n\{g\}\, J_n(pr)\, p\,dp.$$

For tables of Hankel transforms and a rigorous discussion of the inversion relations (9.29) we refer the reader to the references [1, 9], and [10].

Example. A Hankel transform of order 0 (i.e., using the eigenfunction J_0) may be used to solve the following electrostatics problem.

Suppose that it is desired to find the electric potential in the region $z > 0$ resulting from the imposition of a radially symmetric potential in the plane $z = 0$ The governing equations are therefore

$$\frac{\partial^2 \phi}{\partial r^2} + \frac{1}{r}\frac{\partial \phi}{\partial r} + \frac{\partial^2 \phi}{\partial z^2} = 0,\ z > 0 \tag{9.30}$$

$$\phi(r, 0) = g(r).$$

Define

$$\Phi(p, z) = \mathcal{H}_0\{\phi\}(p) \tag{9.31}$$

$$= \int_0^\infty \phi(r, z)\, J_0(pr)\, r\,dr.$$

Then (transforming the governing equation)

$$-p^2\, \Phi(p, z) + \frac{d^2}{dz^2}\, \Phi(p, z) = 0, \tag{9.32}$$

$$\Phi(p, 0) = G(p) = \mathcal{H}_0\{g\}(p).$$

The solution of (9.32) bounded as $z \to \infty$ is

$$\Phi(p, z) = G(p)\, e^{-pz}.$$

Using the inversion formula gives

$$\phi(r, z) = \int_0^\infty G(p) \, e^{-pz} \, J_0(pr) \, p \, dp. \tag{9.33}$$

Problems 9.6.

1. Compute the sine transform of

   a. $f(t) = e^{-t}$,
   b. $f(t) = t e^{-t}$.

2. Invent the Fourier cosine transform, with the associated inversion formula.
3. Compute the cosine transform of $\frac{d^2 f}{dx^2}$ (under appropriate assumptions).
4. Use the cosine transform to solve the heat equation on a half-axis, with prescribed flux as the boundary condition. That is,

$$\frac{\partial u}{\partial t} = \kappa \frac{\partial^2 u}{\partial x^2}, \, x > 0,$$

$$\frac{\partial u}{\partial x}(0, t) = F, \, u(x, 0) = 0.$$

5. Evaluate sine transform

$$\frac{2 \, T_0}{\pi} \int_0^\infty \left[1 - e^{-\kappa p^2 t} \right] \frac{\sin px}{p} \, dp.$$

Hint: The integrand is an even function of p, so we may as well consider

$$\frac{2}{\pi} \int_0^\infty e^{-\kappa p^2 t} \frac{\sin px}{p} \, dp = \frac{2}{2\pi} \int_{-\infty}^\infty e^{-\kappa \omega^2 t} \frac{\sin \omega x}{\omega} e^{i\omega y} \, d\omega \Big|_{y=0}.$$

One of the transform factors is a Gaussian curve, and the other is the transform of a square pulse. The expression can be evaluated using the Fourier transform convolution theorem.

6. Compute the Mellin transform of

   a. x^a,
   b. $\cos x$.

7. Compute

   a. $\mathcal{M} \{x^a f(x)\}$,
   b. $\mathcal{M} \{\ln(x) f(x)\}$,
   c. $\mathcal{M} \{f(a x)\}$.

8. Compute

$$\mathcal{M}\left\{f''(x)\right\}.$$

9. Compute

$$\mathcal{M}\left\{\int_0^\infty f(xu)\, g(u)\, du\right\}$$

and

$$\mathcal{M}\left\{\int_0^\infty f(\frac{x}{u})\, g(u)\, \frac{du}{u}\right\}.$$

10. Use the Mellin transform to find the (formal) solution to the problem of heat conduction in a wedge with prescribed heat flux on the sides.

$$\frac{\partial^2 u}{\partial r^2} + \frac{1}{r}\frac{\partial u}{\partial r} + \frac{1}{r^2}\frac{\partial^2 u}{\partial \theta^2} = 0,$$

$$\frac{\partial u}{\partial \theta}(r,0) = f(r), \quad \frac{\partial u}{\partial \theta}(r,\alpha) = g(r).$$

11. Carry out the details of the change of contour and variable involved in the derivation of equation (9.24) from equation (9.23).
12. Verify the residue calculation in equation (9.22).
13. The symmetrical vibrations of a thin elastic plate are governed by the equation

$$\frac{\partial^2 u}{\partial t^2} = -\omega_0^2\left(\frac{\partial^2 u}{\partial r^2} + \frac{1}{r}\frac{\partial u}{\partial r}\right)^2 u(r,t)$$

$$u(r,0) = f(r), \quad \frac{\partial u}{\partial t}(r,0) = 0.$$

Use a Hankel transform of order 0 to obtain an integral representation of the solution.
14. Formulate the problem of symmetrical (i.e., no angular dependence) oscillations of a very large circular elastic membrane. Find an integral form of the solution of the resulting problem, using a Hankel transform of order 0.
15. Repeat Problem 14 without assuming circular symmetry. Separate the angular variable first in order to obtain a solution.

References

1. E.A. Coddington, N. Levinson, *Theory of Ordinary Differential Equations* (McGraw-Hill, New York, 1955)
2. I. Daubechies, *Ten Lectures on Wavelets* (Society for Industrial and Applied Mathematics, Philadelphia, 1992)
3. N.J. Fine, On the walsh functions. Trans. Am. Math. Soc. **65**, 372–414 (1949)
4. H. Harmuth, *Transmission Of Information by Orthogonal Functions* (Springer-Verlag, New York, 1970)
5. B.W. Kernighan, R. Pike, *The UNIX Programming Environment* (Prentice-Hall, Englewood Cliffs, 1984)
6. S. Mallat, *A Wavelet Tour of Signal Processing* (Academic, San Diego, 1998)
7. Mathworks Incorporated, *Signal Processing Toolbox User's Guide* (Mathworks Incorporated, Natick, 2000)
8. S. Pinker, *The Language Instinct* (William Morrow and Sons, New York, 1995)
9. I.H. Sneddon, *The Use of Integral Transforms* (McGraw-Hill, New York, 1972)
10. C.J. Tranter, *Integral Transforms in Mathematical Physics* (Methuen and Company, London, 1966)

Further Reading

C.S. Burrus, R.A. Gopinath, H. Guo, *Wavelets and Wavelet Transforms* (Prentice-Hall, Upper Saddle River, 1998)

E.U. Condon, *Quantum Mechanics* (McGraw-Hill, New York, 1929)

I. Daubechies (ed.), *Different Perspectives on Wavelets* (American Mathematical Society, Providence, 1993)

J.H. Davis, *Foundations of Deterministic and Stochastic Control* (Birkhäuser, Boston, 2002)

L.I. Schiff, *Quantum Mechanics*, 2nd edn. (McGraw-Hill, New York, 1955)

A. Teolis, *Computational Signal Processing with Wavelets* (Birkhäuser, Boston, 1998)

B. Van der Pol, H. Bremmer, *Operational Calculus Based on the Two-Sided Laplace Integral*, 2nd edn. (Cambridge University Press, Cambridge, 1955)

J.S. Walker, *Wavelets and Their Scientific Applications* (CRC Press LLC, Boca Raton, 1999)

D.F. Walnut, *An Introduction to Wavelet Analysis* (Birkhäuser, Boston, 2002)

D.V. Widder, *The Laplace Transform* (Princeton University Press, Princeton, 1946)

ERRATUM

Chapter 2
Fourier Series

© Springer International Publishing Switzerland 2016
J.H. Davis, *Methods of Applied Mathematics with a Software Overview*,
Applied and Numerical Harmonic Analysis, DOI 10.1007/978-3-319-43370-7

DOI 10.1007/978-3-319-43370-7_10

In Chapter 2 titled "Fourier Series", the equation in page 13 has been corrected to read as follows:

$$c_n = \frac{1}{2\pi}\left[\int_0^{\pi} (1)e^{-inx}\,dx + \int_{\pi}^{2\pi} (-1)e^{-inx}\,dx\right]$$

The updated original online version for this chapter can be found at
http://dx.doi.org/10.1007/978-3-319-43370-7_2

© Springer International Publishing Switzerland 2017
J.H. Davis, *Methods of Applied Mathematics with a Software Overview*,
Applied and Numerical Harmonic Analysis, DOI 10.1007/978-3-319-43370-7_10

Appendix A
Linear Algebra Overview

Some of the topics in this text are approached from the point of view of generalizing fundamental notions from linear algebra. In addition, some of the examples and problems assume a certain level of familiarity with this subject. In this appendix we collect the relevant definitions and basic results for reference.

A.1 Vector spaces

Definition (Vector Space). A vector space V over a field of scalars F is defined to be a set on which there is a (vector) addition operation defined, together with an operation of scalar multiplication of vectors by elements of F. The operations and objects are required to satisfy the following conditions: [1]

1. $u + v = v + u$, for all $u, v \in V$.
2. There exists a vector 0 such that $v + 0 = v$ for all $v \in V$.
3. For each v there exists an additive inverse $(-v)$ such that $v + (-v) = 0$.
4. For scalars $\alpha, \beta \in F$

$$(\alpha + \beta)v = \alpha\, v + \beta\, v,$$

$$\alpha\,(u + v) = \alpha\, u + \alpha\, v.$$

In many applications (including those in this text) the scalar field is either the real or complex numbers.

---

[1] There is of course a distinction between addition of scalars and addition of vectors. However, the symbol "+" is used for both, as the intent is determinable from the context.

© Springer International Publishing Switzerland 2016
J.H. Davis, *Methods of Applied Mathematics with a Software Overview*,
Applied and Numerical Harmonic Analysis, DOI 10.1007/978-3-319-43370-7

Examples of vector spaces include R^n and C^n (n-tuples of real or complex numbers), sets of polynomials of degree N or less, as well as various spaces of functions defined on some domain. In the latter case, one may consider "arbitrary" real valued functions, although for applications in analysis spaces of functions satisfying some constraint of analytical regularity are of interest. Instances of the latter are spaces of continuous functions on some interval, functions continuously differentiable on some interval, or functions square-integrable on some interval. What is essential in these examples is that the desired analytical property is preserved under the operation of the formation of linear combinations of the functions in the set.

Basic notions in the context of vector spaces are those of linear independence, basis, and dimension.

Definition (Linear independence). A set of vectors $\{v_i\}_{i=1}^N$ in a vector space V is linearly independent if the equality

$$c_1 v_1 + c_2 v_2 + \cdots + c_n v_n = 0$$

holds only when $c_1 = c_2 = \cdots = c_N = 0$ (so that all the coefficients vanish).

Definition (Span). Given a set of vectors $U = \{v_i\}_{i=1}^N$ the set of all linear combinations of elements of U,

$$S = \{v = c_1 v_1 + c_2 v_2 + \cdots + c_n v_n\},$$

is called the span of $U = \{v_i\}_{i=1}^N$. This is a subspace of the vector space V.

Definition (Basis, dimension). In the case when there exists a linearly independent set $\{v_i\}_{i=1}^N$ with the property that the vector space V is the span of $\{v_i\}_{i=1}^N$, then V is said to have dimension N, and $\{v_i\}_{i=1}^N$ is called a basis of V.

There exist vector spaces for which it is not possible to find a finite set of vectors such that every vector is expressible as a linear combination of this set. Such vector spaces are called infinite-dimensional. Typical examples are the function spaces mentioned above. Even in this context, it is useful to consider the spans of certain finite sets of vectors. These produce finite-dimensional subspaces (of an infinite-dimensional vector space) to which results of (finite-dimensional) linear algebra may be usefully applied.

If the finite-dimensional vector space V has a basis $B = \{v_i\}_{i=1}^N$, then it is easy to see that the coefficients of the expansion of a given vector v in terms of the basis are unique. If v is expanded in a basis B,

$$v = c_1 v_1 + c_2 v_2 + \cdots + c_n v_n,$$

then we call the array

$$[v]_B = \begin{bmatrix} c_1 \\ c_2 \\ \vdots \\ c_N \end{bmatrix}_B$$

the column vector of coordinates of v with respect to the basis $B = \{v_i\}_{i=1}^N$. Recall that in the above it is useful to emphasize the dependence of the coordinates on the choice of basis B.

A central result of linear algebra deals with the relationship of coordinate vectors computed with respect to two different sets of basis vectors $B = \{v_i\}_{i=1}^N$ and $C = \{u_i\}_{i=1}^N$. This result is the statement that the coordinates are related by multiplication by a non-singular (the so-called change of basis) matrix: in the above notation for coordinates

$$[v]_B = \mathbf{P}[v]_C$$

is the form of the relation.

A.2   Linear Mappings

Linear mappings with domain the vector space U and range in the vector space V are those mappings which respect the operations of vector addition and scalar multiplication. $\mathbf{T} : U \mapsto V$ is a linear mapping from U into V if and only if

$$\mathbf{T}(\alpha u_1 + \beta u_2) = \alpha \, \mathbf{T}(u_1) + \beta \, \mathbf{T}(u_2)$$

holds for all $u_1, u_2 \in V$ and all scalars $\alpha, \beta \in F$.

The prototypical linear mapping is the operation of matrix multiplication acting on column vectors of real or complex numbers. This example is central, since it is possible to show that an arbitrary linear mapping $\mathbf{T}$ from one finite-dimensional vector space U to another finite-dimensional vector space V has a unique representation as matrix multiplication acting on coordinate column vectors with respect to bases $C = \{u_i\}_{i=1}^N$, and $B = \{v_i\}_{i=1}^N$ for U and V. In our previous notation, this takes the form

$$[\mathbf{T}\,u]_B = [\mathbf{T}]_B^C \, [u]_C .$$

The matrix $[\mathbf{T}]_B^C$ is called the matrix of the linear transformation $\mathbf{T}$ with respect to the bases B and C.

In the case that T maps the finite-dimensional vector space U into itself, it is of interest to determine the effect of a change of basis on the matrix of the transformation $\mathbf{T}$. From the matrix representation of $\mathbf{T}$

$$[\mathbf{T}\,u]_B = [\mathbf{T}]_B^B\,[u]_B$$

and the change of basis relations

$$[v]_B = \mathbf{P}[v]_C,$$

we compute

$$[\mathbf{T}\,u]_B = \mathbf{P}\,[\mathbf{T}\,u]_C$$
$$= \mathbf{P}\,[\mathbf{T}]_C^C\,[u]_C$$
$$= \mathbf{P}\,[\mathbf{T}]_C^C\,\mathbf{P}^{-1}\,\mathbf{P}\,[u]_C$$
$$= \mathbf{P}\,[\mathbf{T}]_C^C\,\mathbf{P}^{-1}\,[u]_B$$

which identifies the matrix of $\mathbf{T}$ with respect to the new basis B as the matrix

$$[\mathbf{T}]_B^B = \mathbf{P}\,[\mathbf{T}]_C^C\,\mathbf{P}^{-1}.$$

This matrix is said to be similar to the matrix

$$[\mathbf{T}]_C^C$$

computed with respect to C.

A.3   Inner Products

Vector spaces equipped with an inner product structure find applications in a wide variety of problems. They play a central role in the solution of boundary value problems (see Chapters 2, 3, and 4), and find application in other problems ranging from optimal control to statistical estimation of random processes.

Definition. An inner product space is a vector space V (in general with complex numbers as the field of scalars) equipped with an inner product function $(.,.)$ defined from pairs of vectors $V \times V$ and taking values in C (the complex numbers), satisfying (for all $u, v, w \in V$, $\alpha, \beta \in C$) the conditions

1. $(\alpha u + \beta v,\, w) = \alpha\,(u, w) + \beta\,(v, w)$,
2. $(u, w) = \overline{(w, u)}$,
3. $(u, u) > 0$,
4. $(u, u) = 0$ if and only if $u = 0$.

This definition allows the introduction of a distance function (or norm) according to $\|u\| = (u, u)^{\frac{1}{2}}$. In addition, the geometrically based intuition about mutually orthogonal vectors in R^2 or R^3 can be extended to general inner product spaces by declaring vectors to be orthogonal whenever $(u, v) = 0$.

It can be shown that a finite-dimensional inner product space U always has a basis consisting of mutually orthogonal vectors of unit norm. This is called an orthonormal basis for U.

Examples and further properties of inner product spaces are discussed in Chapter 2.

A.4   Linear Functionals and Dual Spaces

It might at first be thought that (since the real or complex numbers form a one-dimensional vector space) linear mappings from a vector space to the scalar field over which the vector space is defined would be of comparatively little interest. If has turned out (especially in the case of infinite-dimensional vector spaces arising in mathematical analysis, see Section 7.8) that consideration of such mappings is quite profitable.

In the special case of inner product spaces, mappings of the form (f a fixed vector)

$$f(u) = (u, f)$$

are examples of such mappings. For finite-dimensional inner product spaces (and some infinite-dimensional situations) it is also true that a linear functional (a linear mapping $U \mapsto C$, the complex numbers) is representable in the indicated form for some fixed vector.

For vector spaces U without an inner product, linear mappings $U \mapsto C$ may still be considered. Since such objects may be added and scalar multiplied, they form a vector space themselves. This vector space of linear mappings is referred to as the *dual space* to the vector space U.

A.5   Canonical Forms

One of the central problems of linear algebra is to understand and classify the structure of linear mappings from the finite-dimensional vector space $U \mapsto U$. In this pursuit, subspaces of U which are mapped into themselves under the linear mapping **T** (the so-called invariant subspaces) play a central role.

The simplest linear mappings are those for which there exists a basis of U, with each basis vector a basis for a (one-dimensional) invariant subspace of T This means that each basis vector is mapped into a scalar multiple of itself under T, so that

$$\mathbf{T}(u_i) = \lambda_i u_i, \; i = 1, 2, \cdots, \dim U$$

must hold. In short, the basis consists of eigenvectors of the mapping $\mathbf{T}$.

If one has a basis consisting of eigenvectors, it is easy in this situation to compute the matrix of T with respect to the eigenvector basis $B = \{u_i\}_{i=1}^{\dim U}$. Trivially, the matrix is diagonal

$$[\mathbf{T}]_B^B = \begin{bmatrix} \lambda_1 & 0 & \cdots & 0 \\ 0 & \lambda_2 & \cdots & 0 \\ 0 & \ddots & \cdots & 0 \\ 0 & 0 & \cdots & \lambda_{\dim U} \end{bmatrix}_B^B ,$$

since the image of an eigenvector has only a single component in the eigenvector direction.

In general, the linear mapping $\mathbf{T}$ is likely to fail to have enough eigenvectors to form a basis, and the above simple form is unattainable. The situation then divides according to whether or not the eigenvalues of $\mathbf{T}$ lie in the field over which the vector space U is defined. This is not a problem if the scalar field for U is the complex numbers (due to the fundamental theorem of algebra, to the effect that polynomials with complex coefficients have complex roots), but the distinction arises already in the case of the real number field.

If eigenvalues lie outside of the field, then rational canonical forms may be constructed. If the eigenvalues lie within the field, then it is possible to determine a basis with respect to which the matrix of the mapping is in Jordan canonical form.

A basic Jordan block is a matrix of the form

$$\begin{bmatrix} \lambda & 1 & \cdots & 0 \\ 0 & \lambda & \cdots & 0 \\ 0 & \ddots & \cdots & 1 \\ 0 & 0 & \cdots & \lambda \end{bmatrix}$$

with 1's on the super-diagonal, λ along the diagonal, and zeroes elsewhere. The Jordan canonical form is a block diagonal matrix, whose blocks consist of basic Jordan blocks. The computation of a Jordan canonical form is difficult, in both the analytical and numerical sense. However, the form is useful in classifying the form of possible solutions in problems involving linear algebra. The fact that constant coefficient differential equations have solutions representable as a sum of polynomials multiplied by exponential functions is a consequence of the Jordan canonical form.

More information on canonical forms is made available by imposing further structure on the mappings in question The class of mappings for which the most complete description can be obtained is the class of self-adjoint linear mappings.

The notion of an adjoint mapping is defined for linear mappings from an inner product space U to an inner product space V. If $\mathbf{T}$ is such a mapping from the finite-dimensional inner product space U to the finite-dimensional inner product space V, then the expression

$$\{\mathbf{T}u, v\}_V$$

defines a linear mapping $U \mapsto C$ (i.e., a linear functional on U). From the result quoted earlier, it must be representable as an inner product, so

$$\{\mathbf{T}u, v\}_V = (u, w)_U$$

for some $w \in U$. Since the expression depends on $v \in V$, we write

$$w = \mathbf{T}^*(v).$$

Using the properties of inner products, it is possible to deduce from the expression

$$\{\mathbf{T}u, v\}_V$$

that $\mathbf{T}^*$ is a linear mapping $V \mapsto U$. $\mathbf{T}^*$ is called the (inner product space) adjoint of the linear mapping $\mathbf{T}$.

In terms of the matrix of $\mathbf{T}$ with respect to orthonormal bases for U and V, it is possible to show that the matrix of $\mathbf{T}^*$ is the complex-conjugate transpose of that of $\mathbf{T}$.

If one specializes to mappings $\mathbf{T} : U \mapsto U$ which map an inner product space into itself, then the resulting matrix representation is square. Mappings $\mathbf{T} : U \mapsto U$ for which

$$\mathbf{T}^* = \mathbf{T}$$

are called self-adjoint.

Self-adjoint linear mappings are a constant component of problems arising in mathematical physics and mechanical problems in engineering. Chapter 4 uses self-adjointness properties to discuss the general structure of solutions of typical boundary value problems.

Appendix B
Software Resources

B.1 Octave

Octave provides an environment for technical computing and graphics. Because of this orientation, and also because of the range of the included subroutine libraries, it is a useful adjunct to learning and practicing applied and engineering mathematics.

Octave is in origin and at heart an interactive interpreter providing ready access to a variety of mathematical calculations.

The Octave program is not the only project that has sought to provide an interactive scripting language on top of a package of subroutine libraries. The experience of such facilities in word processing and spreadsheet programs is familiar to most computer users.

There are a number of mathematically oriented programs with an interactive scripting language interface. Symbolic processing is provided by Maple $^{(TM)}$ and Mathematica $^{(TM)}$ among commercial programs, and by the Free Software Maxima and SymPy projects.

Maple, Mathematica, and Maxima have support for graphics, and so compete with Octave and Matplotlib in that area.

Graphics are also a feature of interactive programs with a statistics emphasis. The commercial SPlus $^{(TM)}$ comes to mind, and similar functionality is in the GNU R program release. There are also programs with functional areas that overlap those of Octave. Both Octave and Scilab are freely available examples. Octave is distributed under GNU GPL, while Scilab has a unique license agreement.

© Springer International Publishing Switzerland 2016
J.H. Davis, *Methods of Applied Mathematics with a Software Overview*,
Applied and Numerical Harmonic Analysis, DOI 10.1007/978-3-319-43370-7

Octave Overview

Octave is widely used for scientific computation and has been in existence and under development for a long time. The long development implies that many facilities have been added over time. Some of this is in the form of area specific "packages," and this is outlined below. Generally Octave has incorporated improvements in basic numerical analysis routines as these have become available, so that the numerical quality of the routines is high. Octave syntax attempts to be compatible with the commercial program MATLAB, although there are no "toolbox" additions in the Octave context. In a sense, that is replaced by a combination of "packages" and the Graphical UI elements that are available.

Octave graphics flexibility and facilities are highly developed. This makes Octave useful for processing measured data, as well as presenting the output of simulations and other computations. We describe some aspects of the graphics below.

The Octave program is also used as a programming environment. Part of this is "scripting" for driving computations using subroutine libraries and graphics, but it also contains facilities for "user" interface construction [5]. These are associated with the graphical user interface, and can be used to develop a custom interface for a data collection and analysis project. Finally, the program also supports dynamically linking user written and compiled modules. This can be used to overcome speed limitations of the interpreter, if that becomes an issue in a dedicated application.

In this text our interest is in scripting computations and graphics, so we slant our discussion below in those directions.

Octave Data Structures

When Octave development was begun the emphasis was on providing interactive use of linear algebra libraries. In subsequent development, Octave has added syntax expansions and library support for various data processing and file formats. The syntax improvements are present "out of the box" with the standard releases, while libraries for file formats and specialized processing are in separate packages that need to be explicitly loaded for use.

The basic Octave data structures are multidimensional numeric arrays, structures, and cell arrays. For some time after its release numeric arrays were the only data type, and as a result handling of character strings was awkward. The following examples illustrate the basic numeric character of Octave arrays. Arrays constructed with string values are basically numeric with the ascii character values appearing. This appears with an array element-wise addition.

OCTAVE

```
>> b = [[1 2 3]; [4 5 6]]
b =

   1   2   3
   4   5   6

>> b = [[1, 2, 3]; [4, 5, 6]]
b =

   1   2   3
   4   5   6

>> a = ['abc'; 'def']
a =

abc
def

>> a = ['abc'; 'defg']
a =

abc
defg

>> size(a)
ans =

   2   4

>> b(1,1)
ans =  1
>> a(1,1)
ans = a
>> aa = ['abc', 'defg']
aa = abcdefg
>> bb = [[1, 2, 3], [4, 5, 6]]
bb =

   1   2   3   4   5   6

>> aa .+ 2
ans =

    99   100   101   102   103   104   105
```

Octave cell arrays use "curly braces" in place of the square notation of numeric arrays. The curly notation also can be used for subscripting. Examples below use cell arrays of strings in a plotting application. Some example of cell variables follow.

OCTAVE
```
  a = {'abc'; 'defg'}
```

a =

```
    'abc'
    'defg'
```

```
  size(a)
```

ans =

```
      2      1
```

```
  a{1  }
```

ans =

abc

```
  a(1)
```

ans =

```
    'abc'
```

Virtually any Octave data type can be stored in a cell array, so it is very useful as a container type.

The conventional container type in other programming languages is the structure with named fields. This also is used in Octave and is capable of "universal storage."

OCTAVE
```
  a.first_string  =   'abc';
  a.second_string =   'defg';
  a.float = pi;
  a.number = exp(1);
```

```
  a
```

a =

```
      first_string:  'abc'
     second_string:  'defg'
             float:  3.1416
            number:  2.7183
```

Octave also supports passing functions as variables in a function call. There are actually two acceptable versions of the syntax. A quoted name has been used and is embedded in much "legacy code," if that term applies to Octave scripts. More recently "@" read as "address of" has been introduced. Thus both of the code fragments

OCTAVE

```
smelly_call('myfunc', 1, 2);

smelly_call(@myfunc, 1, 2);
```

should work. Our calculations with Fourier series operate by this mechanism. Fourier series differ from one another because the term coefficient formula changes. The plotting function calls can pass a "term function" as an argument. Other such examples involving operations of zero finding or numerical integration also can utilize this facility.

Octave Operators and Syntax

Octave has an online help system which includes information useful in writing programs and scripts. The help system will provide a list of available operators, provided that one can guess one likely instance. For example,

OCTAVE
```
>> help *
```

Operators and special characters.

Arithmetic operators.
plus	− Plus	+
uplus	− Unary plus	+
minus	− Minus	−
uminus	− Unary minus	−
mtimes	− Matrix multiply	*
times	− Array multiply	.*
mpower	− Matrix power	^
power	− Array power	.^
mldivide	− Backslash or left matrix divide	\
mrdivide	− Slash or right matrix divide	/
ldivide	− Left array divide	.\
rdivide	− Right array divide	./
kron	− Kronecker tensor product	kron

Relational operators.
eq	− Equal	==
ne	− Not equal	~=
lt	− Less than	<
gt	− Greater than	>
le	− Less than or equal	<=
ge	− Greater than or equal	>=

Logical operators.
	Short−circuit logical AND	&&
	Short−circuit logical OR	||
and	− Element−wise logical AND	&
or	− Element−wise logical OR		
not	− Logical NOT	~
xor	− Logical EXCLUSIVE OR	
any	− True if any element of vector is nonzero	
all	− True if all elements of vector are nonzero	

Special characters.
colon	− Colon	:
paren	− Parentheses and subscripting	()
paren	− Brackets	[ ]
paren	− Braces and subscripting	{ }
punct	− Function handle creation	@
punct	− Decimal point	.
punct	− Structure field access	.
punct	− Parent directory	..

```
punct        – Continuation                                       . . .
punct        – Separator                                          ,
punct        – Semicolon                                          ;
punct        – Comment                                            %
punct        – Invoke operating system command    !
punct        – Assignment                                         =
punct        – Quote                                              ,
transpose    – Transpose                                          .,
ctranspose   – Complex conjugate transpose        ,
horzcat      – Horizontal concatenation           [,]
vertcat      – Vertical concatenation             [;]
subsasgn     – Subscripted assignment             (), {  },.
subsref      – Subscripted reference              (), {  },.
subsindex    – Subscript index
```

Bitwise operators.
```
bitand       – Bit–wise AND.
bitcmp       – Complement bits.
bitor        – Bit–wise OR.
bitmax       – Maximum floating point integer.
bitxor       – Bit–wise XOR.
bitset       – Set bit.
bitget       – Get bit.
bitshift     – Bit–wise shift.
```

Set operators.
```
union        – Set union.
unique       – Set unique.
intersect    – Set intersection.
setdiff      – Set difference.
setxor       – Set exclusive–or.
ismember     – True for set member.
```

See also ARITH, RELOP, SLASH, FUNCTION_HANDLE.

Octave Programming Structures

As a programming language, Octave provides the usual amenities. This can be deduced by listing the language keywords.

```
OCTAVE
>> iskeyword
ans =
{
  [1,1]  =  __FILE__
  [2,1]  =  __LINE__
  [3,1]  =  break
  [4,1]  =  case
  [5,1]  =  catch
  [6,1]  =  classdef
  [7,1]  =  continue
  [8,1]  =  do
  [9,1]  =  else
  [10,1] =  elseif
  [11,1] =  end
  [12,1] =  end_try_catch
  [13,1] =  end_unwind_protect
  [14,1] =  endclassdef
  [15,1] =  endenumeration
  [16,1] =  endevents
  [17,1] =  endfor
  [18,1] =  endfunction
  [19,1] =  endif
  [20,1] =  endmethods
  [21,1] =  endparfor
  [22,1] =  endproperties
  [23,1] =  endswitch
  [24,1] =  endwhile
  [25,1] =  enumeration
  [26,1] =  events
  [27,1] =  for
  [28,1] =  function
  [29,1] =  global
  [30,1] =  if
  [31,1] =  methods
  [32,1] =  otherwise
  [33,1] =  parfor
  [34,1] =  persistent
  [35,1] =  properties
  [36,1] =  return
  [37,1] =  switch
  [38,1] =  try
  [39,1] =  until
  [40,1] =  unwind_protect
  [41,1] =  unwind_protect_cleanup
  [42,1] =  while
}
```

Generally speaking the controlled blocks are terminated by the corresponding end statement. The for loops are a little unconventional in appearance since the controlling variable steps through an array. This is actually more flexible than conventional instances.

OCTAVE
```
for x = 0: .1: 1
   ... x ...

endfor
```

Octave provides break and continue for early loop exit in the manner of the
C family of languages. The break is not used with the Octave switch/case
structures, since the code exits the block on a successful match (there is no C-like
fall through). The punctuation is different, probably because of the use of ":" in
array indexing. Because of the built-in help, one does not really have to worry about
the differences. Note the use of Octave cell structures for enclosing multiple case
values.

OCTAVE
```
>> help case

-- Keyword: case VALUE
-- Keyword: case {VALUE, ...}
A case statement in a switch.  Octave cases are exclusive
and do not fall-through as do C-language cases.  A switch
statement must  have at least one case.  See 'switch' for
an example.

     See also: switch.


>> help switch

-- Keyword: switch STATEMENT
     Begin a switch block.

          yesno = "yes"

          switch yesno
            case {"Yes" "yes" "YES" "y" "Y"}
              value = 1;
            case {"No" "no" "NO" "n" "N"}
              value = 0;
            otherwise
              error ("invalid value");
          endswitch

     See also: if, case, otherwise.
```

Additional help for built-in functions and operators is
available in the online version of the manual. Use the command
'doc <topic>' to search the manual index.

Help and information about Octave is also available on the WWW
at \cite{web:octave} and via the octave
mailing list, which can be reached through links on the site.

The other "features" that bear notice are the `global` declarations and `function` declarations.

Global variables have to be declared as such both within the function definition that uses them and at the level of the interactive interpreter. A `persistent` declaration within a function body is the same as a C *static* variable, effectively a global variable invisible at the outer level.

Octave Programs and Scripts

Functions are typically defined by writing them in external files. As long as the file is on the `loadpath` with a ".m" suffix, it will execute as a command interactively, or within another function. The `loadpath` can be set, but includes the current directory, at least by default. Many examples of function definitions appear elsewhere in the text, using a standard format of leading with comment lines. These leading comment lines are used to generate online help for the function, so they should be formatted with that fact in mind. The help text listed in the example above for the `case` statement is actually identifiable as the leading comment lines from the file `case.m` in the Octave system directory. The `case` function is actually a compiled entity, so the corresponding ".m" file exists for the sake of the help system.

Functions (by definition) return a value. One can also construct commands by filling a ".m" file on the `loadpath` with lines which would execute if typed at the keyboard. Typing the command name redirects the keyboard input to come from the file as long as it lasts. This makes it possible to save a portion of the interactive history as a command, perhaps editing out the "false steps" of the interactive session.

Common Octave Idioms

Much of the use of Octave involves handling arrays. The syntax features of Octave make possible easy handling of arrays.

Division and multiplication provide *matrix* versions transparently, and what are usually thought of as scalar-valued functions provide element-by-element evaluation without the need to code loops to step through the values of the array. Since array components can be complex, this also works with complex arguments.

There is a small caveat associated with this facility. Computations can occasionally produce results with very small imaginary components in a problem where the values theoretically should be real, typically because the included complex conjugate terms cancel. In practice, they do not exactly cancel because of roundoff, and the small complex values can propagate through subsequent calculations. The errors produced in this situation can come as a surprise, and it may not be evident what caused the problem.

The ":" operator appears in several contexts in Octave as a very concise shorthand. It is very useful for producing (mostly) linearly spaced arrays.

OCTAVE
```
x0 : delta : xf
```

is an array with values starting from $x0$, stepping by *delta*, up to and including xf.[1]

Arrays of this kind are often used as the controlling element in a `for` loop. In this case the fact that an omitted increment value defaults to 1

OCTAVE
```
...
for k = 1 : length (x)
  ...
endfor
```

is often used as a shorthand to obtain a loop where the variable steps by 1.

Another use of the colon operator ":" is as a dimension wildcard. Hence we see statement like

OCTAVE
```
% get the first row
  x = a(1, :);
% pull out the seventh column
  x = a(:, 7);
```

Perhaps the most mysterious use of ":" is as a column creating agent. The expression

OCTAVE
```
x = a(:)
```

will turn a row vector a into a column. More generally, it will create a column out of the components of a listed in column-major order. Because arrays are stored internally in column-major order by Octave, this is effectively a free operation. The presence of this operation also leads to heavy use of columns as the conventional vector form in many Octave codes, since it provides an easy way to force a standard format, and handle inadvertent row vector input at the same time.

The code fragment above mixes assignment lines that are terminated with a semicolon with some that leave it off. Terminating the line with a semicolon suppresses console echo of the result of the line.

Omitting a semicolon on a ".m" file line will often cause a flood of terminal output, although it provides an easy mechanism for the classical "print statement" method of debugging. It is surprisingly effective in this role, since seeing an impossible value appear is often enough to trigger recognition of a problem in the code.

---

[1]This can produce an "uneven" last step. Use `linspace` if even steps are essential.

Common Octave Gotchas

Like any programming or operating environment, Octave takes some use before one
is accustomed to the conventions, syntax, and facilities. One thing you have to get
used to is the difference between scalar and array operations. The general pattern
appears to be that

```
OCTAVE
      a .oper b
% as in
      a .* b
```

represents a component by component operation between the elements of *a* and
those of *b*. This pattern applies to multiplication, division, and powers. If one of the
participants is a numerical value, either the element by element form or the array
form may be used. What appears to happens with use of the scalar syntax is that
Octave promotes the scalar to a scalar multiple of an appropriately sized array of 1s,
and then does the indicated element by element operation.[2]

   The .+ expression is an exception to the "dot-operation" pattern in the operation
list. The .+ operation works as expected, but a simple + between arrays does
element-wise addition.

   We mention an effect of sloppy shift key use. The result of

```
OCTAVE
t = 0:.01;2*pi;
plot(t, sin(t))
```

is surprising to the point of provoking "Whaaaa?," especially if there are a number
of surprisingly rapid intermediate calculations preceding the plot.

   What is happening is that *t* is assigned an array ranging from 0 to .01 stepping
by 1, so that *t* ends up effectively a singleton 0. The $2 * pi$ is echoed to the console,
and silently dropped because of the trailing ";".

Octave User Interface

Although Octave is often used with a command line interface running in a terminal,
it is also possible to interact with it using a graphical user interface. This is illustrated
in Figure B.1. The graphical user interface is particularly useful for beginners
learning to use Octave, since the interface includes editor, file browser, debugger,
command line, and access to the included documentation.

---

[2]There is no intended suggestion that this is what Octave does internally.

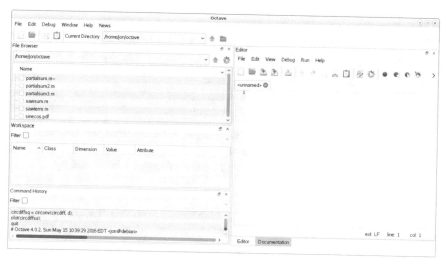

Fig. B.1 Octave graphical interface

The terminal interface can be accessed by typing "`octave ---no-gui`" at the terminal command prompt. Alternatively, essentially the same interface is present as the "command interface" selection on the graphical display version.

Octave Two-Dimensional Plots

Octave has plotting facilities that operate on both high and low levels from the user perspective. The high level plotting facilities operate with arrays of data. In fact, one can ask Octave to plot a single vector (row or column), and the result will be a plot with the array indices indicated as the "x axis" values.

It is more common to want the independent variable value indicated, and for this both arguments are passed to `plot`. Multiple pairs produce overlaid plots.

OCTAVE
```
t = 0:.01:2*pi;
plot(t, sin(t), t, cos(t));
```

Many function examples are given in terms of piecewise formula definitions. It is useful to plot such examples, in connection with investigations of series convergence, for example. To enable this we need a Octave version of the Heaviside step function. This can be constructed on the basis that the Octave relational operators return a 1 for "true" and 0 for "false."

OCTAVE

```
% heaviside(t)
% = 0, for elements of t <  0
% = 1, for elements of t >= 0
function out = heaviside(t)
    out = (t >= 0);
endfunction
```

Combining relational expressions with logic operators actually lets one write down piecewise function definitions such as those used for Laplace transform calculations in Chapter 6 directly. A square pulse of length 1 is defined by

OCTAVE

```
% squarepulse(t)
%       = 1, for elements of in >=0 and <1
function out = squarepulse(t)
    out = (t >= 0) & (t < 1);
endfunction
```

Using this, a function that is 1 between τ and $\tau + T$ is simply

$$squarepulse((t - \tau)/T).$$

In order to plot functions whose graphing data contains very large magnitude values, it is useful to be able to clip the data at some level. If the large values are not suppressed, the default automatic plot scaling tends to obscure useful data.

OCTAVE

```
% clip(x, limits)
% works usefully for real valued things only, probably
function y = clip(x, limits)
    y = min(max(x, limits(1)), limits(2));
endfunction
```

Surface Plots

For plotting surfaces Octave supports both an "altitude over (x,y)" model, and what probably should be considered the "right" approach to surfaces (and multivariable functions). This is referred to as the *parametric* approach in the Octave documentation.

A main benefit of the parametric view is that it extends to situations with a higher number of dimensions than two, although we stick to the two-dimensional case in this Appendix.

A two-dimensional surface in R^3 is defined by a parameter region

$$(u, v) \in P, \text{ usually a rectangle in } R^2,$$

and coordinate functions

$$\begin{bmatrix} X(u, v) \\ Y(u, v) \\ Z(u, v) \end{bmatrix} \in R^3.$$

This model applies to Octave (generally, but in particular with graphics) through the realization that a two-dimensional array really is a function of two integer valued arguments. The mapping involved is

$$(i, j) \mapsto A(i, j).$$

For computations, the u, v parameters are replaced by a finite set of sample values, and the functions of the two parameters then become functions of the two sample subscripts, i.e., two-dimensional arrays.

OCTAVE
```
u = linspace(0, 2, 5)
```

u =

          0   0.5000   1.0000   1.5000   2.0000

```
v = linspace(-2, 1, 6)
```

v =

    -2.0000   -1.4000   -0.8000   -0.2000   0.4000   1.0000

```
[X Y] = meshgrid(u, v)
```

X =

          0   0.5000   1.0000   1.5000   2.0000
          0   0.5000   1.0000   1.5000   2.0000
          0   0.5000   1.0000   1.5000   2.0000
          0   0.5000   1.0000   1.5000   2.0000
          0   0.5000   1.0000   1.5000   2.0000
          0   0.5000   1.0000   1.5000   2.0000

Y =

    -2.0000   -2.0000   -2.0000   -2.0000   -2.0000
    -1.4000   -1.4000   -1.4000   -1.4000   -1.4000
    -0.8000   -0.8000   -0.8000   -0.8000   -0.8000
    -0.2000   -0.2000   -0.2000   -0.2000   -0.2000
     0.4000   0.4000   0.4000   0.4000   0.4000
     1.0000   1.0000   1.0000   1.0000   1.0000

```
% Parametrically this is X(u, v) = u, Y(u, v) = v,
%   Z = F(X(u, v), Y(u, v)).

Z = exp(-(X.^2 + Y.^2));

surf(X, Y, Z);
```

The somewhat coarse surface plot can be seen in Figure B.2.

Much more pleasing plots can be generated by using a finer grid, and setting properties of the graphic objects to interpolate values, introduce transparency, and so on.

There is nothing magical about the use of meshgrid. The argument arrays can be generated using outer products of the data vectors with correctly sized arrays of 1s. The meshgrid function is more convenient, but it obscures what is going on a bit. It should be clear that the arrangement of values in those arrays directly determines the orientation of the plotted surface, and understanding that is an aid in manipulating and interpreting the plot output.

The data generated above by meshgrid can be produced "by hand."

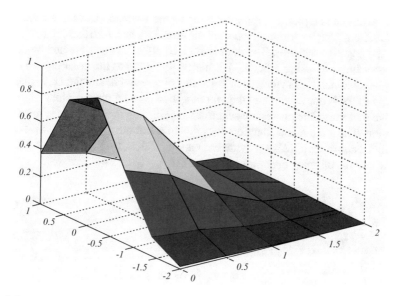

Fig. B.2 Basic surface plot

OCTAVE

```
ones ([length(v)  1]) * u
```

ans =

```
      0      0.5000    1.0000    1.5000    2.0000
      0      0.5000    1.0000    1.5000    2.0000
      0      0.5000    1.0000    1.5000    2.0000
      0      0.5000    1.0000    1.5000    2.0000
      0      0.5000    1.0000    1.5000    2.0000
      0      0.5000    1.0000    1.5000    2.0000
```

```
transpose(v) * ones([1  length(u)])
```

ans =

```
  -2.0000    -2.0000    -2.0000    -2.0000    -2.0000
  -1.4000    -1.4000    -1.4000    -1.4000    -1.4000
  -0.8000    -0.8000    -0.8000    -0.8000    -0.8000
  -0.2000    -0.2000    -0.2000    -0.2000    -0.2000
   0.4000     0.4000     0.4000     0.4000     0.4000
   1.0000     1.0000     1.0000     1.0000     1.0000
```

The following is an example for which a parametric representation of the surface is really essential, since it is not representable as a single-valued function on the (x, y) plane. The parameter range of course is a rectangle in parameter space, but is thought of as a double sweep through the unit circle in planar polar coordinates.

The example is a pictorial representation of the Riemann surface for the square root function. This is a double-sheeted surface, stitched together by a crossover along the negative real axis. The double-sheeted effect can be obtained by defining the Z coordinate by using a sinusoid of period 4π. Then as the polar circle is traced twice, we obtain values symmetrically located above and below the (x, y) plane.

The surface is conventionally thought of as two parallel planes joined along their negative real axes. The planes are suggested by clipping the z values for most of the excursion. The arc tangent function does this in a smooth way. The `clip` function introduced above would also provide flat looking sheets, but also a corner along the lines where the clipping becomes effective. Try it.

```octave
OCTAVE
  r = 0: .05: 1;
  theta = linspace (0, 4*pi, 720);

  [R Theta] = meshgrid (r, theta);

% define the rectangular variables

  X = R .* cos (Theta);
  Y = R .* sin (Theta);

% use  ''(1/pi) * atan (25 *''  for a smooth clipping effect

  Z = (1/pi) * atan (25 * cos (.5 * Theta));

  h = surf (X, Y, Z);

% use the surface handle to set properties

  set (h, 'EdgeColor', 'none', 'FaceColor', 'interp');

% the labels are associated to the current axes handle

  set (gca, 'XTickLabel', []);
  set (gca, 'YTickLabel', []);
  set (gca, 'ZTickLabel', []);

  print -deps riemannsqrt.eps
```

The resulting plot is shown in Figure B.3.

Low Level Graphics

Octave graphics can be produced and manipulated at a lower level than the `plot` and `surf` commands used above. A hint of the mechanisms involved is provided by the `set` commands in the previous examples.

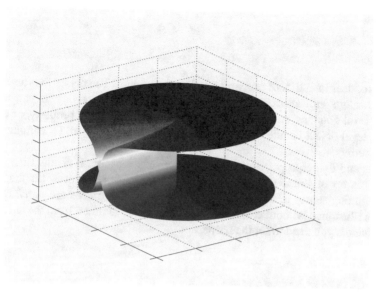

Fig. B.3 Two sheeted surface example

The model is that of an object hierarchy, with all components within a top level figure object. Visually the figure corresponds to the screen window on which the various plots are displayed.

Figures act as a container, while axes are the immediate containers of what most individuals think of as a plot. It is possible to place multiple axes (and hence plots) within a single figure, and produce arbitrarily complicated, oriented, and labeled diagrams in that way. For our purposes it suffices to stick with a single set of axes within the default figure.

True to the object programming dogma, variables associated with the objects are examined and changed through *get* and *set* procedures. The models are

OCTAVE
```
get( object_handle , 'PropertyTag' );

set( object_handle , 'PropertyTag' , property_value );
```

These functions handle variable length argument lists, so arbitrary numbers of properties can be set at once. For that matter, properties can also be set by adding tag-value pairs to plot or surf invocations, instead of "correcting" the values after the plot appears.

The available properties can be examined by means of

OCTAVE
get (gca)

get (gcf)

This produces long lists of properties and current values. The mysterious arguments
stand for "get current axis" and "get current figure" respectively. If you use the
object level graphics, sooner or later you will either misspell a property, or use the
wrong level object handle in trying to access it. Using the get () command is a
useful reminder.

Axis and figure properties are actually not sufficient for all needs. Linear plots
are made up of lines, which are objects on their own, children (in the object
programming sense) of an axis. What we think of as surfaces are *patch* objects,
and (as illustrated without comment earlier) the value returned from the surf call
is the handle you need to set the properties.

OCTAVE

```
fig_axis = newplot;
fig = get(fig_axis, 'Parent');
plot_axis = setup_plot(fig, fig_axis);
%
set(plot_axis, 'YTick', [-1 0 1]);
set(plot_axis, 'YTickLabel', ['-one', 'zero', 'one']);
set(plot_axis, 'XTick', [0]);
set(plot_axis, 'XTickLabel', '0');
%
Epsfile('x', 'fig:2-5-1.eps', ...
        {'sin (x)', 'cos (x)', 'sin(2*x)' } , ...
        {'x=linspace(0, 7,1000);', 'x=linspace(0, 7, 1000);', ...
         'x=linspace(0, 7, 1000);' });
```

The plot setup sets properties in bulk using the option to pass a Octave structure
to set.

OCTAVE
```
% setup_plot(fg, ax)
% sets parameters for a fixed size eps image
 function h = setup_plot(fg, ax)
     fig.PaperUnits = 'inches';
     fig.PaperSize = [4.6063 3.5];
     fig.Color = [.925 .925 .584];
     set(fg, fig)

     axis.Color = [.925 .925 .584];
     axis.FontName = 'times';
     axis.FontAngle = 'italic';
     axis.FontSize = 10;
     axis.LineWidth = 1.0;
     axis.Position = [.1 .1 .8 .8];
     axis.NextPlot = 'add';
     axis.XTickMode = 'manual';
     axis.XTickLabelMode = 'manual';
     axis.YTickMode = 'manual';
     axis.YTickLabelMode = 'manual';
     axis.linewidth = 2;

     set(ax, axis);

     h = ax;
 endfunction
```

In order to generate multiple plots, `Epsfile` takes parallel cell arrays, consisting of a variable range assignment statement and a function expression, both passed as character strings. The *eval* procedure is a Octave built-in that evaluates string expressions. The tricky part of the code is that evaluating the expressions does not give the subroutine the identity of the variable involved. This is handled by also using the string representation of the variable as the first argument. The actual "plotting" is done by the `line` call. If properties of the lines were to be changed from the default, the value returned from `line` call would have to be saved and used for accessing the line properties.

OCTAVE
```
% Epsfile
% write out an eps file containing plots
% other figure content may be set before calling Epsfile
 function y = Epsfile(var, filename, fun_cell, range_cell)

for kk = 1:length(fun_cell)
    eval(range_cell{kk});
    yy = eval(fun_cell{kk});
    tem = eval(var);
    line(tem, yy);
endfor

print(gcf, '-depsc', filename);
endfunction
```

The topic of Octave graphics is full of options, parameters, graph types, and alternative functionality.

Octave Packages

MATLAB is a commercial program, and it is marketed as a base package together with extra cost toolboxes.

The toolboxes add functionality by building on the base to provide routines supporting particular areas of interest. Since the additions are usually ".m" files, the added functionality is available on a command basis, although it seems to be a hallmark of the toolboxes that a menu driven graphical user interface is provided as part of the package.

Octave was developed as free software with the aim of command line compatibility with MATLAB. This was possible because so many of the numerical method software libraries used by both Octave and MATLAB are available as under a free software license. There is an active community behind Octave which works to develop Octave interfaces to additional libraries, and packages are available which add this extra functionality to an Octave installation. The available packages can be viewed at the web site [32].

Whether you download packages from this source or not depends on the system you are using to run Octave. Linux systems typically provide many of the packages through the overall software manager, and this ensures package versions are kept in sync with the installed Octave release. See the discussion of software platforms below in Section B.4.

Octave packages are available for a range of subject areas. The Partial Differential Equation, Signal Processing, and general numeric facilities support the applications discussed in this text.

B.2   Python

Octave is not the only environment that provide software useful for applied mathematics calculations. Maxima is basically a symbolic calculation programming environment, with a lot of support for Fourier analysis calculations, and well as some numerical methods. Python is a widely used programming language, and it has packages supporting scientific calculation ("Scipy"), numerical exploration ("Numpy"), plotting ("Matplotlib") , and even finite element partial differential equation work ("Fenics").

It is hard to overstate the extent to which python is in use for computing projects. Many system configuration programs (think printer management, network setup, ...) are written in python. At the other end of the scale, there are large websites operated entirely through python content management systems. This is all due to the usefulness and power of python in a mixed language programming environment. Scripting with python provides a clean, elegant, object-oriented tool, with access to services of many available library functions.

One example of this is the numerical and plotting facilities described below. Another large python project is the Blender program for computer animation [4].

A capsule history of Python development and licensing is available on the python website [27].

Python Programming

Python is an object-based, block-structured scripting language. The standard warning is that it provides enough rope to allow you to shoot yourself in the foot: there are tools that reveal (and actually allow manipulation of) the internal structure of all the objects in the system. Since everything is an object, the scope for foot damage should be clear.

The aspect of python that surprises users used to other programming languages is that "the white space matters" in python source code. The blocks of code controlled by conditionals, loops, and the like are indicated by indentation in the code text. In conventional languages blocks have delimiters like "begin" ..."end," or matched bracket characters, and the difference is visually striking. A motivation for this can be seen in the International Obfuscated C Contest [14], where the effect of misleadingly formatted code is on display.

Who knows how many spaces a tab key is worth? Writing python code really requires the use of a python-syntax-aware program (such as emacs) for editing, or at the very least having tabbing turned off.

Python is under constant development and currently is present on most systems both in a 2.7.xx version and in a release 3.x.

For the 2.7 series the keyword list shows as:

```
PYTHON
and        del        from       not        while
as         elif       global     or         with
assert     else       if         pass       yield
break      except     import     print
class      exec       in         raise
continue   finally    is         return
def        for        lambda     try
```

For the 3.x series the list looks like:

PYTHON

False	class	finally	is	return
None	continue	for	lambda	try
True	def	from	nonlocal	while
and	del	global	not	with
as	elif	if	or	yield
assert	else	import	pass	
break	except	in	raise	

The later version included True/False , while previously the zero/nonzero distinction was in use. Is a Null object a zero?? Check some common Null objects.

PYTHON

```
>>> def truth_check(p):
...      if p :
...           print 'true'
...      else:
...           print 'false'
...
>>> truth_check(0)
false
>>> truth_check(list())
false
>>> truth_check(dict())
false
>>> truth_check("")
false
>>> truth_check(1.0)
true
>>>
```

With the 3.x series release, the print changed from a keyword to a built-in function, but that synopsis hides a deeper difference. In the 2 series strings were arrays of ascii eight bit characters, while the strings in 3 are changed radically in order to support Unicode. This makes string handling in 3.x a more complicated issue.

The distinction also goes a long way to explain the persistence of 2.x versions. The fact is that many system level libraries are based on the use of ascii text strings for configuration data, and python programs using those libraries are in current use. For such things, the "If it ain't broke, don't fix it" idea rules, and so python 2.x stays around.

Python Structure and Data

Python has built-in data structures , some familiar and some novel. The lists and tuples are standard:

PYTHON

```
>>>
>>> a = (1, 2, 'three')
>>> a
(1, 2, 'three')
>>> b = [1, 2, 3]; b
[1, 2, 3]
>>>
```

The built-in "dir()" and "help()" procedures provide ready information about the state and operations of the current python environment. With Octave, the visibility of available procedures and operations is determined by the directory structure where the script is running. Python controls what is visible in the script scope by use of module import statements. The operating system environment is associated with the "sys" module: the contents of it may be displayed with "dir," and the built in module help with "help."

PYTHON
```
>>> import sys
>>> sys
<module 'sys' (built-in)>
>>> dir(sys)

['__displayhook__', '__doc__', '__excepthook__', '__name__',
 '__package__',
 '__stderr__', '__stdin__', '__stdout__', '_clear_type_cache',
 '_current_frames',
 '_getframe', '_mercurial', '_multiarch', 'api_version',
 'argv',
 'builtin_module_names', 'byteorder', 'call_tracing',
 'callstats',
 'copyright', 'displayhook', 'dont_write_bytecode', 'exc_clear',
 'exc_info',
 'exc_type', 'excepthook', 'exec_prefix', 'executable', 'exit',
```

```
 'flags',
 'float_info', 'float_repr_style', 'getcheckinterval',
 'getdefaultencoding',
 'getdlopenflags', 'getfilesystemencoding', 'getprofile',
 'getrecursionlimit',
 'getrefcount', 'getsizeof', 'gettrace', 'hexversion',
 'last_traceback',
 'last_type', 'last_value', 'long_info', 'maxint', 'maxsize',
 'maxunicode',
 'meta_path', 'modules', 'path', 'path_hooks',
 'path_importer_cache',
 'platform', 'prefix', 'ps1', 'ps2', 'py3kwarning', 'pydebug',
 'setcheckinterval', 'setdlopenflags', 'setprofile',
 'setrecursionlimit',
 'settrace', 'stderr', 'stdin', 'stdout', 'subversion',
 'version', 'version_info', 'warnoptions']
```

```
>>> help(sys)
```

Help on built-in module sys:

NAME
    sys

FILE
    (built-in)

MODULE DOCS
    http://docs.python.org/library/sys

DESCRIPTION
This module provides access to some objects used or
maintained by the interpreter and to functions that
interact strongly with the interpreter.

    Dynamic objects:

    argv -- command line arguments; argv[0] is the
    script pathname if known
    path -- module search path; path[0] is the script
            direectory, else ''
    modules -- dictionary of loaded modules

    displayhook -- called to show results in an interactive
                  session
    excepthook -- called to handle any uncaught exception
                  other than SystemExit
    To customize printing in an interactive session or to
    install a custom top-level exception handler,
    assign other functions to replace these.

    exitfunc -- if sys.exitfunc exists, this routine is called
               when Python exits
    Assigning to sys.exitfunc is deprecated; use the atexit
```
```

module instead.

stdin -- standard input file object; used by raw_input()
        and input()
stdout -- standard output file object; used by the print
        statement
stderr -- standard error object; used for error messages
By assigning other file objects (or objects that behave
like files) to these, it is possible to redirect all of
the interpreter's I/O.

        .....

Any Python object can be inspected, and the instance variables and functions/operations are displayed. "list()" returns an empty list, equipped with all the machinery a self-respecting list would want.

```
PYTHON
>>> bare_list = list ()

>>> dir ( bare_list )
[ '__add__ ', '__class__ ', '__contains__ ', '__delattr__ ',
  '__delitem__ ', '__delslice__ ', '__doc__ ', '__eq__ ',
  '__format__ ', '__ge__ ', '__getattribute__ ',
  '__getitem__ ', '__getslice__ ', '__gt__ ', '__hash__ ',
  '__iadd__ ', '__imul__ ', '__init__ ', '__iter__ ',
  '__le__ ', '__len__ ', '__lt__ ', '__mul__ ', '__ne__ ',
  '__new__ ', '__reduce__ ', '__reduce_ex__ ', '__repr__ ',
  '__reversed__ ', '__rmul__ ', '__setattr__ ', '__setitem__ ',
  '__setslice__ ','__sizeof__ ', '__str__ ', '__subclasshook__ ',
  'append ', 'count ', 'extend ', 'index ', 'insert ', 'pop ',
  'remove ', 'reverse ', 'sort ']

>>> bare_list.append(1); bare_list
[1]
>>>

>>> help( bare_list.append )

Help on built-in function append:

append (...)
    L.append(object) -- append object to end
```

The python language internally makes use of hash tables for storage, and the user level version of these are called a "dict" (for dictionary).

PYTHON

```
>>> my_dict = {'first' : 1.25, 'second' : 2.50}
>>>
>>> my_dict['first']
1.25

>>> my_dict.keys()
['second', 'first']

>>> my_dict['third'] = 'a really big number'

>>> my_dict.viewitems()
dict_items([('second', 2.5), ('third', 'a really big number'),
            ('first', 1.25)])
>>>
```

Object attributes are accessed with a ".name" syntax comment to other object oriented languages. In the case of Python, the attributes can be discovered with "dir," and once the names are known, the "help" will elaborate the properties of the attribute.

PYTHON
```
>>> bare_dict = dict()
dir(bare_dict)
['__class__', '__cmp__', '__contains__', '__delattr__',
 '__delitem__', '__doc__', '__eq__', '__format__',
 '__ge__', '__getattribute__', '__getitem__', '__gt__',
 '__hash__', '__init__', '__iter__', '__le__', '__len__',
 '__lt__', '__ne__', '__new__', '__reduce__',
 '__reduce_ex__', '__repr__', '__setattr__', '__setitem__',
 '__sizeof__', '__str__', '__subclasshook__', 'clear',
 'copy', 'fromkeys', 'get', 'has_key', 'items', 'iteritems',
 'iterkeys', 'itervalues', 'keys', 'pop', 'popitem',
 'setdefault', 'update', 'values', 'viewitems', 'viewkeys',
 'viewvalues']

>>> help(bare_dict.keys)

Help on built-in function keys:

keys(...)
    D.keys() -> list of D's keys
```

Python Libraries and Extensions

The python documentation has always included information about writing python extensions in C, C++, and other languages. As a result there is a very large collection of extension libraries which extend python into use in various areas. Some of this involves numerical and scientific topics and that is discussed further in the following section.

Other extensions involve use of python in internet commerce and website development, and so involve libraries for handling internet protocols, xml and html documents, and database management and website integration.

The development of such software is enabled by tools that support development of the python extensions; one of the most widely used is the SWIG Project [2, 3].

Scipy, Numpy, Matplotlib

Scipy, Numpy, and Matplotlib are interrelated python projects that together cover roughly the same application territory that Octave does. Matplotlib was begun with the idea of providing a plotting api that duplicated the range of Matlab and Octave plotting.

The typical minimal numpy plotting example looks like:

PYTHON

```python
import numpy as np
import matplotlib.pyplot as plt

x = np.linspace(-np.pi, np.pi, 1000)
y = np.sin(x)

plt.plot(x, y)
plt.show()
```

This shows a numerical array x generated with syntax adopted from Octave (Matlab). One might expect that a numerical array in python would just be an ordinary python list, but with numeric entries. That this is not the case comes about because a python list carries the storage overhead of python object-ness. This has consequences in storage requirements, as well as in data access speed that are not good for large numerical calculations.

Octave has data storage that historically derives from the old Fortran numeric libraries originally used. This is reflected in the 1 based array indexing, as well as the row/column/array-shape issues the permeate octave programming.

Numpy has its own data type, based on the idea of storing data in uniform sequential physical memory. This facilitates access as fast as cpu addressing modes can access it, and so aids numerical speed. As well as the uniform storage, numpy arrays include header information which determines the geometry of the access to the stored data. The result of this is that the shape with which the data is viewed and accessed can change without wholesale moving of data within the storage.

The presence of this data structure leads to the need for numerical operation functions that are adapted to this scheme. The *np.sin*() in the above simple plot is an example of such an adapted function. These are known as "ufunc," presumably indicating that they apply to the uniform numpy data arrays. Using such functions with numpy is a matter of importing the numpy namespace (as *np*), and then prefixing with *np.*.

A common octave idiom exploits the fact that 1 represents "true" to mix numerics with logic expressions to get short simple expressions. Octave logic expressions convert to python without much difficulty, but running these in a numpy context produces an exception complaining about use of logic expressions with numpy arrays.

What you need is a ufunc compatible version of the codes causing the problem.

The heaviside case can be handled by using the *np.sign*() that comes with numpy. The octave *clip*() on page 730 is not so easy to fix and needs rewriting as a ufunc. Writing one of those in C and adding it to numpy in theory could happen, but a bit of search with google turns up [30]. With that information "clip" can be rewritten.

The key to this is the use of an "iterator." This is the notion for sequential access to an object (supporting such access). It applies to objects that are iterable, like strings, lists, arrays, . . . Getting an iterator replaces writing for or while loops with explicit object reference syntax. It is useful for code dealing with unicode strings, for example.

PYTHON
```python
import numpy as np

def heaviside(t):

    return(.5*(1   + np.sign(t)))

def clip(a, low, high):
    it = np.nditer([a, None])
    for x, y in it:
        if x <= low:
            y[...] = low
        else:
            if x >= high:
                y[...] = high
            else:
                y[...] = x
    return it.operands[1]


def halfwave(t):
    """ halfwave sine waveform"""

    return clip(np.sin(2 * np.pi * t), 0.0, 1.0)
```

Much of Octave "feels" like its heritage in Fortran Linear Algebra codes. This shows up in the need to keep the shape (row, column) of the variables in mind when programming, and in the convention that the lowest array index is 1, in Octave. On the other hand, the plotting with abstract graphics handles, and getting and setting properties through keywords sounds more object-oriented. Given the thorough object character of python it should be a good fit for building an api that functionally duplicates what Octave graphics can do. Matplotlib is the result of that effort [37].

One can argue about the quality of output from different graphics generating programs, although much of what you see is a result of user settings at the point of production. One can at least say that the amount of documentation for Matplotlib is enormous and readily accessible through a web search.

The examples below are python/matplotlib equivalents to the Octave procedures above. The man differences are between using Octave cells and python lists, and the way the graphic parameters are stored. There is so much overlap in the key names and meanings that it is not difficult to convert from one platform to another.

PYTHON
#

```python
import matplotlib.pyplot as plt
import numpy as np
from mpl_toolkits.mplot3d import Axes3D
import scipy.linalg as slg
import scipy.special as spec

# numpy compatible utility functions

from clip import heaviside, clip, halfwave

# global plot data storage

plot_dict = {}

## matplotlib versions of utility functions used in various plots


def setup_plot(fg, ax):
    """
    sets parameters for a fixed size image
    """
    plt.rcParams['figure.figsize'] = (4.6063, 3)
    plt.rcParams['axes.edgecolor'] = 'b'
    plt.rcParams['text.usetex'] = True
    plt.rcParams['font.serif'] = 'Times'
    plt.rcParams['font.family'] = 'serif'
    plt.rcParams['font.size'] = '10.0'
    plt.rcParams['ps.usedistiller'] = 'xpdf'

    plt.rcParams['axes.facecolor'] = 'ECEC95'
    plt.rcParams['figure.facecolor'] = 'ECEC95'
    plt.rcParams['savefig.facecolor'] = 'ECEC95'


def Epsfile(plot_dict, var, filename, fun_list, range_list):
    """ write out an eps file containing plots
    other figure content may be set before calling Epsfile
    """

    for kk in range(len(fun_list)):
        exec(range_list[kk])
        yy = eval(fun_list[kk])
        tem = eval(var)
        plt.plot(tem, yy)
```

```
        # label the axes

    plt.xticks(plot_dict['xtick'], plot_dict['xticklabel'])
    plt.yticks(plot_dict['ytick'], plot_dict['yticklabel'])

    s = .05

    # get [xmin, xmax, ymin, ymax]

    limits = plt.axis()

    xlims = [limits[0], limits[1]]
    ylims = [limits[2], limits[3]]

    lrange = xlims[1] - xlims[0]
    xlims[0] = xlims[0] - s * lrange
    xlims[1] = xlims[1] + s * lrange

    lrange = ylims[1] - ylims[0]
    ylims[0] = ylims[0] - s * lrange
    ylims[1] = ylims[1] + s * lrange

    plt.axis([xlims[0], xlims[1], ylims[0], ylims[1]])

    # cut down to simple x -y axes

    fig_axis = plt.gca()

    # hide top and right spines

    fig_axis.spines['right'].set_color('none')
    fig_axis.spines['top'].set_color('none')


    plt.grid(color='#404040')

    # add space around the plot in the figure

    plt.margins(0.25)

    plt.savefig(filename, dpi=300)

    plt.close('all')

#######
```

```
PYTHON
import matplotlib.pyplot as plt
import numpy as np

plot_dict = {}

# bandlimited impression
#
def bandlimited(t):
    """
    bandlimited(t)
    spectrum of typical function bandlimited to [-3, 3]
    """
    y = (heaviside(t+3)-heaviside(t-3))*(np.sinc(t))**2

    return y


fig_axis = plt.gca()
fig = plt.gcf()
setup_plot(fig, fig_axis)
#
plot_dict['ytick'] = [0]
plot_dict['yticklabel'] = [''0'']
plot_dict['xtick'] = [-4, 4]
plot_dict['xticklabel'] = ["-B", "B"]
#
Epsfile(plot_dict, 't', 'bandlimited.pdf', \
        ['bandlimited(t)'], \
        ['t = np.linspace(-4, 4, 1000)'])
```

Python Surface Plots

Surface plots may be made with numpy in a manner entirely parallel to methods used writing in Octave. The numpy and matplotlib namespaces have to be imported and used to invoke functions and data values. The same "linspace" and "meshgrid" functionality exists in numpy. The only difference is that TeX mathematical characters in the labels are easy with matplotlib. For Octave there are several different graphic toolkits, and not all readily support mathematical fonts. The Latex psfrag.sty package is a possibility, but that is tricky to use with pdflatex.

PYTHON

```python
from mpl_toolkits.mplot3d import Axes3D
import matplotlib
import numpy as np
from matplotlib import cm
from matplotlib import pyplot as plt

step = 0.04
maxval = 1.0

fig = plt.figure()
ax = fig.add_subplot(111, projection='3d')

# create polar coordinates

r = np.linspace(0, 1.25, 50)
p = np.linspace(0, 2*np.pi, 50)

R, P = np.meshgrid(r, p)

# transform them to cartesian system

X, Y = R*np.cos(P), R*np.sin(P)

Z = ((R**2 - 1)**2)

ax.plot_surface(X, Y, Z, rstride=1, cstride=1, cmap=cm.YlGnBu_r)

ax.set_zlim3d(0, 1)
ax.set_xlabel(r'$\phi_\mathrm{real}$')
ax.set_ylabel(r'$\phi_\mathrm{im}$')
ax.set_zlabel(r'$V(\phi)$')

plt.show()
```

Python, Fenics, and Finite Elements

The FEniCS project [35] develops software for finite element solutions of partial differential equations. The work is interesting on several levels, the software "automatically" provides solutions to problems that historically involved tedious work with various pieces of software dealing with finite elements.

At another level, the FEniCS project provides documentation for the process of building a python extension to give a python interface to a large finite element library written in C++.

The documentation is on the website [34]. This includes a user guide for using the software, as well as a book including the user guide, descriptions of the programs, and a detailed account of the process of building the python interface with SWIG [2, 3]. The glitches that arise are due to the differences in memory management between compiled C++ and interactive garbage-collected python scripts. The discussion in the FEniCS book is instructive in the extreme.

The fenics system includes demo scripts that can be used to learn the capabilities of the system. As fenics is updated it is imperative to update the sample programs to keep them in phase. That there is a demo update process was not evident until the occurrence of an implausible error message. Dolfin is the core of fenics. The process of installing fenics brings dolfin along with it.

BASH

```
jon@debian:~ $ dolfin-get-demos
This script will install the DOLFIN demos in your home directory:

    /home/jon/dolfin-demos

Do you want to install the DOLFIN demos in your home directory?
[y] y

Installing demos...

The demos have now been installed in /home/jon/dolfin-demos.

_____
To get started, enter the following command:

    cd /home/jon/dolfin-demos/pde

The documentation can be found at the following address:

    http://fenicsproject.org/documentation/doc_1.6.0.html
_____
```

The listing below is one of the fenics demo scripts, illustrating that the user interface basically consists of a python description of the variational formulation of the pde problem at issue.

An issue that is avoided in the demos is that of generating the mesh for the model being analyzed. The demo scripts tend to use the built-in geometries and meshes, while meshes for complicated objects probably have to be generated by CAD programs. The FreeCAD program [7] seems to contain some mesh export operations.

PYTHON

```
"""This demo program solves the mixed formulation of Poisson's
equation:

    sigma + grad(u) = 0       in Omega
           div(sigma) = f       in Omega
              du/dn = g         on Gamma_N
                  u = u_D       on Gamma_D

The corresponding weak (variational problem)

    <sigma, tau> + <grad(u), tau>    = 0
                                              for all tau
                - <sigma, grad(v)> = <f, v> + <g, v>
                                        for all v

is solved using DRT (Discontinuous Raviart-Thomas) elements
of degree k for (sigma, tau) and CG (Lagrange) elements
of degree k + 1 for (u, v) for k >= 1.
"""

# Copyright (C) 2014 Jan Blechta
```

```
#
# This file is part of DOLFIN.
#
# DOLFIN is free software: you can redistribute it and/or modify
# it under the terms of the GNU Lesser General Public License
# as published by the Free Software Foundation, either version
# 3 of the License, or (at your option) any later version.
#
# DOLFIN is distributed in the hope that it will be useful,
# but WITHOUT ANY WARRANTY; without even the implied warranty of
# MERCHANTABILITY or FITNESS FOR A PARTICULAR PURPOSE. See the
# GNU Lesser General Public License for more details.
#
# You should have received a copy of the GNU Lesser General Public
# License along with DOLFIN. If not,
# see <http://www.gnu.org/licenses/>.
#
# First added:   2014-01-27
# Last changed:  2014-01-29

# Begin demo

from dolfin import *

# Create mesh
mesh = UnitSquareMesh(32, 32)

# Define function spaces and mixed (product) space
DRT = FunctionSpace(mesh, "DRT", 2)
CG  = FunctionSpace(mesh, "CG", 3)
W = DRT * CG

# Define trial and test functions
(sigma, u) = TrialFunctions(W)
(tau, v) = TestFunctions(W)

# Define source functions
f = Expression("10*exp(-(pow(x[0] - 0.5, 2)\
    + pow(x[1] - 0.5, 2)) / 0.02)")
g = Expression("sin(5.0*x[0])")

# Define variational form
a = (dot(sigma, tau) + dot(grad(u), tau)\
    + dot(sigma, grad(v)))*dx
L = - f*v*dx - g*v*ds

# Define Dirichlet BC
def boundary(x):
    return x[0] < DOLFIN_EPS or x[0] > 1.0 - DOLFIN_EPS
bc = DirichletBC(W.sub(1), 0.0, boundary)

# Compute solution
w = Function(W)
solve(a == L, w, bc)
```

```
(sigma, u) = w. split ()

# Plot sigma and u
plot (sigma)
plot (u)
interactive ()
```

Python Resources

The python website [28] contains a complete range of resources on python as a language, python extensions, python tutorials, and books.

With a Linux desktop system, the "python-doc" package can be installed, and the website information becomes available in the standard documentation directory on the local machine. An image of this is in Figure B.4.

The documentation provides links to books about python and python-based application development. One of these (of note because it is distributed under the GNU Free Documentation License) is the book [26]. This is intended for an audience with some programming experience, but not familiarity with python. One

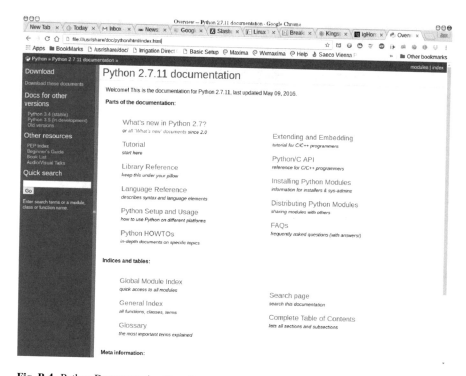

Fig. B.4 Python Documentation (Local)

example in the book is the utility "info," which essentially automates the process suggested above using the python "dir" and "help" procedures to document the functions and data variables of python objects.

B.3   Maxima

The Maxima program has a long and complicated history. It was started at Project MAC at MIT, close to the start of work on artificial intelligence. At one time it was developed as a commercial product by Symbolics Inc., a company spun out of MIT to develop hardware machines running Lisp. A version was eventually released under the Gnu Copyright License and called "Maxima." The details of the history involve many individuals and organizations, and different sources and documents leave the impression that this topic should be handled carefully.

The manual page distributed with the current Maxima says this:

```
BASH
AUTHORS
          MACSYMA (Project MAC's SYmbolic MAnipulation System)
          was developed by the Mathlab group of the MIT Labora-
          tory for Computer Science (originally known as Project
          MAC), during the years 1969-1972. Their work was sup-
          ported by grants NSG 1323 of the National Aeronautics
          and Space Administration, N00014-77-C-0641 of the
          Office of Naval Research, ET-78-C-02-4687 of the U.S.
          Department of Energy, and F49620-79-C-020 of the U.S.
          Air Force. MACSYMA was further modified for use under
          the UNIX operating system (for use on DEC VAX computers
          and Sun workstations), by Richard Fateman and col-
          leagues at the University of California at Berkeley;
          this version of MACSYMA is known as VAXIMA. The
          present version stems from a re-working of the public
          domain MIT MACSYMA for GNU Common Lisp, prepared by
          William Schelter, University of Texas at Austin until
          his passing away in 2001. It contains numerous addi-
          tions, extensions and enhancements of the original.
          The original version of this manual page was written by
          R. P. C. Rodgers, UCSF School of Pharmacy, San Fran-
          cisco, CA 94143 (rodgers@maxwell.mmwb.ucsf.edu) in
          1989. It was extensively revised by James Amundson in
          2002.

          Maxima is now developed and maintained by the Maxima
          project at <http://maxima.sourceforge.net>.
```

Maxima is now developed and maintained by the Maxima project at the website [22].

Maxima Structure and Interfaces

Maxima was started at a time when terminal based interfaces were the rule, and editors that matched the parentheses in lisp code were a stunning innovation. The current Maxima still runs as a large Lisp program, and the text based interface still exposes the underlying Lisp for anyone curious enough to learn enough to take advantage of that.

In fact, the usual user interface to maxima is an interactive interpreter, that just happens to have been written in Lisp. The commands are typed, and the program responds. If it is run in a terminal, the output is formatted as "typewriter math display." At any time the display can be put out in LaTeXformatting, although that is only useful to writers of mathematical texts. On most platforms this interface is provided by an "xmaxima" binary. This has a virtual terminal interface, together with access to the help file documents.

A nicer appearing interface is provided by wxMaxima. The "wx" refers to a cross platform user interface toolkit "wxWindows." The interface is basically a worksheet bearing cells that can be "sent to maxima" on menu command. Other common maxima commands are available on the menus, and anything can be entered as well.

The main interface improvement is in the output: wxMaxima uses the javascript HTML Math fonts, they are certainly a step up over "typewriter math."

The help menu provides access to wxMaxima web documentation, as well as the standard Maxima info formatted files (Figure B.5). The maxima-doc package on Linux downloads the pdf version of the maxima user documentation, and that may be viewed and searched with any one of various pdf viewer programs. It is also made available through the menus on the maxima graphical interface programs.

Maxima Syntax and Scripting

In addition to the overall documentation, help and usage examples for individual commands can be found with the "??" interface command. The example on page 760 shows the help for the "residue" command.

Breadth of Maxima

Maxima "knows" how to solve a large variety of mathematical manipulation problems. Looking at the documentation gives some indication of the range, as well as a discussion of the procedures to manage all sorts of expression manipulations. This is a reflection of the large amount of work that went into the development of Maxima over the years.

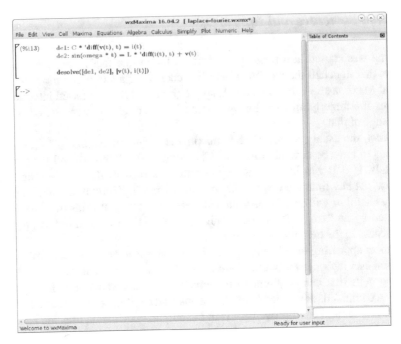

Fig. B.5 The wxMaxima User Interface

In this text we have mentioned mainly the facilities useful for Fourier analysis and Laplace transform manipulation.

There is also support for symbolic solutions to a wide range of differential equations: Maxima knows all of the "standard tricks" leading to closed form solutions. There are also facilities for calculations with physical units and the differential geometry quantities that come up in General Relativity. These and other packages reflect the interests of past and present contributors to the Maxima project.

Graphics in Maxima

Maxima has graphing capacity, and in fact usually uses gnuplot for that purpose. Gnuplot is one of the graphic toolkit options for Octave, although the output has some drawbacks compared to some of the other choices.

Our first choice for plotting is matplotlib, using scipy and numpy with python scripting. The graphics control is as flexible as octave permits, and the labeling easily supports using LaTeXfor labels and annotations.

The plotting for maxima is available, and examples are in the provided documentation.

Maxima Resources

The source for current information on Maxima is the website [22], captured in the appearance on page 759 (Figure B.6).

Documentation and support for using wxMaxima can be found on the developer website [38], illustrated on page 759 (Figure B.7).

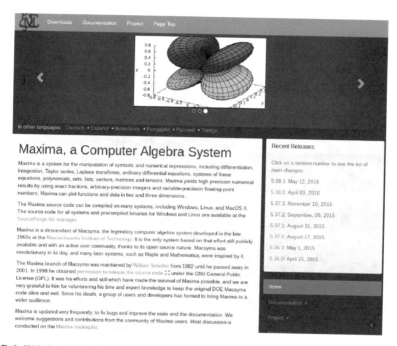

Fig. B.6 Website for Maxima Information

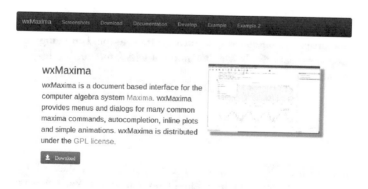

Fig. B.7 Information on wxMaxiam

B.4   Computing Platforms

Doing mathematical calculations requires a programming environment equipped
with a current compatible set of numerical and/or symbolic software libraries.
Looking at the support mailing lists for the software used here, it becomes clear that
a large part of the problems people have arise from incompatible library versions
on the system they are trying to use. These programs are "third-party" products as
far as the commercial system vendors are concerned, and you are basically on your
own searching the web for problem support.

WXMAXIMA

```
(%i1)       ?? residue;
 -- Function: residue (<expr>, <z>, <z_0>)
    Computes the residue in the complex plane of the expression <expr>
    when the variable <z> assumes the value <z_0>.  The residue is the
    coefficient of '(<z> - <z_0>)^(-1)' in the Laurent series for
    <expr>.
        (%i1) residue (s/(s**2+a**2), s, a*%i);
                          1
        (%o1)             -
                          2
        (%i2) residue (sin(a*x)/x**4, x, 0);
                          3
                          a
        (%o2)            - --
                          6
(%o1)       true
```

The best way to solve problems that arise is actually to copy the exact text of the
error message that appeared, and paste it into a Google search. If somebody else has
had the same problem, that is likely to turn up the solution. If nothing turns up, just
hope the error message reveals which version of which module hit the problem, and
continue the search from there.

Windows

It is possible to find Octave and WxMaxima packaged as stand-alone units for
Windows. You get whatever the individual who built the package chose to include,

which means anything not there will have to be installed "by hand" by the user. Python also runs on Windows, and it is possible to add packages to the installation.

All told, it is probably easier to install Linux in a Virtualbox environment, and exploit the package managers to install what is needed with a few clicks.

Macintosh

The Macintosh OS-X is a Unix system under the hood, not that the casual user would realize that. There are versions of Octave and WxMaxima available for Macintosh, and these are subject to the same sort of caveats as those that apply to Windows.

There is another route available for Macintosh, arising from its heritage as a BSD Unix. BSD Unix distributes software on the basis of a source code distribution system known as "ports." On Macintosh, there is a project called "macports"[36]. In principle, all you have to do is install "macports," and then type "sudo port install whatever" to get whatever you want. Of course, it will take a visit to the "macports" website to discover the names of the port packages you may want to install. You may also search for packages with a "port search" command, although the organization of the packages is more evident from looking at the website.

Macs come with a working version of (a somewhat old) python. Things you want require a later version, so you have to build an updated python as part of the process, and keep everything up to date with "macports."

This process is an actual way Mac versions can get generated. There are other avenues to building third party packages on a Mac: I mention "macports" because I have actually used it to install things on a Mac. If you want to distribute a "macports" built version of something to someone else, it is a somewhat involved process. See the "macports" website [36]. The short description of the procedure is that you need a separate parallel macports installation in order to avoid having your package interfere with a macports installation on the target machine. The issue is that the update state of the build machine will not be in sync with the user target. All of the programs need to be installed together with the matching library versions. Keeping all that in sync is the rationale for the ports system.

It is hard to suggest this route to someone who is not fairly "software development literate." A command line XCode development environment has to be installed, and that is what you are using when programs are installed with "macports." A classical reference to command line development is [13], although that contains more information than is actually required to use "macports" successfully.

Linux

The easiest way to get installations of Octave, Python, Matplotlib, and WxMaxima is to run them on a Debian project [12] based Linux distribution. You can use one of the Debian versions itself, or one of the distributions based on those. The reason for this choice is the wide range of configuration possibilities, as well as a well-maintained software selection.

The most widely used distribution derived from Debian is Ubuntu [15]. Ubuntu comes in various flavors and is based on a snapshot of a Debian Testing distribution. This makes it easy to install a choice of user desktop interfaces (or actually all of the choices at once, since this is a Debian-based distribution.) Ubuntu attracts a lot of former Windows users, and as a result there is a large and active online support community. It comes as a revelation that well-maintained applications are readily available from a central source, and not chased down from various websites of varying reliability.

The default Ubuntu interface feels "tablet-ish," but versions using "xfce," "mate," or even "kde" operate more like a conventional user interface with task bars, menus, and multiple desktops. That facility is much more congenial for running multiple programs like Octave, wxMaxima, and Python on separate desktops. There is no need to crowd things.

With a downloaded installation dvd image, you can either install it on spare disk drive space and end up with a dual-boot system, or you can use the dvd image to create a Virtualbox image and run a virtual Ubuntu in a window on your usual desktop.

More information about Linux and other free software is available at the GNU website [9]. Of particular interest is the GNU Public License, under which Linux and Octave are released.

One could write (and authors have written) books about Free Software and the efforts of commercial software vendors to compete with "Free" by political, legal, and astroturf-based public relations efforts. The fact that Google, Facebook, and Amazon are basing their business on large Linux-running data centers is an indication of the way computing technology is evolving.

Virtualbox

Virtualbox [25] is available for virtually all desktop systems at no cost. In the case of Free Software programs, it provides an easy way to run them in a virtual environment, independent of all the software that may or may not be present on the host system.

Duplicating the functionality of my Debian desktop on a Macbook actually took around a week of research, messing with XCode and "macports," then finally making "macports" download, compile, and install everything required. It works, and is an instructive computing experience, but it is not what anyone would call a quick process.

Installing Virtualbox on the Macbook, and then the Ubuntu Mate desktop version on the Virtualbox program took around half an hour. Installing the applied math packages took around another five minutes. Admittedly, the software manager on the Ubuntu system is the same as the one used on my usual desktop. The command line "sudo apt-get install python-scipy python-numpy python-matplotlib python-doc fenics wxmaxima octave" will install most things except for some of the octave extra packages. The package names will lead to other relevant downloads when used in a package manager search.

I don't have a Windows installation to try the same Virtualbox approach, but we can be fairly sure it would proceed roughly as the Mac OS-X version did. The explicit instructions are to first install Virtualbox as an app on your machine, and then download, say, an Ubuntu xfce desktop installation DVD to the desktop. Then use Virtualbox to get a virtual Ubuntu system running. Type the magic apt-get invocation for the mathematics programs, and you are good to go. I simplify slightly, as the virtual Ubuntu will probably insist on updating itself after it is installed, and that will slow the process a little.

Appendix C
Transform Tables

C.1 Laplace Transforms

$f(t)$	$\mathcal{L}\{f(t)\} = F(s)$
$f(t)$	$\int_0^\infty e^{-st} f(t)dt$
$\frac{1}{2\pi i} \int_{\sigma-i\infty}^{\sigma+i\infty} F(s)e^{st}ds$	$F(s)$
$e^{at}f(t)$	F(s-a)
$U(t-T)f(t-T)$	$e^{-sT}F(s)$
$\frac{df}{dt}$	$sF(s) - f(0^+)$
$tf(t)$	$-\frac{d}{ds}F(s)$
$(g*f)(t) = \int_0^t g(t-\tau)f(\tau)d\tau$	$F(s)G(s)$
e^{at}	$\frac{1}{s-a}$
$U(t)$	$\frac{1}{s}$
$\delta(t)$	1
t	$\frac{1}{s^2}$
$\frac{t^2}{2}$	$\frac{1}{s^3}$
$\frac{t^{n-1}}{(n-1)!}$	$\frac{1}{s^n}$
$\sin(\Omega t)$	$\frac{\Omega}{s^2+\Omega^2}$
$\cos(\Omega t)$	$\frac{s}{s^2+\Omega^2}$
$\sinh(at)$	$\frac{a}{s^2-a^2}$
$\cosh(at)$	$\frac{s}{s^2-a^2}$

$U(\cdot)$= Heaviside unit step function

$$U(t) = \begin{cases} 1 & t \geq 0 \\ 0 & t < 0 \end{cases}$$

© Springer International Publishing Switzerland 2016
J.H. Davis, *Methods of Applied Mathematics with a Software Overview*,
Applied and Numerical Harmonic Analysis, DOI 10.1007/978-3-319-43370-7

C.2   Fourier Transforms

$f(t)$	$\mathcal{F}\{f(t)\} = \hat{f}(\omega)$
$f(t)$	$\int_{-\infty}^{\infty} e^{-i\omega t} f(t)dt$
$\frac{1}{2\pi} \int_{-\infty}^{\infty} e^{i\omega t} \hat{f}(\omega)d\omega$	$\hat{f}(\omega)$
$f(t - t_0)$	$e^{-i\omega t_0} \hat{f}(\omega)$
$e^{i\omega_0 t} f(t)$	$\hat{f}(\omega - \omega_0)$
$\frac{df}{dt}$	$i\omega \hat{f}(\omega)$
$tf(t)$	$\frac{-1}{i} \frac{d}{d\omega} \hat{f}(\omega)$
$(g * f)(t) = \int_{-\infty}^{\infty} g(t - \tau)f(\tau)d\tau$	$\hat{g}(\omega)\hat{f}(\omega)$
$2\pi f(t)g(t)$	$(\hat{g} * \hat{f})(\omega) = \int_{-\infty}^{\infty} \hat{g}(\omega - \lambda)\hat{f}(\lambda)d\lambda$
$e^{-at}U(t)$	$\frac{1}{i\omega + a}, (Re(a) > 0)$
$e^{at}U(-t)$	$\frac{1}{-i\omega + a}, (Re(a) > 0)$
$te^{-at}U(t)$	$\frac{1}{(i\omega + a)^2}, (Re(a) > 0)$
$U(t + a) - U(t - a)$	$\frac{2sin(\omega a)}{\omega}$
$tent(t)$	$(\frac{2sin(\omega a)}{\omega})^2$
$\hat{f}(t)$	$2\pi f(-\omega)$
$\frac{sin(\Omega_0 t)}{\Omega_0 t}$	$\frac{\pi}{\Omega_0}(U(\omega + \Omega_0) - U(\omega - \Omega_0))$
$\delta(t)$	1
$\delta'(t)$	$i\omega$
1	$2\pi\delta(\omega)$
t	$2\pi i\delta'(\omega)$
$e^{i\omega_0 t}$	$2\pi\delta(\omega - \omega_0)$
$\int_{-\infty}^{\infty} |f(t)|^2 dt$	$\frac{1}{2\pi} \int_{-\infty}^{\infty} |\hat{f}(\omega)|^2 d\omega$

Here,

$$tent(t) = \begin{cases} 2a - t & 0 \le t \le 2a \\ t + 2a & -2a \le t \le 0 \end{cases}$$

C.3   Z Transforms

$U(\cdot)$ = Heaviside unit step function

$$U(n) = \begin{cases} 1 & n \ge 0 \\ 0 & n < 0 \end{cases}$$

$x(n)$	$\mathcal{Z}\{x(n)\} = X(z)$
$x(n)$	$\sum_{n=0}^{\infty} x(n)z^{-n}$
$\frac{1}{2\pi i} \int_{|z|=R} X(z)z^{n-1}\,dz$	$X(z)$
$a^n x(n)$	$X(\frac{z}{a})$
$U(n-N)x(n-N)$	$\frac{1}{z^N}X(z)$
$x(n+N)$	$z^N X(z) - z^N x(0) - z^{N-1}x(1) - \cdots - zx(N-1)$
$nx(n)$	$-z\frac{d}{dz}X(z)$
$(x*y)(n) = \sum_{k=0}^{n} x(n-k)y(k)$	$X(z)Y(z)$
a^n	$\frac{z}{z-a}$
$U(n)$	$\frac{z}{z-1}$
$\delta(n)$	1
n	$\frac{z}{(z-1)^2}$
$\sin(\Omega n)$	$\frac{z\sin(\Omega)}{z^2 - 2z\cos(\Omega)+1}$

$\delta(\cdot) =$ discrete delta function

$$\delta(n) = \begin{cases} 1 & n = 0 \\ 0 & n \neq 0 \end{cases}$$

C.4   Discrete Fourier Transforms

$f(n)$	$\mathcal{F}\{f(n)\} = \hat{f}(\theta)$
$f(n)$	$\sum_{n=-\infty}^{\infty} f(n)e^{-in\theta}$
$\frac{1}{2\pi}\int_0^{2\pi} \hat{f}(\theta)e^{in\theta}\,d\theta$	$\hat{f}(\theta)$
$e^{in\Omega_0}f(n)$	$\hat{f}(\theta - \Omega_0)$
$f(n-N)$	$e^{-iN\theta}\hat{f}(\theta)$
$nf(n)$	$-\frac{1}{i}\frac{d}{d\theta}\hat{f}(\theta)$
$(f*y)(n) = \sum_{k=-\infty}^{\infty} f(n-k)y(k)$	$\hat{f}(\theta)\hat{y}(\theta)$
$2\pi f(n)g(n)$	$\int_0^{2\pi} \hat{f}(\theta - \phi)\hat{g}(\phi)\,d\phi$
$U(n)a^n$	$\frac{e^{i\theta}}{e^{i\theta}-a}, |a| < 1$
$a^{|n|}$	$\frac{1-a^2}{1-2a\cos(\theta)+a^2}, |a| < 1$
$\delta(n)$	1
$\sum_{-\infty}^{\infty} |f(n)|^2$	$\frac{1}{2\pi}\int_0^{2\pi} |\hat{f}(\theta)|^2\,d\theta$

$U(\cdot)$= Heaviside unit step function

$$U(n) = \begin{cases} 1 & n \ge 0 \\ 0 & n < 0 \end{cases}$$

$\delta(\cdot)$ = discrete delta function

$$\delta(n) = \begin{cases} 1 & n = 0 \\ 0 & n \ne 0 \end{cases}$$

References

1. H. Anton, C. Rorres, *Applications of Linear Algebra* (Wiley, New York, 1979)
2. D.M. Beazley, Swig master (2008). http://www.dabeaz.com/SwigMaster/index.html
3. D.M. Beazley, Swig (2015). http://www.swig.org/Doc1.3/Python.html
4. Blender.org, Blender (2016). https://www.blender.org/. Accessed 25 May 2016
5. J.W. Eaton, Gnu octave ui-elements. https://www.gnu.org/software/octave/doc/interpreter/UI-Elements.html
6. J.W. Eaton, Gnu octave (2016). https://www.gnu.org/software/octave/. Accessed 25 May 2016
7. FreeCAD Developers, Freecad (2016). http://www.freecadweb.org/. Accessed 25 May 2016
8. Free Software Foundation, Free software foundation (2016). http://www.fsf.org/. Accessed 25 May 2016
9. Free Software Foundation, Gnu operating system (2016). https://www.gnu.org/. Accessed 25 May 2016
10. Free Software Foundation, Gnu public license faq (2016). http://www.gnu.org/licenses/gpl-faq.html. Accessed 25 May 2016
11. K. Hoffman, R. Kunze, *Linear Algebra* (Prentice-Hall, Englewood Cliffs, 1960)
12. Inc. Software in the Public Interest, Debian (2016). https://www.debian.org/. Accessed 25 May 2016
13. B.W. Kernighan, R. Pike, *The UNIX Programming Environment* (Prentice-Hall, Englewood Cliffs, 1984)
14. Landon Curt Noll Leo Broukhis, Simon Cooper, The international obfuscated c code contest (2016). http://www.ioccc.org/. Accessed 25 May 2016
15. Ltd. Canonical, Ubuntu for desktops (2016). http://www.ubuntu.com/desktop. Accessed 25 May 2016
16. Mathworks Incorporated, *External Interfaces* (Mathworks Incorporated, Natick, 2000)
17. Mathworks Incorporated, *Partial Differential Equation Toolbox User's Guide* (Mathworks Incorporated, Natick, 2000)
18. Mathworks Incorporated, *Signal Processing Toolbox User's Guide* (Mathworks Incorporated, Natick, 2000)
19. Mathworks Incorporated, *Using MATLAB* (Mathworks Incorporated, Natick, 2000)
20. Mathworks Incorporated, *Using MATLAB Graphics* (Mathworks Incorporated, Natick, 2000)
21. Mathworks Incorporated, *Wavelet Toolbox User's Guide* (Mathworks Incorporated, Natick, 2000)
22. Maxima Developers, Maxima, a computer algebra system (2016). http://maxima.sourceforge.net. Accessed 25 May 2016
23. Numpy Developers, Numpy (2016). http://www.numpy.org/. Accessed 25 May 2016

24. OPEN CASCADE SAS, Open cascade (2016). http://www.opencascade.com/. Accessed 31 May 2016
25. Oracle Inc., Virtualbox (2016). https://www.virtualbox.org/. Accessed 25 May 2016
26. M. Pilgrim, Dive into python (2016). http://www.diveintopython.net/. Accessed 25 May 2016
27. Python Software Foundation, History and license (2016). https://docs.python.org/2/license. html. Accessed 25 May 2016
28. Python Software Foundation, Python (2016). https://www.python.org/. Accessed: 25 May 2016
29. Python Software Foundation, Python (2016). https://docs.python.org/2/extending/extending. html. Accessed 25 May 2016
30. SciPy Developers, Scipy.org (2016). http://docs.scipy.org/doc/numpy-1.10.1/reference/arrays. nditer.html. Accessed 25 May 2016
31. SciPy Developers, Scipy.org (2016). https://www.scipy.org/. Accessed 25 May 2016
32. sourceforge.net, Octave forge (2016). http://octave.sourceforge.net/packages.php. Accessed 25 May 2016
33. The Document Foundation, Libre office. https://www.libreoffice.org/. Accessed 26 May 2016
34. The FEniCS Project, Automated solution of differential equations by the finite element method (2016). http://fenicsproject.org/book/index.html#book. Accessed 25 May 2016
35. The FEniCS Project, Fenics project (2016). http://fenicsproject.org/. Accessed 25 May 2016
36. The MacPorts Project, The macports project official homepage (2016). https://www.macports. org/. Accessed 25 May 2016
37. The Matplotlib Deverlopment Team, Matplotlib (2016). http://matplotlib.org/. Accessed 25 May 2016
38. wxMaxima Developers, wxmaxima (2016). http://andrejv.github.io/wxmaxima/.

Index

A
abscissa of convergence, 353
adjoint mapping, 715
admissible variation, 276
AM modulation, 555
analytic function, 287
approximate delta function, 466, 522
architecture
    DSP, 657
Argand diagram, 285
argument, 286
associated Legendre equation, 249

B
Banach space, 31
band-limited function, 554
basis, 710
Bessel function
    integral representation, 701
Bessel's equation, 178
    order n, 191
Bessel's inequality, 32
besselj(n,x), 234
bilinear transformation, 350
boundary conditions, 89, 98
    general heat equation, 175
    heat flux, 124
    heat transfer, 90
    homogeneous, 89
    insulated, 89
    ladder network, 103
    potential equation, 98
    vibrating beam, 153

branch cut, 294, 296
butterfly pattern, 657

C
CAD, 279
Cauchy Criterion
    continuous functions, 29
    real numbers, 26
    uniform convergence, 28
Cauchy sequence
    Hilbert space, 32
    normed vector space, 30
Cauchy's Integral Formula, 310
Cauchy–Riemann equations, 289
Cauchy–Schwarz inequality, 16
central difference, 266
Cesaro sum, 48
Cesaro sum, 48
change of basis, 711
clipped sine wave, 68
clipping, 730
command
    besselj(n,x), 234
    find, 236
    fzero, 234
complete
    normed vector space, 30
    orthonormal set, 34
complete inner product space, 31
complex plane, 285
computer aided design, 279
computing
    platforms, 760

© Springer International Publishing Switzerland 2016
J.H. Davis, *Methods of Applied Mathematics with a Software Overview*,
Applied and Numerical Harmonic Analysis, DOI 10.1007/978-3-319-43370-7

conformal mapping, 332, 335
convergence
    mean-square, 36
    with discontinuities, 54
convolution, 214, 390
    theorem, 392
convolution law
    z-transform, 580
    discrete Fourier transform, 609
    finite Fourier transform, 643
    Fourier transform, 457
    Laplace transform, 392
cooling
    Newton's law, 86
coordinates, 711
critical point, 335

D
Debian, 761
delay factor
    Fourier series, 62
delay law
    z-transform, 582
    discrete Fourier transform, 609
    finite Fourier transform, 641
    Fourier series, 61
    Fourier transform, 441
    Laplace transform, 359
delayed function
    Fourier series, 61
delta function, 516
DFT, 603
differentiable function, 287
differential equations
    periodic solutions, 69
differentiation
    termwise, 65
differentiation law
    Fourier series, 63
        complex, 64
        real, 66
    Fourier transform, 442
    Laplace transform, 361
diffusion
    minority carrier, 92
    pollution, 90
dimension, 710
discrete Fourier transform, 603
distribution, 516
    convolution, 522
    definition , 516
    derivative, 517
    Fourier transform, 518

distribution theory, 65
distributions, 514
DSP architecture, 657
dual space, 713

E
eigenvectors
    basis, 714
elastic membrane, 176
equation
    Bessel's, 178
    diffusion, 86
    discrete Poissson, 176
    elastic beam, 96
    elastic membrane, 175
    evolution, 158
    fluid flow, 100
    heat, 158
    heat flow, 98
    ladder network, 102
    Laplace's, 97
    Maxwell's, 95
    population, 96
    potential, 97
    random walk, 570
    singular differential, 192
    sound wave, 154
    spring mass, 104
    Sturm–Liouville, 190
    thin ring, 110
    variational, 274
    vibrating beam, 151
    wave, 92
equation of evolution, 158
error
    mean square, 19
essential singularity, 315
Euler equations, 6
Euler's formula, 8, 46
Euler-Mascheroni constant, 226
even function
    periodic, 67
exponential
    matrix, 409

F
fast Fourier transform, 649
fenics, 280
FFT, 649
    algorithm, 652
Fick's law, 88
figure, 736

filter
    low pass, 557
find command, 236
finite Fourier transform, 640
finite mean-square derivative, 39
flow
    about circle, 330, 332
    about ellipse, 346
    ideal fluid, 328
forcing function, 403
formant analysis, 680
Fourier transform
    short time, 675
Fourier coefficient formulas, 19
Fourier inversion, 465
    contour integral, 477
    theorem, 474
Fourier inversion theorem, 474
Fourier series
    complex form, 8
    finite, 7
    multiple, 240
    pointwise convergence, 40
    real form, 8
    uniform convergence, 44
Free Software, 762
frequency
    carrier, 555
    fundamental, 58
Frobenius method, 217, 218
Frobenius' method, 216
full-wave rectifier, 60
fundamental frequency, 58
fzero, 234

G
gca, 736
gcf, 736
generalized functions, 514
Gibb's phenomenon, 45
GNU, 762
    GNU Public LIcense, 762
Green's function, 170, 524

H
half-range sine series, 131
half-wave rectifier, 58
Hankel transform, 700, 704
    inversion, 704
Hilbert space, 31
holomorphic function, 287

I
ideal fluid, 324
ill-posed, 157
impedance
    Fourier series, 75
    Fourier transform, 531
    Laplace transform, 417
impedance relations, 76
impulse response, 399, 524
impulse response matrix, 412
indicial equation, 121, 213, 215
inequality
    Bessel's, 32
    Cauchy–Schwarz, 16
    triangle, 16
inner product, 16, 712
inner product space, 16, 712
integrals
    Lebesgue, 17
    Riemann, 17
inverse Laplace transform, 355
inversion integral
    z-transform, 589, 590
        complex, 590
        real, 589
    complex variable, 477
    discrete Fourier transform, 604
    finite Fourier transform, 640
    Fourier transform, 465
    Laplace transform, 499
irrotational flow, 100, 324

J
Jordan form, 714

K
kde, 762
Kirchoff's laws, 75

L
ladder network, 102
Laplace inversion
    contour integral, 499
Laplace transform, 352
    inversion integral, 499
    one sided, 352
    two sided, 685
Laplace's equation, 98
Laurent series, 315
Lebesgue integrals, 17

Legendre equation
    associated, 249
Legendre equation, 249
limit
    normed vector space, 30
    real number sequence, 26
linear independence, 710
Linux, 761
    Debian, 761
    kde, 762
    mate, 762
    Ubuntu, 762
    xfce, 762
Liouville's theorem, 322, 331
low pass filter, 659
    finite Fourier transform, 659

M
Macintosh, 761
macports, 761
matched filter, 83
mate, 762
MATLAB
    toolbox, 738
matplotlib, 745
matrix
    sparse, 106
matrix exponential, 409
Maxima, 759
    history, 756
    wxMaxima, 759
maxima
    Fourier coefficient, 12
mean-square convergence, 36
mean-square error, 26
method of lines, 270
modulation
    AM, 555
    phase, 701
modulus, 286

N
nmupy
    ufunc, 746
norm, 16
normal mode, 243
normed vector space, 29
numpy, 745
Nyquist rate, 562

O
Octave, 717
    axis, 736
    cell arrays, 719
    cells, 720
    data structures, 718
    fenics, 280
    figure, 736
    graphics, 729
    keywords, 723
    operations, 723
    operator syntax, 721
    programming, 724
    programs, 726
    scripts, 726
    structures, 720
    user interface, 728
odd function
    periodic, 67
Ohm's law, 76
orthogonality, 18
orthogonality conditions, 6
orthonormal basis, 34
orthonormal expansion theorem, 34
orthonormal set, 18

P
parallelogram law, 24
Parseval relations, 23
Parseval's Theorem
    general, 35
Parseval's theorem
    discrete Fourier transform, 617
    finite Fourier transform, 648
    Fourier transform, 461
partial fractions, 380
    Octave, 384
partial sum
    Fourier series, 21
passive system, 70
phase modulation, 701
pointwise convergence, 26
    definition, 27
Poisson kernel, 127
Poisson's equation, 98
    discrete, 176
polar form, 285
pole, 315
pole of order k, 315
power spectral density, 78
power spectrum, 76, 78

principal part, 315
principal value, 299
principle of the argument, 339, 348, 349
projection
    orthogonal, 19
projection coefficients, 20
Projection Theorem, 19, 26
properties
    Fourier series, 56
Pythagorean Theorem, 21
Python
    documentation, 755
    fenics, 280
    introduction, 741

Q
quarter-range expansion, 141

R
random walk, 573
regular function, 287
regular singular point, 210
residue, 316
    function, 316
Riemann integrals, 17
Riemann Mapping Theorem, 344
Riemann surface, 294
Riesz–Fischer theorem, 38
Riesz–Fisher theorem, 184
RLC circuit, 74

S
sampling expansion, 560
sampling rate, 562
sampling theorem, 557, 560
sawtooth function, 44
    pointwise convergence, 44
scaling law
    Fourier transform, 439
    Laplace transform, 357
Scipy, 745
second difference, 569
semi-group property, 159
separable solution, 113
sequency, 693
shift law
    z-transform, 579, 584
    discrete Fourier transform, 608
    finite Fourier transform, 641
    Fourier series, 62

    Fourier transform, 439
    Laplace transform, 358
short time Fourier transform, 675
simple function, 341
sinc function, 435
sine transform, 695
singular point, 315
software, 717
solution
    general, 71
    particular, 71
sparse matrix, 106
special functions, 210
spectrogram, 679
spherical harmonics, 250
stable system, 70
state variables, 73
state vector, 403
STFT, 675
stream function, 326
streamlines, 326
string
    vibrating, 93
sum
    Abel, 56
    Cesaro, 48
summation methods, 47
surface
    parametric, 732
    plot, 734
    Riemann, 734

T
Taylor series, 5
termwise differentiation, 65
test function, 516
theorem
    Riesz–Fischer, 38
time multiplication
    Laplace transform, 361
time multiplication law
    discrete Fourier transform, 608
time multiplication law
    Fourier transform, 443
    z transform, 579
toolbox, 738
transfer function, 394
transfer function matrix, 412, 413
transform pairs, 433

U
Ubuntu, 762
uncertainty principle, 666, 668

uniform convergence
    functions, 28
univalent function, 341

V
variation of constants, 412
vector space
    review, 709
velocity potential, 324
Virtualbox, 762
visualization programs, 717
vorticity, 332

W
weighting pattern, 390

well-posed, 157
Windows, 760
WKB approximation, 204, 205
wxMaxima, 759

X
XCode, 761
xfce, 762

Z
z-transform, 575
z-transforms
    one sided, 575
    two-sided, 687

Applied and Numerical Harmonic Analysis
(68 volumes)

A. Saichev and W.A. Woyczyński: *Distributions in the Physical and Engineering Sciences* (ISBN 978-0-8176-3924-2)

C.E. D'Attellis and E.M. Fernandez-Berdaguer: *Wavelet Theory and Harmonic Analysis in Applied Sciences* (ISBN 978-0-8176-3953-2)

H.G. Feichtinger and T. Strohmer: *Gabor Analysis and Algorithms* (ISBN 978-0-8176-3959-4)

R. Tolimieri and M. An: *Time-Frequency Representations* (ISBN 978-0-8176-3918-1)

T.M. Peters and J.C. Williams: *The Fourier Transform in Biomedical Engineering* (ISBN 978-0-8176-3941-9)

G.T. Herman: *Geometry of Digital Spaces* (ISBN 978-0-8176-3897-9)

A. Teolis: *Computational Signal Processing with Wavelets* (ISBN 978-0-8176-3909-9)

J. Ramanathan: *Methods of Applied Fourier Analysis* (ISBN 978-0-8176-3963-1)

J.M. Cooper: *Introduction to Partial Differential Equations with MATLAB* (ISBN 978-0-8176-3967-9)

A. Procházka, N.G. Kingsbury, P.J. Payner, and J. Uhlir: *Signal Analysis and Prediction* (ISBN 978-0-8176-4042-2)

W. Bray and C. Stanojevic: *Analysis of Divergence* (ISBN 978-1-4612-7467-4)

G.T. Herman and A. Kuba: *Discrete Tomography* (ISBN 978-0-8176-4101-6)

K. Gröchenig: *Foundations of Time-Frequency Analysis* (ISBN 978-0-8176-4022-4)

L. Debnath: *Wavelet Transforms and Time-Frequency Signal Analysis* (ISBN 978-0-8176-4104-7)

J.J. Benedetto and P.J.S.G. Ferreira: *Modern Sampling Theory* (ISBN 978-0-8176-4023-1)

D.F. Walnut: *An Introduction to Wavelet Analysis* (ISBN 978-0-8176-3962-4)

A. Abbate, C. DeCusatis, and P.K. Das: *Wavelets and Subbands* (ISBN 978-0-8176-4136-8)

O. Bratteli, P. Jorgensen, and B. Treadway: *Wavelets Through a Looking Glass* (ISBN 978-0-8176-4280-80)

H.G. Feichtinger and T. Strohmer: *Advances in Gabor Analysis* (ISBN 978-0-8176-4239-6)

O. Christensen: *An Introduction to Frames and Riesz Bases* (ISBN 978-0-8176-4295-2)

L. Debnath: *Wavelets and Signal Processing* (ISBN 978-0-8176-4235-8)

G. Bi and Y. Zeng: *Transforms and Fast Algorithms for Signal Analysis and Representations* (ISBN 978-0-8176-4279-2)

J.H. Davis: *Methods of Applied Mathematics with a MATLAB Overview* (ISBN 978-0-8176-4331-7)

J.J. Benedetto and A.I. Zayed: *Modern Sampling Theory* (ISBN 978-0-8176-4023-1)

E. Prestini: *The Evolution of Applied Harmonic Analysis* (ISBN 978-0-8176-4125-2)

L. Brandolini, L. Colzani, A. Iosevich, and G. Travaglini: *Fourier Analysis and Convexity* (ISBN 978-0-8176-3263-2)

W. Freeden and V. Michel: *Multiscale Potential Theory* (ISBN 978-0-8176-4105-4)

O. Christensen and K.L. Christensen: *Approximation Theory* (ISBN 978-0-8176-3600-5)

O. Calin and D.-C. Chang: *Geometric Mechanics on Riemannian Manifolds* (ISBN 978-0-8176-4354-6)

J.A. Hogan: *Time?Frequency and Time?Scale Methods* (ISBN 978-0-8176-4276-1)

C. Heil: *Harmonic Analysis and Applications* (ISBN 978-0-8176-3778-1)

K. Borre, D.M. Akos, N. Bertelsen, P. Rinder, and S.H. Jensen: *A Software-Defined GPS and Galileo Receiver* (ISBN 978-0-8176-4390-4)

T. Qian, M.I. Vai, and Y. Xu: *Wavelet Analysis and Applications* (ISBN 978-3-7643-7777-9)

G.T. Herman and A. Kuba: *Advances in Discrete Tomography and Its Applications* (ISBN 978-0-8176-3614-2)

M.C. Fu, R.A. Jarrow, J.-Y. Yen, and R.J. Elliott: *Advances in Mathematical Finance* (ISBN 978-0-8176-4544-1)

O. Christensen: *Frames and Bases* (ISBN 978-0-8176-4677-6)

P.E.T. Jorgensen, J.D. Merrill, and J.A. Packer: *Representations, Wavelets, and Frames* (ISBN 978-0-8176-4682-0)

M. An, A.K. Brodzik, and R. Tolimieri: *Ideal Sequence Design in Time-Frequency Space* (ISBN 978-0-8176-4737-7)

S.G. Krantz: *Explorations in Harmonic Analysis* (ISBN 978-0-8176-4668-4)

B. Luong: *Fourier Analysis on Finite Abelian Groups* (ISBN 978-0-8176-4915-9)

G.S. Chirikjian: *Stochastic Models, Information Theory, and Lie Groups, Volume 1* (ISBN 978-0-8176-4802-2)

C. Cabrelli and J.L. Torrea: *Recent Developments in Real and Harmonic Analysis* (ISBN 978-0-8176-4531-1)

M.V. Wickerhauser: *Mathematics for Multimedia* (ISBN 978-0-8176-4879-4)

B. Forster, P. Massopust, O. Christensen, K. Gröchenig, D. Labate, P. Vandergheynst, G. Weiss, and Y. Wiaux: *Four Short Courses on Harmonic Analysis* (ISBN 978-0-8176-4890-9)

O. Christensen: *Functions, Spaces, and Expansions* (ISBN 978-0-8176-4979-1)

J. Barral and S. Seuret: *Recent Developments in Fractals and Related Fields* (ISBN 978-0-8176-4887-9)

O. Calin, D.-C. Chang, and K. Furutani, and C. Iwasaki: *Heat Kernels for Elliptic and Sub-elliptic Operators* (ISBN 978-0-8176-4994-4)

C. Heil: *A Basis Theory Primer* (ISBN 978-0-8176-4686-8)

J.R. Klauder: *A Modern Approach to Functional Integration* (ISBN 978-0-8176-4790-2)

J. Cohen and A.I. Zayed: *Wavelets and Multiscale Analysis* (ISBN 978-0-8176-8094-7)

D. Joyner and J.-L. Kim: *Selected Unsolved Problems in Coding Theory* (ISBN 978-0-8176-8255-2)

G.S. Chirikjian: *Stochastic Models, Information Theory, and Lie Groups, Volume 2* (ISBN 978-0-8176-4943-2)

J.A. Hogan and J.D. Lakey: *Duration and Bandwidth Limiting* (ISBN 978-0-8176-8306-1)

G. Kutyniok and D. Labate: *Shearlets* (ISBN 978-0-8176-8315-3)

P.G. Casazza and P. Kutyniok: *Finite Frames* (ISBN 978-0-8176-8372-6)

V. Michel: *Lectures on Constructive Approximation* (ISBN 978-0-8176-8402-0)

D. Mitrea, I. Mitrea, M. Mitrea, and S. Monniaux: *Groupoid Metrization Theory* (ISBN 978-0-8176-8396-2)

T.D. Andrews, R. Balan, J.J. Benedetto, W. Czaja, and K.A. Okoudjou: *Excursions in Harmonic Analysis, Volume 1* (ISBN 978-0-8176-8375-7)

T.D. Andrews, R. Balan, J.J. Benedetto, W. Czaja, and K.A. Okoudjou: *Excursions in Harmonic Analysis, Volume 2* (ISBN 978-0-8176-8378-8)

D.V. Cruz-Uribe and A. Fiorenza: *Variable Lebesgue Spaces* (ISBN 978-3-0348-0547-6)

W. Freeden and M. Gutting: *Special Functions of Mathematical (Geo-)Physics* (ISBN 978-3-0348-0562-9)

A. Saichev and W.A. Woyczyński: *Distributions in the Physical and Engineering Sciences, Volume 2: Linear and Nonlinear Dynamics of Continuous Media* (ISBN 978-0-8176-3942-6)

S. Foucart and H. Rauhut: *A Mathematical Introduction to Compressive Sensing* (ISBN 978-0-8176-4947-0)

G. Herman and J. Frank: *Computational Methods for Three-Dimensional Microscopy Reconstruction* (ISBN 978-1-4614-9520-8)

A. Paprotny and M. Thess: *Realtime Data Mining: Self-Learning Techniques for Recommendation Engines* (ISBN 978-3-319-01320-6)

A. Zayed and G. Schmeisser: *New Perspectives on Approximation and Sampling Theory: Festschrift in Honor of Paul Butzer's 85th Birthday* (978-3-319-08800-6)

R. Balan, M. Begue, J. Benedetto, W. Czaja, and K.A Okoudjou: *Excursions in Harmonic Analysis, Volume 3* (ISBN 978-3-319-13229-7)

S. Dahlke, F. De Mari, P. Grohs, and D. Labate: *Harmonic and Applied Analysis: From Groups to Signals* (ISBN 978-3-319-18862-1)

G. Pfander: *Sampling Theory, a Renaissance* (ISBN 978-3-319-19748-7)

R. Balan, M. Begue, J. Benedetto, W. Czaja, and K.A Okoudjou: *Excursions in Harmonic Analysis, Volume 4* (ISBN 978-3-319-20187-0)

O. Christensen: *An Introduction to Frames and Riesz Bases, Second Edition* (ISBN 978-3-319-25611-5)

J.H. Davis: *Methods of Applied Mathematics with a Software Overview, Second Edition* (ISBN 978-3-319-43369-1)

For an up-to-date list of ANHA titles, please visit http://www.springer.com/series/4968